Springer Proceedings in Mathematics & Statistics

Volume 451

This book series features volumes composed of selected contributions from workshops and conferences in all areas of current research in mathematics and statistics, including data science, operations research and optimization. In addition to an overall evaluation of the interest, scientific quality, and timeliness of each proposal at the hands of the publisher, individual contributions are all refereed to the high quality standards of leading journals in the field. Thus, this series provides the research community with well-edited, authoritative reports on developments in the most exciting areas of mathematical and statistical research today.

Shuji Machihara
Editor

Mathematical Physics and Its Interactions

In Honor of the 60th Birthday of Tohru Ozawa, Tokyo, Japan, August 2021

Springer

Editor
Shuji Machihara
Department of Mathematics
Saitama University
Saitama, Japan

ISSN 2194-1009 ISSN 2194-1017 (electronic)
Springer Proceedings in Mathematics & Statistics
ISBN 978-981-97-0363-0 ISBN 978-981-97-0364-7 (eBook)
https://doi.org/10.1007/978-981-97-0364-7

Mathematics Subject Classification: 35-XX, 42-XX, 81-XX

This Springer imprint is published by the registered company Springer Nature Singapore Pte Ltd.
The registered company address is: 152 Beach Road, #21-01/04 Gateway East, Singapore 189721, Singapore

Preface

This volume is a collection of the contents of lectures that were to be given at the meeting entitled *Mathematical Physics and Its Interactions*, originally scheduled to be held 25–27 August 2021. The meeting would have a special significance, as it was to be a celebration of Professor Tohru Ozawa's 60th birthday.

Preparations had been underway for about 2 years prior to the planned meeting, before the news of COVID-19. In March 2020, many schools, including universities, were closed in Japan. For a long time afterward, many events such as research meetings were canceled. We are grateful to the authors of this volume for their willingness to participate as speakers in 2021 even though news of COVID-19 continued to appear in the media and many people were reluctant to attend. In actuality, however, the reality of vaccines became an issue that could not be ignored, and on 23 April of that year, we announced the cancellation of the meeting. This collection of reports is therefore the only evidence of the existence of this meeting that would have taken place.

Saitama, Japan
July 2023

Shuji Machihara

Contents

Positive Solutions of Superlinear Elliptic Equations with Respect to the Schrödinger Operator

Kentaro Hirata

Dedicated to Professor Tohru Ozawa on the occasion of his sixtieth birthday.

Abstract We study positive solutions of semilinear elliptic equations with respect to the Schrödinger operator in the superlinear case. In a bounded smooth domain, a priori estimates for solutions and their gradients, the Harnack inequality and the boundary Harnack principle are presented. As applications, we also investigate the asymptotic behavior of positive solutions with isolated boundary singularities and the removability of isolated boundary singularities.

Keywords Semilinear elliptic equation · Schrödinger operator · Superharmonic function · Boundary behavior · Boundary Harnack principle · Apriori estimate

2020 Mathematics Subject Classification Primary 35J91 · Secondary 31B10 · 31B35 · 35B09 · 35B45 · 35J10

1 Introduction

This paper is motivated by the following nonlinear Schrödinger equation. Let Ω be a domain in $\mathbb{R}^N$, let Δ be the usual Laplacian on $\mathbb{R}^N$ and let $p > 1$. The standing wave solution $\psi = \psi(x, t) : \Omega \times \mathbb{R} \to \mathbb{C}$ of the nonlinear Schrödinger equation with potential functions $U_1, U_2 : \Omega \to \mathbb{R}$ of the form

This work was supported in part by JSPS KAKENHI Grant Number JP18K03333.

K. Hirata (✉)
Department of Mathematics, Graduate School of Advanced Science and Engineering, Hiroshima University, Higashi-Hiroshima 739-8526, Japan
e-mail: hiratake@hiroshima-u.ac.jp

S. Machihara (ed.), *Mathematical Physics and Its Interactions*, Springer Proceedings in Mathematics & Statistics 451, https://doi.org/10.1007/978-981-97-0364-7_1

$$i\frac{\partial \psi}{\partial t} = -\Delta\psi + U_1\psi - U_2|\psi|^{p-1}\psi \quad \text{in } \Omega \times \mathbb{R}$$

is given by $\psi(x,t) = e^{i\omega t}u(x)$ with a pair of $\omega \in \mathbb{R}$ and a solution $u : \Omega \to \mathbb{R}$ of the elliptic equation $-\Delta u + (U_1 + \omega)u = U_2|u|^{p-1}u$ in Ω. Thus we are interested in revealing properties and the behavior of positive solutions of such elliptic equations.

Throughout this paper, we suppose that Ω is a bounded domain in $\mathbb{R}^N$ ($N \geq 3$) with a $C^{1,1}$-boundary $\partial\Omega$; roughly speaking, there exists a constant $r_\Omega > 0$ such that, in an r_Ω-neighborhood of each point in $\partial\Omega$, we can describe $\partial\Omega$ as the graph of a $C^{1,1}$-function of $(N-1)$-variables by retaking a Cartesian coordinate system if necessary (see Definition 1). We deal with the superlinear elliptic equation with respect to the Schrödinger operator $-\Delta + V_1$ of the form

$$-\Delta u + V_1 u = V_2|u|^{p-1}u \quad \text{in } \Omega, \tag{1}$$

where V_1 and V_2 are real-valued measurable functions on Ω possibly having singularities on $\partial\Omega$. By $\delta_\Omega(x)$, we denote the distance from $x \in \Omega$ to $\partial\Omega$. The letter c stands for a generic positive constant whose value may vary at each occurrence. If necessary, we add the subscript like $c_1, c_2, \ldots$ to specify constants.

Unless otherwise stated explicitly, we always assume that

$$1 < p < \frac{N}{N-2}, \quad \alpha_1 < 2 \quad \text{and} \quad \alpha_2 < N + 1 - p(N-1) \tag{2}$$

and there exist positive constants c_1, c_2 such that

(V1) $|V_1(x)| \leq c_1\,\delta_\Omega(x)^{-\alpha_1}$ for a.e. $x \in \Omega$,
(V2) $0 \leq V_2(x) \leq c_2\,\delta_\Omega(x)^{-\alpha_2}$ for a.e. $x \in \Omega$.

Note that $V_1, V_2 \in L^\infty_{\mathrm{loc}}(\Omega)$. We say that $u \in L^p_{\mathrm{loc}}(\Omega)$ is a *solution* of (1) if

$$\int_\Omega u(-\Delta\phi)\,dx = \int_\Omega \{V_2|u|^{p-1}u - V_1 u\}\phi\,dx$$

for all $\phi \in C_c^\infty(\Omega)$; the set of all C^∞-functions with compact support in Ω. Let $c_3 > 0$ and let x_0 be a fixed point in Ω such that $\delta_\Omega(x_0) > r_\Omega$. By $\mathscr{S}(\Omega)$, we denote the collection of all positive solutions $u \in C(\Omega)$ of (1) such that

$$u(x_0) \leq c_3.$$

Note that $\mathscr{S}(\Omega)$ depends on c_3, p, V_1, V_2, etc., but we do not express them in the symbol $\mathscr{S}(\Omega)$ for simplicity.

In the statements of theorems, the notation $c = c(a, b, \ldots)$ means that the constant c depends at most on the parameters $a, b, \ldots$, while $c = c(\Omega)$ means its dependence on the dimension N and some characters regarding Ω.

Now, let us state main results of this paper. Since we impose no boundary condition, the collection $\mathscr{S}(\Omega)$ may include positive solutions of (1) with singularities on $\partial\Omega$. The following theorem reveals the possible growth rate of such solutions.

Theorem 1 (Boundary growth estimate) *There exists* $c = c(c_1, c_2, c_3, \alpha_1, \alpha_2, p, V_1^+, \Omega) > 0$ *such that*

$$u(x) \le c\,\delta_\Omega(x)^{1-N} \quad \textit{for all } x \in \Omega, \tag{3}$$

whenever $u \in \mathscr{S}(\Omega)$.

Corollary 1 $\mathscr{S}(\Omega) \subset C^1(\Omega)$. *Moreover, there exists* $c = c(c_1, c_2, c_3, \alpha_1, \alpha_2, p, V_1^+, \Omega) > 0$ *such that*

$$|\nabla u(x)| \le c\,\delta_\Omega(x)^{-N} \quad \textit{for all } x \in \Omega, \tag{4}$$

whenever $u \in \mathscr{S}(\Omega)$.

Theorem 1 is relevant to the result due to Bidaut-Véron–Vivier [4] for the Lane–Emden equation $-\Delta u = |u|^{p-1}u$ with $1 < p < (N+1)/(N-1)$: for any positive solution u on Ω, there exists a nonnegative harmonic function h on Ω such that

$$h(x) \le u(x) \le c\{h(x) + \delta_\Omega(x)\} \quad \text{for all } x \in \Omega. \tag{5}$$

Taking account of the possible growth rate near $\partial\Omega$ of positive harmonic functions on Ω, we can derive (3) from (5). In our setting of V_1 and V_2, it is hard to show a sharper estimate like (5) for positive solutions of (1). But estimate (3) is sufficient to obtain a variety of nice properties of solutions.

Also, as a special case of Serrin–Zou [19], we know that if $0 \le V_1 \in L^\infty(\Omega)$ and $V_2 \in L^\infty(\Omega)$ has a positive lower bound on Ω, then all positive weak solutions u of (1) with $1 < p < N/(N-2)$ satisfy

$$u(x) \le c\,\delta_\Omega(x)^{-\frac{2}{p-1}} \quad \text{for all } x \in \Omega, \tag{6}$$

where the constant c is independent of c_3: the upper bound of $u(x_0)$. See also Dancer [7] and Poláčik–Quittner–Souplet [16] for the case $V_1 \equiv 0$, $V_2 \equiv 1$ and $1 < p < (N+2)/(N-2)$. Note that

$$1 - N > -\frac{2}{p-1} \iff p < \frac{N+1}{N-1},$$

and also that the growth rate of a positive solution of the Lane–Emden equation with an isolated singularity on $\partial\Omega$ differs among the cases $1 < p < (N+1)/(N-1)$, $p = (N+1)/(N-1)$ and $(N+1)/(N-1) < p < (N+2)/(N-2)$ (see Bidaut-Véron–Ponce–Véron [3] for details). Moreover, since our setting includes the case that V_2 is the indicator function of a set K in Ω (that is, positive solutions are

$(-\Delta + V_1)$-harmonic on $\Omega \setminus \overline{K}$), estimate (3) is sharp even when $(N+1)/(N-1) \leq p < N/(N-2)$. Thus the boundary behavior of positive solutions of (1) depends on both of p and V_2.

Remark 1 If V_2 has a positive lower bound on some $B(x_0, r) \subset \Omega$, then we can take the constants c in (3) and (4) independently of c_3. This can be seen by applying (6) with $\Omega = B(x_0, r)$ to u or to $\widetilde{u}$ defined by (36) when V_1 is not nonnegative: In fact, $u(x_0) \leq c\widetilde{u}(x_0) \leq cr^{-2/(p-1)}$.

Remark 2 Theorem 1 and Corollary 1 are valid for the case $\alpha_2 = N + 1 - p(N-1)$ as well (see their proof). But we see from [8, Theorem 1.5] that Theorem 1 does not hold for $\alpha_2 > N + 1 - p(N-1)$ in general.

The following theorem reveals the vanishing rate of $u \in \mathscr{S}(\Omega)$ vanishing on a portion of $\partial\Omega$. For $0 < r < r_\Omega$ and $\xi \in \partial\Omega$, we let $\xi_r := \xi + r\boldsymbol{n}_\xi$, where $\boldsymbol{n}_\xi$ is the inward unit normal vector to $\partial\Omega$ at ξ.

Theorem 2 *There exist* $c_4 = c(\alpha_1, \alpha_2, p, N) \geq 4$, $c = c(c_1, c_2, c_3, \alpha_1, \alpha_2, p, V_1^+, \Omega) \geq 1$ *and* $r_1 = r(\alpha_1, \alpha_2, p, \Omega) > 0$ *with the following property: Let* $\xi \in \partial\Omega$ *and* $0 < r < r_1$. *If* $u \in \mathscr{S}(\Omega)$ *vanishes continuously on* $\partial\Omega \cap B(\xi, c_4 r)$, *then*

$$\frac{1}{c}\frac{u(\xi_r)}{r}\,\delta_\Omega(x) \leq u(x) \leq c\frac{u(\xi_r)}{r}\,\delta_\Omega(x) \quad \text{for all } x \in \Omega \cap B(\xi, r). \tag{7}$$

Corollary 2 *There exists* $c = c(c_1, c_2, c_3, \alpha_1, \alpha_2, p, V_1^+, \Omega) \geq 1$ *such that*

$$u(x) \leq c\,\delta_\Omega(x) \quad \text{for all } x \in \Omega \tag{8}$$

and

$$|\nabla u(x)| \leq c \quad \text{for all } x \in \Omega, \tag{9}$$

whenever $u \in \mathscr{S}(\Omega)$ *vanishes continuously on the whole of* $\partial\Omega$.

Remark 3 By the same reasoning as in Remark 1, we can take the constants c, appearing in all the estimates in Theorem 2 and Corollary 2, independently of c_3 if V_2 has a positive lower bound on some $B(x_0, r) \subset \Omega$.

As another application of Theorem 2, we present results concerning the asymptotic behavior of solutions in $\mathscr{S}(\Omega)$ having isolated singularities on $\partial\Omega$ and the removability of isolated boundary singularities in Sect. 4.

The above results are new at least for the case $V_1 \not\equiv 0$, whereas those in the case $V_1 \equiv 0$ were proved by the author [8, 10] (Theorem 2 was obtained under $\alpha_2 = 0$), using potential theory and an iteration method.

The plan of this paper is as follows. In Sect. 2, we prove the above theorems in the case $V_1 \leq 0$ by using superharmonicity of positive solutions of (1) and modifying the arguments in the case $V_1 \equiv 0$. When V_1 takes positive values, we cannot apply our previous argument. We try to find a positive superharmonic function $\widetilde{u}$ on Ω such that it is comparable to a given $u \in \mathscr{S}(\Omega)$ and the above theorems hold for $\widetilde{u}$, by using the perturbation theory of the Green function and potential theory for the linear Schrödinger equation. Proofs of the above theorems are given in Sect. 3, and their applications and some remarks are given in Sect. 4.

2 The Case $V_1 \leq 0$

This section presents proofs of our results in the case $V_1 \leq 0$. In light of an application to the case that V_1 takes positive values, we address the problems for a slightly general class which includes $\mathscr{S}(\Omega)$ when $V_1 \leq 0$.

2.1 Positive Superharmonic Functions Satisfying a Nonlinear Inequality

A function $u : \Omega \to \mathbb{R} \cup \{+\infty\}$ is called *superharmonic* on Ω if the following conditions are fulfilled:

(i) $u \not\equiv +\infty$,
(ii) u is lower semicontinuous on Ω,
(iii) for any $x \in \Omega$ and $0 < r < \delta_\Omega(x)$, we have

$$u(x) \geq \fint_{B(x,r)} u(y)\, dy,$$

where the symbol $\fint_{B(x,r)}$ stands for the integral average over the open ball $B(x, r)$ of center x and radius r.

For $v \in L^1_{\mathrm{loc}}(\Omega)$, it is known that there exists a superharmonic function u on Ω such that $u = v$ a.e. on Ω if and only if v is a distributional supersolution of the Laplace equation in Ω in the sense that the *distributional Laplacian* of v defined by

$$(-\Delta v)(\phi) := \int_\Omega v(-\Delta \phi)\, dx$$

is nonnegative for all nonnegative $\phi \in C_c^\infty(\Omega)$. Observe that, for continuous functions, the notions of superharmonic functions and distributional supersolutions are equivalent. Also, it is known that a superharmonic function on Ω is in $L^1_{\mathrm{loc}}(\Omega)$ and its distributional Laplacian is regarded as a Radon measure on Ω.

We consider the collection $\mathscr{U}(\Omega)$ of all positive superharmonic functions u on Ω satisfying

(U1) $u(x_0) \leq c_3$,

(U2) the distributional Laplacian $-\Delta u \in L^1_{\mathrm{loc}}(\Omega)$,
(U3) $-\Delta u(x) \le c_1\, \delta_\Omega(x)^{-\alpha_1} u(x) + c_2\, \delta_\Omega(x)^{-\alpha_2} u(x)^p$ for a.e. $x \in \Omega$.

Though $\mathscr{U}(\Omega)$ depends on the parameters $c_1, c_2, c_3, \alpha_1, \alpha_2, p$ (these are the same as in the introduction), we do not express them in the symbol $\mathscr{U}(\Omega)$ for simplicity. Observe from (V1) and (V2) that $\mathscr{S}(\Omega) \subset \mathscr{U}(\Omega)$ if $V_1 \le 0$.

2.2 Boundary Growth Estimate

In this subsection, we prove the following.

Theorem 3 *There exist $c = c(c_1, c_2, \alpha_1, \alpha_2, p, \Omega) > 0$ and $\beta = \beta(p, N) \ge 1$ such that*

$$u(x) \le c\, \delta_\Omega(x)^{1-N} \max\{u(x_0), 1\}^\beta \quad \text{for all } x \in \Omega, \tag{10}$$

whenever $u \in \mathscr{U}(\Omega)$.

Corollary 3 *$\mathscr{U}(\Omega) \subset C^1(\Omega)$. Moreover, there exist $c = c(c_1, c_2, \alpha_1, \alpha_2, p, \Omega) > 0$ and $\beta = \beta(p, N) \ge 1$ such that*

$$|\nabla u(x)| \le c\, \delta_\Omega(x)^{-N} \max\{u(x_0), 1\}^{\beta p} \quad \text{for all } x \in \Omega,$$

whenever $u \in \mathscr{U}(\Omega)$.

Remark 4 All the results regarding $\mathscr{U}(\Omega)$ in Sects. 2.2 and 2.3 are valid under $\alpha_1 \le 2$ and $\alpha_2 \le N + 1 - p(N-1)$.

In the arguments below, we suppose $u \in \mathscr{U}(\Omega)$. A proof of Theorem 3 is based on the author's previous work [8]. The next two lemmas follow from the Riesz decomposition theorem for superharmonic functions, the Martin (Poisson) integral representation of harmonic functions and the Green function estimate.

Lemma 1 ([8, Lemma 2.3]) *There exists $c = c(\Omega) > 0$ such that*

$$\int_\Omega \delta_\Omega(y)(-\Delta u(y))\, dy \le c u(x_0).$$

Lemma 2 ([8, Lemma 2.4]) *Let $j \in \mathbb{N}$. Then there exists $c_j = c(j, \Omega) > 0$ such that*

$$u(x) \le c_j \left\{ \delta_\Omega(z)^{1-N} u(x_0) + \int_{B(z, 2^{-j}\delta_\Omega(z))} \frac{-\Delta u(y)}{|x-y|^{N-2}}\, dy \right\} \tag{11}$$

for all $z \in \Omega$ and $x \in B(z, 2^{-j-1}\delta_\Omega(z))$.

For simplicity, we use the symbol $v_1 \lesssim v_2$ if two positive functions v_1 and v_2 satisfy $v_1(x) \le c v_2(x)$ for some constant c independent of variable x. If both of $v_1 \lesssim v_2$ and $v_2 \lesssim v_1$ hold, then we write $v_1 \approx v_2$.

In the arguments below, we fix $z \in \Omega$ and show (10) for $x = z$. For simplicity, we write $B_j := B(0, 2^{-j})$ and

$$\psi(\eta) := \delta_\Omega(z)^{N+1}(-\Delta u(z + \delta_\Omega(z)\eta)) \quad \text{for } \eta \in B_0.$$

Since $-\Delta u \geq 0$ a.e. on Ω and $\delta_\Omega(y) \approx \delta_\Omega(z)$ for all $y \in B(z, \delta_\Omega(z)/2)$, it follows from Lemma 1 and the change of variables that

$$\|\psi\|_{L^1(B_1)} \lesssim u(x_0). \tag{12}$$

Also, we note from (2) and the boundedness of Ω that

$$\delta_\Omega(x)^{-\alpha_1} \lesssim \delta_\Omega(x)^{-2} \quad \text{and} \quad \delta_\Omega(x)^{-\alpha_2} \lesssim \delta_\Omega(x)^{p(N-1)-N-1} \quad \text{for all } x \in \Omega. \tag{13}$$

Lemma 3 *Let*

$$s := \frac{1}{2}\left(p + \frac{N}{N-2}\right) \quad \textit{and} \quad \ell := \left[\frac{\log(s/(s-1))}{\log(s/p)}\right] + 1,$$

where $[\,\cdot\,]$ *is the Gauss symbol. Let* $\kappa \geq 1$. *Then there exists* $c = c(\kappa, c_1, c_2, c_3, \alpha_1, \alpha_2, p, \Omega) > 0$ *such that*

$$\|\psi\|_{L^{\frac{\kappa s}{p}}(B_{j+1})} \leq c\left(\max\{u(x_0), 1\}^p + \|\psi\|^p_{L^\kappa(B_j)}\right) \quad \textit{for all } j \in \{1, \ldots, \ell\}.$$

Proof Let $j \in \{1, \ldots, \ell\}$ and $c_0 := \max\{c_1, \ldots, c_\ell\}$. We write

$$\Psi_j(\eta) := \int_{B_j} \frac{\psi(\zeta)}{|\eta - \zeta|^{N-2}}\, d\zeta.$$

Making the change of variables $x = z + \delta_\Omega(z)\eta$ and $y = z + \delta_\Omega(z)\zeta$ in (11), we have

$$\delta_\Omega(z)^{N-1} u(z + \delta_\Omega(z)\eta) \leq c_0\{u(x_0) + \Psi_j(\eta)\} \quad \text{for all } \eta \in B_{j+1}. \tag{14}$$

It follows from (U3), (13) and (14) that for a.e. $\eta \in B_{j+1}$,

$$\begin{aligned}
\psi(\eta) &\lesssim \delta_\Omega(z)^{N-1} u(z + \delta_\Omega(z)\eta) + \delta_\Omega(z)^{p(N-1)} u(z + \delta_\Omega(z)\eta)^p \\
&\lesssim \{u(x_0) + \Psi_j(\eta)\} + \{u(x_0) + \Psi_j(\eta)\}^p \\
&\lesssim \left(\max\{u(x_0), 1\} + \Psi_j(\eta)\right)^p,
\end{aligned}$$

and therefore

$$\|\psi\|_{L^{\frac{\kappa s}{p}}(B_{j+1})} \lesssim \max\{u(x_0), 1\}^p + \|\Psi_j\|^p_{L^{\kappa s}(B_j)}. \tag{15}$$

To estimate the right hand side, we apply the Jensen inequality. Then

$$\Psi_j(\eta)^\kappa \le \left(\int_{B_0} \frac{d\zeta}{|\eta-\zeta|^{N-2}}\right)^{\kappa-1} \int_{B_j} \frac{\psi(\zeta)^\kappa}{|\eta-\zeta|^{N-2}}\,d\zeta \lesssim \int_{B_j} \frac{\psi(\zeta)^\kappa}{|\eta-\zeta|^{N-2}}\,d\zeta.$$

By the Minkowski inequality for integrals and $s < N/(N-2)$, we have

$$\|\Psi_j\|_{L^{\kappa s}(B_j)} \lesssim \|\psi\|_{L^\kappa(B_j)}.$$

Substituting this into (15), we obtain the desired inequality. □

Proof (*Proof of Theorem* 3) Let s and ℓ be as in Lemma 3, and let $\gamma := s/p$. Then $\gamma > 1$ and $\gamma^\ell/(\gamma^\ell - 1) \le s < N/(N-2)$. By Lemma 2 with $x = z$ and the Hölder inequality,

$$\begin{aligned}\delta_\Omega(z)^{N-1}u(z) &\le c_{\ell+1}\left(u(x_0) + \int_{B_{\ell+1}} \frac{\psi(\zeta)}{|\zeta|^{N-2}}\,d\zeta\right)\\ &\lesssim u(x_0) + \|\psi\|_{L^{\gamma^\ell}(B_{\ell+1})}.\end{aligned}$$

By the way, applying Lemma 3 ℓ times and (12), we have

$$\|\psi\|_{L^{\gamma^\ell}(B_{\ell+1})} \lesssim A + \|\psi\|_{L^1(B_1)}^{p^\ell} \lesssim A,$$

where $A := \max\{u(x_0), 1\}^p + \cdots + \max\{u(x_0), 1\}^{p^\ell}$. Hence we obtain the desired estimate. □

Since $-\Delta u \in L^\infty_{\mathrm{loc}}(\Omega)$ by (U3) and Theorem 3, the classical regularity theorem ensures $u \in C^1(\Omega)$. As a consequence of Theorem 3, we obtain the following estimate.

Lemma 4 *There exists* $c = c(c_1, c_2, c_3, \alpha_1, \alpha_2, p, \Omega) > 0$ *such that*

$$\int_{B(x,r)} |\nabla u|^2\,dy \le cr^{-N} \max\{u(x_0), 1\}^{\beta(p+1)},$$

whenever $u \in \mathscr{U}(\Omega)$ *and* $B(x, 3r) \subset \Omega$.

Proof Since $u \in C^1(\Omega)$ and $-\Delta u \in L^1_{\mathrm{loc}}(\Omega)$, we see from the Riesz representation theorem and (U3) that for all nonnegative $\phi \in C_c^\infty(\Omega)$,

$$\begin{aligned}\int_\Omega \nabla u \cdot \nabla\phi\,dy &= \int_\Omega u(-\Delta\phi)\,dy = (-\Delta u)(\phi) = \int_\Omega (-\Delta u)\phi\,dy\\ &\lesssim \int_\Omega \{\delta_\Omega(y)^{-\alpha_1}u(y) + \delta_\Omega(y)^{-\alpha_2}u(y)^p\}\phi(y)\,dy.\end{aligned} \tag{16}$$

Observe from the standard approximation that (16) holds for all nonnegative $\phi \in C_c^1(\Omega)$. Let $B(x, 3r) \subset \Omega$. Take $\psi \in C_c^\infty(B(x, 2r))$ satisfying $0 \le \psi \le 1$, $|\nabla\psi| \le 2/r$ on $B(x, 2r)$ and $\psi = 1$ on $B(x, r)$. Substituting $\phi := u\psi^2 \in C_c^1(\Omega)$ into (16), we have by (13) that

$$\int_{B(x,2r)} |\nabla u|^2\psi^2\,dy \le 2\int_{B(x,2r)} u\psi|\nabla u||\nabla\psi|\,dy + c\left(r^{-2}\int_{B(x,2r)} u^2\,dy + r^{p(N-1)-N-1}\int_{B(x,2r)} u^{p+1}\,dy\right).$$

Using the Young inequality

$$u\psi|\nabla u||\nabla\psi| \le \frac{(\psi|\nabla u|)^2}{4} + (u|\nabla\psi|)^2 \le \frac{(\psi|\nabla u|)^2}{4} + 4\left(\frac{u}{r}\right)^2,$$

we have by (10) that

$$\begin{aligned}\int_{B(x,2r)} |\nabla u|^2\psi^2\,dy &\lesssim r^{-2}\int_{B(x,2r)} u^2\,dy + r^{p(N-1)-N-1}\int_{B(x,2r)} u^{p+1}\,dy \\ &\lesssim r^{-N}\max\{u(x_0), 1\}^{\beta(p+1)}.\end{aligned} \tag{17}$$

Thus the lemma follows. □

Then we notice that the following gradient estimate established by Kuusi–Mingione [12] leads to our pointwise estimate for $|\nabla u|$.

Lemma 5 ([12, Theorem 1.1]) *Let $v \in C^1(\Omega)$ be a weak solution of $-\Delta v = \mu$ in Ω, where μ is a finite signed measure on Ω. Then there exists $c = c(N) > 0$ such that*

$$|\nabla v(x)| \le c\left(\int_0^r \frac{|\mu|(B(x,t))}{t^{N-1}}\frac{dt}{t} + \fint_{B(x,r)} |\nabla v|\,dy\right),$$

whenever $B(x, r) \subset \Omega$.

Proof (*Proof of Corollary* 3) As mentioned above, $u \in C^1(\Omega)$. Applying the Hölder inequality and Lemma 4 with $r := \delta_\Omega(x)/3$, we have

$$\fint_{B(x,r)} |\nabla u|\,dy \le \left(\fint_{B(x,r)} |\nabla u|^2\,dy\right)^{1/2} \lesssim \delta_\Omega(x)^{-N}\max\{u(x_0), 1\}^{\frac{\beta(p+1)}{2}}.$$

Also, (U3), (13) and Theorem 3 yield

$$\int_0^r t^{1-N}\left(\int_{B(x,t)} (-\Delta u)\,dy\right)\frac{dt}{t} \lesssim \delta_\Omega(x)^{-N}\max\{u(x_0), 1\}^{\beta p}.$$

Therefore the conclusion follows from Lemma 5. □

2.3 Harnack Inequality

The Harnack inequality is the interior estimate that

$$\sup_{B(x,r)} u \leq c_5 \inf_{B(x,r)} u \quad \text{whenever } B(x, 2r) \subset \Omega, \tag{18}$$

where the constant c_5 is independent of x, r and u. Following the previous work [9], let us show this for the class $\mathscr{U}(\Omega)$. Our approach is based on the boundary growth estimate (Theorem 3) and classical potential theory. As a consequence, we see that the lower constraint like $-\Delta u \geq cu^p$ is not need.

Let $u \in \mathscr{U}(\Omega)$. By superharmonicity, we have

$$u(x) \geq \fint_{B(x,r)} u(y)\, dy \geq \fint_{\partial B(x,r)} u\, d\sigma, \tag{19}$$

whenever $\overline{B(x,r)} \subset \Omega$. Here σ is surface area measure on $\partial B(x,r)$. We need the opposite inequalities in some sense to obtain the Harnack inequality.

Let $u \in \mathscr{U}(\Omega)$. By Theorem 3 and (13), we have

$$\delta_\Omega(x)^{-\alpha_2} u(x)^{p-1} \lesssim \delta_\Omega(x)^{-\alpha_2+(p-1)(1-N)} \lesssim \delta_\Omega(x)^{-2} \quad \text{for all } x \in \Omega.$$

Thus, by (U3), there exists $c_6 = c(c_1, c_2, c_3, \alpha_1, \alpha_2, p, \Omega) > 0$ such that

$$-\Delta u(x) \leq c_6\, \delta_\Omega(x)^{-2} u(x) \quad \text{for a.e. } x \in \Omega. \tag{20}$$

Theorem 4 (Strong Harnack inequality) *Let* $0 < \kappa < 1$ *and let* c_6 *be as in* (20). *Put*

$$c_\kappa := \kappa \min\left\{\sqrt{\frac{N\kappa}{8c_6}}, \frac{1}{8}\right\}.$$

If u *is a positive superharmonic function on* Ω *with the distributional Laplacian* $-\Delta u \in L^1_{\mathrm{loc}}(\Omega)$ *satisfying* (20), *then*

$$\frac{1-\kappa}{(1+\kappa)^N} u(x) \leq u(y) \leq \frac{(1+\kappa)^N}{1-\kappa} u(x) \tag{21}$$

for any pair of $x, y \in \Omega$ *satisfying* $|x-y| \leq c_\kappa\, \delta_\Omega(x)$. *In particular, the conclusion is true for all* $u \in \mathscr{U}(\Omega)$.

Proof Let $x \in \Omega$. We first show that

$$(1-\kappa)u(x) \leq (1+\kappa)^N u(y) \quad \text{for all } y \in B(x, 2c_\kappa\, \delta_\Omega(x)). \tag{22}$$

To this end, let $r := 2\kappa^{-1} c_\kappa\, \delta_\Omega(x)$. Since $\delta_\Omega(z) \geq \delta_\Omega(x)/2$ for all $z \in B(x,r)$, it follows from the classical theorem by Riesz, (20) and (19) that

$$u(x) - \fint_{\partial B(x,r)} u\, d\sigma = \frac{1}{N} \int_0^r t \left(\fint_{B(x,t)} (-\Delta u(z))\, dz \right) dt$$
$$\leq \frac{2c_6}{N\, \delta_\Omega(x)^2} u(x) \leq \kappa u(x),$$

whence

$$(1-\kappa)u(x) \leq \fint_{\partial B(x,r)} u\, d\sigma. \tag{23}$$

By the way, if $y \in B(x, \kappa r)$, then $B(x,r) \subset B(y, r + |x-y|) \subset B(x, 3\,\delta_\Omega(x)/4) \subset \Omega$, and so

$$\begin{aligned} \fint_{B(x,r)} u(z)\, dz &\leq \left(\frac{r + |x-y|}{r} \right)^N \fint_{B(y,r+|x-y|)} u(z)\, dz \\ &\leq (1+\kappa)^N u(y). \end{aligned} \tag{24}$$

Combining (19), (23) and (24), we obtain (22).

Let $y \in B(x, c_\kappa\, \delta_\Omega(x))$. Then $x \in B(y, 2c_\kappa\, \delta_\Omega(y))$. Changing the position of x and y in (22) yields $(1-\kappa)u(y) \leq (1+\kappa)^N u(x)$. This completes the proof.

Remark 5 By the standard Harnack chain argument, we can obtain (18) from the conclusion of Theorem 4. Moreover, (21) is stronger than (18), since (21) includes the information that $u(y)$ converges uniformly to $u(x)$ as $\kappa \to 0$. This enables us to obtain the result regarding the existence of nontangential boundary limits of $u \in \mathscr{U}(\Omega)$. See [9, Theorem 6.1].

2.4 Boundary Harnack Principle

The aim of this section is to prove the following.

Theorem 5 *There exist $c_4 = c(\alpha_1, \alpha_2, p, N) \geq 4$, $c = c(c_1, c_2, c_3, \alpha_1, \alpha_2, p, \Omega) \geq 1$ and $r_1 = r(\alpha_1, \alpha_2, p, \Omega) > 0$ with the following property: Let $\xi \in \partial\Omega$ and $0 < r < r_1$. If $u \in \mathscr{U}(\Omega)$ vanishes continuously on $\partial\Omega \cap B(\xi, c_4 r)$, then (7) holds.*

By $\mathscr{H}_+(\Omega)$, we denote the collection of all positive harmonic functions on Ω. The following is an immediate consequence of Theorem 5.

Corollary 4 (Boundary Harnack principle) *The constants c_4, r_1 are the same as in Theorem 5. Let $\xi \in \partial\Omega$ and $0 < r < r_1$. If $u, v \in \mathscr{U}(\Omega) \cup \mathscr{H}_+(\Omega)$ vanish continuously on $\partial\Omega \cap B(\xi, c_4 r)$, then*

$$\frac{1}{c}\frac{u(x)}{v(x)} \leq \frac{u(y)}{v(y)} \leq c\frac{u(x)}{v(x)} \quad \text{for all } x, y \in \Omega \cap B(\xi, r), \tag{25}$$

where $c = c(c_1, c_2, c_3, \alpha_1, \alpha_2, p, \Omega) > 1$.

In [10], the author could obtain the above results under the assumption that $c_1 = 0$, $\alpha_2 = 0$ and Ω is a bounded Lipschitz domain satisfying the uniform exterior ball condition. Using Lemma 9, we modify our previous arguments.

Let us recall the result for harmonic functions.

Lemma 6 (e.g. [2]) *There exist $c = c(\Omega) \geq 1$ and $r_2 = r(\Omega) > 0$ with the following property: Let $\xi \in \partial\Omega$ and $0 < r < r_2$. If h is a positive harmonic function on $\Omega \cap B(\xi, 4r)$ vanishing continuously on $\partial\Omega \cap B(\xi, 4r)$, then*

$$\frac{1}{c}\frac{h(\xi_r)}{r}\,\delta_\Omega(x) \leq h(x) \leq c\frac{h(\xi_r)}{r}\,\delta_\Omega(x) \quad \textit{for all } x \in \Omega \cap B(\xi, r).$$

Remark 6 In [2], the intersections with a bigger ball were considered instead of $B(\xi, 4r)$, but we can modify it by the use of Lemma 18 if $r > 0$ is sufficiently small. For $-\Delta u + Vu = 0$ with an unbounded potential V, estimate (7) is known to hold for positive solutions in the whole of Ω, not in the restricted place $\Omega \cap B(\xi, 4r)$ (see [6, 18]), but this is not applicable to our argument.

The Green function on an open set D for the Dirichlet–Laplacian is denoted by G_D. Recall that $\xi_r := \xi + r\boldsymbol{n}_\xi$ for $\xi \in \partial\Omega$ and $0 < r < r_\Omega$, where $\boldsymbol{n}_\xi$ is the inward unit normal vector to $\partial\Omega$ at ξ.

Lemma 7 *Assume that u is a positive superharmonic function on Ω enjoying the Harnack inequality* (18) *with constant c_5. Then there exist $c = c(c_5, \Omega) > 0$ and $r_3 = r(\Omega) > 0$ with the following property: Let $\xi \in \partial\Omega$ and $0 < r < r_3$. Then*

$$\frac{u(\xi_r)}{r}\,\delta_\Omega(x) \leq cu(x) \quad \textit{for all } x \in \Omega \cap B(\xi, r).$$

Proof Let $G(x) := G_{\Omega\cap B(\xi,6r)}(x, \xi_{4r})$. The Harnack inequality gives

$$\frac{G(x)}{G(\xi_r)} \approx 1 \approx \frac{u(x)}{u(\xi_r)} \quad \text{for all } x \in \partial B(\xi_{4r}, r).$$

It follows from the minimum principle and Lemma 6 that

$$\frac{u(x)}{u(\xi_r)} \gtrsim \frac{G(x)}{G(\xi_r)} \approx \frac{\delta_\Omega(x)}{r} \quad \text{for all } x \in \Omega \cap B(\xi, r).$$

Thus the lemma is proved. □

In order to obtain Theorem 5 for $\alpha_2 > 0$, we need the following estimate, which is also used in the study of removable isolated singularities on $\partial\Omega$ in Sect. 4.2.

Lemma 8 *Assume $0 \in \partial\Omega$. Let $\beta > -2$ and $\gamma > -N - 1 - \beta$. Then there exists $c = c(\beta, \gamma, \Omega) > 0$ such that*

$$\int_\Omega G_\Omega(x, y)\,\delta_\Omega(y)^\beta |y|^\gamma\, dy \leq c\,\delta_\Omega(x)\{J_1(x) + J_2(x)\} \quad \textit{for all } x \in \Omega,$$

where

$$J_1(x) := \begin{cases} |x|^{1+\beta+\gamma} & \text{if } \beta > -1, \\ |x|^\gamma \log \dfrac{2|x|}{\delta_\Omega(x)} & \text{if } \beta = -1, \\ \delta_\Omega(x)^{1+\beta}|x|^\gamma & \text{if } \beta < -1, \end{cases}$$

and

$$J_2(x) := \begin{cases} |x|^{1+\beta+\gamma} & \text{if } 1+\beta+\gamma < 0, \\ \log \dfrac{\operatorname{diam}\Omega}{|x|} & \text{if } 1+\beta+\gamma = 0, \\ 1 & \text{if } 1+\beta+\gamma > 0. \end{cases}$$

Proof Fix $x \in \Omega$ and split the range of integration into four parts:

$$E_1 := B(x, \delta_\Omega(x)/2), \qquad E_2 := \Omega \cap B(x, |x|/2) \setminus E_1,$$
$$E_3 := \Omega \cap B(0, 2|x|) \setminus E_2, \quad E_4 := \Omega \setminus (E_1 \cup E_2 \cup E_3).$$

We use the Green function estimate:

$$G_\Omega(x, y) \lesssim \min\left\{\frac{1}{|x-y|^{N-2}}, \frac{\delta_\Omega(x)\,\delta_\Omega(y)}{|x-y|^N}\right\}.$$

If $y \in E_1$, then $\delta_\Omega(y) \approx \delta_\Omega(x)$ and $|y| \approx |x|$, and so

$$\int_{E_1} G_\Omega(x, y)\, \delta_\Omega(y)^\beta |y|^\gamma\, dy \lesssim \delta_\Omega(x)^{2+\beta}|x|^\gamma.$$

If $y \in E_2$, then $|y| \approx |x|$. Let $A_j := B(x, 2^j\,\delta_\Omega(x)) \setminus B(x, 2^{j-1}\,\delta_\Omega(x))$ and let m be the smallest integer such that $2^m\,\delta_\Omega(x) \geq |x|/2$. Then $A_j \subset B(\eta, 2^{j+1}\,\delta_\Omega(x))$ for some $\eta \in \partial\Omega$. By $\beta > -2$, we have

$$\begin{aligned} \int_{E_2} G_\Omega(x, y)\, \delta_\Omega(y)^\beta |y|^\gamma\, dy &\lesssim \delta_\Omega(x)|x|^\gamma \sum_{j=0}^{m} \int_{\Omega\cap A_j} \frac{\delta_\Omega(y)^{1+\beta}}{|x-y|^N}\, dy \\ &\lesssim \delta_\Omega(x)|x|^\gamma \sum_{j=0}^{m} (2^j\,\delta_\Omega(x))^{-N} \int_{\Omega\cap B(\eta, 2^{j+1}\,\delta_\Omega(x))} \delta_\Omega(y)^{1+\beta}\, dy \\ &\lesssim \delta_\Omega(x)|x|^\gamma \sum_{j=0}^{m} (2^j\,\delta_\Omega(x))^{1+\beta} \\ &\lesssim \delta_\Omega(x) J_1(x). \end{aligned}$$

If $y \in E_3$, then $|x-y| \approx |x|$. Let $A'_j := B(0, 2^{-j+1}|x|) \setminus B(0, 2^{-j}|x|)$. Then

$$\begin{aligned}
\int_{E_3} G_\Omega(x,y)\,\delta_\Omega(y)^\beta |y|^\gamma\,dy &\lesssim \frac{\delta_\Omega(x)}{|x|^N} \sum_{j=0}^\infty \int_{\Omega\cap A'_j} \delta_\Omega(y)^{1+\beta}|y|^\gamma\,dy \\
&\lesssim \frac{\delta_\Omega(x)}{|x|^N} \sum_{j=0}^\infty (2^{-j}|x|)^\gamma \int_{\Omega\cap B(0,2^{-j+1}|x|)} \delta_\Omega(y)^{1+\beta}\,dy \\
&\lesssim \delta_\Omega(x)|x|^{1+\beta+\gamma} \sum_{j=0}^\infty 2^{-j(N+1+\beta+\gamma)}.
\end{aligned}$$

Since $N+1+\beta+\gamma > 0$, the last series converges.

If $y \in E_4$ (assuming $E_4 \neq \emptyset$), then $|x-y| \approx |y|$. Let $A''_j := B(0, 2^{j+1}|x|) \setminus B(0, 2^j|x|)$ and let n be the smallest integer such that $2^{n+1}|x| \geq \operatorname{diam}\Omega$. Then

$$\begin{aligned}
\int_{E_4} G_\Omega(x,y)\,\delta_\Omega(y)^\beta |y|^\gamma\,dy &\lesssim \delta_\Omega(x) \sum_{j=1}^n \int_{\Omega\cap A''_j} \frac{\delta_\Omega(y)^{1+\beta}}{|y|^{N-\gamma}}\,dy \\
&\lesssim \delta_\Omega(x) \sum_{j=1}^n (2^j|x|)^{\gamma-N} \int_{\Omega\cap B(0,2^{j+1}|x|)} \delta_\Omega(y)^{1+\beta}\,dy \\
&\lesssim \delta_\Omega(x)|x|^{1+\beta+\gamma} \sum_{j=1}^n 2^{j(1+\beta+\gamma)} \\
&\lesssim \delta_\Omega(x) J_2(x).
\end{aligned}$$

Combining all the estimates yields the desired estimate. □

The special case $\gamma = 0$ becomes as follows.

Lemma 9 *Let $\beta > -2$. Then there exists $c = c(\beta, \Omega) > 0$ such that*

$$\int_\Omega G_\Omega(x,y)\,\delta_\Omega(y)^\beta\,dy \leq cJ(x) \quad \text{for all } x \in \Omega,$$

where

$$J(x) := \begin{cases} \delta_\Omega(x) & \text{if } \beta > -1, \\ \delta_\Omega(x) \log \dfrac{2\operatorname{diam}\Omega}{\delta_\Omega(x)} & \text{if } \beta = -1, \\ \delta_\Omega(x)^{\beta+2} & \text{if } \beta < -1. \end{cases}$$

Lemma 10 *There exist $c = c(c_1, c_2, c_3, \alpha_1, \alpha_2, p, \Omega) > 0$ and $\gamma = \gamma(\alpha_1, \alpha_2, p, \Omega) > 0$ with the following property: Let $\xi \in \partial\Omega$ and $0 < r < r_\Omega/2$. Then*

$$u(x) \le c\left(\frac{|x-\xi|}{r}\right)^{\gamma} \sup_{\Omega\cap B(\xi,r)} u \quad \textit{for all } x \in \Omega \cap B(\xi, r),$$

whenever $u \in \mathscr{U}(\Omega)$ *vanishes continuously on* $\partial\Omega \cap B(\xi, r)$.

Proof We may assume $\xi = 0$. Let $x \in \Omega_r := \Omega \cap B(0, r)$. By the Riesz decomposition theorem, there exists a nonnegative harmonic function h on Ω_r such that

$$u(x) = h(x) + \int_{\Omega_r} G_{\Omega_r}(x, y)(-\Delta u(y))\, dy.$$

As is well known, by comparing h with the Martin kernel $|x|^{\gamma_1} f(x/|x|)$ with pole at ∞ of the complement of the exterior cone of Ω with vertex at 0, we have

$$h(x) \lesssim \left(\frac{|x|}{r}\right)^{\gamma_1} \sup_{\Omega_r} u.$$

Let $\beta := -\max\{\alpha_1, \alpha_2 + (N-1)(p-1)\}$. Then $\beta > -2$ by (2). Also, by (U3) and Theorem 3,

$$-\Delta u(y) \lesssim \delta_\Omega(y)^\beta \sup_{\Omega_r} u \quad \text{for a.e. } y \in \Omega_r.$$

Since $G_{\Omega_r} \le G_\Omega$ on $\Omega_r \times \Omega_r$, it follows from Lemma 9 that

$$\int_{\Omega_r} G_{\Omega_r}(x, y)(-\Delta u(y))\, dy \lesssim \delta_\Omega(x)^{\gamma_2} \sup_{\Omega_r} u \lesssim \left(\frac{|x|}{r}\right)^{\gamma_2} \sup_{\Omega_r} u$$

for some $\gamma_2 > 0$. Thus the lemma follows. □

Lemma 11 (Carleson estimate) *There exist* $c = c(c_1, c_2, c_3, \alpha_1, \alpha_2, p, \Omega) > 0$ *and* $r_4 = r(\Omega) > 0$ *with the following property: Let* $\xi \in \partial\Omega$ *and* $0 < r < r_4$. *If* $u \in \mathscr{U}(\Omega)$ *vanishes continuously on* $\partial\Omega \cap B(\xi, 2r)$, *then*

$$u(x) \le c u(\xi_r) \quad \textit{for all } x \in \Omega \cap B(\xi, r). \tag{26}$$

Proof An idea of the proof is based on [5, 11]. First, we observe from Theorem 4 (Remark 5) and the standard Harnack chain argument (Lemma 18) that there exists a constant $\lambda > 1$ such that

$$u(x) \le \left(\frac{4r}{\delta_\Omega(x)}\right)^{\lambda} u(\xi_r) \quad \text{for all } x \in \Omega \cap B(\xi, 2r),$$

or equivarently

$$\delta_\Omega(x) \le 4\left(\frac{u(\xi_r)}{u(x)}\right)^{\frac{1}{\lambda}} r \quad \text{for all } x \in \Omega \cap B(\xi, 2r). \tag{27}$$

Also, by Lemma 10, there exists $c_7 > 1$ such that

$$u(x) < \frac{1}{2^\lambda} \sup_{\Omega \cap B(\eta_x, c_7\, \delta_\Omega(x))} u, \tag{28}$$

whenever $x \in \Omega \cap B(\xi, 2r)$ satisfies $c_7\, \delta_\Omega(x) < r$ and $\eta_x \in \partial\Omega$ is such that $\delta_\Omega(x) = |\eta_x - x|$ (a similar notation is used in the argument below to denote the nearest boundary point). Put $c_8 := 1/(33c_7)$.

Let us show (26) with $c = 1/c_8^\lambda$. To this end, we suppose to the contrary that there exists $x_1 \in \Omega \cap B(\xi, r)$ such that $c_8^\lambda u(x_1) > u(\xi_r)$. Then $\delta_\Omega(x_1) < 4c_8 r < r/c_7$ by (27). Therefore, by (28), we find $x_2 \in \Omega \cap B(\eta_{x_1}, c_7\, \delta_\Omega(x_1))$ with $2^\lambda u(x_1) \le u(x_2)$. Then

$$\begin{aligned} |x_2 - \xi| &\le |x_2 - \eta_{x_1}| + |\eta_{x_1} - x_1| + |x_1 - \xi| \le 2c_7\, \delta_\Omega(x_1) + r \\ &\le (8c_7c_8 + 1)r \le \frac{3}{2}r \end{aligned}$$

and $c_8^\lambda u(x_2) > 2^\lambda u(\xi_r)$, which implies $\delta_\Omega(x_2) < (4c_8 r)/2 < r/c_7$ by (27). Repeating this process, we obtain a sequence $\{x_k\}$ in Ω such that

$$\begin{aligned} c_8^\lambda u(x_k) &> 2^{\lambda(k-1)} u(\xi_r), \\ \delta_\Omega(x_k) &< \frac{4c_8}{2^{k-1}} r < \frac{r}{c_7} \end{aligned}$$

and

$$\begin{aligned} |x_k - \xi| &\le \sum_{j=2}^{k} |x_j - x_{j-1}| + |x_1 - \xi| \le \sum_{j=2}^{k} 2c_7\, \delta_\Omega(x_{j-1}) + r \\ &\le \sum_{j=2}^{\infty} \frac{8c_7c_8}{2^{j-2}} r + r \le \frac{3}{2} r. \end{aligned}$$

But this contradicts that u vanishes continuously on $\partial\Omega \cap B(\xi, 2r)$. Thus the lemma is proved. □

We are now in a position to prove Theorem 5.

Proof (*Proof of Theorem* 5) Let $c_4 := 4^{m+3}$, where $m \in \mathbb{N}$ is chosen later, and let r_2, r_3, r_4 be as in Lemmas 6, 7, 11, respectively. Put $r_1 := \min\{r_2, r_3, r_4\}/c_4$.

Let $0 < r < r_1$ and $\xi \in \partial\Omega$. For $k \in \{0, 1, \ldots, m + 2\}$, we write

$$\Omega_k := \Omega \cap B(\xi, 4^k r) \quad \text{and} \quad G_k := G_{\Omega_k}.$$

Let $u \in \mathscr{U}(\Omega)$ vanish continuously on $\partial\Omega \cap B(\xi, c_4 r)$. In light of the Harnack inequality (Theorem 4) and Lemma 7, it suffices to show the upper estimate:

$$u(x) \lesssim \frac{u(\xi_r)}{r} \delta_\Omega(x) \quad \text{for all } x \in \Omega_0. \tag{29}$$

Let $k \in \{1, \ldots, m+2\}$. By the Riesz decomposition theorem, there exists a positive harmonic function h_k on Ω_k such that

$$u(x) = h_k(x) + \int_{\Omega_k} G_k(x, y)(-\Delta u(y))\, dy \quad \text{for all } x \in \Omega_k. \tag{30}$$

Since h_k vanishes continuously on $\partial\Omega \cap B(\xi, 4^k r)$, it follows from Lemma 6 and the Harnack inequality that for all $x \in \Omega_{k-1}$,

$$h_k(x) \lesssim \frac{h_k(\xi_{4^{k-1}r})}{r} \delta_\Omega(x) \lesssim \frac{h_k(\xi_r)}{r} \delta_\Omega(x) \lesssim \frac{u(\xi_r)}{r} \delta_\Omega(x). \tag{31}$$

Also, letting $\beta := -\max\{\alpha_1, \alpha_2 + (N-1)(p-1)\} > -2$, we have from (U3) and Theorem 3 that

$$-\Delta u(y) \lesssim \delta_\Omega(y)^\beta u(y) \lesssim \delta_\Omega(y)^\beta u(\xi_r) \quad \text{for a.e. } y \in \Omega_k, \tag{32}$$

since $u(y) \lesssim u(\xi_{4^k r}) \lesssim u(\xi_r)$ by Lemma 11 and the Harnack inequality.

Case 1: $\beta > -1$. Let $x \in \Omega_0$. By (32) and Lemma 9,

$$\int_{\Omega_1} G_1(x, y)(-\Delta u(y))\, dy \lesssim u(\xi_r)\, \delta_\Omega(x) \lesssim \frac{u(\xi_r)}{r} \delta_\Omega(x).$$

This, together with (30) and (31), yields (29).

Case 2: $\beta = -1$. Let $z \in \Omega_1$. By (32) and Lemma 9,

$$\int_{\Omega_2} G_2(z, y)(-\Delta u(y))\, dy \lesssim u(\xi_r)\, \delta_\Omega(z) \log \frac{2 \operatorname{diam} \Omega}{\delta_\Omega(z)} \lesssim \frac{u(\xi_r)}{r} \delta_\Omega(z)^{\frac{1}{2}}.$$

Substituting this and (31) into (30), we have

$$u(z) \lesssim \frac{u(\xi_r)}{r} \delta_\Omega(z)^{\frac{1}{2}}.$$

Therefore

$$-\Delta u(z) \lesssim \delta_\Omega(z)^\beta u(z) \lesssim \frac{u(\xi_r)}{r} \delta_\Omega(z)^{\beta + \frac{1}{2}} \quad \text{for a.e. } z \in \Omega_1.$$

Again, we apply Lemma 9 to obtain

$$\int_{\Omega_1} G_1(x,z)(-\Delta u(z))\,dz \lesssim \frac{u(\xi_r)}{r}\delta_\Omega(x) \quad \text{for all } x \in \Omega_1.$$

Then (30) and (31) yield (29).

Case 3: $\beta < -1$. Let $\gamma_1 := \beta + 2$. Since $0 < \gamma_1 < 1$, we take the largest $m \in \mathbb{N}$ with $m\gamma_1 < 1$. Define

$$\gamma_k := k\gamma_1 \quad \text{for } k \in \{0, \ldots, m+2\}.$$

Note that $\gamma_0 = 0$, $0 < \gamma_k < 1$ for all $k \in \{1, \ldots, m\}$ and $\gamma_{m+1} \geq 1$.

We claim that, for each $k \in \{0, \ldots, m\}$, we have

$$u(x) \lesssim \frac{u(\xi_r)}{r}\delta_\Omega(x)^{\gamma_k} \quad \text{for all } x \in \Omega_{m+2-k}. \tag{33}$$

Let us show this by induction. The case $k = 0$ follows from Lemma 11 and the Harnack inequality. In fact, for all $x \in \Omega_{m+2}$,

$$u(x) \lesssim u(\xi_{4^{m+2}r}) \lesssim u(\xi_r) \lesssim \frac{u(\xi_r)}{r}.$$

Assume that (33) holds for some $k \in \{0, \ldots, m-1\}$. Then, for a.e. $y \in \Omega_{m+2-k}$,

$$-\Delta u(y) \lesssim \delta_\Omega(y)^\beta u(y) \lesssim \frac{u(\xi_r)}{r}\delta_\Omega(y)^{\beta+\gamma_k}. \tag{34}$$

Since $\beta + \gamma_k = \gamma_{k+1} - 2 \in (-2, -1)$, it follows from (34) and Lemma 9 that

$$\int_{\Omega_{m+2-k}} G_{m+2-k}(x,y)(-\Delta u(y))\,dy \lesssim \frac{u(\xi_r)}{r}\delta_\Omega(x)^{\gamma_{k+1}} \quad \text{for all } x \in \Omega_{m+2-k}.$$

This, together with (30) and (31), yields that

$$u(x) \lesssim \frac{u(\xi_r)}{r}\delta_\Omega(x)^{\gamma_{k+1}} \quad \text{for all } x \in \Omega_{m+1-k}.$$

Thus the claim holds.

By (33) with $k = m$, we have

$$-\Delta u(y) \lesssim \delta_\Omega(y)^\beta u(y) \lesssim \frac{u(\xi_r)}{r}\delta_\Omega(y)^{\beta+\gamma_m} \quad \text{for a.e. } y \in \Omega_2$$

and $\beta + \gamma_m = \gamma_{m+1} - 2 \geq -1$. If $\beta + \gamma_m > -1$, then we can apply Lemma 9 to obtain

$$u(x) \lesssim \frac{u(\xi_r)}{r}\,\delta_\Omega(x) \quad \text{for all } x \in \Omega_1 \supset \Omega_0.$$

If $\beta + \gamma_m = -1$, then we apply Lemma 9 twice to obtain (29), as in the Case 2.
This completes the proof. □

3 Proofs of Theorems 1, 2 and Corollaries 1, 2

We develop some results in Sect. 2 to a general case of V_1. But we should note that the results and arguments in Sect. 2 cannot be applied to positive solutions of (1), since they are not necessarily superharmonic on Ω if V_1 takes positive values. Moreover, it seems that there is no counterpart of Lemma 6 for positive solutions of $-\Delta u + V_1^+ u = 0$ in the restricted place $\Omega \cap B(\xi, 4r)$, not in the whole of Ω, whence the argument given in the proof of Theorem 5 does not work well even if we use the decomposition (35) below. We borrow an idea from Cranston–Fabes–Zhao [6] in the study of the linear Schrödinger equation.

For $a \in \mathbb{R}$, we write

$$a^+ := \max\{a, 0\} \quad \text{and} \quad a^- := -\min\{a, 0\}.$$

Using the estimate: if $0 \le \varepsilon \le 1$, then

$$G_\Omega(x, y) \lesssim \frac{\delta_\Omega(x)\,\delta_\Omega(y)^\varepsilon}{|x-y|^{N-1+\varepsilon}} \quad \text{for all } x, y \in \Omega,$$

we see from (V1) and $\alpha_1 < 2$ that for all $x \in \Omega$,

$$\begin{aligned}\int_\Omega \frac{\delta_\Omega(y)}{\delta_\Omega(x)} G_\Omega(x, y) V_1^+(y)\,dy &\lesssim \int_\Omega \frac{\delta_\Omega(y)}{\delta_\Omega(x)} G_\Omega(x, y)\,\delta_\Omega(y)^{-\alpha_1}\,dy \\ &\lesssim \int_\Omega |x-y|^{1-N-(\alpha_1-1)^+}\,dy \lesssim 1.\end{aligned}$$

Thus $V_1^+ \in \mathcal{K}(\Omega)$ in the sense of Riahi [17, 18], by which the Green function $G_{V_1^+,\Omega}$ for the Dirichlet–Schrödinger operator $-\Delta + V_1^+$ on Ω has the following comparison (see also [1, 6, 14, 15, 21]). Note also that $G_{V_1^+,\Omega}$ is positive and continuous on $\Omega \times \Omega$ in the extended sense ([20]). Let M_Ω and $M_{V_1^+,\Omega}$ denote the Martin kernels on Ω (normalized at x_0) for $-\Delta$ and $-\Delta + V_1^+$, respectively.

Lemma 12 *There exists $c_9 = c(V_1^+, \Omega) \ge 1$ such that*

$$\frac{1}{c_9} G_\Omega(x, y) \le G_{V_1^+,\Omega}(x, y) \le G_\Omega(x, y) \quad \textit{for all } x, y \in \Omega$$

and

$$\frac{1}{c_9}M_\Omega(x,y) \le M_{V_1^+,\Omega}(x,y) \le c_9 M_\Omega(x,y) \quad \textit{for all } x \in \Omega \textit{ and } y \in \partial\Omega.$$

Lemma 13 *Assume that* $0 \le f \in L^\infty_{\mathrm{loc}}(\Omega)$ *for which equation* $-\Delta u + V_1^+ u = f$ *in* Ω *has a positive solution* $u \in C(\Omega)$. *Then there exists a nonnegative solution* $h \in C(\Omega)$ *of* $-\Delta h + V_1^+ h = 0$ *in* Ω *such that*

$$u(x) = h(x) + \int_\Omega G_{V_1^+,\Omega}(x,y) f(y)\,dy \quad \textit{for all } x \in \Omega.$$

Proof This may be known as the Riesz decomposition. For completeness, we give a sketch of the proof. Let $\{\Omega_k\}$ be a sequence of smooth domains such that $\overline{\Omega_k} \subset \Omega_{k+1}$ and $\bigcup_{k=1}^\infty \Omega_k = \Omega$. For $k \in \mathbb{N}$, let $h_k := u - v_k$, where

$$v_k(x) := \int_{\Omega_k} G_{V_1^+,\Omega_k}(x,y) f(y)\,dy \quad \text{for } x \in \Omega_k.$$

By the Fubini theorem, we see that h_k satisfies $-\Delta h_k + V_1^+ h_k = 0$ in Ω_k in the sense of distributions. Moreover, $h_k \in C(\Omega_k)$ by [20]. Since v_k vanishes continuously on $\partial\Omega_k$, the minimum principle ensures that $h_k > 0$ in Ω_k. Also, from the argument in [13, p. 873], we know that $G_{V_1^+,\Omega_k}$ converges increasingly to $G_{V_1^+,\Omega}$ on Ω as $k \to \infty$. By the monotone convergence,

$$\int_\Omega G_{V_1^+,\Omega}(x,y) f(y)\,dy = \lim_{k\to\infty} v_k(x) \le u(x) < \infty \quad \text{for all } x \in \Omega$$

and $\{h_k\}$ converges decreasingly to some nonnegative solution h of $-\Delta h + V_1^+ h = 0$ in Ω. Again, $h \in C(\Omega)$ by [20]. Thus the lemma is proved. □

Let $u \in \mathscr{S}(\Omega)$. By Lemma 13, there is a nonnegative solution $h \in C(\Omega)$ of $-\Delta h + V_1^+ h = 0$ in Ω such that

$$u(x) = h(x) + \int_\Omega G_{V_1^+,\Omega}(x,y)\{V_1^-(y)u(y) + V_2(y)u(y)^p\}\,dy \quad \text{for all } x \in \Omega. \tag{35}$$

By the Martin representation theorem (cf. [18, Theorem 6.1]), there exists a Radon measure μ on $\partial\Omega$ such that

$$h(x) = \int_{\partial\Omega} M_{V_1^+,\Omega}(x,y)\,d\mu(y) \quad \text{for all } x \in \Omega.$$

Using this μ, we define

$$\widetilde{u}(x) := \int_{\partial\Omega} M_\Omega(x,y)\,d\mu(y) + \int_\Omega G_\Omega(x,y)\{V_1^-(y)u(y) + V_2(y)u(y)^p\}\,dy \tag{36}$$

for $x \in \Omega$. Then $\widetilde{u}$ is a positive superharmonic function on Ω with the distributional Laplacian $-\Delta\widetilde{u} = V_1^- u + V_2 u^p \in L^\infty_{\mathrm{loc}}(\Omega)$. Moreover, by Lemma 12, we have

$$\frac{1}{c_9}\widetilde{u}(x) \le u(x) \le c_9\widetilde{u}(x) \quad \text{for all } x \in \Omega$$

and

$$-\Delta\widetilde{u}(x) \le c_9 V_1^-(x)\widetilde{u}(x) + c_9^p V_2(x)\widetilde{u}(x)^p \quad \text{for a.e. } x \in \Omega.$$

These, together with (V1) and (V2), signify that $\widetilde{u} \in \mathscr{U}(\Omega)$ if replacing the constants c_1, c_2, c_3 in (U1) and (U3) by bigger one, and that $\widetilde{u}$ behaves like u. This is a reason why we considered the class $\mathscr{U}(\Omega)$. Then the results for $u \in \mathscr{S}(\Omega)$ stated in the introduction can be obtained by applying the results in Sect. 2 to $\widetilde{u}$ as follows.

Proof (*Proof of Theorem* 1) Applying Theorem 3 to $\widetilde{u}$, we have

$$u(x) \lesssim \widetilde{u}(x) \lesssim \delta_\Omega(x)^{1-N} \quad \text{for all } x \in \Omega,$$

as desired. □

Proof (*Proof of Corollary* 1) Let $u \in \mathscr{S}(\Omega)$. Since $-\Delta u \in L^\infty_{\mathrm{loc}}(\Omega)$ by (V1), (V2) and Theorem 1, the classical regularity theorem ensures $u \in C^1(\Omega)$. Note that we have no information about $|\nabla u| \approx |\nabla\widetilde{u}|$. But, since $-\Delta u$ satisfies (U3) by (V1) and (V2), we observe that (16) holds. Repeating the same argument as in the proof of Lemma 4 with the help of Theorem 1, we can obtain

$$\int_{B(x,r)} |\nabla u|^2\,dy \lesssim r^{-N} \quad \text{whenever } B(x,3r) \subset \Omega.$$

Thus (4) can be proved in the same manner as in the proof of Corollary 3. □

Theorem 6 (Harnack inequality) *There exists* $c_5 = c(c_1, c_2, c_3, \alpha_1, \alpha_2, p, V_1^+, \Omega) \ge 1$ *such that* (18) *holds for all* $u \in \mathscr{S}(\Omega)$.

Proof Let $u \in \mathscr{S}(\Omega)$ and $B(x,2r) \subset \Omega$. Applying Theorem 4 (Remark 5) to $\widetilde{u}$, we have

$$\sup_{B(x,r)} u \lesssim \sup_{B(x,r)} \widetilde{u} \lesssim \inf_{B(x,r)} \widetilde{u} \lesssim \inf_{B(x,r)} u,$$

as desired. □

Proof (*Proof of Theorem* 2) By the assumption on $u \in \mathscr{S}(\Omega)$, $\widetilde{u}$ vanishes continuously on $\partial\Omega \cap B(\xi, c_4 r)$. Applying Theorem 5 to $\widetilde{u}$, we obtain

$$\frac{u(x)}{u(\xi_r)} \approx \frac{\widetilde{u}(x)}{\widetilde{u}(\xi_r)} \approx \frac{\delta_\Omega(x)}{r} \quad \text{for all } x \in \Omega \cap B(\xi, r),$$

as desired. □

The next corollary follows immediately from Theorem 2.

Corollary 5 *The constants c_4 and r_1 are the same as in Theorem 2. Let $\xi \in \partial\Omega$ and $0 < r < r_1$. If $u, v \in \mathscr{S}(\Omega) \cup \mathscr{H}_+(\Omega)$ vanish continuously on $\partial\Omega \cap B(\xi, c_4 r)$, then* (25) *holds with $c = c(c_1, c_2, c_3, \alpha_1, \alpha_2, p, V_1^+, \Omega) > 1$.*

Proof (*Proof of Corollary* 2) Using Theorem 2 with $r = r_1/2$ and the Harnack inequality, we have $u(x) \lesssim \delta_\Omega(x) u(x_0) \lesssim c_3\, \delta_\Omega(x)$ for all $x \in \Omega$, whence (8) follows.

To show (9), let $r := \delta_\Omega(x)/3$. Since u satisfies (16), it follows from (17) and (8) that

$$\int_{B(x,r)} |\nabla u|^2\, dy \lesssim r^{-2} \int_{B(x,2r)} u^2\, dy + r^{p(N-1)-N-1} \int_{B(x,2r)} u^{p+1}\, dy$$
$$\lesssim r^N + r^{pN} \lesssim r^N.$$

Also, by (V1), (V2), (2) and (8), we have

$$\int_0^r t^{1-N} \int_{B(x,t)} |-\Delta u(y)|\, dy \frac{dt}{t} \lesssim 1.$$

Appealing to Lemma 5, we can obtain (9). This completes the proof. □

4 Applications

This section gives some applications of our main results stated in the introduction.

4.1 Positive Solutions with Isolated Boundary Singularities

Let us show the asymptotic behavior of a positive solution of (1) near its isolated singularity on $\partial\Omega$. For $p \geq 1$, $\alpha \in \mathbb{R}$ and $\xi \in \partial\Omega$, we let

$$F_{p,\alpha}(x) := \begin{cases} |x-\xi|^{N+1-p(N-1)-\alpha} & \text{if } 1 - p(N-1) - \alpha < 0, \\ |x-\xi|^N \log \dfrac{2\operatorname{diam}\Omega}{|x-\xi|} & \text{if } 1 - p(N-1) - \alpha = 0, \\ |x-\xi|^N & \text{if } 1 - p(N-1) - \alpha > 0. \end{cases}$$

Observe that $F_{p,\alpha}$ vanishes continuously at ξ if $\alpha < N + 1 - p(N-1)$.

Theorem 7 *Let $\xi \in \partial\Omega$. If $u \in \mathscr{S}(\Omega)$ vanishes continuously on $\partial\Omega \setminus \{\xi\}$, then there exist $\lambda = \lambda(u) \geq 0$ and $c_{10} = c(c_1, c_2, c_3, \alpha_1, \alpha_2, p, V_1^+, \Omega) > 0$ such that*

$$0 \leq \frac{u(x)}{M_{V_1^+,\Omega}(x,\xi)} - \lambda \leq c_{10} \max\{F_{1,\alpha_1}(x), F_{p,\alpha_2}(x)\} \quad \textit{for all } x \in \Omega. \tag{37}$$

Moreover, there exist $\lambda' = \lambda(u) \geq 0$ *and* $c = c(c_1, \alpha_1, V_1^+, \Omega) > 0$ *such that*

$$-c\lambda F_{1,\alpha_1}(x) \leq \frac{u(x)}{M_\Omega(x,\xi)} - \lambda' \leq c_{10} \max\{F_{1,\alpha_1}(x), F_{p,\alpha_2}(x)\} \quad \textit{for all } x \in \Omega. \tag{38}$$

Remark 7 If $\lambda = 0$, then $\lambda' = 0$.

We need the following modification of [4, Theorem 3.1].

Lemma 14 *Let* $p \geq 1$ *and* $\alpha < N + 1 - p(N-1)$. *Then there exists* $c = c(p, \alpha, \Omega) > 0$ *such that*

$$\int_\Omega G_\Omega(x,y) M_\Omega(y,\xi)^p \, \delta_\Omega(y)^{-\alpha} \, dy \leq c F_{p,\alpha}(x) M_\Omega(x,\xi)$$

for all $x \in \Omega$ *and* $\xi \in \partial\Omega$.

Proof To shorten the notation, we may let $\xi = 0$. We use the estimates for the Green function and the Martin kernel: if $0 \leq \varepsilon \leq 1$, then

$$G_\Omega(x,y) \lesssim \min\left\{\frac{1}{|x-y|^{N-2}}, \frac{\delta_\Omega(x)\,\delta_\Omega(y)^{1-\varepsilon}}{|x-y|^{N-\varepsilon}}\right\} \quad \text{for all } x, y \in \Omega$$

and

$$M_\Omega(x,0) \approx \frac{\delta_\Omega(x)}{|x|^N} \quad \text{for all } x \in \Omega.$$

Let $x \in \Omega$. We split the range of integration into three parts:

$$E_1 := \Omega \cap B(0, |x|/2), \quad E_2 := \Omega \cap B(x, |x|) \setminus E_1, \quad E_3 := \Omega \setminus (E_1 \cup E_2).$$

If $y \in E_1$, then $|x-y| \approx |x|$. Since $1 + p - \alpha > 0$, we have

$$\begin{aligned}\frac{G_\Omega(x,y) M_\Omega(y,0)^p \, \delta_\Omega(y)^{-\alpha}}{M_\Omega(x,0)} &\lesssim \frac{\frac{\delta_\Omega(x)\,\delta_\Omega(y)}{|x-y|^N}\left(\frac{\delta_\Omega(y)}{|y|^N}\right)^p \delta_\Omega(y)^{-\alpha}}{\frac{\delta_\Omega(x)}{|x|^N}} \\ &\lesssim \frac{\delta_\Omega(y)^{1+p-\alpha}}{|y|^{Np}} \lesssim |y|^{1-p(N-1)-\alpha},\end{aligned}$$

whence

$$\int_{E_1} G_\Omega(x,y) M_\Omega(y,0)^p \, \delta_\Omega(y)^{-\alpha} \, dy \lesssim |x|^{N+1-p(N-1)-\alpha} M_\Omega(x,0).$$

If $y \in E_2$, then $|y| \approx |x|$. Take $\varepsilon > 0$ so that $1 + p - \alpha - \varepsilon > 0$. Then

$$\frac{G_\Omega(x,y)M_\Omega(y,0)^p\,\delta_\Omega(y)^{-\alpha}}{M_\Omega(x,0)} \lesssim \frac{\frac{\delta_\Omega(x)\,\delta_\Omega(y)^{1-\varepsilon}}{|x-y|^{N-\varepsilon}}\left(\frac{\delta_\Omega(y)}{|y|^N}\right)^p\delta_\Omega(y)^{-\alpha}}{\frac{\delta_\Omega(x)}{|x|^N}}$$
$$\lesssim \frac{\delta_\Omega(y)^{1+p-\alpha-\varepsilon}|x|^N}{|x-y|^{N-\varepsilon}|y|^{Np}} \lesssim \frac{|x|^{N+1-p(N-1)-\alpha-\varepsilon}}{|x-y|^{N-\varepsilon}},$$

whence

$$\int_{E_2} G_\Omega(x,y)M_\Omega(y,0)^p\,\delta_\Omega(y)^{-\alpha}\,dy \lesssim |x|^{N+1-p(N-1)-\alpha}M_\Omega(x,0).$$

If $y \in E_3$, then $|x-y| \geq |x| \geq |y| - |x-y|$, and so $|x-y| \gtrsim |y|$. By $1+p-\alpha > 0$, we have

$$\frac{G_\Omega(x,y)M_\Omega(y,0)^p\,\delta_\Omega(y)^{-\alpha}}{M_\Omega(x,0)} \lesssim \frac{\frac{\delta_\Omega(x)\,\delta_\Omega(y)}{|x-y|^{N}}\left(\frac{\delta_\Omega(y)}{|y|^N}\right)^p\delta_\Omega(y)^{-\alpha}}{\frac{\delta_\Omega(x)}{|x|^N}}$$
$$\lesssim \frac{|x|^N}{|y|^{N-1+p(N-1)+\alpha}},$$

whence

$$\frac{1}{M_\Omega(x,0)}\int_{E_3} G_\Omega(x,y)M_\Omega(y,0)^p\,\delta_\Omega(y)^{-\alpha}\,dy$$
$$\lesssim \int_{\Omega\setminus B(0,|x|/2)} \frac{|x|^N}{|y|^{N-1+p(N-1)+\alpha}}\,dy$$
$$\lesssim \begin{cases} |x|^{N+1-p(N-1)-\alpha} & \text{if } N-1+p(N-1)+\alpha > N, \\ |x|^N \log\dfrac{2\operatorname{diam}\Omega}{|x|} & \text{if } N-1+p(N-1)+\alpha = N, \\ |x|^N & \text{if } N-1+p(N-1)+\alpha < N. \end{cases}$$

Combining all the estimates, we obtain the desired one. □

Proof (*Proof of Theorem* 7) Thanks to Theorems 1, 6 and Corollary 5, the same argument as in [10, Proof of Corollary 2.3] leads to

$$u(x) \lesssim M_\Omega(x,\xi) \lesssim M_{V_1^+,\Omega}(x,\xi) \quad \text{for all } x \in \Omega. \tag{39}$$

This and the minimality of $M_{V_1^+,\Omega}(\cdot,\xi)$ imply that h in (35) is proportional to $M_{V_1^+,\Omega}(\cdot,\xi)$, so that

$$u(x) = \lambda M_{V_1^+,\Omega}(x,\xi) + \int_\Omega G_{V_1^+,\Omega}(x,y)\{V_1^-(y)u(y) + V_2(y)u(y)^p\}\,dy \tag{40}$$

for all $x \in \Omega$ and some constant $\lambda \geq 0$. By Lemmas 12, 14 and (39), we have

$$\begin{aligned} 0 \leq \frac{u(x)}{M_{V_1^+,\Omega}(x,\xi)} - \lambda &\lesssim \int_\Omega \frac{G_\Omega(x,y)}{M_\Omega(x,\xi)}\{V_1^-(y)M_\Omega(y,\xi) + V_2(y)M_\Omega(y,\xi)^p\}\,dy \\ &\lesssim \max\{F_{1,\alpha_1}(x), F_{p,\alpha_2}(x)\} \end{aligned}$$

for all $x \in \Omega$. Thus (37) follows.

Since the subharmonic function $\lambda M_{V_1^+,\Omega}(\cdot,\xi)$ has a harmonic majorant of the form $cM_\Omega(\cdot,\xi)$ on Ω by Lemma 12 and $M_\Omega(\cdot,\xi)$ is minimal (i.e., every positive harmonic function on Ω not bigger than $M_\Omega(\cdot,\xi)$ must be proportional to $M_\Omega(\cdot,\xi)$), it follows from the Riesz decomposition theorem for subharmonic functions that there exists $\lambda' \geq 0$ such that

$$\lambda M_{V_1^+,\Omega}(x,\xi) = \lambda' M_\Omega(x,\xi) - \lambda \int_\Omega G_\Omega(x,y)V_1^+(y)M_{V_1^+,\Omega}(y,\xi)\,dy \tag{41}$$

for all $x \in \Omega$. After substituting this into (40), we have by Lemma 12 and (39) that

$$\frac{u(x)}{M_\Omega(x,\xi)} - \lambda' \gtrsim -\lambda \int_\Omega \frac{G_\Omega(x,y)}{M_\Omega(x,\xi)} V_1^+(y)M_\Omega(y,\xi)\,dy$$

and

$$\frac{u(x)}{M_\Omega(x,\xi)} - \lambda' \lesssim \int_\Omega \frac{G_\Omega(x,y)}{M_\Omega(x,\xi)}\{V_1^-(y)M_\Omega(y,\xi) + V_2(y)M_\Omega(y,\xi)^p\}\,dy.$$

Thus (38) follows from Lemma 14. This completes the proof. □

4.2 Removable Boundary Singularities

Let us consider the case $\lambda = 0$ in Theorem 7. The reader is referred to [3, 4] for the Lane–Emden equation $-\Delta u = u^p$.

Theorem 8 *Let $\xi \in \partial\Omega$. If $u \in \mathcal{S}(\Omega)$ vanishes continuously on $\partial\Omega \setminus \{\xi\}$ and*

$$\liminf_{x\to\xi} \frac{u(x)}{M_\Omega(x,\xi)} = 0, \tag{42}$$

then ξ is a removable singularity of u in the sense that u has a continuous extension to $\overline{\Omega}$.

Proof We may let $\xi = 0$. By (42) and Lemma 12, we have

$$\lambda = \liminf_{x \to 0} \frac{u(x)}{M_{V_1^+,\Omega}(x,0)} = 0,$$

where λ is as in Theorem 7. Therefore (40) implies that for all $x \in \Omega$,

$$\begin{aligned} u(x) &= \int_\Omega G_{V_1^+,\Omega}(x,y)\{V_1^-(y)u(y) + V_2(y)u(y)^p\}\,dy \\ &\lesssim \int_\Omega G_\Omega(x,y)\{\delta_\Omega(y)^{-\alpha_1}u(y) + \delta_\Omega(y)^{-\alpha_2}u(y)^p\}\,dy. \end{aligned}$$

Moreover, by Theorem 7 with $\lambda' = 0$, there exists a constant $\tau \in (0, N]$ such that

$$u(x) \lesssim |x|^\tau M_\Omega(x,0) \lesssim |x|^{\tau-N}\,\delta_\Omega(x) \lesssim \delta_\Omega(x)^{\tau+1-N} \quad \text{for all } x \in \Omega,$$

and therefore

$$u(x) \lesssim \int_\Omega G_\Omega(x,y)\,\delta_\Omega(y)^{\beta_0}u(y)\,dy \tag{43}$$

$$\lesssim \int_\Omega G_\Omega(x,y)\,\delta_\Omega(y)^{\beta_0+1}|y|^{\tau-N}\,dy \tag{44}$$

for all $x \in \Omega$, where $\beta_0 := \min\{-\alpha_1, -\alpha_2 + (p-1)(\tau+1-N)\} > -2$ by (2).

Let us show that u vanishes continuously at 0 by using Lemma 8 with $\beta = \beta_0 + 1$ and $\gamma \le 0$: for all $x \in \Omega$,

$$\int_\Omega G_\Omega(x,y)\,\delta_\Omega(y)^{\beta_0+1}|y|^\gamma\,dy \lesssim \begin{cases} \delta_\Omega(x) & \text{if } \gamma > -\beta_0 - 2, \\ \delta_\Omega(x)\log\dfrac{\operatorname{diam}\Omega}{|x|} & \text{if } \gamma = -\beta_0 - 2, \\ \delta_\Omega(x)|x|^{\beta_0+2+\gamma} & \text{if } \gamma < -\beta_0 - 2. \end{cases} \tag{45}$$

Take $m \in \mathbb{Z}$ so that

$$m(\beta_0 + 2) < N - \tau \le (m+1)(\beta_0 + 2).$$

Then $m \ge -1$. If $m = -1$ or $m = 0$, then $\tau - N \ge -\beta_0 - 2$, and thus we see from (44) and (45) that u vanishes continuously at 0.

We claim that if $m \ge 1$ and $k \in \{1, \ldots, m\}$, then

$$u(x) \lesssim \delta_\Omega(x)|x|^{k(\beta_0+2)+\tau-N} \quad \text{for all } x \in \Omega. \tag{46}$$

This is true for $k = 1$. In fact, (44) and (45) with $\gamma = \tau - N < -\beta_0 - 2$ give

$$u(x) \lesssim \delta_\Omega(x)|x|^{\beta_0+2+\tau-N} \quad \text{for all } x \in \Omega.$$

Assume that $m \geq 2$ and (46) holds for some $k \leq m - 1$. Then, by (43) and (45) with $\gamma = k(\beta_0 + 2) + \tau - N < -\beta_0 - 2$, we have

$$u(x) \lesssim \int_\Omega G_\Omega(x, y)\, \delta_\Omega(y)^{\beta_0+1}|y|^{k(\beta_0+2)+\tau-N}\, dy \lesssim \delta_\Omega(x)|x|^{(k+1)(\beta_0+2)+\tau-N}$$

for all $x \in \Omega$. Thus the claim holds.

Using (46) with $k = m$ and repeating the same argument, we deduce that

$$u(x) \lesssim \begin{cases} \delta_\Omega(x) & \text{if } (m+1)(\beta_0+2) > N - \tau, \\ \delta_\Omega(x) \log \dfrac{\operatorname{diam} \Omega}{|x|} & \text{if } (m+1)(\beta_0+2) = N - \tau. \end{cases}$$

Hence u vanishes continuously on the whole of $\partial\Omega$. This completes the proof. □

4.3 Dirichlet Problem

In this final section, we give some remarks on the sharpness of the vanishing estimate (8) and the uniqueness of the Dirichlet problem

$$\begin{cases} -\Delta u + V_1 u = V_2 |u|^{p-1} u & \text{in } \Omega, \\ u = 0 & \text{on } \partial\Omega. \end{cases} \tag{47}$$

In addition, we impose

(V3) $\|V_1^-\|_{G_{V_1^+,\Omega}} := \sup_{x\in\Omega} \int_\Omega G_{V_1^+,\Omega}(x, y) V_1^-(y)\, dy < 1.$

Note from Lemma 9 that (V3) is satisfied whenever c_1 is small enough, and from (49) that, in this case, the existence of a positive solution of (47) implies $V_2 > 0$ on a set of positive Lebesgue measure.

Lemma 15 *Assume that* V_1 *and* V_2 *satisfy* (V1)–(V3). *Then there exists* $c = c(c_1, c_2, c_3, \alpha_1, \alpha_2, p, V_1^+, \|V_1^-\|_{G_{V_1^+,\Omega}}, \|V_2\|_{G_{V_1^+,\Omega}}, \Omega) \geq 1$ *such that*

$$\frac{1}{c}\, \delta_\Omega(x) \leq u(x) \leq c\, \delta_\Omega(x) \quad \textit{and} \quad |\nabla u(x)| \leq c \quad \textit{for all } x \in \Omega, \tag{48}$$

whenever $u \in C(\overline{\Omega})$ *is a positive solution of* (47).

Proof The upper estimates for u and $|\nabla u|$ were proved in Corollary 2.

Let us show the lower estimate for u. By Theorem 2 with $r = r_1/2$ and the Harnack inequality, we have $u(x) \gtrsim \delta_\Omega(x)u(x_0)$ for all $x \in \Omega$ and $\|u\|_\infty \lesssim u(x_0)$. Thus it suffices to show $\|u\|_\infty \gtrsim 1$. Since u vanishes on $\partial\Omega$, we see from (35) that

$$u(x) = \int_\Omega G_{V_1^+,\Omega}(x,y)\{V_1^-(y)u(y) + V_2(y)u(y)^p\}\,dy \quad \text{for all } x \in \Omega.$$

Then $\|u\|_\infty \le \|u\|_\infty \|V_1^-\|_{G_{V_1^+,\Omega}} + \|u\|_\infty^p \|V_2\|_{G_{V_1^+,\Omega}}$, and so

$$0 < 1 - \|V_1^-\|_{G_{V_1^+,\Omega}} \le \|u\|_\infty^{p-1} \|V_2\|_{G_{V_1^+,\Omega}}. \tag{49}$$

Since $\alpha_2 < 2$, it follows from (V2) and Lemmas 12 and 9 that

$$\|V_2\|_{G_{V_1^+,\Omega}} \lesssim \sup_{x\in\Omega} \int_\Omega G_\Omega(x,y)\,\delta_\Omega(y)^{-\alpha_2}\,dy \lesssim 1.$$

Therefore

$$\|u\|_\infty \ge \left(\frac{1 - \|V_1^-\|_{G_{V_1^+,\Omega}}}{\|V_2\|_{G_{V_1^+,\Omega}}}\right)^{\frac{1}{p-1}} > 0.$$

Thus the lemma is proved. □

Remark 8 We do not know whether $u \in C^1(\overline{\Omega})$ or not, so that (48) does not follow from the Hopf maximum principle. Moreover, the constant c is independent of u.

A positive solution $u \in C(\overline{\Omega})$ of (47) is called *minimal* if $u \le v$ on Ω for all positive solutions $v \in C(\overline{\Omega})$ of (47).

Theorem 9 *Assume that V_1 and V_2 satisfy* (V1)–(V3). *If* (47) *has a minimal positive solution in $C(\overline{\Omega})$, then it is a unique solution.*

Proof We first claim that every positive solution $u \in C(\overline{\Omega})$ of (47) satisfies

$$\int_\Omega \nabla u \cdot \nabla\phi\,dx = \int_\Omega \{V_2 u^p - V_1 u\}\phi\,dx \tag{50}$$

for all positive $\phi \in C^1(\Omega)$ satisfying

$$\phi(x) \lesssim \delta_\Omega(x) \quad \text{and} \quad |\nabla\phi(x)| \lesssim 1 \quad \text{for all } x \in \Omega.$$

In fact, taking a sequence of compact subsets E_k of Ω such that $E_k \subset E_{k+1}$ and $\bigcup_k E_k = \Omega$ and considering the convolution of a mollifier and $\phi\chi_{E_k}$ for such a function ϕ, we see that there exists a sequence $\{\phi_k\}$ in $C_c^\infty(\Omega)$ converging to ϕ on Ω such that

$$\left\| \frac{\phi_k}{\delta_\Omega} \right\|_\infty \le \left\| \frac{\phi}{\delta_\Omega} \right\|_\infty \lesssim 1 \quad \text{and} \quad \|\nabla \phi_k\|_\infty \le \|\nabla \phi\|_\infty \lesssim 1 \quad \text{for all } k \in \mathbb{N},$$

since $\delta_\Omega(y) \approx \delta_\Omega(x)$ for all $y \in B(x, \delta_\Omega(x)/2)$. Thanks to (48), we see from (V1) and (V2) that for a.e. $x \in \Omega$ and all $k \in \mathbb{N}$,

$$|\nabla u(x) \cdot \nabla \phi_k(x)| \lesssim 1$$

and

$$\left|\{V_2(x)u(x)^p - V_1(x)u(x)\}\phi_k(x)\right| \lesssim \delta_\Omega(x)^{\alpha_0} \frac{|\phi_k(x)|}{\delta_\Omega(x)} \lesssim 1,$$

where $\alpha_0 := \min\{2 - \alpha_1, 1 + p - \alpha_2\} > 0$ by (2). Applying (50) to $\phi = \phi_k$, we see from the bounded convergence theorem that the above claim holds.

Let $u \in C(\overline{\Omega})$ be a minimal positive solution of (47) and let $w \in C(\overline{\Omega})$ be any positive solution of (47). By Lemma 15 and the above claim, we can use u and w as test functions for each other. Then

$$\int_\Omega \nabla u \cdot \nabla w \, dx = \int_\Omega \{V_2 u^p - V_1 u\} w \, dx,$$
$$\int_\Omega \nabla w \cdot \nabla u \, dx = \int_\Omega \{V_2 w^p - V_1 w\} u \, dx,$$

whence

$$\int_\Omega V_2 u w \{w^{p-1} - u^{p-1}\} \, dx = 0.$$

Since $w \ge u > 0$ and $V_2 \ge 0$ a.e. on Ω, this implies that $u = w$ a.e. on the set where $V_2 \ne 0$. Therefore, using the representation (35) with $h \equiv 0$, we have for all $x \in \Omega$,

$$\begin{aligned} 0 &\le w(x) - u(x) \\ &= \int_\Omega G_{V_1^+,\Omega}(x, y)\big(V_1^-(y)\{w(y) - u(y)\} + V_2(y)\{w(y)^p - u(y)^p\}\big) \, dy \\ &= \int_\Omega G_{V_1^+,\Omega}(x, y) V_1^-(y)\{w(y) - u(y)\} \, dy \\ &\le \|w - u\|_\infty \int_\Omega G_{V_1^+,\Omega}(x, y) V_1^-(y) \, dy, \end{aligned}$$

whence

$$\|w - u\|_\infty \le \|w - u\|_\infty \|V_1^-\|_{G_{V_1^+,\Omega}}.$$

Taking (V3) and continuity into account, we conclude that $w \equiv u$ in the whole of Ω. This completes the proof. □

We give an example regarding a relation between the validity of (48) and the upper bounds of α_1 and α_2.

Example 1 We may assume that $0 \in \partial\Omega$ and $e = (0, \ldots, 0, 1)$ is the inward normal vector to $\partial\Omega$ at 0. Let $\kappa > 1$ and write

$$\Gamma := \{x = (x_1, \ldots, x_N) \in \mathbb{R}^N : |x| < \kappa x_N\}.$$

Let $\alpha_1 < 2$. Consider $V_1, V_2 \in L^1_{\mathrm{loc}}(\Omega)$ satisfying $V_1^+(x) \lesssim \delta_\Omega(x)^{-\alpha_1}$ and $V_2(x) \geq 0$ for a.e. $x \in \Omega$. Assume either

$$V_1^-(x) \gtrsim \delta_\Omega(x)^{-2} \quad \text{for a.e. } x \in \Gamma \cap B(0, r) \tag{51}$$

or

$$V_2(x) \gtrsim \delta_\Omega(x)^{-p-1} \quad \text{for a.e. } x \in \Gamma \cap B(0, r), \tag{52}$$

where $r > 0$ is sufficiently small. Let $u \in C(\overline{\Omega})$ be a positive solution of (47). Then at least one of the following fails to hold:

$$\frac{u(x)}{\delta_\Omega(x)} = O(1) \quad \text{as } \Gamma \ni x \to 0, \tag{53}$$

or

$$\frac{\delta_\Omega(x)}{u(x)} = O(1) \quad \text{as } \Gamma \ni x \to 0. \tag{54}$$

To show this, we have only to prove that (53) fails to hold if (54) is true. Without loss of generality, we may assume that

$$u(x) \gtrsim \delta_\Omega(x) \quad \text{for all } x \in \Gamma \cap B(0, r).$$

Let $x \in \Gamma \cap B(0, r/4)$. By Lemma 12 and (51) or (52), we have for all $x \in \Omega$,

$$\begin{aligned} u(x) &= \int_\Omega G_{V_1^+,\Omega}(x, y)\{V_1^-(y)u(y) + V_2(y)u(u)^p\}\, dy \\ &\gtrsim \int_{\Gamma\cap B(0,r)} G_\Omega(x, y)\, \delta_\Omega(y)^{-1}\, dy. \end{aligned}$$

By the way, we see from the Green function estimate that

$$G_\Omega(x, y)\, \delta_\Omega(y)^{-1} \approx \delta_\Omega(x)|y|^{-N} \quad \text{for all } y \in \Gamma \cap B(0, r) \setminus B(0, 2|x|).$$

Therefore

$$u(x) \gtrsim \delta_\Omega(x) \int_{\Gamma\cap B(0,r)\setminus B(0,2|x|)} |y|^{-N}\, dy \gtrsim \delta_\Omega(x) \log \frac{r}{2|x|},$$

which implies that (53) fails to hold.

Note that there is a gap between the exponent $-p-1$ in (52) and the lower bound $-N-1+p(N-1)$ of $-\alpha_2$. Thus it remains open whether or not the upper estimate $u(x) \le c\,\delta_\Omega(x)$ holds for all positive solutions u of (47) in the case that V_2 satisfies (V2) for some $\alpha_2 \in [N+1-p(N-1), p+1)$.

5 Appendix

We give some supplementary remarks for the reader's convenience.

Definition 1 A bounded domain $\Omega \subset \mathbb{R}^N$ $(N \ge 2)$ is called a $C^{1,1}$*-domain* if there exist $r_\Omega > 0$ and $c_{11} > 0$ such that to each point $\xi \in \partial\Omega$ there corresponds a Cartesian coordinate system $(x', x^N) = (x^1, \dots, x^{N-1}, x^N)$ and a C^1-function $\psi : \mathbb{R}^{N-1} \to \mathbb{R}$ such that

$$|\nabla\psi(x') - \nabla\psi(y')| \le c_{11}|x' - y'| \quad \text{for all } x', y' \in \mathbb{R}^{N-1} \tag{55}$$

and $\Omega \cap B(\xi, r_\Omega) = \{(x', x^N) : x^N > \psi(x')\} \cap B(\xi, r_\Omega)$.

Definition 2 We say that Ω satisfies the *uniform ball condition* if there exists $r_5 > 0$ satisfying the following property: for each $\xi \in \partial\Omega$, there exist $\xi^i, \xi^e \in \mathbb{R}^N$ such that $B(\xi^i, r_5) \subset \Omega$, $B(\xi^e, r_5) \subset \mathbb{R}^N \setminus \overline{\Omega}$ and $\xi \in \partial B(\xi^i, r_5) \cap \partial B(\xi^e, r_5)$.

Lemma 16 ([2, Lemma 2.2]) *Let Ω be a bounded domain. Then Ω is a $C^{1,1}$-domain if and only if Ω satisfies the uniform ball condition.*

In what follows, we suppose that Ω is a bounded $C^{1,1}$-domain. For $\xi \in \partial\Omega$ and $0 < \theta < \pi$, we denote the circular cone of vertex ξ, axis $\{\xi + t\boldsymbol{n}_\xi : t \ge 0\}$ and aperture θ by

$$\Gamma_\theta(\xi) := \{x \in \mathbb{R}^N : \angle x\xi\xi_1 < \theta\},$$

where $\xi_1 = \xi + \boldsymbol{n}_\xi$ with the inward unit normal vector $\boldsymbol{n}_\xi$ to $\partial\Omega$ at ξ.

Let x, y be points in an open set D. We say that $\{B(x_k, r_k)\}_{k=1}^m$ is a *Harnack chain* from x to y in D if

$$x_1 = x, \quad |x_{k+1} - x_k| \le r_k \ (k = 1, \dots, m-1), \quad |y - x_m| \le r_m$$

and $B(x_k, 2r_k) \subset D$ for $k = 1, \dots, m$.

Lemma 17 *Let $\xi \in \partial\Omega$, $r > 0$ and $0 < \theta < \widehat{\theta} < \pi$. If u is a function on $\Gamma_{\widehat{\theta}}(\xi) \cap B(\xi, 2r) \subset \Omega$ satisfying the Harnack inequality* (18) *there, then*

$$\frac{1}{c}\left(\frac{|x-\xi|}{|y-\xi|}\right)^\beta u(y) \le u(x) \le c\left(\frac{|y-\xi|}{|x-\xi|}\right)^\beta u(y)$$

for all $x, y \in \Gamma_\theta(\xi) \cap B(\xi, r)$ satisfying $|x-\xi| \le |y-\xi|$, where $c > 1$ and $\beta > 0$ depend only on θ, $\widehat{\theta} - \theta$ and the Harnack constant c_5 in (18).

Proof We may assume that $\xi = 0$, $\xi_1 = (0', 1)$ and $x, y \in \Gamma_\theta(0)$ satisfy $|x| < |y| < r$. Let $z := (0', |x|)$ and $w := (0', |y|)$. Also, let

$$r_k := \left(1 + \frac{\sin\theta}{2}\right)^{k-1} \frac{\sin\theta}{2}|x| \quad \text{for } k \in \mathbb{N}.$$

Take the smallest $m \in \mathbb{N}$ with $|x| + \sum_{k=1}^m r_k \geq |y|$. Then

$$m \leq \frac{1}{\log\left(1 + \frac{\sin\theta}{2}\right)} \log\frac{|y|}{|x|} + 1.$$

Put $x_1 := z$ and $x_{k+1} := (0', |x| + \sum_{j=1}^k r_j)$ for $k = 1, \dots, m-1$. Then $\{B(x_k, r_k)\}_{k=1}^m$ is a Harnack chain from z to w in $\Gamma_\theta(0) \cap B(0, 2r)$. Using (18) repeatedly, we have

$$\frac{1}{c_5^m} u(w) \leq u(z) \leq c_5^m u(w),$$

and so

$$\frac{1}{c_5}\left(\frac{|x|}{|y|}\right)^\beta u(w) \leq u(z) \leq c_5\left(\frac{|y|}{|x|}\right)^\beta u(w) \tag{56}$$

for some $\beta = \beta(\theta, c_5) > 0$. Also, it is easy to see that

$$\frac{1}{c}u(z) \leq u(x) \leq cu(z) \quad \text{and} \quad \frac{1}{c}u(y) \leq u(w) \leq cu(y) \tag{57}$$

for some $c = c(\widehat{\theta} - \theta, c_5) > 1$. Combining (56) and (57) yields

$$\frac{1}{c}\left(\frac{|x|}{|y|}\right)^\beta u(y) \leq u(x) \leq c\left(\frac{|y|}{|x|}\right)^\beta u(y).$$

This completes the proof. □

Lemma 18 *There exist $r_6 = r(\Omega) > 0$ and $\lambda = \lambda(c_5, \Omega) > 0$ with the following property: Let $\xi \in \partial\Omega$, $1 \leq \alpha \leq 2$ and $0 < r < r_6$. If u is a function on $\Omega \cap B(\xi, 4\alpha r)$ satisfying the Harnack inequality* (18) *there, then*

$$u(x) \leq \left(\frac{4r}{\delta_\Omega(x)}\right)^\lambda u(\xi_r) \quad \textit{for all } x \in \Omega \cap B(\xi, \alpha r).$$

Proof Without loss of generality, we may assume that $\xi = 0$ and $|\nabla\psi(0)| = 0$, where ψ is as in Definition 1. Note from Lemma 16 that

$$\left\{(y', \psi(y')) : |y'| < \frac{12}{37}r_5\right\} \subset \left\{(y', y_N) : |y_N| \le \frac{1}{6}|y'|\right\} \tag{58}$$

and $\Gamma_{\widehat{\theta}}(0) \cap B(0, 12r_5/37)) \subset \Omega$, where $\widehat{\theta} := \tan^{-1} 6 \in (0, \pi/2)$.

Let $0 < 4\alpha r \le 8r < \min\{12r_5/37, 1/c_{11}\}$ and $x \in \Omega \cap B(0, \alpha r)$. Take $\eta = (\eta, \psi(\eta')) \in \partial\Omega$ with $|\eta - x| = \delta_\Omega(x)$. Then $|\eta'| \le |\eta| \le |\eta - x| + |x| \le 2|x| < 2\alpha r$, and so $\psi(\eta') + \alpha r \ge -|\eta'|/6 + \alpha r > 2\alpha r/3$ by (58). Thus

$$(\eta', \psi(\eta') + \alpha r) \in \Gamma_\theta(0) \cap B(0, 3\alpha r) \setminus B(0, 2\alpha r/3)$$

with $\theta := \tan^{-1} 3 \in (0, \pi/2)$. Obviously, $\theta < \widehat{\theta}$.

By the way, since $\boldsymbol{n}_\eta = (\nabla\psi(\eta'), 1)/|(\nabla\psi(\eta'), 1)|$ and $\boldsymbol{n}_0 = (0', 1)$, we have

$$|\boldsymbol{n}_\eta - \boldsymbol{n}_0| \le 2\min\{1, c_{11}|\eta'|\} \le 2\min\{1, 2\alpha c_{11} r\} = 4\alpha c_{11} r$$

by (55), whence the angle between $\boldsymbol{n}_\eta$ and $\boldsymbol{n}_0$ is less than $2\sin^{-1}(2\alpha c_{11} r)$. Thus

$$(\eta', \psi(\eta') + \alpha r) = \eta + \alpha r \boldsymbol{n}_0 \in \Gamma_\theta(\eta)$$

whenever $r > 0$ is sufficiently small such that $2\sin^{-1}(4c_{11} r) \le \theta$.

By the same reasoning as above, we have $\Gamma_{\widehat{\theta}}(\eta) \cap B(\eta, 2\alpha r) \subset \Omega \cap B(0, 4\alpha r)$. Applying Lemma 17, we have

$$u(x) \le c\left(\frac{\alpha r}{\delta_\Omega(x)}\right)^\beta u((\eta', \psi(\eta') + \alpha r)) \le c\left(\frac{12r}{\delta_\Omega(x)}\right)^\beta u(0_r).$$

After taking λ sufficiently large so that $c6^\beta \le 2^\lambda$, we can obtain the desired estimate, since $\delta_\Omega(x) < 2r$. □

References

1. H. Aikawa, Norm estimate of Green operator, perturbation of Green function and integrability of superharmonic functions. Math. Ann. **312**(2), 289–318 (1998)
2. H. Aikawa, T. Kilpeläinen, N. Shanmugalingam, X. Zhong, Boundary Harnack principle for p-harmonic functions in smooth Euclidean domains. Potential Anal. **26**(3), 281–301 (2007)
3. M.F. Bidaut-Véron, A.C. Ponce, L. Véron, Isolated boundary singularities of semilinear elliptic equations. Calc. Var. Partial Differ. Equ. **40**(1–2), 183–221 (2011)
4. M.F. Bidaut-Véron, L. Vivier, An elliptic semilinear equation with source term involving boundary measures: the subcritical case. Rev. Mat. Iberoamericana **16**(3), 477–513 (2000)
5. L. Caffarelli, E. Fabes, S. Mortola, S. Salsa, Boundary behavior of nonnegative solutions of elliptic operators in divergence form. Indiana Univ. Math. J. **30**(4), 621–640 (1981)
6. M. Cranston, E. Fabes, Z. Zhao, Conditional gauge and potential theory for the Schrödinger operator. Trans. Amer. Math. Soc. **307**(1), 171–194 (1988)
7. E.N. Dancer, Superlinear problems on domains with holes of asymptotic shape and exterior problems. Math. Z. **229**(3), 475–491 (1998)

8. K. Hirata, The boundary growth of superharmonic functions and positive solutions of nonlinear elliptic equations. Math. Ann. **340**(3), 625–645 (2008)
9. K. Hirata, Existence and nonexistence of a positive solution of the Lane-Emden equation having a boundary singularity: the subcritical case. Monatsh. Math. **186**(4), 635–652 (2018)
10. K. Hirata, Boundary estimates for superharmonic functions and solutions of semilinear elliptic equations with source. Collect. Math. **72**(1), 43–61 (2021)
11. D.S. Jerison, C.E. Kenig, Boundary behavior of harmonic functions in nontangentially accessible domains. Adv. in Math. **46**(1), 80–147 (1982)
12. T. Kuusi, G. Mingione, Linear potentials in nonlinear potential theory. Arch. Ration. Mech. Anal. **207**(1), 215–246 (2013)
13. M. Murata, Structure of positive solutions to $(-\Delta + V)u = 0$ in R^n. Duke Math. J. **53**(4), 869–943 (1986)
14. M. Murata, Semismall perturbations in the Martin theory for elliptic equations. Israel J. Math. **102**, 29–60 (1997)
15. Y. Pinchover, Criticality and ground states for second-order elliptic equations. J. Differ. Equ. **80**(2), 237–250 (1989)
16. P. Poláčik, P. Quittner, P. Souplet, Singularity and decay estimates in superlinear problems via Liouville-type theorems. I. Elliptic equations and systems. Duke Math. J. **139**(3), 555–579 (2007)
17. L. Riahi, Comparison of Green functions for generalized Schrödinger operators on $C^{1,1}$-domains. JIPAM. J. Inequal. Pure Appl. Math. **4**(1), Article 22, 14 (2003)
18. L. Riahi, The 3G-inequality for general Schrödinger operators on Lipschitz domains. Manuscripta Math. **116**(2), 211–227 (2005)
19. J. Serrin, H. Zou, Cauchy-Liouville and universal boundedness theorems for quasilinear elliptic equations and inequalities. Acta Math. **189**(1), 79–142 (2002)
20. G. Stampacchia, Le problème de Dirichlet pour les équations elliptiques du second ordre à coefficients discontinus. Ann. Inst. Fourier (Grenoble) **15**, 189–258 (1965)
21. Z. Zhao, Green function for Schrödinger operator and conditioned Feynman-Kac gauge. J. Math. Anal. Appl. **116**, 309–334 (1986)

Convexity Phenomena Arising in an Area-Preserving Crystalline Curvature Flow

Tetsuya Ishiwata and Shigetoshi Yazaki

Abstract An area-preserving crystalline curvature flow is regarded as a simple model of the deformation process of a negative crystal. In the present paper, behavior of polygonal curves by area-preserving crystalline curvature flow is discussed. We show "convexity phenomena", that is, the solution polygon from a non-convex initial polygon becomes convex in a finite time. In order to show this assertion, we classify edge-disappearing patterns completely and prove that all zero-curvature edges disappear in a finite time, and we also show that evolution process of the flow can be continued beyond such edge-disappearing singularities.

Keywords Crystalline curvature flow · Crystalline motion · Area-preserving property · Convexity phenomena · Edge-disappearing · Negative ice crystal

MSC Classification 53C44 · 53A04 · 82B26 · 82C26

1 Introduction

When a block of ice crystal is exposed to sunlight or strong beams, the ice crystal starts to melt from not only on the surface but also inside the crystal and each water region forms a snowflake-like-pattern which has six petals, called "Tyndall figure" (Fig. 1a).

Shigetoshi Yazaki—These authors contributed equally to this work.

T. Ishiwata (✉)
College of Systems Engineering and Science, Shibaura Institute of Technology,
307 Fukasaku, Minuma-ku, Saitama-shi, Saitama 337-8570, Japan
e-mail: tisiwata@shibaura-it.ac.jp

S. Yazaki
School of Science and Technology, Meiji University, 1-1-1 Higashi-Mita,
Tama-ku, Kanagawa 214-8571, Japan
e-mail: syazaki@meiji.ac.jp

S. Machihara (ed.), *Mathematical Physics and Its Interactions*, Springer Proceedings in Mathematics & Statistics 451, https://doi.org/10.1007/978-981-97-0364-7_2

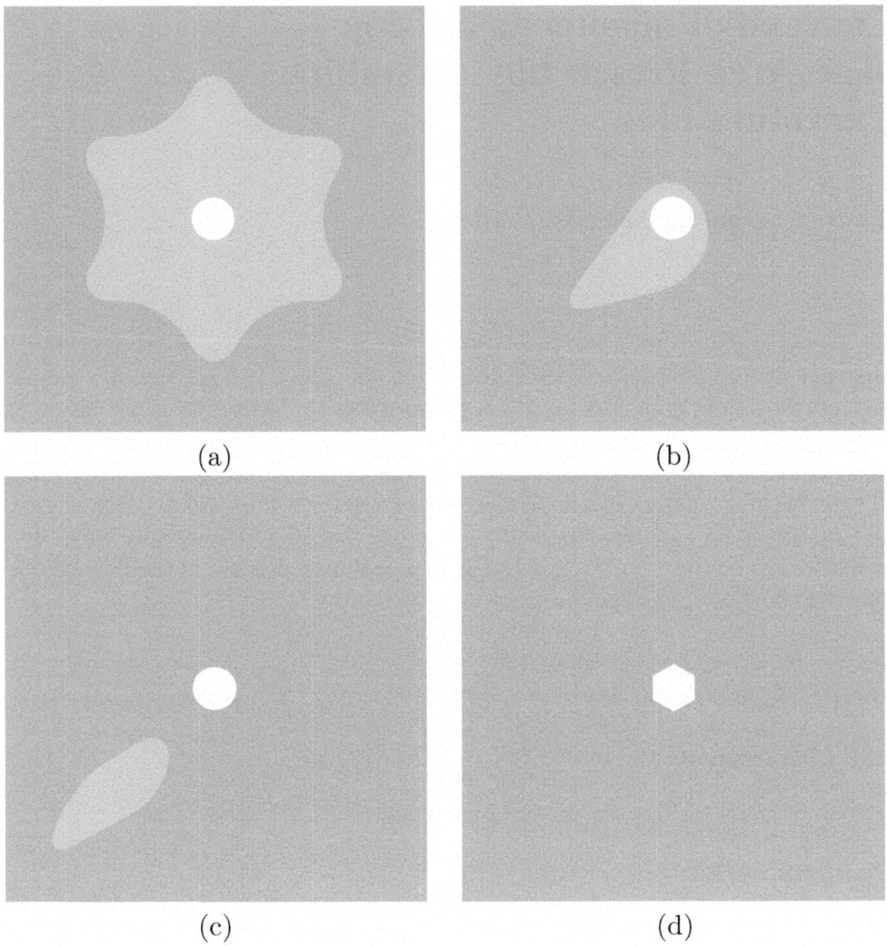

Fig. 1 In all figures, the dark gray region is ice portion. **a** Tyndall figure: vapor bubble (white) is floating in a water region (light gray), **b** Freezing of water proceeds, **c** Water part is separated and bubble part is isolated in the ice, and **d** Deformation process from an "apparently circular" region to a hexagonal disk

The figure is filled with water except for a vapor bubble. This phenomenon was first observed by Tyndall [20]. When Tyndall figure is refrozen, the vapor bubble remains in the ice as a hexagonal disk (see Fig. 1c, d). This hexagonal disk is filled with water vapor saturated at that temperature and surrounded by ice. McConnel found these disks in the ice of Davos lake [14]. Nakaya called this hexagonal disk "vapor figure" and investigated its properties precisely [16]. Adams and Lewis (1934) called it "negative crystal". (Although Nakaya said "this term does not seem adequate" with a certain reason, hereafter we use the term negative crystal for avoiding confusion.) Furukawa and Kohata made hexagonal prisms experimentally in a single ice crystal, and investigated the habit change of negative crystals with respect to the temperatures and the evaporation mechanisms of ice surfaces [3].

To the best of author's knowledge, after the Furukawa and Kohata's experimental research, there have been no published results on negative crystals, and there are no dynamical model equations describing the process of formation of negative crystals. In this paper, we will focus on the process of formation of negative crystals after Tyndall figures are refrozen, and propose a model equation of interfacial motion which tracks the deformation of negative crystals in time.

Nakaya's observation. Figure 1 shows how the Tyndall figure is refrozen from its initial state to the eventual formation of negative crystals. This process may be described such that negative crystal changes the shape from an apparently circular shape to the hexagon (Fig. 1c, d). Nakaya pointed out the apparently circular shape is not a circle, it should be a stepped structure such as Fig. 2a from a surface energy point of view [16, Sect. 24]. Note that the stepped structures in Fig. 2a–c and the hexagon (d) are in a same polygon's class which is determined by the surface energy. The class is called admissible which will be defined later. Nakaya also mentioned that the circumference of the stepped structures are longer than that of a hexagon with the same area. Consequently, Nakaya observed that the deformation of vapor figure from the apparently circular shape to a hexagon can be explained as a result of motion of curves minimizing the energy.

The Nakaya's observation can be regarded as the motion of curves in a direction which minimize the total interfacial energy, as follows: Deformation process of the shape of vapor figures (Fig. 1c, d) correspond to Fig. 2a–d. This process is modelled by an area-preserving crystalline curvature flow such as a gradient flow of a surface energy on polygonal curves in some suitable metric [11, 12].

In the present paper, we prove that a solution polygon to an area-preserving crystalline curvature flow becomes convex in a finite time. The process is called "convexity phenomena", since all crystalline curvatures (defined below) of the initial polygon are nonnegative and they become positive at the final stage. After the solution polygon becomes convex, it converges to the shape of the boundary of the Wulff shape at time infinity. (See Proposition 1 from [21].)

The organization of this paper is as follows. In the next section, we will set up the problem of an area-preserving crystalline curvature flow, and give main results. In Sect. 3, the main theorem will be proved.

2 Setting Up and Main Results

In this section, we will set up the problem of an area-preserving crystalline curvature flow, and introduce the known results. Finally, the main results will be given.

Crystalline energy. Assume that an interfacial energy density γ is a continuous convex function of $\boldsymbol{x} \in \mathbb{R}^2$ and positive for $\boldsymbol{x} \neq \boldsymbol{0}$, and satisfies $\gamma(\lambda \boldsymbol{x}) = \lambda\gamma(\boldsymbol{x})$ for all $\lambda \geq 0$ and $\boldsymbol{x} \in \mathbb{R}^2$. In this paper, we consider the case where the Wulff shape

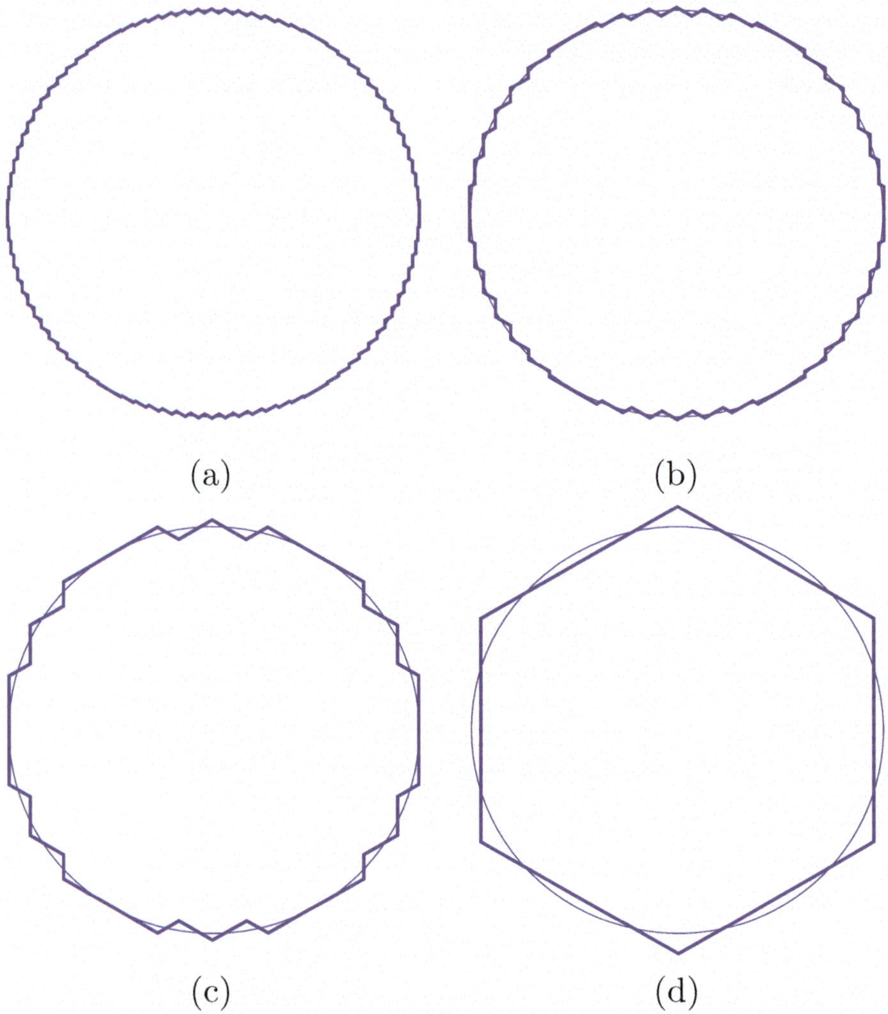

Fig. 2 Examples of stepped structured shapes. All shapes have the same area as the circle in **d**. From right to left, the structure changes finer and finer, and surface energy becomes bigger and bigger. **d** hexagon and **a** an apparently circular shape, which is a very fine stepped structure indeed

of γ, $W_\gamma = \bigcap_{\theta \in S^1} \left\{ \boldsymbol{x} \in \mathbb{R}^2;\ \boldsymbol{x} \cdot \boldsymbol{n}(\theta) \leq \gamma(\boldsymbol{n}(\theta)) \right\}$, is a convex polygon, where $S^1 = \mathbb{R}/2\pi\mathbb{Z}$, $\boldsymbol{n}(\theta) = (\cos\theta, \sin\theta)$ and $\boldsymbol{a}_1 \cdot \boldsymbol{a}_2$ is the Euclidean inner product of vectors $\boldsymbol{a}_1$ and $\boldsymbol{a}_2$. Such γ is called *crystalline energy*, and if W_γ is an N_γ-sided convex polygon ($N_\gamma \geq 3$), then W_γ is expressed as

$$W_\gamma = \bigcap_{i=1}^{N_\gamma} \left\{ \boldsymbol{x} \in \mathbb{R}^2;\ \boldsymbol{x} \cdot \boldsymbol{\nu}_i \leq \gamma(\boldsymbol{\nu}_i) \right\},$$

where $\boldsymbol{\nu}_i = \boldsymbol{n}(\phi_i)$ and ϕ_i is the exterior normal angle of the i-th edge with $\phi_i \in (\phi_{i-1}, \phi_{i-1} + \pi)$ for all i ($\phi_0 = \phi_{N_\gamma} - 2\pi$, $\phi_{N_\gamma+1} = \phi_1 + 2\pi$). We define a set of normal vectors of W_γ by

$$\mathcal{V}_\gamma = \{\boldsymbol{\nu}_1, \boldsymbol{\nu}_2, \ldots, \boldsymbol{\nu}_{N_\gamma}\}.$$

Polygons. Let Γ be an N-sided polygonal Jordan curve in the plane $\mathbb{R}^2$ and Ω a bounded interior region surrounded by Γ. We assume that Ω is simply connected, that is, $\Gamma = \partial\Omega$, and use the terminology "polygon" for Γ as well as for Ω. Let us label the position vector of vertices $\boldsymbol{p}_i$ ($i = 1, 2, \ldots, N$) in an anticlockwise order:

$$\Gamma = \bigcup_{i=1}^{N} \Gamma_i, \quad \Gamma_i = \left\{(1-\lambda)\boldsymbol{p}_i + \lambda \boldsymbol{p}_{i+1};\ \lambda \in [0, 1]\right\},$$

where Γ_i is the i-th edge ($\boldsymbol{p}_0 = \boldsymbol{p}_N$, $\boldsymbol{p}_{N+1} = \boldsymbol{p}_1$). We denote the length of Γ_i by $d_i = |\boldsymbol{p}_{i+1} - \boldsymbol{p}_i|$, and then the i-th unit tangent vector is $\boldsymbol{t}_i = (\boldsymbol{p}_{i+1} - \boldsymbol{p}_i)/d_i$ and the i-th unit outward normal vector is $\boldsymbol{n}_i = -\boldsymbol{t}_i^\perp$, where $(a, b)^\perp = (-b, a)$. We define a set of normal vectors of Γ by

$$\mathcal{N} = \{\boldsymbol{n}_1, \boldsymbol{n}_2, \ldots, \boldsymbol{n}_N\}.$$

Let θ_i be the exterior normal angle of Γ_i. Then $\boldsymbol{n}_i = \boldsymbol{n}(\theta_i)$ and $\boldsymbol{t}_i = \boldsymbol{t}(\theta_i)$ hold ($\theta_0 = \theta_N - 2\pi$, $\theta_{N+1} = \theta_1 + 2\pi$), where $\boldsymbol{t}(\theta) = (-\sin\theta, \cos\theta)$. We define the i-th height function $h_i = \boldsymbol{p}_i \cdot \boldsymbol{n}_i = \boldsymbol{p}_{i+1} \cdot \boldsymbol{n}_i$ ($h_0 = h_N$, $h_{N+1} = h_1$). By using $\{h_{i-1}, h_i, h_{i+1}\}$ and $\{\boldsymbol{n}_{i-1}, \boldsymbol{n}_i, \boldsymbol{n}_{i+1}\}$, the length of i-th edge d_i is described as follows

$$d_i = \frac{\chi_{i-1,i}(h_{i-1} - (\boldsymbol{n}_{i-1} \cdot \boldsymbol{n}_i)h_i)}{\sqrt{1 - (\boldsymbol{n}_{i-1} \cdot \boldsymbol{n}_i)^2}} + \frac{\chi_{i,i+1}(h_{i+1} - (\boldsymbol{n}_i \cdot \boldsymbol{n}_{i+1})h_i)}{\sqrt{1 - (\boldsymbol{n}_i \cdot \boldsymbol{n}_{i+1})^2}}, \quad i = 1, 2, \ldots, N,$$

where $\chi_{i,j} = \mathrm{sgn}(\det(\boldsymbol{n}_i, \boldsymbol{n}_j))$, and $\det(\boldsymbol{a}_1, \boldsymbol{a}_2)$ is the determinant of the 2 by 2 matrix with column vectors $\boldsymbol{a}_1$ and $\boldsymbol{a}_2$. Since $\boldsymbol{n}_i \cdot \boldsymbol{n}_j = \cos(\theta_i - \theta_j)$, we have another expression

$$d_i = -(\cot\vartheta_i + \cot\vartheta_{i+1})h_i + h_{i-1}\mathrm{cosec}\vartheta_i + h_{i+1}\mathrm{cosec}\vartheta_{i+1}, \quad i = 1, 2, \ldots, N, \tag{1}$$

where $\vartheta_i = \theta_i - \theta_{i-1}$. Note that $0 < |\vartheta_i| < \pi$ holds for all i. Furthermore, the i-th vertex $\boldsymbol{p}_i$ is described as follows

$$\boldsymbol{p}_i = h_i \boldsymbol{n}_i + \frac{h_{i-1} - (\boldsymbol{n}_{i-1} \cdot \boldsymbol{n}_i)h_i}{\boldsymbol{n}_{i-1} \cdot \boldsymbol{t}_i} \boldsymbol{t}_i. \quad i = 1, 2, \ldots, N. \tag{2}$$

Admissibility and crystalline curvature. We call polygon Γ *admissible* (associated with W_γ) if and only if the following conditions hold:

(i) The normal vector of each edge of the curve Γ is included in the set of all normal vectors $\mathcal{V}_\gamma$, that is, $\boldsymbol{n}_i \in \mathcal{V}_\gamma$ for all i.
(ii) $\boldsymbol{n}_{i-1}, \boldsymbol{n}_{i+1} \in \{\boldsymbol{\nu}_{j-1}, \boldsymbol{\nu}_{j+1}\}$ while $\boldsymbol{n}_i = \boldsymbol{\nu}_j$ for all i.

The latter condition means that the adjacency of each edge of Γ is the same as that of each edge of $\mathcal{V}_\gamma$.

Note that if Γ is admissible and convex, then the number of edges of Γ is same as that of the Wulff shape. On the other hand, if Γ is admissible and non-convex, then the number of edges of Γ is larger than that of the Wulff shape. In any case, $\mathcal{N} = \mathcal{V}_\gamma$ holds.

Let Γ be an N-sided admissible polygon. For each edge Γ_i a *crystalline curvature* is defined by

$$H_i = \frac{\chi_i l_\gamma(\boldsymbol{n}_i)}{d_i}, \quad i = 1, 2, \ldots, N,$$

where $\chi_i = (\chi_{i-1,i} + \chi_{i,i+1})/2$ is the transition number and it takes $+1$ (resp. -1) if Γ is convex (resp. concave) around Γ_i in the direction of $-\boldsymbol{n}_i$, otherwise $\chi_i = 0$; and $l_\gamma(\boldsymbol{n}_i)$ is the length of the j-th edge of W_γ if $\boldsymbol{n}_i = \boldsymbol{\nu}_j$. If Γ is an N-sided admissible convex polygon, then $\boldsymbol{n}_i = \boldsymbol{\nu}_i$ and $\chi_i = 1$ for all $i = 1, 2, \ldots, N = N_\gamma$; and moreover, if $\Gamma = \partial W_\gamma$, then all the crystalline curvatures are 1. The total interfacial crystalline energy on Γ is

$$E_\gamma = \sum_{i=1}^{N} \gamma(\boldsymbol{n}_i) d_i, \tag{3}$$

and the i-th crystalline curvature H_i is characterized as the first variation of E_γ on Γ at Γ_i with a suitable norm.

Area-preserving crystalline curvature flow. The normal velocity on Γ_i in the direction $\boldsymbol{n}_i$ is $V_i = \dot{h}_i$. Here and hereafter, we denote that the derivative of a function $u = u(t)$ with respect to time t by $\dot{u}$. The area-preserving crystalline curvature flow is the gradient flow of E_γ along Γ which encloses a fixed area, and it is described as follows

$$V_i = \overline{H} - H_i, \quad i = 1, 2, \ldots, N, \tag{4}$$

where

$$\overline{H} = \frac{\sum_{i=1}^{N} \chi_i l_\gamma(\boldsymbol{n}_i)}{L} = \frac{L_\gamma}{L}$$

is the average of the crystalline curvature, L_γ is the total length of ∂W_γ, and

$$L = \sum_{k=1}^{N} d_k$$

is the total length of Γ.

Remark 1 The equivalence $\sum_{i=1}^N H_i d_i = L_\gamma$ is not obvious. It will be proved in Appendix B.

From (1), we have

$$\dot{d}_i = -(\cot\vartheta_i + \cot\vartheta_{i+1})V_i + V_{i-1}\mathrm{cosec}\vartheta_i + V_{i+1}\mathrm{cosec}\vartheta_{i+1}, \quad i = 1, 2, \ldots, N. \tag{5}$$

Furthermore, by (2) we have

$$\dot{\boldsymbol{p}}_i = V_i \boldsymbol{n}_i + \frac{V_{i-1} - (\boldsymbol{n}_{i-1} \cdot \boldsymbol{n}_i)V_i}{\boldsymbol{n}_{i-1} \cdot \boldsymbol{t}_i} \boldsymbol{t}_i, \quad i = 1, 2, \ldots, N. \tag{6}$$

Note that (4), (5) and (6) are equivalent. It is easy to check that the enclosed area $A = \sum_{i=1}^N h_i d_i/2$ is preserved in time

$$\dot{A} = \sum_{i=1}^N V_i d_i = 0.$$

Moreover, the total interfacial energy E_γ is decreasing in time

$$\dot{E}_\gamma = \sum_{i=1}^N H_i V_i d_i \le 0. \tag{7}$$

Remark 2 The proof of the left equality in (7) can be found in [24, Remark 1.43], and the right inequality is given by Cauchy-Bunyakovsky-Schwarz inequality. From this energy dissipation and the positivity of γ, we have the boundedness of the length of each edge during time evolution.

Our problem is stated as follows.

Problem. For a given admissible polygon Γ_0, find a family of admissible polygons $\{\Gamma(t)\}_{0\le t<T}$ satisfying (4) and (5) (or (6)) with $\Gamma(0) = \Gamma_0$.

We note that the maximal existence time is positive $T > 0$, since (5) are the system of ordinary differential equations.

Known results. What might happen to $\Gamma(t)$ as t tends to $T \le \infty$? When polygon is convex, we have the following result.

Proposition 1 *Let the crystalline energy be $\gamma > 0$. Assume that the initial polygon Γ_0 is an N_γ-sided admissible convex polygon. Then a solution admissible polygon $\Gamma(t)$ exists globally in time, keeping the area enclosed by the polygon equal to constant A, and $\Gamma(t)$ converges to the shape of the boundary of the Wulff shape ∂W_{γ_*} in the Hausdorff metric as t tends to infinity, where $\gamma_*(\boldsymbol{n}_i) = \gamma(\boldsymbol{n}_i)/W$, $W = \sqrt{A_\gamma/A}$ for all $i = 1, 2, \ldots, N_\gamma$ and $A_\gamma = \sum_{k=1}^N \gamma(\boldsymbol{n}_k) l_\gamma(\boldsymbol{n}_k)/2$ is the area of W_γ.*

This proposition is proved in [24] by using the anisoperimetric inequality or Brünn and Minkowski's inequality and the theory of dynamical systems which is the similar technique as in the paper of isotropic version [21]. The results were extended when Γ_0 was essentially admissible polygon [22].

Main results. Our main results are extension of Proposition 1 under the following assumptions.

(A0) W_γ is an N_γ-sided symmetric polygon with respect to the origin.

Under the assumption (A0), the number of edges of W_γ is even, and $-\mathcal{V}_\gamma = \mathcal{V}_\gamma$ holds, i.e., $\gamma(\boldsymbol{n}(\phi_i + \pi)) = \gamma(\boldsymbol{n}(\phi_i))$ holds for all i.

(A1) Γ_0 is an N-sided admissible simple closed polygon and $N > N_\gamma$.

(A2) Transition numbers of Γ_0 are all nonnegative: $\chi_i \geq 0 (i = 1, 2, \ldots, N)$.

The assumption (A1) means that Γ_0 is not convex. Moreover, under the assumption (A2), crystalline curvatures of Γ_0 are also all nonnegative and then Γ_0 is "almost convex" polygon. In addition, under the assumption (A1) and (A2), Γ_0 has at least one edge with zero transition number.

Theorem 2 *Assume the assumptions* (A0), (A1) *and* (A2). *Then there exists* $T_1 > 0$ *such that the solution polygon* $\Gamma(t)$ *is an* N*-sided admissible polygon for* $0 \leq t < T_1$ *and there exists at least one edge with zero transition number whose length tends to zero as* $t \to T_1$. *Moreover,* $\Gamma(t)$ *converges to an admissible* N'*-sided polygon* Γ' $(N_\gamma \leq N' < N)$ *in the Hausdorff metric as* $t \to T_1$ *and enclosed area of* Γ' *is equal to enclosed area of* Γ_0.

In the case where $N' > N_\gamma$ holds, the result of Theorem 2 will be repeated, that is, there exists $T_2 > T_1$ such that a solution polygon $\Gamma(t)$ starting with the initial polygon Γ' evolves until time T_2 and the same phenomena (as in Theorem 2) occur as t tends to T_2. After the time T_2, since number of edges is finite, edge-disappearing occurs at most finitely many instants

$$0 < T_1 < T_2 < \cdots < T_m,$$

and eventually the solution polygon becomes an N_γ-sided admissible convex polygon at the final time $t = T_m$. In the case $N' = N_\gamma$, the final time T_m is T_1 and the polygon Γ' is an N_γ-sided admissible convex polygon. In any case, after the above "convexity phenomena" occurred, the solution polygon follows the theory as in Proposition 1.

At the end of this section, we touch on a general small history of crystalline curvature flow. The crystalline curvature flow is the gradient flow of total interfacial crystalline energy in the case where the Frank diagram, dual shape of the Wulff shape, is a convex polygon. To handle such flow, smooth curves are restricted in an admissible class and the crystalline curvature is defined on each edge of admissible polygonal curve, as we have seen above. This crystalline strategy was proposed by Taylor [18] and independently by Angenent and Gurtin [2]. To select just a few, we refer to [1, 6, 8–10, 17, 19, 21–24]. Besides this crystalline strategy, other strategies by subdifferential and level-set method have been extensively studied [5].

Convexity phenomenon is one of interesting issue in the research of curvature flows, not only the crystalline. Grayson [7] proved convexity phenomena of smooth plane curves for a classical curvature flow, and the first author [10] proved it for a crystalline curvature flow. These results can not be extended for the area-preserving version. For examples, in smooth case, a self-intersection phenomenon is conjectured in [4] and is proved [15]. For the crystalline case, we will prove that Theorem 2: all almost convex polygons become convex eventually. If the initial polygon is NOT "almost convex", i.e., in the case where a negative transition number exists, self-touching phenomenon or edge-edge contact may occur. See Appendix A.

3 Proof of Theorem 2

In what follows, we say that a polygon is *convex* if all transition numbers are positive, and that a polygon is *almost convex* if all transition numbers are nonnegative and there is at least one edge whose transition number is zero, and also that a Γ_j is *inflection edge* if $\chi_j = 0$. We first mention the following properties for almost convex polygons.

Proposition 3 *An almost convex polygon has at least one edge whose transition number is 1.*

Proof Assume that the polygon has N sides and that $\chi_1 = 0$ with $\chi_{N,1} = 1$ and $\chi_{1,2} = -1$, and also that

$$\boldsymbol{p}_1 = \boldsymbol{0},\ \boldsymbol{p}_2 = d_1\boldsymbol{t}_1 + \boldsymbol{p}_1,\ \boldsymbol{t}_1 = \begin{pmatrix} 0 \\ 1 \end{pmatrix},\ \boldsymbol{t}_2 = \begin{pmatrix} \alpha \\ \beta \end{pmatrix} = \boldsymbol{t}_N,$$

without loss of generality. Suppose that $\chi_j = 0$ for all $j = 1, 2, \ldots, N$. Then N is even, and $\boldsymbol{t}_3 = \boldsymbol{t}_5 = \cdots = \boldsymbol{t}_1$ and $\boldsymbol{t}_4 = \boldsymbol{t}_6 = \cdots = \boldsymbol{t}_2$ hold. Since $\chi_{N,1} = \mathrm{sgn}\,(\det(\boldsymbol{t}_N, \boldsymbol{t}_1)) = \mathrm{sgn}(\alpha) = 1$, $\alpha > 0$ holds. Therefore, it holds that

$$\sum_{i=1}^{N} d_i\boldsymbol{t}_i = \sum_{i:\text{even}} d_i \begin{pmatrix} \alpha \\ \beta \end{pmatrix} + \sum_{i:\text{odd}} d_i \begin{pmatrix} 0 \\ 1 \end{pmatrix} \neq \boldsymbol{0}.$$

This is a contradiction to closedness of the polygon. Hence, there is at least one edge whose transition number is 1. ■

Proposition 4 *The number of consecutive inflection edges of an almost convex polygon is even.*

Proof We first assume that the k-th edge is inflection edge and the $(k-1)$-th edge is not the one, that is, assume that $\chi_{k-1} = 1$ and $\chi_k = 0$. This assumption is guaranteed by Proposition 3. From

$$\chi_{k-1} = \frac{1}{2}(\chi_{k-2,k-1} + \chi_{k-1,k}) = 1, \quad \chi_k = \frac{1}{2}(\chi_{k-1,k} + \chi_{k,k+1}) = 0,$$

we have $\chi_{k-1,k} = 2 - \chi_{k-2,k-1} \geq 1$ and $\chi_{k,k+1} = -\chi_{k-1,k}$. Hence, $\chi_{k-1,k} = 1$ holds by $\chi_{k-1,k} \in \{1, -1\}$ and then $\chi_{k,k+1} = -1$ holds. Since the polygon is almost convex, that is, $\chi_i \geq 0$ holds for all i, we have

$$\chi_{k+1} = \frac{1}{2}(\chi_{k,k+1} + \chi_{k+1,k+2}) = \frac{1}{2}(-1 + \chi_{k+1,k+2}) \geq 0,$$

and $\chi_{k+1,k+2} \geq 1$. Therefore, $\chi_{k+1,k+2} = 1$ holds and we obtain $\chi_{k+1} = 0$. Therefore, an almost convex polygon has at least two consecutive inflection edges.

Next, we assume that there are odd number of consecutive inflection edges such as $\chi_k = \chi_{k+1} = \cdots = \chi_{k+2m} = 0$ for some $m = 0, 1, 2, \ldots$, and $\chi_{k-1} = 1$. By the same argument as the above, we have

$$\begin{cases} \chi_{k-1,k} = \chi_{k+1,k+2} = \cdots = \chi_{k+2m-1,k+2m} = 1, \\ \chi_{k,k+1} = \chi_{k+2,k+3} = \cdots = \chi_{k+2m,k+2m+1} = -1. \end{cases}$$

From the almost convexity assumption and

$$\chi_{k+2m+1} = \frac{1}{2}(\chi_{k+2m,k+2m+1} + \chi_{k+2m+1,k+2m+2}) \geq 0,$$

we obtain $\chi_{k+2m+1,k+2m+2} \geq -\chi_{k+2m,k+2m+1} = 1$, and then $\chi_{k+2m+1,k+2m+2} = 1$. Hence, $\chi_{k+2m+1} = 0$ holds. Therefore, the number of consecutive inflection edges is even for almost convex polygons. ■

The following proposition is obvious but will be useful later.

Proposition 5 *Assume* (A2). *If* $\theta_j = \phi_k$, *then* $\theta_i \geq \phi_{k-1}$ *holds for all* $i > j$.

Let the duration of all edges be

$$T_1 = \sup\left\{t > 0; \min_{j=1,2,\ldots,N} d_j(t) > 0\right\}.$$

Proposition 6 *Assume* (A0), (A1) *and* (A2). *Then the solution polygon* $\Gamma(t)$ *has no self-tangencies nor self-intersections for* $0 \leq t < T_1$.

Proof Possible cases of self-contact (self-tangency or self-intersection) of Γ are classified into three cases as follows: (1) vertex-edge contact, (2) edge-edge contact and (3) vertex-vertex contact. We, however, will show no possibilities of any contacts.

Suppose that the first self-contact occurred at the time $T \in (0, T_1)$, and before the time T, no self-contacts have occurred. Then no edges disappear at the time T: $d_k > 0$ $(\forall k)$.

(1) Firstly, we will show that vertices never touch edges. The assertion will be proved by a contradiction. Suppose that a vertex $\boldsymbol{p}_i$ touches an edge Γ_j $(j \neq i-1, i)$

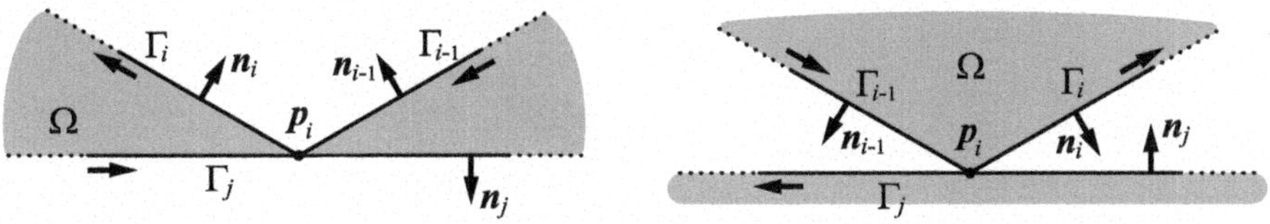

Fig. 3 Possible two cases of vertex-edge contact

Fig. 4 Possible one case of edge-edge contact

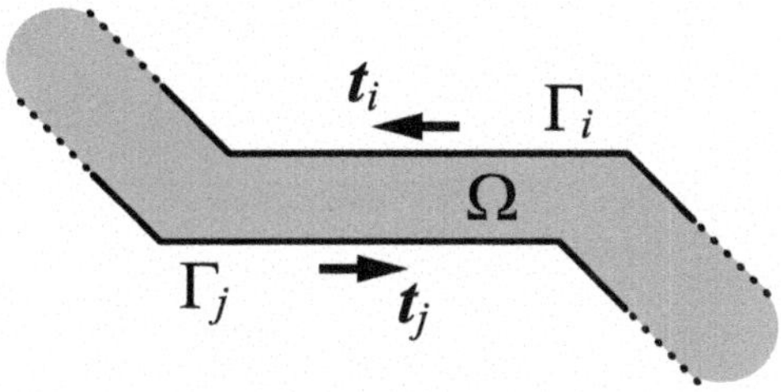

at the time T. Then $\boldsymbol{p}_i \in \Gamma_j$ and $\Gamma_{i-1} \cap \Gamma_i \cap \Gamma_j = \{\boldsymbol{p}_i\}$ are satisfied at the time T. Moreover, we assume that Γ_j lies on the x-axis and $\Gamma_{i-1} \cup \Gamma_i \backslash \{\boldsymbol{p}_i\}$ is in the upper half plane at the time T, without loss of generality. Figure 3 depicts possible two cases of vertex-edge contact, where the shadow region indicates Ω.

Then there exists $s \in (0, 1)$ such that $-\boldsymbol{n}_j = \boldsymbol{n}_*/|\boldsymbol{n}_*|$, $\boldsymbol{n}_* = (1-s)\boldsymbol{n}_{i-1} + s\boldsymbol{n}_i$ holds. This is a contradiction, since by the assumption (A0) and the admissibility there should exist an edge between Γ_{i-1} and Γ_i whose exterior normal vector is $-\boldsymbol{n}_j \in \mathcal{V}_\gamma = \mathcal{N}$. The case where $\boldsymbol{p}_j$ touches an edge Γ_i $(i \neq j-1, j)$ is also applied the same argument as above. Therefore, the vertex-edge contact does not occur.

(2) Secondly, we will show that edges never touch edges. Suppose that Γ_i touches Γ_j at the time T. Then Γ_i and Γ_j are parallel (denote $\Gamma_i \mathbin{/\!\!/} \Gamma_j$ in short) and $\chi_i, \chi_j \geq 0$ hold. Assume that Γ_i and Γ_j lie on the x-axis at the time T, without loss of generality. Then the unit tangent vectors on the edges are $\boldsymbol{t}_k = (t_{k,1}, t_{k,2}) = (1, 0)$ or $(-1, 0)$ for $k = i, j$, and all possible combinations for Γ_i are 8 cases such as $(\chi_{i-1,i}, \chi_{i,i+1}, t_{i,1}) = (\pm 1, \pm 1, \pm 1)$ except the two cases $(\chi_{i-1,i}, \chi_{i,i+1}, t_{i,1}) = (-1, -1, \pm 1)$, i.e., six cases for each edge are possible. Hence, the total amount of 36 cases are possible as in Fig. 9. But under the assumptions (A0), (A1) and (A2), possible edge-edge contact case is essentially only one. See Appendix D in detail. Figure 4 indicates the case at the time just before T, where the shadow region indicates Ω.

In this case, the time derivative of the distance between Γ_i and Γ_j is $V_i + V_j = 2\overline{H} > 0$. Therefore, the edge-edge contact does not occur.

(3) Finally, we will show that vertices never touch vertices. Suppose that $\boldsymbol{p}_i$ and $\boldsymbol{p}_j$ contact each other at the time T. In this case, the total amount of 8 cases are possible. See Fig. 10. However, under the assumption, two cases are possible. See Appendix E in detail. Figure 5 depicts possible two cases of vertex-vertex contact at the time just before T, where $\{a, b\} = \{i, j\}$.

Note that $\Gamma_i \mathbin{/\!\!/} \Gamma_{j-1}$, $\Gamma_{i-1} \mathbin{/\!\!/} \Gamma_j$ (Fig. 5a) or $\Gamma_i \mathbin{/\!\!/} \Gamma_j$, $\Gamma_{i-1} \mathbin{/\!\!/} \Gamma_{j-1}$ (Fig. 5b) hold. If $\chi_{i-1,i}\chi_{j-1,j} = -1$ (Fig. 5a for $\{a, b\} = \{i, j\}$) holds, then the edge-edge con-

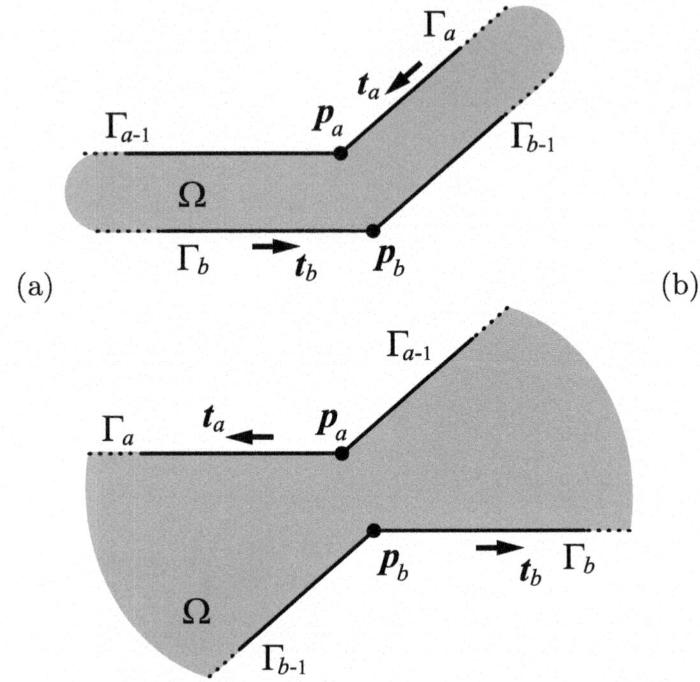

Fig. 5 Possible two cases of vertex-vertex contact

tact occurs at the time T, which was denied in the above. In the other case $(\chi_{i-1,i}, \chi_{j-1,j}) = (-1, -1)$ (Fig. 5b for $\{a, b\} = \{i, j\}$), under the assumption (A2), $\chi_k = 0$ holds for $k = i-1, i, j-1, j$. Hence, we have $V_k = \overline{H} > 0$, which means that the distance between $\boldsymbol{p}_i$ and $\boldsymbol{p}_j$ increase in time. Therefore, the vertex-vertex contact does not occur. Hence, we have the assertion. ■

Lemma 7 *Assume* (A0), (A1) *and* (A2). *Then at least one edge of solution polygon disappears in a finite time, that is,* $T_1 < \infty$.

Proof By Proposition 3, there exists j_0 such that $(\chi_{j_0-1}, \chi_{j_0}, \chi_{j_0+1}) = (1, 0, 0)$. Note that $\Gamma_{j_0-1} /\!/ \Gamma_{j_0+1}$. From (7), $d_j(t) \le E_\gamma(0)/\gamma(\boldsymbol{n}_j)$ holds for all j. Hence, we have $V_{j_0-1} - V_{j_0+1} = -l_\gamma(\boldsymbol{n}_{j_0-1})/d_{j_0-1}(t) \le -l_\gamma(\boldsymbol{n}_{j_0-1})\gamma(\boldsymbol{n}_{j_0-1})/E_\gamma(0) < 0$. Therefore, by (5), there exist $T_1 > 0$ such that $\liminf_{t\to T_1} d_{j_0}(t) = 0$. On the other hand, we can easily obtain that there exist the limit $\lim_{t\to T_1} d_j(t)$ for all j by the same arguments as in [9, 10]. Hence, $\lim_{t\to T_1} d_{j_0}(t) = 0$ holds and we have the assertion. ■

From the argument in the above proof, we can define $d_j(T_1) = \lim_{t\to T_1} d_j(t)$ for all j and the limit $\Omega(T_1) = \lim_{t\to T_1} \Omega(t)$ exists in the Hausdorff topology. If $\Omega(T_1)$ is admissible, then the flow of admissible polygons can be continued restarting from the initial polygon $\Omega(T_1)$.

In the following, we will characterize disappearing edges at $t = T_1$, and eventually show that $\Omega(T_1)$ is admissible.

Proposition 8 *Assume* (A0), (A1) *and* (A2). *Then only inflection edges may disappear at* $t = T_1$.

We will prepare two lemmas to prove this proposition.

Lemma 9 *Assume* (A0), (A1) *and* (A2). *Suppose that there exist* k_1 *and* k_2 *satisfying* $k_2 \geq k_1 + 2$, $0 < \theta_{k_2} - \theta_{k_1} < \pi$ *and* $d_{k_1}(T_1) > 0$, *and that* $\chi_j = 1$ *and* $d_j(t) > 0$ *hold for* $j \in Q := \{k_1 + 1, k_1 + 2, \ldots, k_2 - 1\}$ *and* $t \in [0, T_1)$. *If* $d_{k_2}(T_1) > 0$ *or* $\chi_{k_2} = 0$, *it is impossible that* $d_j(T_1) = 0$ *holds for all* $j \in Q$.

Hereafter, $A \sim B$ means $c_1 B \leq A \leq c_2 B$ for some positive constants c_1 and c_2.

Proof Note that the assumption $k_2 \geq k_1 + 2$ implies $Q \neq \emptyset$. The lemma will be proved by a contradiction. Assume that $d_j(T_1) = 0$ holds for all $j \in Q$. Then we can define the meeting point $\boldsymbol{p}^* = \lim_{t \to T_1} \boldsymbol{p}_i(t)$ for $i \in Q \cup \{k_2\}$. Let $L_k(t)$ be the straight line extending the k-th edge:

$$L_k(t) = \{\boldsymbol{x} \in \mathbb{R}^2;\ (\boldsymbol{p}_k - \boldsymbol{x}) \cdot \boldsymbol{n}_k = 0\},$$

and let $\boldsymbol{q}(t)$ be the intersection point of L_{k_1} and L_{k_2}. We take any fixed $i \in Q$ and define two functions $a_1(t)$ and $a_2(t)$ by $a_1(t) = (\boldsymbol{q}(t) - \boldsymbol{p}^*) \cdot \boldsymbol{n}_i$ and $a_2(t) = (\boldsymbol{p}_i(t) - \boldsymbol{p}^*) \cdot \boldsymbol{n}_i$, respectively. Note that $a_1(T_1) = a_2(T_1) = 0$ holds. Since $d_i(T_1) = 0$, we have $V_i(t) \to -\infty$ as $t \to T_1$. Hence, $V_i(t) < 0$ holds for $t \in (T_1 - \delta, T_1)$ with a sufficiently small $\delta > 0$. We define the half plane $\Omega_j(t)$ by

$$\Omega_j(t) = \{\boldsymbol{x} \in \mathbb{R}^2;\ (\boldsymbol{p}_j(t) - \boldsymbol{x}) \cdot \boldsymbol{n}_j \geq 0\}.$$

Then $\boldsymbol{p}^* \in \bigcap_{t \in (T_1 - \delta, T_1)} \bigcap_{k_1 \leq j \leq k_2} \Omega_j(t)$ holds. Hence, we have $a_2(t) > 0$.

Claim 10 $\Gamma_i \subset \Omega_{k_1} \cap \Omega_{k_2}$ *holds.*

Proof We assume that $\theta_{k_1} = 0$ and $0 < \theta_j < \pi$ $(k_1 + 1 \leq j \leq k_2)$ without loss of generality. Then we have $\boldsymbol{n}_{k_1} = (1, 0)$ and L_{k_1} is the y-axis. Let the entry of $\boldsymbol{p}_j$ be (x_j, y_j). Since $x_{k_1+1} = 0$, $x_i = -\sum_{j=k_1}^{i-1} d_j \sin\theta_j < 0$. Hence, $x_i \leq 0$ holds, and we have $\boldsymbol{p}_i \in \Omega_{k_1}$. In the same way, we have $\boldsymbol{p}_i \in \Omega_{k_2}$. Therefore, $\boldsymbol{p}_i \in \Omega_{k_1} \cap \Omega_{k_2}$ holds. Similarly, we have $\boldsymbol{p}_{i+1} \in \Omega_{k_1} \cap \Omega_{k_2}$. Hence, the assertion of the claim is proved. □

Claim 11 $a_1(t) > a_2(t)$ *holds.*

Proof By the same calculation as in [23, p. 166], we obtain

$$\boldsymbol{q}(t) = \boldsymbol{p}_{k_1+1} + s\boldsymbol{t}_{k_1},$$

$$s = \frac{(\boldsymbol{p}_{k_1} - \boldsymbol{p}_{k_1+1}) \cdot (\boldsymbol{t}_{k_1} - \mu \boldsymbol{t}_{k_2})}{1 - \mu^2} = \frac{(\boldsymbol{p}_{k_1} - \boldsymbol{p}_{k_1+1}) \cdot \boldsymbol{n}_{k_2}}{\sin(\theta_{k_2} - \theta_{k_1})} = \sum_{j=k_1+1}^{k_2-1} d_j \frac{\sin(\theta_{k_2} - \theta_j)}{\sin(\theta_{k_2} - \theta_{k_1})},$$

where $\mu = \cos(\theta_{k_2} - \theta_{k_1})$. Note that $s > 0$ holds, since we have

$$\theta_{k_2} - \pi < \theta_{k_1} < \theta_{k_1+1} < \cdots \theta_{k_2-1} < \theta_{k_2} < \theta_{k_1} + \pi.$$

Hence, it holds that

$$
\begin{aligned}
a_1 - a_2 = (\boldsymbol{q} - \boldsymbol{p}_i) \cdot \boldsymbol{n}_i &= - \sum_{j=k_1+1}^{i-1} d_j \sin(\theta_i - \theta_j) + s \sin(\theta_i - \theta_{k_1}) \\
&= - \sum_{j=k_1+1}^{i-1} d_j \sin(\theta_i - \theta_j) + \frac{\sin(\theta_i - \theta_{k_1})}{\sin(\theta_{k_2} - \theta_{k_1})} \sum_{j=k_1+1}^{k_2-1} d_j \sin(\theta_{k_2} - \theta_j) \\
&> - \sum_{j=k_1+1}^{k_2-1} d_j \sin(\theta_i - \theta_j) + \frac{\sin(\theta_i - \theta_{k_1})}{\sin(\theta_{k_2} - \theta_{k_1})} \sum_{j=k_1+1}^{k_2-1} d_j \sin(\theta_{k_2} - \theta_j) \\
&= \sum_{j=k_1+1}^{k_2-1} d_j \frac{\sin(\theta_{k_2} - \theta_i) \sin(\theta_j - \theta_{k_1})}{\sin(\theta_{k_2} - \theta_{k_1})} > 0.
\end{aligned}
$$

Hence, the assertion of the claim is proved. □

The time derivative of $\boldsymbol{q}$ can be calculated as

$$
\dot{\boldsymbol{q}}(t) = V_{k_1} \boldsymbol{n}_{k_1} + \frac{V_{k_2} - V_{k_1} \cos(\theta_{k_2} - \theta_{k_1})}{\sin(\theta_{k_2} - \theta_{k_1})}
$$

via the same calculation as in [23, p. 166]. Note that $\overline{H} \sim 1$. Since $d_{k_1}(T_1) > 0$, $\sup_{T_1-\delta<t<T_1} |V_{k_1}(t)|$ is bounded. Moreover, $\sup_{T_1-\delta<t<T_1} |V_{k_2}(t)|$ is also bounded, since $d_{k_2}(T_1) > 0$ or $\chi_{k_2} = 0$ are assumed. Therefore, there exists a constant $C_* > 0$ such that $|\dot{\boldsymbol{q}}(t)| \leq C_*$ and then $|\dot{a}_1(t)| \leq C_*$ hold for $t \in (T_1 - \delta, T_1)$. Hence there exists $\eta \in (T_1 - \delta, T_1)$ such that we have

$$
0 < - \int_{T_1-\delta}^{T_1} V_i(t)\, dt = a_2(T_1 - \delta) < a_1(T_1 - \delta) - a_1(T_1) = -\dot{a}_1(\eta)\delta \leq C_*\delta.
$$

This contradicts the fact $V_i(t) \to -\infty$ as $t \to T_1$. Hence, we have the assertion. ■

The following result is proved in [6, 10] for crystalline curvature flow without area-preserving constraint. We need to modify the proof as a nonlocal version.

Lemma 12 *Assume* (A0), (A1) *and* (A2). *Suppose that there exist k_1 and k_2 satisfying $k_2 \geq k_1 + 2$, $\theta_{k_2} - \theta_{k_1} = \pi$, $d_{k_1}(T_1) > 0$ and $\chi_{k_1} = \chi_{k_1+1} = 1$, and that $d_j(t) > 0$ holds for $j \in Q := \{k_1 + 1, k_1 + 2, \ldots, k_2 - 1\}$ and $t \in [0, T_1)$. If $d_{k_2}(T_1) > 0$ or $\chi_{k_2} = 0$, it is impossible that $d_j(T_1) = 0$ holds for all $j \in Q$.*

Hereafter, we denote various positive constants by C.

Proof Note that the assumption $k_2 \geq k_1 + 2$ implies $Q \neq \emptyset$. The lemma will be proved by a contradiction. Assume that $d_j(T_1) = 0$ holds for all $j \in Q$. Let $w(t)$ be the distance in the normal direction between Γ_{k_1} and Γ_{k_2}, which have the same normal direction. Then, we have $w(t) = (\boldsymbol{p}_{k_1} - \boldsymbol{p}_{k_2}) \cdot \boldsymbol{n}_{k_1} = h_{k_1} + h_{k_2}$ from $-\boldsymbol{n}_{k_1} = \boldsymbol{n}_{k_2}$. By the assumptions, we have $w(t) > 0$ for $0 \leq t < T_1$ and $\lim_{t \to T_1} w(t) = 0$

holds. Also, $V_{k_1-1} \leq \overline{H} \sim 1$ holds from (A2). Note that $|V_{k_1}| \sim 1$ holds, both $m := \sin(\theta_{k_1+1} - \theta_{k_1})$ and $\sin(\theta_{k_1} - \theta_{k_1-1})$ are positive constants, and $|\cot(\theta_{k_1+1} - \theta_{k_1}) + \cot(\theta_{k_1} - \theta_{k_1-1})|$ is bounded. Using (5), we have

$$\dot{d}_{k_1} \leq C + \frac{1}{m} V_{k_1+1}. \tag{8}$$

From $\chi_{k_1+1} = 1$ and $d_{k_1+1}(T_1) = 0$, we have $\dot{d}_{k_1}(t) \to -\infty$ as $t \to T_1$, and there exists $\delta > 0$ such that $\dot{d}_{k_1}(t) < 0$ for $t \in (T_1 - \delta, T_1)$.

We put $W = \dot{w} - 2\overline{H}$. Note that $\dot{w}(t) = V_{k_1} + V_{k_2} = 2\overline{H} - l_\gamma(\boldsymbol{n}_{k_1})/d_{k_1} - \chi_{k_2} l_\gamma(\boldsymbol{n}_{k_2})/d_{k_2}$. Hence, we have $W \sim -1$, since $d_{k_2}(T_1) > 0$ or $\chi_{k_2} = 0$ are assumed. Therefore, it holds that

$$I := \int_{T_1-\delta}^{T_1} \dot{d}_{k_1} W \, dt \leq C \int_{T_1-\delta}^{T_1} (-\dot{d}_{k_1}) dt = C\big(d_{k_1}(T_1 - \delta) - d_{k_1}(T_1)\big) < \infty.$$

On the other hand, we have $m d_{k_1+1}(t) \leq w(t)$ by geometry. Then, $V_{k_1+1} \leq \overline{H} - m l_\gamma(\boldsymbol{n}_{k_1+1})/w$ holds. By virtue of (8), we have

$$\begin{aligned}
I &\geq \int_{T_1-\delta}^{T_1} \big(C + \frac{1}{m} V_{k_1+1}\big) W \, dt \\
&\geq \int_{T_1-\delta}^{T_1} \big(C + \frac{1}{m}\overline{H} - \frac{1}{w} l_\gamma(\boldsymbol{n}_{k_1+1})\big) W \, dt \\
&\geq \int_{T_1-\delta}^{T_1} \big(C - \frac{1}{w} l_\gamma(\boldsymbol{n}_{k_1+1})\big)(\dot{w} - 2\overline{H}) \, dt \\
&\geq \int_{T_1-\delta}^{T_1} \big(C\dot{w} - 2C\overline{H} - l_\gamma(\boldsymbol{n}_{k_1+1})\frac{\dot{w}}{w} + l_\gamma(\boldsymbol{n}_{k_1+1})\frac{2\overline{H}}{w}\big) \, dt \\
&\geq -C - l_\gamma(\boldsymbol{n}_{k_1+1})\big[\log w\big]_{w(T_1-\delta)}^{0} = +\infty.
\end{aligned}$$

This leads to a contradiction, and completes the proof. ■

The symmetric result of Lemma 12 also holds.

Corollary 13 *Assume* (A0), (A1) *and* (A2). *Suppose that there exist* k_1 *and* k_2 *satisfying* $k_2 \geq k_1 + 2$, $\theta_{k_2} - \theta_{k_1} = \pi$, $d_{k_2}(T_1) > 0$ *and* $\chi_{k_2} = \chi_{k_2-1} = 1$, *and that* $d_j(t) > 0$ *holds for* $j \in Q := \{k_1 + 1, k_1 + 2, \ldots, k_2 - 1\}$ *and* $t \in [0, T_1)$. *If* $d_{k_1}(T_1) > 0$ *or* $\chi_{k_1} = 0$, *it is impossible that* $d_j(T_1) = 0$ *holds for all* $j \in Q$.

Now we will show Proposition 8.

Proof of Proposition 8 By virtue of the isoperimetric inequality $L(t)^2 \geq CA(t)$ and the area-preserving property $A(t) \equiv A(0)$, neither a single-point-extinction nor a line-segment-extinction occur at $t = T_1$, and at least three edges remain at $t = T_1$. We can find the two edges Γ_{j_0} and Γ_{j_1} $(j_0 + 2 \leq j_1)$ such that $d_{j_0}(T_1),\ d_{j_1}(T_1) > 0$,

$d_j(T_1) = 0$ for all $j \in Q := \{j_0 + 1, \dots, j_1 - 1\}$ and Q is not empty. Here we can define $\boldsymbol{p}^* = (x^*, y^*) := \lim_{t \to T_1} \boldsymbol{p}_i(t)$ for $i \in Q \cup \{j_1\}$, since $d_i(T_1) = 0$ for $i \in Q$.

Let the set of the zero-transition numbers be $Q_0 := \{j \in Q;\ \chi_j = 0\}$. We will show that $Q_0 = Q$ by a contradiction. Assume that $Q_0 \subsetneq Q$. Then there exists $j_2 \in Q$ such that

$$\chi_{j_2} = 1 \quad \text{and} \quad \chi_j = 0 \text{ for } j \in \{j_0 + 1, \dots, j_2 - 1\} \subset Q \tag{9}$$

Claim 14 *$\theta_{j_0} \le \theta_{j_1}$ holds.*

Proof If $\theta_{j_0} = \phi_k$, from Proposition 5 and (9), we have

$$\begin{cases} \theta_j \ge \phi_{k-1} & \text{for } j \in \{j_0 + 1, \dots, j_2 - 1\}, \\ \theta_{j_2} = \theta_{j_0} = \phi_k, \\ \theta_{j_2+1} = \phi_{k+1}. \end{cases}$$

Again from Proposition 5, $\theta_i \ge \phi_k = \theta_{j_0}$ holds for all $k \ge j_2 + 1$, especially we have $\theta_{j_1} \ge \theta_{j_0}$. □

Claim 15 *$\theta_{j_0} < \theta_{j_1}$ holds.*

Proof Suppose that $\theta_{j_1} = \theta_{j_0}$. From (9) and the previous claim, we have $\chi_{j_2} = 1$ and $\theta_{j_2} = \theta_{j_0}$, and hence the number of edges between Γ_{j_0} and Γ_{j_1} should be at least three, i.e., $j_0 + 2 \le j_2 \le j_1 - 2$ holds.

If $\chi_{j_2+1} = 1$, then from $\chi_{j_2} = 1$, $\theta_{j_2} = \theta_{j_0} = \phi_k$ and Proposition 5, $\theta_i \ge \phi_{k+1}$ holds for all $i \ge j_2 + 1$ including $i = j_1$. This contradicts $\theta_{j_1} = \theta_{j_0}$, and hence $\chi_{j_2 \pm 1} = 0$ holds.

Since $\chi_{j_2} = 1$ and $\chi_{j_2 \pm 1} = 0$, we have $\chi_{j_2 \pm 2} = 0$. Indeed, if $\chi_{j_2-2} = 1$, then $\chi_{j_2-2, j_2-1} = 1$, thus $\chi_{j_2-1} = 1$ holds, since $\chi_{j_2-1, j_2} = 1$. Hence four edges Γ_{j_2-2}, Γ_{j_2-1}, Γ_{j_2+1} and Γ_{j_2+2} are inflection edges and move toward outside with normal velocity $\overline{H} > 0$. Therefore, the vertex $\boldsymbol{p}_{j_2-1}$ (resp. $\boldsymbol{p}_{j_2+2}$) moves along the half line $\widetilde{L}_{j_2-1}$ (resp. $\widetilde{L}_{j_2+1}$), where

$$\widetilde{L}_j = \{\boldsymbol{x} \in \mathbb{R}^2 \mid \boldsymbol{x} = \boldsymbol{p}_j(0) + r(\boldsymbol{n}_j + \boldsymbol{n}_{j-1}),\ r \ge 0\}, \quad j = 1, 2, \dots, N.$$

Note that "$\boldsymbol{p}_{j_2-1}(0) \notin \Omega_{j_2+1}(0)^c$, $\widetilde{L}_{j_2-1}(0) \not\subset \Omega_{j_2+1}(0)^c$" or "$\boldsymbol{p}_{j_2+2}(0) \notin \Omega_{j_2-1}(0)^c$, $\widetilde{L}_{j_2+2}(0) \not\subset \Omega_{j_2-1}(0)^c$" hold, since $\Gamma(0)$ has no self-intersection. Thus, the above two half lines do not have an intersection point, that is, $\boldsymbol{p}_{j_2-1}(t)$ and $\boldsymbol{p}_{j_2+2}(t)$ can not meet at the same point $\boldsymbol{p}^*$, as $t \to T_1$, which is a contradiction.

Therefore $\theta_{j_1} = \theta_{j_0}$ does not hold. □

Without loss of generality, we may assume $\theta_{j_0} = 0$. If $\pi < \theta_{j_1} \le 2\pi$, from $\lim_{t \to T_1} \sum_{j=j_0+1}^{j_1-1} d_j(t) = 0$, we have $\lim_{t \to T_1} \mathrm{dist}(\boldsymbol{p}_{j_0+1}(t), \boldsymbol{p}_{j_1}(t)) = 0$. Then, self-intersection of the curve $\sum_{j=j_0}^{j_1} \Gamma_j(t)$ occurs before $t = T_1$, since $d_{j_0}(T_1), d_{j_1}(T_1) > 0$. Hence, we have $0 < \theta_{j_1} \le \pi$.

Hereafter, $\sharp S$ stands for the number of element of the set S.

Claim 16 $\sharp Q_0 \geq 1$ *holds.*

Proof We first assume that $\sharp Q_0 = 0$, i.e., $\chi_j = 1$ for all $j \in Q$, especially $\chi_{j_0+1} = \chi_{j_1-1} = 1$ holds. In the case where $\theta_{j_1} = \pi$, if $(\chi_{j_0}, \chi_{j_1}) \neq (0, 0)$, we have a contradiction by Lemma 12. Hence $(\chi_{j_0}, \chi_{j_1}) = (0, 0)$ holds.

From $V_{j_0} = V_{j_1} = \overline{H} > 0$ and $\boldsymbol{n}_{j_0} = -\boldsymbol{n}_{j_1}$, the distance between Γ_{j_0} and Γ_{j_1} increases in time. Therefore, we have $0 < \theta_{j_1} < \pi$. By Lemma 9, this situation does not appear. Therefore, Q_0 is not empty, and $\sharp Q_0 \geq 1$ holds. □

Claim 17 *Only one pair of inflection edges exists on the curve* $\sum_{j=j_0}^{j_1} \Gamma_j(t)$.

Proof By Proposition 4, there are several pairs of two consecutive inflection edges. Suppose that there are two pairs $(k_1 - 1, k_1)$ and $(k_2 - 1, k_2)$ $(k_1 + 1 < k_2)$ in $Q \cup \{j_0, j_1\}$. Then, $V_k = \overline{H}(t) > 0$ for $k = k_1 - 1, k_1, k_2 - 1, k_2$ and thus $\boldsymbol{p}_k(t)$ $(k = k_1, k_2)$ moves toward outside region of the solution polygon along the half line $L_k = \{\boldsymbol{p} \in \mathbb{R}^2 \mid \boldsymbol{p} = \boldsymbol{p}_k(0) + \lambda(\boldsymbol{n}_{k-1} + \boldsymbol{n}_k), \lambda \geq 0\}$, which tangential angle α_k is given by $\alpha_k = (\theta_k + \theta_{k-1})/2$. Since $\chi_i \geq 0$ for all j, we have $0 \leq \alpha_{k_2} - \alpha_{k_1} < 2\pi$, thus, from almost convexity of $\Omega(t)$, L_{k_1} and L_{k_2} do not have a intersection point (see Appendix C), and this leads a contradiction to the fact $\lim_{t \to T_1} \boldsymbol{p}_{k_1}(t) = \lim_{t \to T_1} \boldsymbol{p}_{k_2}(t) = \boldsymbol{p}^*$. Thus, only one pair of inflection edges can exist on the curve $\sum_{j=j_0}^{j_1} \Gamma_j(t)$. □

Without loss of generality, we consider the case that $j_0 + 2 < k \leq j_1$ and $\chi_{k-1} = \chi_k = 0$ and $\chi_j = 1$ for $j = j_0 + 1, \ldots, k - 2$. We first consider the case $0 < \theta_{j_1} < \pi$. From $\chi_{k-1,k} = -1$, we have $0 < \theta_{k-1} \leq \pi$. By using Lemma 9 and Lemma 12, we have a contradiction. Therefore, only the case $k = j_1$ and $\theta_{j_1} = \pi$ remain. By $\chi_{j_1} \geq 0$ and $\chi_{j_1-1,j_1} = -1$, we have $\chi_{j_1} = 0$ and $\chi_{j_1,j_1+1} = 1$.

Claim 18 $\chi_{j_0} = 1$ *holds.*

Proof We denote $\boldsymbol{p}_i(t) = (x_i(t), y_i(t))$. Suppose that $\chi_{j_0} = 0$. Then $V_{j_0} = V_{j_1} = \overline{H}(t) > 0$ holds, from which it deduces that $\dot{x}_{j_0}(t) > 0$ and $\dot{x}_{j_1}(t) < 0$. From $\boldsymbol{p}_{j_0+1}(T_1) = \boldsymbol{p}_{j_1}(T_1) = \boldsymbol{p}^* := (x^*, y^*)$, we have

$$x_{j_0}(t) = x_{j_0+1}(t) < x^* < x_{j_1}(t) = x_{j_1+1}(t)$$

for $0 \leq t < T_1$. Since $\chi_j \geq 0$ for all j, by *Claim* 17, the intersection set of the extended line L_{j_0} and the arc $\mathcal{C}_1 = \sum_{j=j_0+1}^{j_1-1} \Gamma_j$ consists of one point, say $\boldsymbol{p}^{**} = (x_{j_0}(t), y^{**}(t))$, satisfying $y^{**}(t) < y_{j_0}(t)$. Then the length of the subarc of $\mathcal{C}_1$ from $\boldsymbol{p}_{j_0}$ to $\boldsymbol{p}^{**}$ is greater than $d_{j_0}(t)$, and thus $\sum_{j=j_0+1}^{j_1-1} d_j(t) \geq \inf_{t \in [0,T_1]} d_{j_0}(t)$ holds for $0 \leq t < T_1$. This contradicts to the assumptions $\inf_{t \in [0,T_1]} d_{j_0}(t), \inf_{t \in [0,T_1]} d_{j_1}(t) > 0$ and $d_j(T_1) = 0$ for $j \in Q$. □

Claim 19 $\chi_{j_0-1} = 1$ *holds.*

Proof The curve $\mathcal{C}(t) := \sum_{j \notin Q} \Gamma_j(t)$ is contained in the region $\{(x, y) \in \mathbb{R}^2 \mid x_{j_1}(t) \leq x \leq x_{j_0}(t)\}$, since $\chi_{j_1, j_1+1} = 1$, $\chi_{j_0} = 0$ and $\chi_j \geq 0$ for all j. By $x_{j_1}(T_1) = x_{j_0}(T_1) = x^*$, $x_j(T_1) = x^*$ holds for $j \notin Q$.

Suppose that $\chi_{j_0-1} = 0$. Then, $\boldsymbol{n}_{j_0-2} = \boldsymbol{n}_{j_0}$ and $\chi_{j_0-2} = 0$ and thus, $\dot{x}_{j_0-2} = \overline{H}(t) > 0$. From $\dot{x}_{j_1} = -\overline{H}(t) < 0$, we have $\dfrac{d}{dt}(x_{j_0-2} - x_{j_1}) = 2\overline{H}(t) > 0$. Therefore, we have $x_{j_0-2}(t) - x_{j_1}(t) > x_{j_0-2}(0) - x_{j_1}(0) \geq 0$, which leads a contradiction, and hence $\chi_{j_0-1} = 1$ holds. □

Now we apply Lemma 12 to $\mathcal{C}(t)$, we obtain that this situation does not occur. Hence, we have $Q_0 = Q$. ■

Proof of Theorem 2. Here we use the same notation in the previous proof. We already know that the number of inflection edges are locally two at disappearing part of $\Gamma(T_1)$. The case $\sharp Q = 2$ means $j_1 = j_0 + 3$ and Γ_{j_0+1} and Γ_{j_0+2} disappear at $t = T_1$. In this case, the admissibility is preserved since $\boldsymbol{n}_{j_0+2} = \boldsymbol{n}_{j_0}$ and $\boldsymbol{n}_{j_0+1} = \boldsymbol{n}_{j_1}$. The case $\sharp Q = 1$ means $j_1 = j_0 + 2$ and Γ_{j_0+1} disappears at $t = T_1$. In this case, $\boldsymbol{n}_{j_0} = \boldsymbol{n}_{j_1}$. That is, when Γ_{j_0+1} disappears, Γ_{j_0} and Γ_{j_1} merge into one edge and thus the admissibility is preserved. Therefore, $\Omega(T_1)$ is admissible. The proof is completed. ■

Appendix A: Example of Self-touching Starting from Non-convex Polygon

Here, to construct the example of self-touching phenomenon during time evolution, such as Fig. 6b starting from (a), we solve the evolution in inverse time direction from a initial polygon which has overlap edges. Set the initial polygon as Fig. 6b. We see that the i-th and j-th edges have overlap and their transition numbers are zero. By the

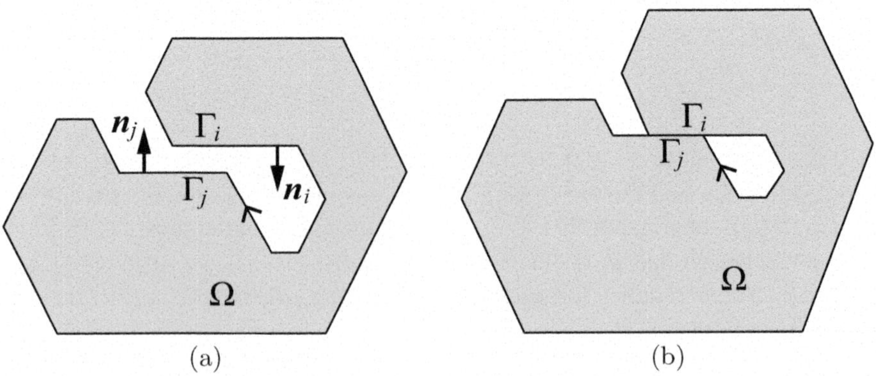

Fig. 6 **a** a non-convex admissible polygon **b** self-touching phenomenon

standard theory of ODEs, we can obtain a time local solution for $t < 0$ and any edge-disappearings and any self-intersections never occur in a short time interval $(-T, 0)$ for a small $T > 0$. Note that both the i-th and the j-th edges move to inward direction with an inward normal velocity $L_\gamma/L(t)$ and thus lose the overlap instantaneously. Hence the solution polygon $\widetilde{\Omega}(t)$ is admissible without any overlaps in the short time interval $(-T, 0)$. Therefore, if we set the initial polygon $\Omega(0) = \widetilde{\Omega}(-T')$ with $-T' \in (-T, 0)$, then the solution polygon $\Omega(T')$ creates a hole.

Appendix B: Proof of the Remark 1

Proposition 20 *For an admissible polygon we have*

$$\sum_{i=1}^{N} H_i d_i = L_\gamma.$$

This assertion is the crystalline version of

$$\int_C H\, ds = L_\gamma,$$

for smooth Jordan curve C, where $H = (\sigma(\theta) + \sigma''(\theta))\kappa$, $\sigma(\theta) = \gamma(\boldsymbol{n}(\theta))$, κ is the curvature, L_γ is the total length of W_γ. In particular, if $\gamma(\boldsymbol{x}) = |\boldsymbol{x}|$, then we have $\sigma = 1$ and ∂W_γ is the unit circle, and the above equality is nothing but the definition of winding number of C such as

$$\int_C \kappa\, ds = 2\pi.$$

To show Proposition 20, we prepare the following lemma.

Lemma 21 *For N-sided admissible polygon, we have*

$$\#I^+(\boldsymbol{n}) - \#I^-(\boldsymbol{n}) = 1 \qquad (\forall \boldsymbol{n} \in \mathcal{N}_\gamma),$$

where

$$I^+(\boldsymbol{n}) = \{i; \boldsymbol{n}_i = \boldsymbol{n}, \chi_i = 1, 1 \le i \le N\},$$
$$I^-(\boldsymbol{n}) = \{i; \boldsymbol{n}_i = \boldsymbol{n}, \chi_i = -1, 1 \le i \le N\}$$

are defined as the set of indices for each $\boldsymbol{n} \in \mathcal{N}_\gamma$.

Using this Lemma, we can easily show the proposition.

Proof of Proposition 20 Let $I^0(\boldsymbol{n}) = \{i \,|\, \boldsymbol{n}_i = \boldsymbol{n}, \chi_i = 0, 1 \le i \le N\}$. From the definition of H_i and χ_i and Lemma 21, we have

$$\begin{aligned}\sum_{i=1}^{N} H_i d_i &= \sum_{i=1}^{N} \chi_i l_\gamma(\boldsymbol{n}_i)\\ &= \sum_{\boldsymbol{n}\in\mathcal{N}_\gamma} (1\cdot \#I^+(\boldsymbol{n}) + 0\cdot \#I^0(\boldsymbol{n}) + (-1)\cdot \#I^-(\boldsymbol{n}))l_\gamma(\boldsymbol{n})\\ &= \sum_{\boldsymbol{n}\in\mathcal{N}_\gamma} (\#I^+(\boldsymbol{n}) - \#I^-(\boldsymbol{n}))l_\gamma(\boldsymbol{n})\\ &= \sum_{\boldsymbol{n}\in\mathcal{N}_\gamma} l_\gamma(\boldsymbol{n})\\ &= L_\gamma.\end{aligned}$$

■

To show Lemma 21, we prepare some properties of admissible polygons $\Gamma = \bigcup_{i=1}^{N} \Gamma_i$. Let θ_i be the exterior normal angle of the i-th edge Γ_i for $i = 1, 2, \ldots, N$ and define $\theta_{i+mN} = \theta_i + 2m\pi$ for $m \in \mathbb{Z}$. Let $\Theta := \{\theta_i \mid i \in \mathbb{Z}\}$.

Let us consider the graph which describes piecewise linear curve connecting all adjacent points of the set $\{(i, \theta_i) \in \mathbb{R}^2\}_{i\in\mathbb{Z}}$ such as $\bigcup_{i\in\mathbb{Z}}[(i-1, \theta_{i-1}), (i, \theta_i)]$. We call this graph (i, θ_i)-graph. For instance, if Γ is the regular N-polygon, then (i, θ_i)-graph describes a straight line with the slope being $2\pi/N$.

Let ϕ_k be the normal angle of the k-th edge of the Wulff shape W_γ for $k = 1, 2, \ldots, N_\gamma$ and define $\phi_{k+mN_\gamma} = \phi_k + 2m\pi$ for $m \in \mathbb{Z}$. Let $\Phi_\gamma := \{\phi_k \mid k \in \mathbb{Z}\}$.

If $\theta_i = \phi_k$, then we have $\theta_{i-1}, \theta_{i+1} \in \{\phi_{k+1}, \phi_{k-1}\}$ because of the admissibility of the curve. Thus, connecting patterns of (i, θ_i)-graph of admissible polygon are classified the only following four cases:

$$(*)\begin{cases}(i-1, \phi_{k-1}) - (i, \phi_k) - (i+1, \phi_{k+1}) & (\chi_i = 1),\\ (i-1, \phi_{k+1}) - (i, \phi_k) - (i+1, \phi_{k-1}) & (\chi_i = -1),\\ (i-1, \phi_{k-1}) - (i, \phi_k) - (i+1, \phi_{k-1}) & (\chi_i = 0),\\ (i-1, \phi_{k+1}) - (i, \phi_k) - (i+1, \phi_{k+1}) & (\chi_i = 0).\end{cases}$$

Using this property, we can show the following discrete version of intermediate value theorem.

Lemma 22 *Let* $\theta_{j_0}, \theta_{j_1} \in \Theta$. *For any* $\phi \in \Phi_\gamma$ *between* θ_{j_0} *and* θ_{j_1} *there exists* k *between* j_0 *and* j_1 *such that* $\theta_k = \phi$.

Proof Without loss of generality, we can assume that $j_0 < j_1$ and $\theta_{j_0} < \theta_{j_1}$. If the assertion does not hold, there exists $\phi \in (\theta_{j_0}, \theta_{j_1}) \cap \Phi_\gamma$ and $k \in (j_0, j_1)$ such that

$\theta_k < \phi < \theta_{k+1}$ or $\theta_k > \phi > \theta_{k+1}$. This contradicts the property $(*)$. Thus we have the assertion. ∎

Moreover, we can show the convexity/concavity of the edge. Hereafter, we assume $j_0 < j_1$ and $\theta_j \in \Theta$ for all $j \in Q := \{j_0 + 1, \ldots, j_1 - 1\}$.

Lemma 23 *Let $\theta_{j_0} < \theta_{j_1}$ (resp. $\theta_{j_0} > \theta_{j_1}$). For any ϕ between θ_{j_0} and θ_{j_1}, there exists $k \in Q$ such that $\theta_k = \phi$ and $\chi_k = 1$ (resp. $\chi_k = -1$).*

Proof Let $\theta_{j_0} < \theta_{j_1}$. By Lemma 22, there exists $K := \{k_1 < k_2 < \cdots < k_m\} \subset Q$ such that $\theta_k = \phi$ for $k \in K$ and $\theta_j \neq \phi$ for $j \in Q \setminus K$. Suppose $\chi_k \leq 0$ holds for all $k \in K$. Then by Lemma 22, we have $\theta_j < \phi$ for $j = j_0, \ldots, k_1 - 1$, since $k_1 = \min K$. From $\chi_{k_1} \leq 0$ and $\theta_{k_1 - 1} < \phi$, we have $\theta_{k_1+1} < \phi$. By the definition of K, $\theta_j < \phi$ holds for $j = k_1 + 1, \ldots, k_2 - 1$, and from $\chi_{k_2} \leq 0$ and $\theta_{k_2 - 1} < \phi$, we have $\theta_{k_2+1} < \phi$. Applying repeatedly the above argument, we have $\theta_j \leq \phi$ for $j_0 \leq j \leq k_m$ and $\phi_{k_m+1} < \phi$. Again by the definition of K, $\theta_j < \phi$ holds for $j = k_m + 1, \ldots, j_1$, especially $\theta_{j_1} < \phi$. This contradicts to $\phi < \theta_{j_1}$. The other case $\theta_{j_0} > \theta_{j_1}$ can be proved by the same argument. ∎

Corollary 24 *Assume $\theta_{j_0} = \theta_{j_1}$ and $\chi_{j_0} = \chi_{j_1} = 1$ (resp. -1). Then, there exists at least one number $k \in Q$ such that $\theta_k = \theta_{j_0}$ and $\chi_k = -1$ (resp. 1).*

Proof Suppose $\chi_{j_0} = \chi_{j_1} = 1$. We have $\theta_{j_0+1} > \theta_{j_0} > \theta_{j_1-1}$. By applying Lemma 23 for the case $\theta_{j_0+1} > \theta_{j_1-1}$, there exists at least one $k \in \{j_0 + 2, \ldots, j_1 - 2\}$ such that $\theta_k = \theta_{j_0}$ and $\chi_k = -1$. The other case $\chi_{j_0} = \chi_{j_1} = -1$ can be proved by the same argument. ∎

Using this corollary, we can easily show the following proposition.

Lemma 25 *Assume $\theta_{j_0} = \theta_{j_1}$ and $\chi_{j_0} = \chi_{j_1} = 1$ (resp. -1). Suppose that $\chi_j \leq 0$ (resp. ≥ 0) for all $j \in Q' := \{j \in Q \mid \theta_j = \theta_{j_0}\}$. Then, there exists a unique $k \in Q'$ such that $\chi_k = -1$ (resp. 1) and $\chi_j = 0$ for all $j \in Q' \setminus \{k\}$.*

Proof We prove the case $\chi_{j_0} = \chi_{j_1} = 1$. (The other case can be proved in the same way.)

By Corollary 24, suppose that there exist $k_1, k_2 \in Q$ such that $k_1 < k_2$, $\theta_{k_1} = \theta_{k_2} = \theta_{j_0}$ and $\chi_{k_1} = \chi_{k_2} = -1$. Applying Corollary 24 for the case $j_0 = k_1$, $j_1 = k_2$, there exists k_3 such that $k_1 < k_3 < k_2$, $\theta_{k_3} = \theta_{j_0}$ and $\chi_{k_3} = 1$. This contradicts the assumption $\chi_k \leq 0$ for $k \in Q'$. Thus we have the assertion. ∎

The following two lemmas can be proved by a contradiction, respectively.

Lemma 26 *Assume $j_0 < j_1$, $\theta_{j_1} = \theta_{j_0} + 2\pi$ and $\chi_{j_0} = \chi_{j_1} = 1$ (resp. -1). Suppose that (i, θ_i)-subgraph $\bigcup_{i=j_0+1}^{j_1} [(i-1, \theta_{i-1}), (i, \theta_i)]$ is included in the rectangle $[j_0, j_1] \times [\theta_{j_0}, \theta_{j_1}]$. Then, either $\boldsymbol{n}_k \neq \boldsymbol{n}(\theta_{j_0}) = \boldsymbol{n}(\theta_{j_1})$ or $\chi_k = 0$ hold for $k = j_0 + 1, \ldots, j_1 - 1$.*

Lemma 27 *Assume* $j_0 < j_1$, $\theta_{j_0} - 2\pi = \theta_{j_1}$ *and* $\chi_{j_0} = \chi_{j_1} = 1$. *Suppose that* $\chi_j \leq 0$ *for all* $j \in \{j \in Q \mid \boldsymbol{n}(\theta_j) = \boldsymbol{n}(\theta_{j_0})\}$. *Then, there exist exactly two numbers* j_0', j_1' *such that* $j_0 < j_0' < j_1' < j_1$, $\chi_{j_0'} = \chi_{j_1'} = -1$, $\theta_{j_0'} = \theta_{j_0}$ *and* $\theta_{j_1'} = \theta_{j_1}$ *hold.*

Now we are ready to show Lemma 21.

Proof Let Γ be N-sided admissible polygon. Taking any $\boldsymbol{n} \in \mathcal{N}_\gamma$, we obtain $I^+(\boldsymbol{n}) = \{i_1 < \cdots < i_M\}$ and it can be divided such as $I^+(\boldsymbol{n}) = \bigoplus_{l=1}^{m} J_l$ (J_l takes consecutive numbers in $I^+(\boldsymbol{n})$). Here for $l = 1, 2, \ldots, m$ there exist $\psi_l \in \Phi_\gamma$ such that $|\psi_l - \psi_{l+1}| = 2\pi$ and $\theta_i = \psi_l$ holds for all $i \in J_l$. Since Γ is Jordan curve, $\theta_{i_1+N} = \theta_{i_1} + 2\pi$ holds. Then we have $\psi_m = \psi_1$ from $\chi_{i_1+N} = \chi_{i_1} = 1$. Let $Z_\pm := \#\{l \mid \psi_{l+1} = \psi_l \pm 2\pi\}$. Hence $Z_+ = Z_-$ holds. Obviously, $Z_+ + Z_- = m - 1$. Therefore, $Z_+ = Z_- = \dfrac{m-1}{2}$. By Lemma 25, for $l = 1, 2, \ldots, m$

$$\#\left(\{\min J_l, (\min J_l) + 1, \ldots, \max J_l\} \cap I^-(\boldsymbol{n})\right) = \#J_l - 1.$$

By Lemma 26, for any $l \in \{l \mid \psi_{l+1} = \psi_l + 2\pi\}$ we have

$$\#\left(\{(\max J_l) + 1, (\max J_l) + 2, \ldots, (\min J_{l+1}) - 1\} \cap I^-(\boldsymbol{n})\right) = 0,$$

and by Lemma 27, for any $l \in \{l \mid \psi_{l+1} = \psi_l - 2\pi\}$ we have

$$\#\left(\{(\max J_l) + 1, (\max J_l) + 2, \ldots, (\min J_{l+1}) - 1\} \cap I^-(\boldsymbol{n})\right) = 2.$$

Therefore we obtain

$$\#I^+(\boldsymbol{n}) - \#I^-(\boldsymbol{n}) = \sum_{l=1}^{m} \#J_l - \left(\sum_{l=1}^{m} (\#J_l - 1) + 0 \times Z_+ + 2 \times Z_-\right) = m - 2\frac{m-1}{2} = 1.$$

■

Appendix C: Non-intersection Lemma for Half Lines

Let Γ be an almost convex admissible polygon. We define the half straight line starting from each vertex of Γ such as

$$\widetilde{L}_j = \{\boldsymbol{x} \in \mathbb{R}^2;\ \boldsymbol{x} = \boldsymbol{p}_j + r\widetilde{\boldsymbol{n}}_j,\ r \geq 0\}, \quad j = 1, 2, \ldots, N,$$

where $\widetilde{\boldsymbol{n}}_j = \boldsymbol{n}_j + \boldsymbol{n}_{j-1}$.

Lemma 28 *Any lines* $\widetilde{L}_i$ *and* $\widetilde{L}_j$ *do not intersect for* $i \neq j$ *and* $i, j = 0, 1, 2, \ldots, N$ ($\widetilde{L}_0 = \widetilde{L}_N$, $\widetilde{L}_{N+1} = \widetilde{L}_1$).

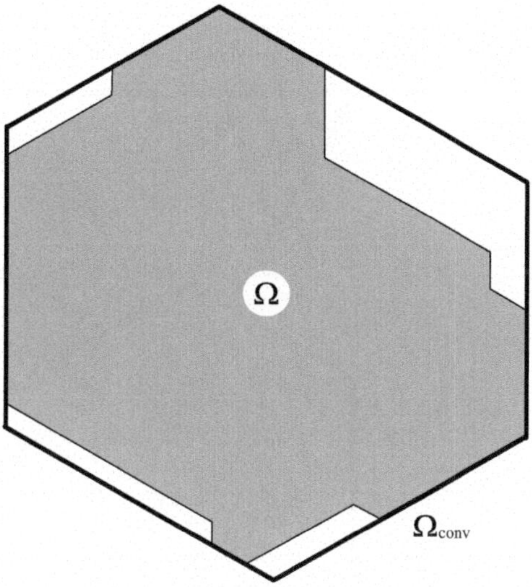

Fig. 7 An almost convex admissible polygon Ω and its convex admissible cover Ω_{conv}

Proof Let Ω be an almost convex admissible polygon. For each $\boldsymbol{\nu}_j \in \mathcal{V}_\gamma$, the maximum of the height function in the direction $\boldsymbol{\nu}_j$ can be determined such as

$$\widetilde{h}_j = \max_{\boldsymbol{n}_i \in \mathcal{N}} \{\boldsymbol{p}_i \cdot \boldsymbol{n}_i;\ \boldsymbol{n}_i = \boldsymbol{\nu}_j\}.$$

Then a convex admissible polygon is uniquely determined:

$$\Omega_{\text{conv}} = \bigcap_{j=1}^{\mathcal{V}_\gamma} \{\boldsymbol{x} \in \mathbb{R}^2;\ \boldsymbol{x} \cdot \boldsymbol{\nu}_j \le \widetilde{h}_j\},$$

which includes Ω, that is, $\Omega \subset \Omega_{\text{conv}}$. See Fig. 7. The k-th edge of Ω_{conv} includes the j_k-th edge of Ω, from which it follows that the k-th normal vector is $\boldsymbol{n}_{j_k} = \boldsymbol{\nu}_k$.

We define the half straight line such as

$$\widehat{L}_k = \{\boldsymbol{x} \in \mathbb{R}^2;\ \boldsymbol{x} = \widehat{\boldsymbol{p}}_{j_k} + r\boldsymbol{\nu}_k,\ r \ge 0\}, \quad k = 1, 2, \ldots, \mathcal{V}_\gamma,$$

where $\widehat{\boldsymbol{p}}_{j_k} = (\boldsymbol{p}_{j_k} + \boldsymbol{p}_{j_k+1})/2$. By convexity of Ω_{conv}, $\widehat{L}_k$'s never intersect each other and $\bigcup_{k=1}^{J} \widehat{L}_k$ divide $\mathbb{R}^2 \setminus \Omega$ into J regions, $R_1, R_2, \ldots, R_J$, where R_k (resp. R_J)is an outer region of Ω between $\widehat{L}_k$ and $\widehat{L}_{k+1}$ (resp. $\widehat{L}_J$ and $\widehat{L}_1$) (Fig. 8).

Consider the region R_1 and the lines $\widetilde{L}_i$ for $i \in Q := \{j_1 + 1, \ldots, j_2 - 1\}$.

By almost convexity of Ω, we have $\chi_i = 0$ for $i \in Q$ and thus $\boldsymbol{n}_i = \boldsymbol{n}_{j_1}$ for $i = j_1 + 2, j_1 + 4, \ldots$ and $\boldsymbol{n}_i = \boldsymbol{n}_{j_2}$ for $i = j_1 + 1, j_1 + 3, \ldots$. Therefore, $\widetilde{\boldsymbol{n}}_i = \boldsymbol{n}_{j_1} + \boldsymbol{n}_{j_2}$ for all $i \in Q$, from which it follow that all $\widetilde{L}_j$'s ($j \in Q$) never intersect each other.

Claim 29 *All* $\widetilde{L}_j$*'s (*$j \in Q$*) is in* R_1.

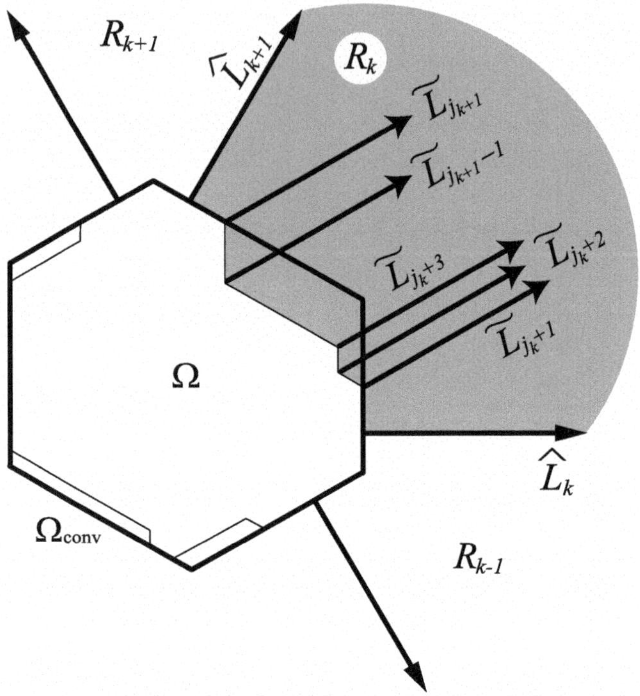

Fig. 8 Division of outer region of Ω

Proof We show that $\widetilde{L}_j$ does not intersect $\widehat{L}_1$ and $\widehat{L}_2$. Since all $\widetilde{L}_j$'s ($j \in Q$) are parallel each other, it is enough to show that $\widetilde{L}_{j_1+1}$ (resp. $\widetilde{L}_{j_2-1}$) does not intersect $\widehat{L}_1$ (resp. $\widehat{L}_2$).

Without loss of generality, we set $\widehat{\boldsymbol{p}}_{j_1} = \boldsymbol{0}$ and $\boldsymbol{n}_{j_1} = (1, 0)$. Then, $\widehat{L}_1 = \{(x, 0) | x > 0\}$, $\boldsymbol{p}_{j_1+1} = (0, d_{j_1}/2)$ and $\widetilde{\boldsymbol{n}}_{j_1+1} = (1 + \cos\theta_{j_1}, \sin\theta_{j_1})$. Thus, two straight lines which include $\widetilde{L}_{j_1+1}$ and $\widehat{L}_1$ respectively, intersect at $(-d_{j_1}(1 + \cos\theta_{j_1})/(2\sin\theta_{j_1}), 0) \notin \widehat{L}_1$, since $0 < \theta_{j_1} < \pi$ and $d_{j_1} > 0$. Thus, $\widetilde{L}_{j_1+1}$ never intersect $\widehat{L}_1$. We also show that $\widetilde{L}_{j_2-1}$ never intersect $\widehat{L}_2$ and thus we have the assertion. □

For other regions $R_2, \ldots, R_J$, we can show the same assertion. Hence, all $\widetilde{L}_j$'s never intersect each other. ■

Appendix D: Edge-Edge Contact

As we mentioned proof (2) of Proposition 6, all possible edge-edge contacts are classified into 36 cases. Figure 9 depicts all of it, where the upper edge is Γ_i and the lower edge is Γ_j, the shadow region indicates Ω, and on each edge, the arrow corresponds to tangent vector. In Fig. 9, cases of 18 marks of x are the case where $\boldsymbol{t}_i = \boldsymbol{t}_j$. This contradicts to the "first" contact of Γ_i and Γ_j at the time T, since under

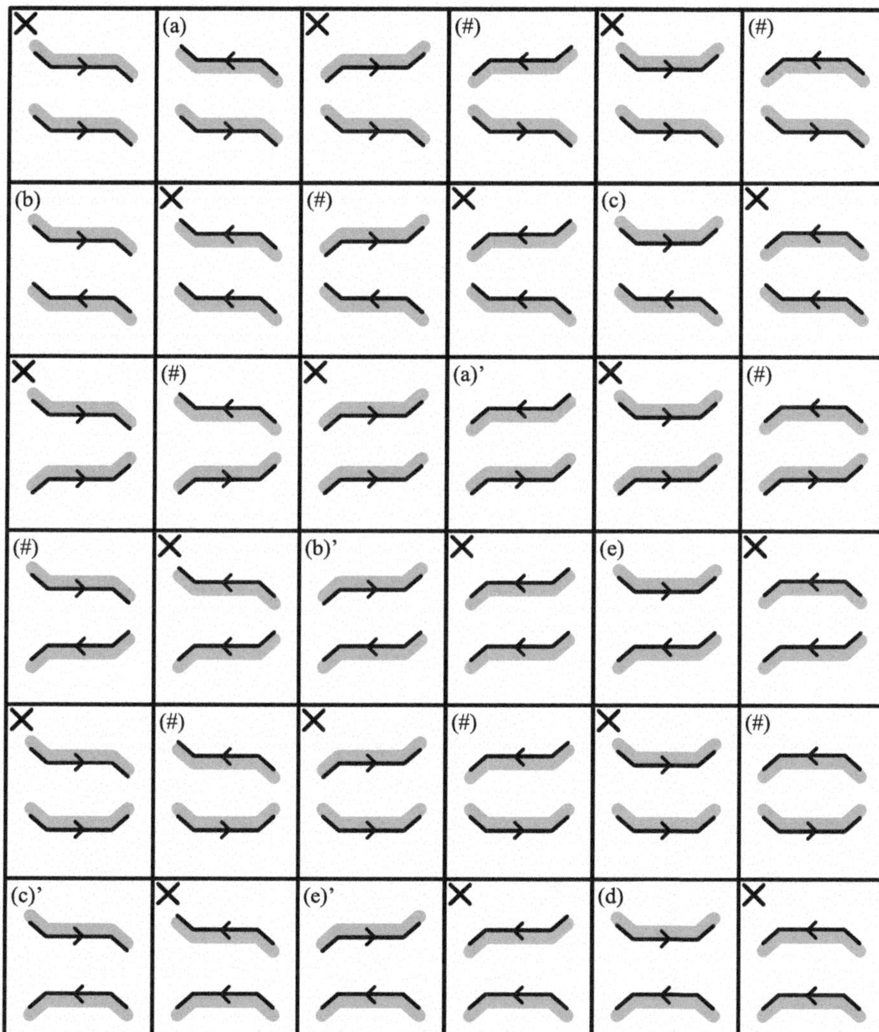

Fig. 9 Possible 36 cases of edge-edge contact

the edge-edge contact situation, one of connected curves between Γ_i and Γ_j should pass through the sandwiched region between Γ_i and Γ_j for making polygonal Jordan curve Γ. In the other 18 cases in Fig. 9, case of 9 marks of (#) also contradicts to the "first" contact, since the distance of two of four adjacent edges of Γ_i and Γ_j are smaller than the distance between Γ_i and Γ_j. In the rest 9 cases, (a)' and (b)' are reflection of (a) and (b) with respect to a vertical line, respectively, and (c)' and (e)' are 180 °C-rotation of (c) and (e), respectively. Moreover, (e) is regarded the same as (c). Therefore, the rest 4 cases (a), (b), (c) and (d) remain as essentially difference cases.

However, under the assumption (A0), (A1) and (A2), cases (b), (c) and (d) are impossible by the following reason. In the figures (b), (c) and (d), let us assume that the upper (resp. lower) edge is Γ_i (resp. Γ_j) and they are parallel to the x-axis. Then outward normal vectors on the edges are $\boldsymbol{n}_i = (0, -1)$ and $\boldsymbol{n}_j = -\boldsymbol{n}_i = (0, 1)$, respectively. In all cases, there exists at least one edge Γ_k satisfying $\chi_{i,k} = \chi_{k,j} = -1$ for connecting Γ_i and Γ_j by an admissible polygonal curve (If not, the polygon does not close), and the first entry of $\boldsymbol{n}_k$ is negative. Hence there are consecutive three edges Γ_{l-1}, Γ_l and Γ_{l+1} in between Γ_i and Γ_j, and $\chi_{l-1,l} = \chi_{l,l+1} = -1$ hold. This means $\chi_l = -1$ which contradicts to the assumption (A2).

Thus the only case (a) remains, which corresponds to the one in Fig. 4.

Appendix E: Vertex-Vertex Contact

As we mentioned proof (3) of Proposition 6, all possible edge-edge contacts are classified into 8 cases. Figure 10 depicts all of it, where the upper vertex is $\boldsymbol{p}_i$ and the lower vertex is $\boldsymbol{p}_j$, the shadow region indicates Ω, and on each edge, the arrow corresponds to tangent vector. In Fig. 10, cases of 4 marks of x contradict to the

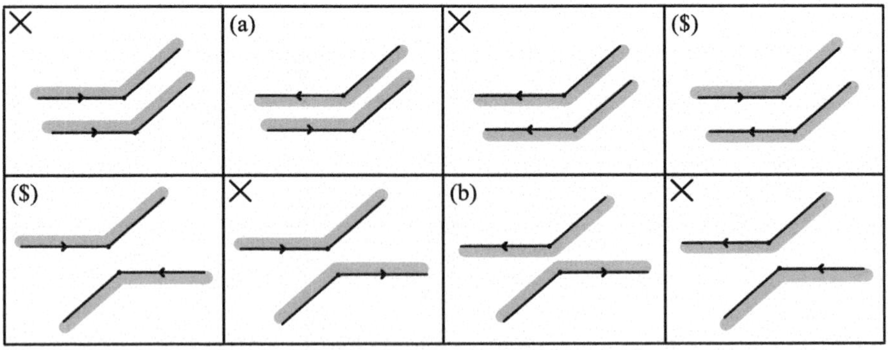

Fig. 10 Possible 8 cases of vertex-vertex contact

"first" contact of $\boldsymbol{p}_i$ and $\boldsymbol{p}_j$ at the time T, which is the same reason as in Sect. D. In the other 4 cases, case of 2 marks of (\$) is also a contradiction, which is the same reason as mentioned in Sect. D.

Thus the two cases (a) and (b) remain, which correspond to the ones in Fig. 5.

References

1. F. Almgren, J.E. Taylor, Flat flow is motion by crystalline curvature for curves with crystalline energies. J. Diff. Geom. **42**, 1–22 (1995)
2. S. Angenent, M.E. Gurtin, Multiphase thermomechanics with interfacial structure. 2. Evolution of an isothermal interface. Arch. Rational Mech. Anal. **108**, 323–391 (1989)
3. Y. Furukawa, S. Kohata, Temperature dependence of the growth form of negative crystal in an ice single crystal and evaporation kinetics for its surfaces. J. Crystal Growth **129**, 571–581 (1993)
4. M. Gage, *On an Area-Preserving Evolution Equation for Plane Curves*, Nonlinear Problems in Geometry, Contemp. Math., vol. 51, ed. by D.M. DeTurck (1986), pp. 51–62
5. Y. Giga, Anisotropic curvature effects in interface dynamics, Sūgaku 52 (2000) 113–127. English transl. Sugaku Expositions **16**, 135–152 (2003)
6. M.-H. Giga, Y. Giga, Crystalline and level set flow – convergence of a crystalline algorithm for a general anisotropic curvature flow in the plane, in *Free boundary problems: theory and applications, I (Chiba, 1999)*, GAKUTO Internat. Ser. Math. Sci. Appl., vol. 13 (Gakkōtosho, Tokyo, 2000), pp. 64–79
7. M.A. Grayson, The heat equation shrinks emmbedded plane curves to round points. J. Diff. Geom. **26**, 285–314 (1987)
8. M.E. Gurtin, *Thermomechanics of Evolving Phase Boundaries in the Plane* (Clarendon Press, Oxford, 1993)
9. K. Ishii, H.M. Soner, Regularity and convergence of crystalline motion. SIAM J. Math. Anal. **30**, 19–37 (1999)
10. T. Ishiwata, Motion of non-convex polygons by crystalline curvature and almost convexity phenomena. Jpn. J. Ind. Appl. Math. **25**, 233–253 (2008)
11. T. Ishiwata, S. Yazaki, Towards modelling the formation of negative ice crystals or vapor figures produced by freezing of internal melt figures. RIMS Kōkyūroku **1542**, 1–11 (2007)
12. T. Ishiwata, S. Yazaki, Interface motion of a negative crystal and its analysis. RIMS Kōkyūroku **1588**, 23–29 (2008)
13. T. Ishiwata, S. Yazaki, Structure-preserving numerical scheme for a generalized area-preserving crystalline curvature flow. Comput. Methods Mater. Sci. **17**, 122–135 (2017)
14. J.C. McConnel, The crystallization of lake ice. Nature **39**, 367 (1889)
15. U.F. Mayer, G. Simonett, Self-intersections for the surface diffusion and the volume-preserving mean curvature flow. Differ. Integral Equ. **13**, 1189–1199 (2000)
16. U. Nayaka, *Properties of Single Crystals of Ice, Revealed by Internal Melting*, SIPRE (Snow, Ice and Permafrost Research Establishment) Research Paper **13** (1956)
17. A. Stancu, Uniqueness of self-similar solutions for a crystalline flow. Indiana U. Math. J. **45**, 1157–1174 (1996)
18. J.E. Taylor,*Motion of Curves by Crystalline Curvature, Including Triple Junctions and Boundary Points, Difference Geometry of Partial Differential Equations on Manifolds (Los Angeles, CA, 1990)*, Proceedings of Symposia in Pure Mathematics, vol. 54, Part I (AMS, Providencd, 1993), pp. 417–438
19. J.E. Taylor, J. Cahn, C. Handwerker, Geometric models of crystal growth. Acta Metall. **40**, 1443–1474 (1992)
20. J. Tyndall, On some physical properties of ice. J. Proc. Roy. Soc. London, Ser. A **9** (1858) 76–80

21. S. Yazaki, On an area-preserving crystalline motion. Calc. Var. **14**, 85–105 (2002)
22. S. Yazaki, Asymptotic behavior of solutions to an area-preserving motion by crystalline curvature. Kybernetika **43**, 903–912 (2007)
23. S. Yazaki, Motion of nonadmissible convex polygons by crystalline curvature. Publ. Res. Inst. Math. Sci. **43**, 155–170 (2007)
24. S. Yazaki, *An Area-Preserving Crystalline Curvature Flow Equation*, Topics in Mathematical Modeling, Part **4**, Jindrich Nečas Center for Mathematical Modeling Lecture Notes, vol. 4, ed. by M. Beneš, E. Feireisl (Matfyzpress, 2008), pp. 169–215

Rate of Convergence for Approximate Solutions in Obstacle Problems for Nonlinear Operators

Shigeaki Koike and Takahiro Kosugi

Abstract The rate of convergence of approximate solutions via penalization and regularization for bilateral obstacle problems of parabolic partial differential equations is shown by the nonlinear adjoint method introduced by Evans [6]. In particular, quasilinear operators arising in the level set mean curvature flow and the p-Laplace operator for $p > 2$ with obstacles can be applied.

Keywords Rate of convergence · Obstacle problem · Quasilinear equation · Adjoint method · Viscosity solution

MSC 2020 35K55 · 35R35 · 35K86 · 35A08 · 35D40

1 Introduction

Obstacle problems arise not only in variational problems under certain restrictions, the so-called variational inequalities, but also in optimal stopping time problems in stochastic control theory. In order to show the existence of solutions, it is often useful to introduce approximate equations associated with the original problems. To this end, we need enough estimates on solutions of approximate partial differential equations (PDEs for short). For the details of this issue, we refer to [16] and references therein.

In [17], we studied the rate of convergence of penalized problems for the elliptic PDEs with bilateral obstacles $\Phi, \Psi : \Omega \to \mathbb{R}$ and a given $f : \Omega \to \mathbb{R}$,

$$\min \left\{ \max \left\{ \hat{L}u - f, u - \Phi \right\}, u - \Psi \right\} = 0 \quad \text{in } \Omega \tag{1.1}$$

S. Koike (✉)
Department of Applied Physics, Waseda University, Tokyo 169-8555, Japan
e-mail: skoike@waseda.jp

T. Kosugi
General Education Center, Tottori University of Environmental Studies, Tottori 680-1111, Japan

S. Machihara (ed.), *Mathematical Physics and Its Interactions*, Springer Proceedings in Mathematics & Statistics 451, https://doi.org/10.1007/978-981-97-0364-7_3

under the homogeneous Dirichlet condition, where

$$\hat{L}u := -\mathrm{tr}[A(x)D^2u] + \langle b(x), Du\rangle + c(x)u$$

with given $A : \Omega \to S^n$, $b : \Omega \to \mathbb{R}^n$, $c : \Omega \to \mathbb{R}$, and a bounded domain $\Omega \subset \mathbb{R}^n$. Here, S^n denotes the set of symmetric matrices, $\langle \cdot, \cdot \rangle$ is the standard inner product in $\mathbb{R}^n$.

In order to show the existence of solutions of (1.1), we often introduce approximate PDEs via penalization

$$\hat{L}u^\varepsilon + \beta\left(\frac{u^\varepsilon - \Phi}{\varepsilon}\right) - \beta\left(\frac{\Psi - u^\varepsilon}{\varepsilon}\right) = f \quad \text{in } \Omega, \tag{1.2}$$

where $\beta : \mathbb{R} \to \mathbb{R}$ in the penalty terms will be given below. When $\hat{L}$ is a uniformly elliptic operator, in [17], we showed the rate of convergence $\|u^\varepsilon - u\|_{L^\infty(\Omega)} \leq C\varepsilon$ for some $C > 0$. Here and later, $C > 0$ denotes various constants independent of parameters. We also apply the argument to weakly coupled systems with obstacles in [17].

In this paper, we will generalize the rate of convergence result in [17] to possibly degenerate nonlinear parabolic PDEs with obstacles.

On the other hand, for the Hamilton-Jacobi equation

$$\partial_t v + H(Dv) = 0 \quad \text{in } \mathbb{R}^n \times (0, T), \tag{1.3}$$

where $T > 0$ is fixed, and the associated approximate equation with a vanishing viscosity term

$$\partial_t v^\delta - \delta \Delta v^\delta + H(Dv^\delta) = 0 \quad \text{in } \mathbb{R}^n \times (0, T), \tag{1.4}$$

by letting v and v^δ, respectively, be solutions of (1.3) and (1.4), the rate of convergence $\|v^\delta - v\|_{L^\infty(\mathbb{R}^n \times (0,T))} \leq C\sqrt{\delta}$ was shown for some $C > 0$ in [6]. We refer to [4] for the same rate of convergence by using the viscosity solution theory under appropriate conditions.

Our aim here is to establish the convergence rate of approximate solutions via singular perturbation and penalization with different parameters to the original nonlinear parabolic PDEs with bilateral obstacles. Our result will indicate the best choice of two different parameters in penalization and singular perturbation.

Since the gradient estimate on approximate solutions on the boundary of domains cannot be obtained in general, in order to simplify and unify the argument, we consider the obstacle problems on the torus $\mathbb{T}^n := \mathbb{R}^n/\mathbb{Z}^n$ in the space variables.

We shall denote by ∂_t the partial derivative with respect to the time variable t, and by D the gradient in the space variable $x \in \mathbb{R}^n$. Given the obstacles $\Phi, \Psi : \mathbb{T}^n \times [0, T] \to \mathbb{R}$ and the initial datum $g : \mathbb{T}^n \to \mathbb{R}$, we study the rate of convergence for approximate solutions of bilateral obstacle problems of fully nonlinear parabolic PDEs,

$$\min\{\max\{\partial_t u + G(x, t, u, Du, D^2u),\ u - \Phi\},\ u - \Psi\} = 0 \quad \text{in } \mathbb{T}^n \times (0, T), \tag{1.5}$$

with an initial condition

$$u(x, 0) = g(x) \quad \text{for } x \in \mathbb{T}^n,$$

where $G : \mathbb{T}^n \times [0, T] \times \mathbb{R}^n \times S^n \to \mathbb{R}$ is continuous.

We shall explain our strategy when G depends only on $\mathbb{R}^n \times S^n$. A typical example for this case is as follows:

$$G(\xi, X) := -\operatorname{tr}\left[\left(I - \frac{\xi \otimes \xi}{1 + |\xi|^2}\right) X\right]$$

This corresponds to the mean curvature flow equation of graphs. For a surface Γ_t at time t, the velocity V of the evolving surface is given by

$$V = -\operatorname{div} \boldsymbol{n} \quad \text{on } \Gamma_t$$

with constraints

$$\Gamma_t \subset U_t^1 \cap U_t^2, \quad t \geq 0,$$

where $\boldsymbol{n}$ is the outward unit normal vector field of Γ_t, and

$$\Gamma_t = \{(x_1, \dots, x_n, x_{n+1}) \in \mathbb{R}^{n+1} | x_{n+1} = u(x, t),\ x = (x_1, \dots, x_n) \in \mathbb{T}^n\},$$

$$U_t^1 = \{(x_1, \dots, x_n, x_{n+1}) \in \mathbb{R}^{n+1} | x_{n+1} \geq \Psi(x, t),\ x = (x_1, \dots, x_n) \in \mathbb{T}^n\},$$

$$U_t^2 = \{(x_1, \dots, x_n, x_{n+1}) \in \mathbb{R}^{n+1} | x_{n+1} \leq \Phi(x, t),\ x = (x_1, \dots, x_n) \in \mathbb{T}^n\}.$$

Mercier and Novaga [24] obtained a regularity result when obstacles Ψ and Φ are independent of t. We refer to [13] for bilateral obstacle problems when G is associated with the level set mean curvature flow equation.

To treat obstacle problems, we consider penalized PDEs:

$$\begin{cases} \partial_t u^\varepsilon + G(Du^\varepsilon, D^2u^\varepsilon) + \beta\left(\dfrac{u^\varepsilon - \Phi}{\varepsilon}\right) - \beta\left(\dfrac{\Psi - u^\varepsilon}{\varepsilon}\right) = 0 \text{ in } Q_T, \\ u^\varepsilon(\cdot, 0) = g \text{ in } \mathbb{T}^n, \end{cases} \tag{1.6}$$

where $Q_T := \mathbb{T}^n \times (0, T)$ in what follows. We define $\beta \in C^2(\mathbb{R})$, which satisfies that $0 \leq \beta' \leq 1$, $\beta'' \geq 0$ in $\mathbb{R}$, and

$$\beta(r) = \begin{cases} 0 & \text{for } r \leq 0, \\ r - 1 & \text{for } r \geq 2. \end{cases}$$

For the penalty method and its usage, we refer to [13, 16–18] and references therein.

Expanding u^ε with respect to $\varepsilon > 0$ at $\varepsilon_0 > 0$ formally, we have

$$u^\varepsilon = u^{\varepsilon_0} + (\varepsilon - \varepsilon_0)u^{\varepsilon_0}_\varepsilon + o(|\varepsilon - \varepsilon_0|) \quad \text{for } \varepsilon_0, \varepsilon > 0,\ |\varepsilon - \varepsilon_0| < 1, \tag{1.7}$$

where $u^{\hat{\varepsilon}}_\varepsilon := \left.\dfrac{\partial u^\varepsilon}{\partial \varepsilon}\right|_{\varepsilon=\hat{\varepsilon}}$.

We shall utilize the summation convention for the repeated indices in the proceeding calculations. By expanding the penalized equation (1.6) with respect to $\varepsilon > 0$, it follows that

$$\begin{aligned}
0 &= \partial_t u^\varepsilon + G(Du^\varepsilon, D^2u^\varepsilon) + \beta((u^\varepsilon - \Phi)/\varepsilon) - \beta((\Psi - u^\varepsilon)/\varepsilon)\\
&= \partial_t u^{\varepsilon_0} + G(Du^{\varepsilon_0}, D^2u^{\varepsilon_0}) + \beta((u^{\varepsilon_0} - \Phi)/\varepsilon_0) - \beta((\Psi - u^{\varepsilon_0})/\varepsilon_0)\\
&\quad + (\varepsilon - \varepsilon_0)\Big\{\partial_t u^{\varepsilon_0}_\varepsilon + G_{X_{ij}} u^{\varepsilon_0}_{\varepsilon x_i x_j} + G_{\xi_i} u^{\varepsilon_0}_{\varepsilon x_i} + \varepsilon_0^{-1}\beta'((u^{\varepsilon_0} - \Phi)/\varepsilon_0)u^{\varepsilon_0}_\varepsilon\\
&\qquad + \varepsilon_0^{-1}\beta'((\Psi - u^{\varepsilon_0})/\varepsilon_0)u^{\varepsilon_0}_\varepsilon - \varepsilon_0^{-2}\beta'((u^{\varepsilon_0} - \Phi)/\varepsilon_0)(u^{\varepsilon_0} - \Phi)\\
&\qquad - \varepsilon_0^{-2}\beta'((\Psi - u^{\varepsilon_0})/\varepsilon_0)(u^{\varepsilon_0} - \Psi)\Big\} + o(|\varepsilon - \varepsilon_0|).
\end{aligned}$$

Thus, we may expect $w = u^{\varepsilon_0}_\varepsilon$ to solve the linear equation

$$\begin{aligned}
&\partial_t w + G_{X_{ij}} w_{x_i x_j} + G_{\xi_i} w_{x_i} + \varepsilon_0^{-1}\beta'((u^{\varepsilon_0} - \Phi)/\varepsilon_0)w + \varepsilon_0^{-1}\beta'((\Psi - u^{\varepsilon_0})/\varepsilon_0)w\\
&= \varepsilon_0^{-2}\beta'((u^{\varepsilon_0} - \Phi)/\varepsilon_0)(u^{\varepsilon_0} - \Phi) + \varepsilon_0^{-2}\beta'((\Psi - u^{\varepsilon_0})/\varepsilon_0)(u^{\varepsilon_0} - \Psi).
\end{aligned} \tag{1.8}$$

Once we obtain the estimate

$$\|u^{\varepsilon_0}_\varepsilon\|_{L^\infty(Q_T)} \le C, \quad \text{where } C \text{ is independent of } \varepsilon_0 \in (0, 1), \tag{1.9}$$

from (1.8), by plugging (1.9) in (1.7), there is $C > 0$ such that

$$\|u^\varepsilon - u^{\varepsilon_0}\|_{L^\infty(Q_T)} \le (\varepsilon - \varepsilon_0)\|u^{\varepsilon_0}_\varepsilon\|_{L^\infty(Q_T)} + o(|\varepsilon - \varepsilon_0|) \le C(\varepsilon - \varepsilon_0)$$

for $0 < \varepsilon_0 < \varepsilon < 1$. Hence, sending $\varepsilon_0 \to 0$ in the above, we obtain the rate of convergence is of order ε.

In [6, 17], we applied the nonlinear adjoint method to obtain the rate of convergence for elliptic obstacle problems (1.4) and (1.2), respectively. We refer to [2, 27] for the rate of convergence by the nonlinear adjoint method. We also refer to [25] for related topics.

Our plan is as follows: First, assuming that there exist classical solutions of approximate PDEs,

$$\partial_t u - \delta\Delta u + G(x, t, u, Du, D^2u) + \beta((u - \Phi)/\varepsilon) - \beta((\Psi - u)/\varepsilon) = f \quad \text{in } Q_T, \tag{1.10}$$

we obtain the rate of convergence in ε and δ. The reason why we add $-\delta\Delta$ term is that when we introduce the adjoint operator of the associated linearized problem, it seems difficult to obtain classical solutions of the adjoint problem because G is degenerate parabolic in general.

Secondly, we will propose modifications G^γ for G, where the corresponding PDEs admit classical solutions $u^{\varepsilon,\delta,\gamma}$ of

$$\partial_t u - \delta\Delta u + G^\gamma(x,t,u,Du,D^2u) + \beta((u-\Phi)/\varepsilon) - \beta((\Psi-u)/\varepsilon) = f \quad \text{in } Q_T \tag{1.11}$$

under suitable conditions on G^γ. In fact, if we do not suppose enough regularity on the coefficients in Example 2.9 for instance in Sect. 2.4, we need to regularize those to obtain $C^{3,2}$ solutions.

Letting $u^{\varepsilon,\delta,\gamma}$ be a solution of (1.11), we show that

$$\|u^{\varepsilon,\delta,\gamma} - u\|_{L^\infty(Q_T)} \le \|u^{\varepsilon,\delta,\gamma} - u^{0,0,\gamma}\|_{L^\infty(Q_T)} + \|u^{0,0,\gamma} - u\|_{L^\infty(Q_T)} \le C(\varepsilon + \sqrt{\delta}) + \omega(\gamma)$$

for some $\omega : [0,\infty) \to [0,\infty)$, where $u^{0,0,\gamma} := \lim_{\varepsilon,\delta\to 0} u^{\varepsilon,\delta,\gamma}$. We will propose suitable conditions on G^γ to enable us to derive $\lim_{\gamma\to 0}\omega(\gamma) = 0$.

In order to simplify the argument, we propose a sufficient condition for G, where the uniqueness of viscosity solutions of

$$\partial_t u + G(x,t,u,Du,D^2u) = 0 \quad \text{in } Q_T \tag{1.12}$$

under $u(\cdot,0) = g$ in $\mathbb{T}^n$ holds. Our sufficient condition consists of two properties:

$$\begin{cases} (i) \text{ There is } \nu \in \mathbb{R} \text{ such that } G(x,t,r,\xi,X) - G(x,t,s,\xi,X) \\ \quad \ge \nu(r-s) \text{ for } (x,t) \in Q_T,\ r \ge s,\ (\xi,X) \in \mathbb{R}^n \times S^n. \\ (ii) \text{ For any } R > 0, \text{ there is a function } \omega_R : [0,\infty) \to [0,\infty) \\ \quad \text{satisfying } \lim_{r\to 0+}\omega_R(r) = 0 \text{ such that if } X, Y \in S^n \text{ and} \\ \quad \alpha > 1 \text{ satisfy that } \begin{pmatrix} X & O \\ O & Y \end{pmatrix} \le 3\alpha \begin{pmatrix} I & -I \\ -I & I \end{pmatrix}, \\ \quad \text{then } G(y,s,r,\alpha(x-y),-Y) - G(x,t,r,\alpha(x-y),X) \\ \quad \le \omega_R(\alpha|x-y|^2 + |x-y| + |t-s|) \text{ for } (x,t),(y,s) \in Q_T \\ \quad \text{and } (r,X,Y) \in [-R,R] \times S^n \times S^n. \end{cases} \tag{1.13}$$

Notice that (1.13) (ii) is a modification of the standard "Structure Condition" in [3, 14] for degenerate elliptic PDEs.

Remark 1.1 When $\nu \ge 0$ in (1.13) (i), it is well-known to show that the uniqueness of viscosity solutions of (1.12) holds. We will give a proof for general $\nu \in \mathbb{R}$ in Appendix under an additional hypothesis (4.2).

This paper is organized as follows. In Sects. 2 and 3, we deal with degenerate and uniformly parabolic equations, respectively. Each section gives several examples of operators G and G^γ. In Appendix, we give proofs of some propositions for the reader's convenience.

2 Degenerate Parabolic PDE

In this section, we consider the case when G is degenerate parabolic. We first give our hypotheses of G, and then present examples.

2.1 Assumptions and Penalized Equation

Throughout this paper, we suppose that the uniqueness of viscosity solutions of (1.5) holds true to avoid choosing subsequences of approximate solutions.

$$\begin{cases} \text{If } \Phi, \Psi \in C(\overline{Q}_T), \text{ and } g \in C(\mathbb{T}^n) \text{ satisfy that } \Psi \le \Phi \text{ in } \overline{Q}_T \text{ and} \\ \Psi(x,0) \le g(x) \le \Phi(x,0) \text{ for } x \in \mathbb{T}^n, \text{ then there is at most one} \\ \text{viscosity solution } u \in C(\overline{Q}_T) \text{ of (1.5) such that } u = g \text{ in } \mathbb{T}^n \times \{0\}. \end{cases} \tag{2.1}$$

We will give a sufficient condition for (2.1), and recall the definition of viscosity solutions in Appendix for the reader's convenience.

Besides the standard function spaces, for $Q_T \subset U \subset \overline{Q}_T$ and $k \in \mathbb{N}$, we denote by $C^{k+1,k}(U)$ the set of functions $f : U \to \mathbb{R}$ such that

$$D^\alpha f, D^\beta \partial_t f, \partial_t^\ell f \in C(U) \quad \text{provided } |\alpha| \le k+1, |\beta| \le k, 0 \le \ell \le k.$$

For given data, we assume that Φ, Ψ and g satisfy

$$\begin{cases} \Psi, \Phi \in C^{2,1}(\overline{Q}_T), \\ g \in C^2(\mathbb{T}^n), \\ \Psi \le \Phi & \text{in } \overline{Q}_T, \\ \Psi(x,0) \le g(x) \le \Phi(x,0) & \text{for } x \in \mathbb{T}^n. \end{cases} \tag{2.2}$$

We shall use the following notations of derivatives with respect to each variable $x = (x_k)$, t, r, $\xi = (\xi_k)$, $X = (X_{ij})$:

$$G_{x_k} := \frac{\partial G}{\partial x_k}, \quad G_t := \frac{\partial G}{\partial t}, \quad G_r := \frac{\partial G}{\partial r}, \quad G_{\xi_k} := \frac{\partial G}{\partial \xi_k}, \quad G_{X_{ij}} := \frac{\partial G}{\partial X_{ij}}.$$

We shall also use the summation convention on the repeated indices.

We suppose the following properties for $G \in C(\overline{Q}_T \times \mathbb{R} \times \mathbb{R}^n \times S^n)$.

$$\begin{cases} (i) \quad G \in C^4(Q_T \times \mathbb{R} \times \mathbb{R}^n \times S^n), \\ (ii) \quad \text{there exists a constant } \nu \in \mathbb{R} \text{ such that } G_r(x,t,r,\xi,X) \geq \nu \\ \qquad \text{for } (x,t) \in Q_T \text{ and } (r,\xi,X) \in \mathbb{R} \times \mathbb{R}^n \times S^n, \\ (iii) \text{ for } R > 0, \text{ there is } M = M(R) > 0 \text{ such that} \\ \qquad G_z(x,t,u,\xi,X)^2 \leq M\left(-G_{X_{ij}}(x,t,u,\xi,X)X_{ik}X_{jk} + |\xi|^2 + R^2\right) \\ \qquad \text{for } (x,t) \in Q_T, (r,\xi,X) \in [-R,R] \times \mathbb{R}^n \times S^n, z = x_1, \dots, x_n, t, \\ (iv) \ (G_{X_{ij}}(x,t,u,\xi,X)) \leq O \text{ for } (x,t) \in Q_T, (r,\xi,X) \in \mathbb{R} \times \mathbb{R}^n \times S^n. \end{cases} \tag{2.3}$$

Remark 2.1 For simplicity, we suppose (i) in (2.3) to establish classical solutions of adjoint equations (2.16) below.

The property (iii) in the above is a fully nonlinear version of (2.27) in Example 2.9, which is stated in [26].

We introduce the penalized equation for (1.5) with a vanishing viscosity term.

$$\begin{cases} \partial_t u - \delta \Delta u + G(x,t,u,Du,D^2u) + \beta\left(\dfrac{u-\Phi}{\varepsilon}\right) - \beta\left(\dfrac{\Psi-u}{\varepsilon}\right) = 0 \text{ in } Q_T, \\ \qquad\qquad\qquad\qquad\qquad\qquad\qquad\qquad u(\cdot,0) = g \text{ in } \mathbb{T}^n. \end{cases} \tag{2.4}$$

To simplify the notations, we will write

$$\beta := \beta\left(\frac{u^\varepsilon - \Phi}{\varepsilon}\right), \ \beta' := \beta'\left(\frac{u^\varepsilon - \Phi}{\varepsilon}\right), \ \hat{\beta} := \beta\left(\frac{\Psi - u^\varepsilon}{\varepsilon}\right), \ \hat{\beta}' := \beta'\left(\frac{\Psi - u^\varepsilon}{\varepsilon}\right).$$

Throughout this paper, we will assume that the approximate equation (2.4) via regularization and penalization admits a smooth solution $u^{\varepsilon,\delta} \in C^{3,2}(Q_T) \cap C^{2,1}(\overline{Q}_T)$ for all $\varepsilon, \delta \in (0,1)$. We refer to [1, 5, 7, 23] for estimates on solutions of nonlinear equations. We also refer to [13] for the existence of classical solutions of (2.4) when G is the level set mean curvature flow.

2.2 Estimates

Under the above assumptions (2.1), (2.2), (2.3), assuming $u^{\varepsilon,\delta} \in C^{3,2}(Q_T) \cap C^{2,1}(\overline{Q}_T)$, we first show that β, $\hat{\beta}$, $u^{\varepsilon,\delta}$, $Du^{\varepsilon,\delta}$ and $\partial_t u^{\varepsilon,\delta}$ are uniformly bounded in $\overline{Q}_T$. Because of (2.1), this implies that there is a unique viscosity solution $u \in C(\overline{Q}_T)$ of (1.5) such that $u = g$ in $\mathbb{T}^n \times \{0\}$, and

$$u^{\varepsilon,\delta} \longrightarrow u \quad \text{uniformly in } \overline{Q}_T, \ \text{as } \varepsilon, \delta \to 0.$$

Proposition 2.2 *Assume* (2.1), (2.2) *and* (2.3). *Let* $u^{\varepsilon,\delta} \in C^{3,2}(Q_T) \cap C^{2,1}(\overline{Q}_T)$ *be solutions of* (2.4). *Then, there exists a constant* $C_k > 0$ $(k = 1, 2, 3)$ *such that*

$$\sup_{\varepsilon,\delta\in(0,1)}\sup_{Q_T}\frac{u^{\varepsilon,\delta}-\Phi}{\varepsilon}\le C_1,\qquad \sup_{\varepsilon,\delta\in(0,1)}\sup_{Q_T}\frac{\Psi-u^{\varepsilon,\delta}}{\varepsilon}\le C_1, \tag{2.5}$$

$$\sup_{\varepsilon,\delta\in(0,1)}\sup_{Q_T}\beta((u^{\varepsilon,\delta}-\Phi)/\varepsilon)\le C_2,\qquad \sup_{\varepsilon,\delta\in(0,1)}\sup_{Q_T}\beta((\Psi-u^{\varepsilon,\delta})/\varepsilon)\le C_2, \tag{2.6}$$

and

$$\|u^{\varepsilon,\delta}\|_{L^\infty(Q_T)},\ \|Du^{\varepsilon,\delta}\|_{L^\infty(Q_T)},\ \|\partial_t u^{\varepsilon,\delta}\|_{L^\infty(Q_T)}\le C_3. \tag{2.7}$$

Remark 2.3 For the existence of classical solutions $u^{\varepsilon,\delta}\in C^{2,1}(Q_T)\cap C(\overline{Q}_T)$ of (2.4) to G in the examples of Sect. 2.4, we refer to [11, 19, 20]. We will discuss further regularity for those in Appendix.

Remark 2.4 Although (2.5) directly shows $\|u^{\varepsilon,\delta}\|_{L^\infty(Q_T)}$ estimate in (2.7), we will give a different proof because our argument for L^∞ estimate below can be applied to unilateral obstacle problems:

$$\max\{\partial_t u+G(x,t,u,Du.D^2u),u-\Phi\}=0\quad\text{in } Q_T,$$

$$\min\{\partial_t u+G(x,t,u,Du.D^2u),u-\Psi\}=0\quad\text{in } Q_T.$$

Proof of Proposition 2.2 We shall drop ε and δ of $u^{\varepsilon,\delta}$. We first prove (2.5). For $\mu>0$, set

$$\theta_\mu:=\sup_{(x,t)\in Q_T}\left(\frac{e^{-\nu^-t}(u(x,t)-\Phi(x,t))}{\varepsilon}-\frac{\mu}{T-t}\right),$$

where $\nu^-:=\max\{-\nu,0\}$, and $\nu\in\mathbb{R}$ is from (2.3) (ii). Let us show that there exists a constant $C>0$, independent of $\mu>0$ and $\varepsilon,\delta\in(0,1)$, such that

$$\theta_\mu\le C. \tag{2.8}$$

Assume (2.8) holds. Then, by definition of θ_μ, it follows that

$$\frac{u(x,t)-\Phi(x,t)}{\varepsilon}\le e^{\nu^-t}\theta_\mu+\frac{e^{\nu^-t}\mu}{T-t}\quad\text{for } (x,t)\in\mathbb{T}^n\times(0,T),$$

which implies the first inequality in (2.5) by sending $\mu\to 0$.

We may suppose that $\theta_\mu\ge 2$, and the supremum is attained at $(x_\mu,t_\mu)\in\mathbb{T}^n\times[0,T)$. By (2.2), we note that $t_\mu>0$,

$$u(x_\mu,t_\mu)>\Phi(x_\mu,t_\mu),\quad\text{and}\quad \hat{\beta}=\beta((\Psi(x_\mu,t_\mu)-u(x_\mu,t_\mu))/\varepsilon)=0.$$

Hence, we have

$$\partial_t u(x_\mu, t_\mu) - \partial_t \Phi(x_\mu, t_\mu) - \nu^-(u(x_\mu, t_\mu) - \Phi(x_\mu, t_\mu)) - \frac{\mu\varepsilon e^{\nu^- t_\mu}}{(T - t_\mu)^2} = 0,$$

$$Du(x_\mu, t_\mu) = D\Phi(x_\mu, t_\mu), \quad \text{and} \quad D^2u(x_\mu, t_\mu) \le D^2\Phi(x_\mu, t_\mu).$$

From (2.3) (ii), (iv) and the mean value theorem, it follows that at (x_μ, t_μ),

$$\begin{aligned}
0 &= \partial_t u - \delta\Delta u + G(x, t, u, Du, D^2u) + \beta \\
&\ge \partial_t \Phi + \nu^-(u - \Phi) + \frac{\mu\varepsilon e^{\nu^- t}}{(T - t_\mu)^2} - \delta\Delta\Phi + G(x, t, u, D\Phi, D^2\Phi) + \beta \\
&\ge \partial_t \Phi + \frac{\mu\varepsilon e^{\nu^- t}}{(T - t_\mu)^2} - \delta\Delta\Phi + G(x, t, \Phi, D\Phi, D^2\Phi) + \beta.
\end{aligned}$$

Thus, by recalling $\theta_\mu \ge 2$, it follows that at (x_μ, t_μ),

$$\begin{aligned}
\theta_\mu - 1 + \frac{\mu\varepsilon e^{\nu^- t}}{(T - t_\mu)^2} &= \beta(\theta_\mu) + \frac{\mu\varepsilon e^{\nu^- t}}{(T - t_\mu)^2} \\
&\le \beta + \frac{\mu\varepsilon e^{\nu^- t}}{(T - t_\mu)^2} \\
&\le \|\partial_t \Phi\|_{L^\infty(Q_T)} + \|\Delta\Phi\|_{L^\infty(Q_T)} + \|G(\cdot, \cdot, \Phi, D\Phi, D^2\Phi)\|_{L^\infty(Q_T)}.
\end{aligned}$$

Therefore, we obtain (2.8).

We analogously obtain the other inequality in (2.5). Moreover, from the definition of β, (2.5) yields (2.6).

We next show the bound of u from above in (2.7). For $\mu > 0$, assume that the mapping

$$(x, t) \in \mathbb{T}^n \times [0, T) \longmapsto w(x, t) := e^{(\nu-1)t}u(x, t) - \frac{\mu}{T - t}$$

attains its positive maximum at a point $(x_\mu, t_\mu) \in \mathbb{T}^n \times [0, T)$. If there exists a subsequence $\mu_k \to 0$ such that $t_{\mu_k} = 0$, then it follows that for any $(x, t) \in Q_T$,

$$\begin{aligned}
u(x, t) &= e^{-(\nu-1)t}\left(e^{(\nu-1)t}u(x, t) - \frac{\mu_k}{T - t}\right) + \frac{e^{-(\nu-1)t}\mu_k}{T - t} \\
&\le e^{-(\nu-1)t}\|g^+\|_{L^\infty(\mathbb{T}^n)} + \frac{e^{-(\nu-1)t}\mu_k}{T - t}.
\end{aligned}$$

Hence, sending $\mu_k \to 0$, we have $u \le e^{(\nu^-+1)T}\|g^+\|_{L^\infty(\mathbb{T}^n)}$ in Q_T.

Thus, we may suppose $t_\mu \in (0, T)$. Following the previous argument in the above, we observe that at (x_μ, t_μ),

$$\begin{aligned} 0 &= \partial_t u - \delta \Delta u + G(x, t, u, Du, D^2u) + \beta - \hat{\beta} \\ &\geq -(\nu - 1)u + \frac{\nu e^{-(\nu-1)t}}{(T-t)^2} + G(x, t, u, 0, O) - \hat{\beta} \\ &\geq u + G(x, t, 0, 0, O) - \hat{\beta}. \end{aligned}$$

Since $u(x_\mu, t_\mu) > 0$, we have

$$e^{(\nu-1)t_\mu} u(x_\mu, t_\mu) - \frac{\mu}{T - t_\mu} \leq e^{\nu^- T} u(x_\mu, t_\mu) \leq e^{\nu^- T}\{-G(x_\mu, t_\mu, 0, 0, O) + \hat{\beta}\} \leq C_0$$

for some $C_0 > 0$ by (2.6). Hence, it follows that for any $(x, t) \in Q_T$,

$$u(x,t) = e^{-(\nu-1)t} w(x,t) + \frac{e^{-(\nu-1)t}\mu}{T-t} \leq e^{\nu^- T} C_0 + \frac{e^{-(\nu-1)t}\mu}{T-t},$$

which implies $u \leq e^{\nu^- T} C_0$ in Q_T by sending $\mu \to 0$.

We obtain a uniform bound from below in $\mathbb{T}^n \times [0, T)$ analogously. Therefore, we have the first estimate in (2.7). We fix $R > 0$ satisfying that

$$\sup_{\varepsilon, \delta \in (0,1)} \|u^{\varepsilon,\delta}\|_{L^\infty(Q_T)} \leq R. \tag{2.9}$$

We will use $M = M(R)$ in the hypothesis (2.3) (*iii*) with this $R > 0$.

We next show the estimates on the first derivatives of u. Define the linear operator

$$L_0 v := -\delta \Delta v + G_{X_{ij}}(x, t, u^{\varepsilon,\delta}, Du^{\varepsilon,\delta}, D^2u^{\varepsilon,\delta}) v_{ij} + G_{\xi_i}(x, t, u^{\varepsilon,\delta}, Du^{\varepsilon,\delta}, D^2u^{\varepsilon,\delta}) v_i,$$

where $v_i := \dfrac{\partial v}{\partial x_i}$, $v_{ij} := \dfrac{\partial^2 v}{\partial x_i \partial x_j}$. We shall drop ε and δ of $u^{\varepsilon,\delta}$ again.

We set

$$W := e^{-\tau t}\left(|Du|^2 + (\partial_t u)^2\right) - \frac{\mu}{T-t}$$

for small $\mu > 0$ and $\tau > 0$, which will be chosen later. Let W attain the maximum at $(x_\mu, t_\mu) \in \mathbb{T}^n \times [0, T)$.

We first suppose that $t_\mu \in (0, T)$. Then, it follows that at (x_μ, t_μ),

$$\begin{aligned} 0 &\leq \partial_t W + L_0 W + \frac{\mu}{(T-t)^2} \\ &= e^{-\tau t}\Big\{ -\tau |Du|^2 + 2u_k \partial_t u_k + 2(-\delta\delta^{ij} + G_{X_{ij}})(u_k u_{kij} + u_{ki}u_{kj}) \\ &\quad +2G_{\xi_i} u_k u_{ki} - \tau(\partial_t u)^2 + 2\partial_t u \partial_t^2 u \\ &\quad +2(-\delta\delta^{ij} + G^\delta_{X_{ij}})(\partial_t u \partial_t u_{ij} + \partial_t u_i \partial_t u_j) + 2G_{\xi_i}\partial_t u \partial_t u_i \Big\}, \end{aligned} \tag{2.10}$$

where δ^{ij} denotes the Kronecker delta. Differentiating (2.4) with respect to x_k and t, respectively, we have

$$\partial_t u_k - \delta\Delta u_k + G_{x_k} + G_r u_k + G_{\xi_i} u_{ik} + G_{X_{ij}} u_{ijk} + \frac{\beta'}{\varepsilon}(u_k - \Phi_k) - \frac{\hat{\beta}'}{\varepsilon}(\Psi_k - u_k) = 0 \tag{2.11}$$

and

$$\partial_t^2 u - \delta\Delta\partial_t u + G_t + G_r\partial_t u + G_{\xi_i}\partial_t u_i + G_{X_{ij}}\partial_t u_{ij} + \frac{\beta'}{\varepsilon}(\partial_t u - \partial_t\Phi) - \frac{\hat{\beta}'}{\varepsilon}(\partial_t\Psi - \partial_t u) = 0. \tag{2.12}$$

Using these equalities in (2.10), we have

$$\begin{aligned}
&\tau(|Du|^2 + (\partial_t u)^2) + 2\delta(|D^2u|^2 + |D\partial_t u|^2) \\
&\le 2G_{X_{ij}} u_{ki} u_{kj} - 2G_r|Du|^2 - 2G_{x_k} u_k + \frac{2\beta'}{\varepsilon}(\langle Du, D\Phi\rangle - |Du|^2) \\
&\quad + \frac{2\hat{\beta}'}{\varepsilon}(\langle Du, D\Psi\rangle - |D\Psi|^2) + 2G_{X_{ij}}\partial_t u_i \partial_t u_j - 2G_r(\partial_t u)^2 - 2G_t\partial_t u \\
&\quad + \frac{2\beta'}{\varepsilon}(\partial_t u \partial_t\Phi - (\partial_t u)^2) + \frac{2\hat{\beta}'}{\varepsilon}(\partial_t u\partial_t\Psi - (\partial_t u)^2).
\end{aligned} \tag{2.13}$$

Since we may suppose

$$|Du|^2 + (\partial_t u)^2 > \hat{R} \quad \text{at } (x_\mu, t_\mu),$$

where $\hat{R} := \max\left\{\|D\Phi\|^2_{L^\infty(Q_T)} + \|\partial_t\Phi\|^2_{L^\infty(Q_T)}, \|D\Psi\|^2_{L^\infty(Q_T)} + \|\partial_t\Psi\|^2_{L^\infty(Q_T)}\right\}$, we have

$$\begin{cases} \dfrac{2\beta'}{\varepsilon}(\langle Du, D\Phi\rangle - |Du|^2 + \partial_t u\partial_t\Phi - (\partial_t u)^2) \le 0, \\ \dfrac{2\hat{\beta}'}{\varepsilon}(\langle Du, D\Psi\rangle - |Du|^2 + \partial_t u\partial_t\Psi - (\partial_t u)^2) \le 0. \end{cases} \tag{2.14}$$

Hence, by using (2.3) (ii), and (2.14) in (2.13), the left hand side of (2.13) is estimated from above by

$$2G_{X_{ij}}(u_{ki}u_{kj} + \partial_t u_i\partial_t u_j) - 2\nu(|Du|^2 + (\partial_t u)^2) - 2G_{x_k}u_k - 2G_t\partial_t u.$$

On the other hand, in view of (2.3) (iii), for any $\rho > 0$, Young's inequality yields

$$\begin{aligned}
-G_{x_k}u_k - G_t\partial_t u &\le \frac{\rho}{2}\sum_{z=x_1,\dots,x_n,t}(G_z)^2 + \frac{1}{2\rho}(|Du|^2 + (\partial_t u)^2) \\
&\le \frac{\rho M}{2}\left(-G_{X_{ij}}u_{ik}u_{jk} + |Du|^2 + R^2\right) + \frac{1}{2\rho}(|Du|^2 + (\partial_t u)^2).
\end{aligned}$$

By combining these inequalities with (2.3) (iv) and by setting $\rho = M^{-1}$, it follows that at (x_μ, t_μ),

$$(\tau + 2\nu - M - 1)\,|Du|^2 + (\tau + 2\nu - M)\,(\partial_t u)^2 \le G_{X_{ij}}(u_{ki}u_{kj} + 2\partial_t u_i \partial_t u_j) + R^2.$$

Now, we choose a largely fixed $\tau > 0$ such that

$$\tau + 2\nu - M - 1 \ge 1.$$

Hence, in view of (2.3) (iv), we obtain

$$|Du|^2 + (\partial_t u)^2 \le R^2 \quad \text{at } (x_\mu, t_\mu),$$

which implies that at any $(x, t) \in Q_T$,

$$|Du(x,t)|^2 + (\partial_t u(x,t))^2 = e^{\tau t} W(x,t) + \frac{\mu e^{\tau t}}{T-t} \le e^{\tau T} \max\{R^2, \hat{R}\} + \frac{\mu e^{\tau T}}{T-t}. \tag{2.15}$$

Now, we have fixed $\tau > 0$ as above. When $t_\mu = 0$, by (2.2) and (2.4), it follows that at $(x_\mu, 0)$,

$$\begin{aligned} W &= |Dg|^2 + (\partial_t u)^2 - \frac{\mu}{T} \\ &\le |Dg|^2 + \{-\delta \Delta g + G(x_\mu, 0, g, Dg, D^2 g)\}^2 \le \hat{C}_0, \end{aligned}$$

where $\hat{C}_0 = \hat{C}_0(\|g\|_{W^{2,\infty}(\mathbb{T}^n)})$. Hence, it follows that for any $(x, t) \in Q_T$,

$$|Du(x,t)|^2 + (\partial_t u(x,t))^2 \le e^{\tau T} W(x_\mu, 0) + \frac{e^{\tau T}\mu}{T-t}.$$

Thus, this inequality together with (2.15) implies that

$$|Du|^2 + (\partial_t u)^2 \le e^{\tau t} W(x_\mu, t_\mu) + \frac{e^{\tau t}\mu}{T-t} \le e^{\tau T} \max\{R^2, \hat{R}, \hat{C}_0\} + \frac{e^{\tau T}\mu}{T-t} \quad \text{in } Q_T,$$

which gives $|Du|^2 + (\partial_t u)^2 \le e^{\tau T} \max\{R^2, \hat{R}, \hat{C}_0\}$ in Q_T by sending $\mu \to 0$. □

2.3 Adjoint Operator and Rate of Convergence

For the penalized equation (2.4), we introduce the linearized operator involving $u^{\varepsilon,\delta}$ in the coefficients:

$$\begin{aligned} Lv := -\delta \Delta v + G_{X_{ij}}(x, t, u^{\varepsilon,\delta}, Du^{\varepsilon,\delta}, D^2 u^{\varepsilon,\delta}) v_{ij} + G_{\xi_i}(x, t, u^{\varepsilon,\delta}, Du^{\varepsilon,\delta}, D^2 u^{\varepsilon,\delta}) v_i \\ + \left(G_r(x, t, u^{\varepsilon,\delta}, Du^{\varepsilon,\delta}, D^2 u^{\varepsilon,\delta}) + \frac{\beta'}{\varepsilon} + \frac{\hat{\beta}'}{\varepsilon} \right) v, \end{aligned}$$

and its adjoint operator

$$
\begin{aligned}
L^*\sigma := & -\delta\Delta\sigma + \frac{\partial^2}{\partial x_i \partial x_j}\left(G_{X_{ij}}(x,t,u^{\varepsilon,\delta},Du^{\varepsilon,\delta},D^2u^{\varepsilon,\delta})\sigma\right) \\
& -\frac{\partial}{\partial x_i}\left(G_{\xi_i}(x,t,u^{\varepsilon,\delta},Du^{\varepsilon,\delta},D^2u^{\varepsilon,\delta})\sigma\right) \\
& +\left(G_r(x,t,u^{\varepsilon,\delta},Du^{\varepsilon,\delta},D^2u^{\varepsilon,\delta}) + \frac{\beta'}{\varepsilon} + \frac{\hat{\beta}'}{\varepsilon}\right)\sigma.
\end{aligned}
$$

For $\hat{x} \in \mathbb{T}^n$ and $0 < \hat{t} \le T$, by the standard theory of parabolic equations (e.g. in [11, 19, 20]), there exists a unique solution $\sigma = \sigma^{\varepsilon,\delta} \in C^{3,2}(Q_{\hat{t}}) \cap C(\overline{Q}_{\hat{t}})$ of

$$
\begin{cases} -\partial_t\sigma + L^*\sigma = 0 & \text{in } Q_{\hat{t}}, \\ \sigma(x,\hat{t}) = \rho_{\hat{x}}^N(x) & \text{for } x \in \mathbb{T}^n. \end{cases} \tag{2.16}
$$

Here $\rho_{\hat{x}}^N$ is a nonnegative smooth function, which approximates the Dirac measure at $\hat{x}$; $\rho_{\hat{x}}^N \in C^\infty(\mathbb{T}^n)$ satisfies

$$
\rho_{\hat{x}}^N \ge 0 \quad \text{in } \mathbb{T}^n, \quad \int_{\mathbb{T}^n} \rho_{\hat{x}}^N(x)dx = 1, \quad \lim_{N\to\infty}\int_{\mathbb{T}^n} f(x)\rho_{\hat{x}}^N(x)dx = f(\hat{x}) \text{ for any } f \in C(\mathbb{T}^n).
$$

We show some properties of solutions of the adjoint equation (2.16).

Lemma 2.5 *Assume* (2.1), (2.2), (2.3). *Let* $u^{\varepsilon,\delta} \in C^{3,2}(Q_T) \cap C^{2,1}(\overline{Q}_T)$ *and* $\sigma = \sigma^{\varepsilon,\delta} \in C(\overline{Q}_{\hat{t}}) \cap C^{3,2}(Q_{\hat{t}})$ *be solutions of* (2.4) *and* (2.16), *respectively. Then,* $\sigma \ge 0$ *in* $\overline{Q}_{\hat{t}}$. *Furthermore, there exists a constant* $C_4 > 0$, *independent of* ε, δ *and* $\hat{t}$, *such that*

$$
\sup_{0\le s<\hat{t},\varepsilon,\delta\in(0,1)} \left\{ \int_s^{\hat{t}}\int_{\mathbb{T}^n}\left(\frac{\beta'}{\varepsilon}+\frac{\hat{\beta}'}{\varepsilon}+1\right)\sigma(x,t)dxdt + \int_{\mathbb{T}^n}\sigma(x,s)dx \right\} \le C_4. \tag{2.17}
$$

Proof For any $\eta \in C_0^\infty(Q_T)$ with $\eta \ge 0$, there exists a unique classical solution w of

$$
\begin{cases} \partial_t w + Lw = \eta & \text{in } Q_T, \\ w(\cdot,0) = 0 & \text{in } \mathbb{T}^n. \end{cases}
$$

The maximum principle implies $w \ge 0$ in Q_T. It follows that

$$\int_0^{\hat t}\int_{\mathbb{T}^n}\sigma(x,t)\eta(x,t)dxdt=\int_0^{\hat t}\int_{\mathbb{T}^n}(\partial_t w+Lw)\sigma(x,t)dxdt$$
$$=\int_{\mathbb{T}^n}\big(w(x,\hat t)\sigma(x,\hat t)-w(x,0)\sigma(x,0)\big)\,dx$$
$$+\int_0^{\hat t}\int_{\mathbb{T}^n}(-\partial_t\sigma(x,t)+L^*\sigma)w(x,t)dxdt$$
$$=\int_{\mathbb{T}^n}w(x,\hat t)\rho^N_{\hat x}(x)dx\ge 0.$$

Since η is arbitrary, we verify that $\sigma\ge 0$ in $Q_{\hat t}$.

Next, we shall show (2.17). The adjoint equation (2.16) combined with (2.3) (*ii*) yields

$$0=\int_s^{\hat t}\int_{\mathbb{T}^n}e^{(\nu^-+1)t}(-\partial_t\sigma(x,t)+L^*\sigma(x,t))dxdt$$
$$=\int_{\mathbb{T}^n}(-e^{(\nu^-+1)\hat t}\sigma(x,\hat t)+e^{(\nu^-+1)s}\sigma(x,s))dx$$
$$+\int_s^{\hat t}\int_{\mathbb{T}^n}\sigma(x,t)\left(\frac{\beta'}{\varepsilon}+\frac{\hat\beta'}{\varepsilon}+G_r(x,u^{\varepsilon,\delta},Du^{\varepsilon,\delta},D^2u^{\varepsilon,\delta})+\nu^-+1\right)e^{(\nu^-+1)t}dxdt$$
$$\ge -e^{(\nu^-+1)\hat t}\int_{\mathbb{T}^n}\rho^N_{\hat x}dx+\int_{\mathbb{T}^n}e^{(\nu^-+1)s}\sigma(x,s)dx$$
$$+\int_s^{\hat t}\int_{\mathbb{T}^n}\sigma(x,t)\left(\frac{\beta'}{\varepsilon}+\frac{\hat\beta'}{\varepsilon}+1\right)e^{(\nu^-+1)t}dxdt.$$

Since $\sigma,\beta',\hat\beta'\ge 0$ and $\int_{\mathbb{T}^n}\rho^N_{\hat x}dx=1$, we obtain (2.17). □

The next lemma gives an estimate on the second derivative of $u^{\varepsilon,\delta}$. To this end, we will use the following norm for $f:Q_T\to\mathbb{R}$:

$$\|f\|_{W^{2,1}_\infty(Q_T)}:=\sum_{|\alpha|\le 2}\|D^\alpha f\|_{L^\infty(Q_T)}+\|\partial_t f\|_{L^\infty(Q_T)}.$$

Lemma 2.6 *Under the hypotheses of Lemma 2.5, there exists $C_5>0$ such that*

$$\sup_{\varepsilon\in(0,1)}\int_0^{\hat t}\int_{\mathbb{T}^n}\left(|D^2u^{\varepsilon,\delta}(x,t)|^2+|D\partial_t u^{\varepsilon,\delta}(x,t)|^2\right)\sigma(x,t)\,dxdt\le\frac{C_5}{\delta}.$$

Proof We set $V:=\dfrac{|Du^{\varepsilon,\delta}|^2+(\partial_t u^{\varepsilon,\delta})^2}{2}$. We shall drop ε and δ of $u^{\varepsilon,\delta}$ as before. Using (2.11) and (2.12), we have

$$\begin{aligned}\partial_t V + LV = & u_k\partial_t u_k + \partial_t u\partial_t^2 u - \delta(u_k\Delta u_k + |D^2u|^2 + \partial_t u\Delta\partial_t u + |D\partial_t u|^2) \\ & + G_{X_{ij}}(u_k u_{kij} + u_{ki}u_{kj} + \partial_t u\partial_t u_{ij} + \partial_t u_i\partial_t u_j) \\ & + G_{\xi_i}(u_k u_{ki} + \partial_t u\partial_t u_i) + \left(G_r + \frac{\beta'}{\varepsilon} + \frac{\hat{\beta}'}{\varepsilon}\right)\frac{|Du|^2 + (\partial_t u)^2}{2} \\ = & -\delta(|D^2u|^2 + |D\partial_t u|^2) + G_{X_{ij}}(u_{ki}u_{kj} + \partial_t u_i\partial_t u_j) \\ & - G_{x_k}u_k - G_t\partial_t u - G_r\frac{|Du|^2 + (\partial_t u)^2}{2} \\ & + \frac{\beta'}{2\varepsilon}(-|Du|^2 - (\partial_t u)^2 + 2\langle Du, D\Phi\rangle + 2\partial_t u\partial_t\Phi) \\ & + \frac{\hat{\beta}'}{2\varepsilon}(-|Du|^2 - (\partial_t u)^2 + 2\langle Du, D\Psi\rangle + 2\partial_t u\partial_t\Psi).\end{aligned} \tag{2.18}$$

Proposition 2.2 combined with (2.3) (ii) yields

$$-G_r\frac{|Du|^2 + (\partial_t u)^2}{2} \le -\frac{|Du|^2 + (\partial_t u)^2}{2}\nu \le C_3^2|\nu|, \tag{2.19}$$

where $C_3 > 0$ is from (2.7). From (2.3) (iii) and Young's inequality, it follows that

$$\begin{aligned} & - G_{x_k}u_k - G_t\partial_t u + G_{X_{ij}}(u_{ki}u_{kj} + \partial_t u_i\partial_t u_j) \\ \le & \frac{1}{2M}\sum_{z=x_1,\dots,x_k,t}(G_z)^2 + \frac{M}{2}(|Du|^2 + (\partial_t u)^2) + G_{X_{ij}}(u_{ki}u_{kj} + \partial_t u_i\partial_t u_j) \\ \le & \frac{M+1}{2}(|Du|^2 + (\partial_t u)^2) + \frac{R^2}{2} + G_{X_{ij}}\left(\frac{u_{ki}u_{kj}}{2} + \partial_t u_i\partial_t u_j\right),\end{aligned} \tag{2.20}$$

where $R > 0$ is from (2.9), and $M = M(R) > 0$ in (2.3) (iii). Again, Proposition 2.2 together with (2.3) (iv) yields

$$-G_{x_k}u_k - G_t\partial_t u + G_{X_{ij}}(u_{ki}u_{kj} + \partial_t u_i\partial_t u_j) \le (M+1)C_3^2 + \frac{R^2}{2}.$$

For the penalty terms in (2.18), by Young's inequality, we have

$$\begin{aligned}\frac{\beta'}{2\varepsilon}(-|Du|^2 - (\partial_t u)^2 + 2\langle Du, D\Phi\rangle + 2\partial_t u\partial_t\Phi) \le \frac{\beta'}{2\varepsilon}(|D\Phi|^2 + (\partial_t\Phi)^2), \\ \frac{\hat{\beta}'}{2\varepsilon}(-|Du|^2 - (\partial_t u)^2 + 2\langle Du, D\Psi\rangle + 2\partial_t u\partial_t\Psi) \le \frac{\hat{\beta}'}{2\varepsilon}(|D\Psi|^2 + (\partial_t\Psi)^2).\end{aligned} \tag{2.21}$$

Thus, we see that

$$\partial_t V + LV \le -\delta\left(|D^2u|^2 + |D\partial_t u|^2\right) + C\left(1 + \frac{\beta'}{\varepsilon} + \frac{\hat{\beta}'}{\varepsilon}\right),$$

where $C > 0$ depends only on $\|\Phi\|_{W^{2,1}_\infty(Q_T)}$, $\|\Psi\|_{W^{2,1}_\infty(Q_T)}$ and $\|g\|_{W^{2,\infty}(\mathbb{T}^n)}$. By multiplying by σ and integrating by parts, Lemma 2.5 combined with (2.16) implies

$$\begin{aligned}
&\int_0^{\hat{t}}\int_{\mathbb{T}^n} \delta\left(|D^2u|^2 + |D\partial_t u|^2\right)\sigma dxdt \\
&\le \int_{\mathbb{T}^n} \left(-V(x,\hat{t})\sigma(x,\hat{t}) + V(x,0)\sigma(x,0)\right) dx \\
&\quad + C\int_0^{\hat{t}}\int_{\mathbb{T}^n}\left(1 + \frac{\beta'}{\varepsilon} + \frac{\hat{\beta}'}{\varepsilon}\right)\sigma(x,t)dxdt \\
&\le \int_{\mathbb{T}^n} \left(-V(x,\hat{t})\sigma(x,\hat{t}) + V(x,0)\sigma(x,0)\right) dx + C.
\end{aligned} \tag{2.22}$$

Since $-V(x,\hat{t})\sigma(x,\hat{t}) \le 0$, and

$$\int_{\mathbb{T}^n} V(x,0)\sigma(x,0)dx \le \frac{\|Dg\|_{L^\infty(\mathbb{T}^n)} + \|\partial_t u\|_{L^\infty(Q_T)}}{2}\int_{\mathbb{T}^n}\sigma(x,0)dx \le C,$$

we conclude the proof. □

We obtain the rate of convergence.

Theorem 2.7 *Under the hypotheses of Lemma 2.6, let $u^{\varepsilon,\delta} \in C^{3,2}(Q_T) \cap C^{2,1}(\overline{Q}_T)$ be solutions of* (2.4)*. Then, there exist a unique viscosity solution $u \in C(\overline{Q}_T) \cap W^{1,1}_\infty(Q_T)$ of* (1.5) *satisfying $u(\cdot,0) = g$ in $\mathbb{T}^n$, and a constant $C > 0$, independent of $\varepsilon > 0$ and $\delta > 0$, such that*

$$\|u^{\varepsilon,\delta} - u\|_{L^\infty(Q_T)} \le C(\varepsilon + \sqrt{\delta}) \quad \textit{for } \varepsilon,\delta \in (0,1).$$

Proof Due to the viscosity solution theory, it is easy to find a (unique) viscosity solution $u \in C(\overline{Q}_T)$ of (1.5) such that $u^{\varepsilon,\delta}$ converges to u uniformly in $\overline{Q}_T$ as $(\varepsilon,\delta) \to (0,0)$. Note that (2.7) in Proposition 2.2 yields $u \in W^{1,1}_\infty(Q_T)$.

Fix $(\hat{x},\hat{t}) \in \mathbb{T}^n \times (0,T)$. We will use the solution σ in (2.16) for this $(\hat{x},\hat{t})$.

Differentiating (2.4) with respect to ε, we have

$$\frac{\partial^2 u^{\varepsilon,\delta}}{\partial t\partial\varepsilon} + L\left(\frac{\partial u^{\varepsilon,\delta}}{\partial\varepsilon}\right) = \frac{\beta'}{\varepsilon^2}(u^{\varepsilon,\delta} - \Phi) + \frac{\hat{\beta}'}{\varepsilon^2}(u^{\varepsilon,\delta} - \Psi).$$

Since

$$\operatorname{supp}\beta'_\varepsilon \subset \{(x,t) \in \mathbb{T}^n \times (0,T) | u^{\varepsilon,\delta}(x,t) \ge \Phi(x,t)\} =: \{u^{\varepsilon,\delta} \ge \Phi\},$$

$$\operatorname{supp}\hat{\beta}'_\varepsilon \subset \{(x,t) \in \mathbb{T}^n \times (0,T) | u^{\varepsilon,\delta}(x,t) \le \Psi(x,t)\} =: \{u^{\varepsilon,\delta} \le \Psi\}$$

and

$$\frac{\partial u^{\varepsilon,\delta}}{\partial \varepsilon}(\cdot, 0) = \frac{\partial g}{\partial \varepsilon} = 0 \quad \text{in } \mathbb{T}^n,$$

multiplying by σ and integrating by parts, in view of Proposition 2.2 and Lemma 2.5, we have

$$\begin{aligned}
\left| \int_{\mathbb{T}^n} \frac{\partial u^{\varepsilon,\delta}}{\partial \varepsilon}(x, \hat{t}) \rho_{\hat{x}}^N \, dx \right| &= \left| \int_0^{\hat{t}} \left(\int_{\mathbb{T}^n} \frac{\partial^2 u^{\varepsilon,\delta}}{\partial t \partial \varepsilon} + L\left(\frac{\partial u^{\varepsilon,\delta}}{\partial \varepsilon} \right) \right) \sigma \, dxdt \right| \\
&\leq C_1 \left(\iint_{Q_i \cap \{u^{\varepsilon,\delta} \geq \Phi\}} \frac{\beta'}{\varepsilon} \sigma \, dxdt + \iint_{Q_i \cap \{u^{\varepsilon,\delta} \leq \Psi\}} \frac{\hat{\beta}'}{\varepsilon} \sigma \, dxdt \right) \\
&\leq C_1 C_4.
\end{aligned}$$

Letting $N \to \infty$, we have

$$\left| \frac{\partial u^{\varepsilon,\delta}}{\partial \varepsilon}(\hat{x}, \hat{t}) \right| \leq C_1 C_4. \tag{2.23}$$

On the other hand, differentiating (2.4) with respect to δ, we have

$$\frac{\partial^2 u^{\varepsilon,\delta}}{\partial t \partial \delta} + L\left(\frac{\partial u^{\varepsilon,\delta}}{\partial \delta} \right) = \Delta u^{\varepsilon,\delta}.$$

Hence, in view of Lemmas 2.5 and 2.6 here, we have

$$\begin{aligned}
\left| \int_{\mathbb{T}^n} \frac{\partial u^{\varepsilon,\delta}}{\partial \delta}(x, \hat{t}) \rho_{\hat{x}}^N \, dx \right| &= \left| \int_0^{\hat{t}} \left(\int_{\mathbb{T}^n} \frac{\partial^2 u^{\varepsilon,\delta}}{\partial t \partial \delta} + L\left(\frac{\partial u^{\varepsilon,\delta}}{\partial \delta} \right) \right) \sigma \, dxdt \right| \\
&\leq \int_0^{\hat{t}} \int_{\mathbb{T}^n} |\Delta u^{\varepsilon,\delta}| \sigma \, dxdt \\
&\leq \frac{\sqrt{C_4 C_5}}{\sqrt{\delta}}.
\end{aligned} \tag{2.24}$$

Thus, sending $N \to \infty$, we have

$$\left| \frac{\partial u^{\varepsilon,\delta}}{\partial \delta}(\hat{x}, \hat{t}) \right| \leq \frac{\sqrt{C_4 C_5}}{\sqrt{\delta}}. \tag{2.25}$$

Remember that $u^{\varepsilon,\delta}$ converges to u uniformly in $\overline{Q}_T$ because of the uniqueness of viscosity solutions of (1.5).

Now, by (2.23) and (2.25), for $0 < \varepsilon_0 < \varepsilon_1 < 1$ and $0 < \delta_0 < \delta_1 < 1$, it follows that

$$\begin{aligned}|u^{\varepsilon_1,\delta_1}(\hat{x},\hat{t}) - u^{\varepsilon_0,\delta_0}(\hat{x},\hat{t})| &\le |u^{\varepsilon_1,\delta_1}(\hat{x},\hat{t}) - u^{\varepsilon_1,\delta_0}(\hat{x},\hat{t})| + |u^{\varepsilon_1,\delta_0}(\hat{x},\hat{t}) - u^{\varepsilon_0,\delta_0}(\hat{x},\hat{t})| \\ &\le \int_{\delta_0}^{\delta_1} \left|\frac{\partial u^{\varepsilon_1,\delta}}{\partial \delta}(\hat{x},\hat{t})\right| d\delta + \int_{\varepsilon_0}^{\varepsilon_1} \left|\frac{\partial u^{\varepsilon,\delta_0}}{\partial \varepsilon}(\hat{x},\hat{t})\right| d\varepsilon \\ &\le \frac{\sqrt{C_4C_5}}{2}(\sqrt{\delta_1} - \sqrt{\delta_0}) + C_1C_4(\varepsilon_1 - \varepsilon_0).\end{aligned}$$

Letting $\varepsilon_0, \delta_0 \to 0$, we have

$$|u^{\varepsilon_1,\delta_1}(\hat{x},\hat{t}) - u(\hat{x},\hat{t})| \le C(\varepsilon_1 + \sqrt{\delta_1})$$

for some $C > 0$, which is independent of $(\hat{x},\hat{t}) \in Q_T$. □

In order to generalize our result to merely continuous G, we suppose the following:

$$\begin{cases}\text{There are } G^\gamma \in C(\overline{Q}_T \times \mathbb{R} \times \mathbb{R}^n \times S^n) \text{ for } \gamma \in (0,1) \text{ such that} \\ G^\gamma \text{ satisfy (2.3) with } \nu \in \mathbb{R}, M(R) > 0 \text{ (for each } R > 0\text{), and} \\ G^\gamma \text{ converges } G \text{ uniformly locally in } \overline{Q}_T \times \mathbb{R} \times \mathbb{R}^n \times S^n \text{ as } \gamma \to 0.\end{cases} \tag{2.26}$$

Corollary 2.8 *Assume* (2.1), (2.2). *Assume that* $G \in C(\overline{Q}_T \times \mathbb{R} \times \mathbb{R}^n \times S^n)$ *satisfies* (2.26). *Let* $u^{\varepsilon,\delta,\gamma} \in C^{3,2}(Q_T) \cap C^{2,1}(\overline{Q}_T)$ *be solutions of*

$$\partial_t u - \delta \Delta u + G^\gamma(x,t,u,Du,D^2u) + \beta - \hat{\beta} = 0 \quad \text{in } Q_T$$

under $u^{\varepsilon,\delta,\gamma}(\cdot,0) = g$ *in* $\mathbb{T}^n$. *Then, there exist a unique viscosity solution* $u \in C(\overline{Q}_T) \cap W^{1,1}_\infty(Q_T)$ *of* (1.5) *under* $u(\cdot,0) = g$ *in* $\mathbb{T}^n$, *a constant* $C > 0$, *independent of* $\varepsilon, \delta, \gamma \in (0,1)$, *and a continuous function* $\omega^* : [0,\infty) \to [0,\infty)$ *such that* $\omega^*(0) = 0$ *and*

$$\|u^{\varepsilon,\delta,\gamma} - u\|_{L^\infty(Q_T)} \le C(\varepsilon + \sqrt{\delta}) + \omega^*(\gamma).$$

Proof Adding one more parameter $\gamma \in (0,1)$, we can proceed the same argument in Theorem 2.7 to obtain that for any fixed $(\hat{x},\hat{t}) \in Q_T$,

$$|u^{\varepsilon,\delta,\gamma}(\hat{x},\hat{t}) - u^{\varepsilon_0,\delta_0,\gamma_0}(\hat{x},\hat{t})| \le C(\varepsilon - \varepsilon_0 + \sqrt{\delta} - \sqrt{\delta_0}) + |u^{\varepsilon_0,\delta_0,\gamma}(\hat{x},\hat{t}) - u^{\varepsilon_0,\delta_0,\gamma_0}(\hat{x},\hat{t})|.$$

Sending $\varepsilon_0, \delta_0, \gamma_0 \to 0$, we have

$$|u^{\varepsilon,\delta,\gamma}(\hat{x},\hat{t}) - u(\hat{x},\hat{t})| \le C(\varepsilon + \sqrt{\delta}) + \|u^{0,0,\gamma} - u\|_{L^\infty(Q_T)}.$$

We note that since $u^{\varepsilon,\delta,\gamma}$ converges to $u^{0,0,\gamma}$ uniformly in $\overline{Q}_T$, which is a unique viscosity solution of

$$\min\{\max\{\partial_t u + G^\gamma(x,r,u,Du,D^2u),\ u - \Phi\},\ u - \Psi\} = 0 \quad \text{in } Q_T$$

under $u(\cdot, 0) = g$ in $\mathbb{T}^n$. In view of the standard theory of viscosity solutions, since $u^{0,0,\gamma}$ converges to u uniformly in Q_T as $\gamma \to 0$, we can find a desired ω^* because the uniqueness of viscosity solutions holds true for the limit PDE. See [3] for the details. □

2.4 Examples

We shall present examples G and G^γ, to which Theorem 2.7 and Corollary 2.8 can be applied.

Example 2.9 The first example of G corresponds to the second-order degenerate linear parabolic operator.

$$G(x, t, r, \xi, X) = -\operatorname{tr}[A(x, t)X] + \langle b(x, t), \xi\rangle + c(x, t)r - f(x, t), \tag{2.27}$$

where $A = (a^{ij}) : \overline{Q}_T \to S^n$ is in $C(\overline{Q}_T) \cap W^{2,\infty}(Q_T)$, and $b = (b^i) : \overline{Q}_T \to \mathbb{R}^n$ and $c, f : \overline{Q}_T \to \mathbb{R}$ are Lipschitz continuous functions such that

$$\langle A(x, t)\eta, \eta\rangle \ge 0 \quad \text{for } \eta \in \mathbb{R}^n \text{ and } (x, t) \in \overline{Q}_T,$$

and there exists $\nu \in \mathbb{R}$ such that

$$c(x, t) \ge \nu \quad \text{for } (x, t) \in \overline{Q}_T. \tag{2.28}$$

Letting $A^\gamma, b^\gamma, c^\gamma, f^\gamma$ be smooth functions by mollifications of A, b, c, f, respectively, we set

$$G^\gamma(x, t, r, \xi, X) := -\operatorname{tr}[A^\gamma(x, t)X] + \langle b^\gamma(x, t), \xi\rangle + c^\gamma(x, t)r - f^\gamma(x, t). \tag{2.29}$$

According to Lemma 1.7.1 in [26], it follows that

$$(a_z^{ij} u_{x_i x_j})^2 \le C a^{ij} u_{x_i x_\ell} u_{x_j x_\ell} \quad \text{in } Q_T \tag{2.30}$$

for $z = x_1, \dots, x_n, t$, where the constant $C = C(\|A\|_{W^{2,\infty}(Q_T)}) > 0$. Since A has $n + 1$ variables here, we recall the proof of [26] in Appendix for the reader's convenience. Our hypothesis (2.3) (*iii*) is motivated by this inequality when G is given by (2.27). We refer to [8, 9] for an application of (2.30) to unilateral obstacle problems in bounded domains.

We give some examples of G associated with quasilinear operators.

Example 2.10 For $p \in (2, \infty)$, we define

$$G(\xi, X) = -|\xi|^{p-2} \operatorname{tr}\left[\left(I + (p-2)\frac{\xi \otimes \xi}{|\xi|^2}\right) X\right].$$

In this example, G^γ is given by

$$G^\gamma(\xi, X) = \begin{cases} G(\xi, X) & \text{for } p \in [4, \infty), \\ -(|\xi|^2 + \gamma)^{\frac{p-2}{2}} \operatorname{tr}\left[\left(I + (p-2)\dfrac{\xi \otimes \xi}{|\xi|^2 + \gamma}\right) X\right] & \text{for } p \in (2, 4). \end{cases}$$

We refer to [22] and its references for p Laplace operator Δ_p.

Example 2.11 The last example appears in eikonal equation.

$$G(\xi) = |\xi|, \quad G^\gamma(\xi) = \sqrt{|\xi|^2 + \gamma}$$

Since we do not need regularization in the following case, we will not use G^γ.

Example 2.12 The infinity Laplace operator Δ_∞ is derived from a formal limit of $-(p-2)^{-1}\Delta_p$ as $p \to \infty$. We refer to [21] and references therein for Δ_∞.

$$G(\xi, X) = -\operatorname{tr}[\xi \otimes \xi X].$$

3 Uniformly Parabolic PDE

3.1 Assumptions and Rate of Convergence

In this section, we treat uniformly parabolic equations. We replace the property (2.3) (iv) by (3.1) (iv), i.e., we suppose that the following for $G \in (\overline{Q}_T \times \mathbb{R} \times \mathbb{R}^n \times S^n)$.

$$\begin{cases} (i) \;\; G \in C^4(Q_T \times \mathbb{R} \times \mathbb{R}^n \times S^n), \\ (ii) \;\; \text{there exists a constant } \nu \in \mathbb{R} \text{ such that } G_r(x, t, r, \xi, X) \geq \nu \\ \qquad \text{for } (x, t) \in Q_T \text{ and } (r, \xi, X) \in \mathbb{R} \times \mathbb{R}^n \times S^n, \\ (iii) \;\; \text{for } R > 0, \text{ there is } M = M(R) > 0 \text{ such that} \\ \qquad G_z(x, t, u, \xi, X)^2 \leq M\left(-G_{X_{ij}}(x, t, u, \xi, X)X_{ik}X_{jk} + |\xi|^2 + R^2\right) \\ \qquad \text{for } (x, t) \in Q_T, (r, \xi, X) \in [-R, R] \times \mathbb{R}^n \times S^n, z = x_1, \ldots, x_n, t, \\ (iv) \;\; \text{for any } R > 0, \text{ there exists } \lambda = \lambda(R) > 0 \text{ such that} \\ \qquad (G_{X_{ij}}(x, t, u, \xi, X)) \leq -\lambda I \\ \qquad \text{for } (x, t) \in Q_T, (r, \xi, X) \in \mathbb{R} \times \overline{B_R} \times S^n. \end{cases} \tag{3.1}$$

Here, $B_R := \{\xi \in \mathbb{R}^n \mid |\xi| < R\}$.

It is obvious that (3.1) is stronger than (2.3), and Proposition 2.2 and Lemma 2.5 hold. Furthermore, the following lemma is a modification of Lemma 2.6.

Lemma 3.1 *Assume* (2.1), (2.2) *and* (3.1). *Let* $u^{\varepsilon,\delta} \in C^{3,2}(Q_T) \cap C^{2,1}(\overline{Q}_T)$ *and* $\sigma = \sigma^{\varepsilon,\delta} \in C(\overline{Q}_{\hat{t}}) \cap C^{3,2}(Q_{\hat{t}})$ *be solutions of* (2.4) *and* (2.16), *respectively. Then, there exists* $C_6 > 0$, *independent of* ε *and* δ, *such that*

$$\int_0^{\hat{t}}\int_{\mathbb{T}^n}\left(|D^2u^{\varepsilon,\delta}(x,t)|^2+|D\partial_t u^{\varepsilon,\delta}(x,t)|^2\right)\sigma(x,t)\,dxdt\le C_6.$$

Proof Set $V := \dfrac{|Du^{\varepsilon,\delta}|^2+(\partial_t u^{\varepsilon,\delta})^2}{2}$. In the same manner as the proof of Lemma 2.6, we can see that (2.18), (2.19), (2.20) and (2.21) hold.

In view of Proposition 2.2, we recall $C_3>0$ such that

$$\|Du^{\varepsilon,\delta}\|_{L^\infty(Q_T)}\le C_3.$$

We fix $\lambda=\lambda(C_3)>0$ in (3.1) (iv).

We shall drop ε and δ from $u^{\varepsilon,\delta}$ as before. From (2.20) and (3.1) (iv), it follows that

$$\begin{aligned}
&-u_kG_{x_k}-\partial_t uG_t+G_{X_{ij}}(u_{ki}u_{kj}+\partial_t u_i\partial_t u_j)\\
\le&\frac{M+1}{2}(|Du|^2+(\partial_t u)^2)+\frac{R^2}{2}+G_{X_{ij}}\left(\frac{u_{ki}u_{kj}}{2}+\partial_t u_i\partial_t u_j\right)\\
\le&(M+1)C_3^2+\frac{R^2}{2}-\frac{\lambda}{2}|D^2u|^2-\lambda|D\partial_t u|^2.
\end{aligned}$$

Combining this inequality with (2.18), (2.19) and (2.21) yields

$$\partial_t V+LV\le-\frac{\lambda}{2}|D^2u|^2-\lambda|D\partial_t u|^2+C\left(1+\frac{\beta'}{\varepsilon}+\frac{\gamma'}{\varepsilon}\right).$$

Thus, we can obtain (2.22) with δ replaced by $\lambda/2$, and we conclude the assertion. □

The rate of convergence in the uniformly parabolic case is as follows.

Theorem 3.2 *Under hypotheses of Lemma 3.1, let $u^{\varepsilon,\delta}\in C^{3,2}(Q_T)\cap C^{2,1}(\overline{Q}_T)$ be solutions of* (2.4). *Then, there exist a unique viscosity solution $u\in C(\overline{Q}_T)\cap W^{1,1}_\infty(Q_T)$ of* (1.5) *satisfying $u(\cdot,0)=g$ in $\mathbb{T}^n$, and a constant $C>0$, independent of $\varepsilon>0$ and $\delta>0$, such that*

$$\|u^{\varepsilon,\delta}-u\|_{L^\infty(Q_T)}\le C(\varepsilon+\delta)\quad for\ \varepsilon,\delta\in(0,1).$$

Proof In the same manner as the proof of Theorem 2.7, we have (2.24) with $\sqrt{C_4C_5}/\sqrt{\delta}$ replaced by $\sqrt{C_4C_6}$ thanks to Lemma 3.1 instead of Lemma 2.6. Therefore, we obtain

$$\frac{\partial u^{\varepsilon,\delta}}{\partial\delta}(\hat{x},\hat{t})\le\sqrt{C_4C_6}$$

for any $(\hat{x},\hat{t})\in\overline{Q}_T$, and

$$\|u^{\varepsilon,\delta} - u\|_{L^\infty(Q_T)} \le C(\varepsilon + \delta)$$

for $\varepsilon, \delta \in (0, 1)$. □

As in Sect. 2, it is possible to obtain the same rate of convergence with an error under the following hypotheses.

$$\begin{cases} \text{There are } G^\gamma \in C(\overline{Q}_T \times \mathbb{R} \times \mathbb{R}^n \times S^n) \text{ for } \gamma \in (0, 1) \text{ such that} \\ G^\gamma \text{ satisfy (3.1) with } \nu \in \mathbb{R}, M(R) > 0 \text{ (for each } R > 0), \text{ and} \\ G^\gamma \text{ converges } G \text{ uniformly locally in } \overline{Q}_T \times \mathbb{R} \times \mathbb{R}^n \times S^n \text{ as } \gamma \to 0. \end{cases} \tag{3.2}$$

Corollary 3.3 *Assume* (2.1), (2.2). *Assume that* $G \in C(\overline{Q}_T \times \mathbb{R} \times \mathbb{R}^n \times S^n)$ *satisfies* (3.2). *Let* $u^{\varepsilon,\delta,\gamma} \in C^{3,2}(Q_T) \cap C^{2,1}(\overline{Q}_T)$ *be solutions of*

$$\partial_t u - \delta\Delta u + G^\gamma(x, t, u, Du, D^2u) + \beta - \hat{\beta} = 0 \quad \text{in } Q_T$$

under $u^{\varepsilon,\delta,\gamma}(\cdot, 0) = g$ *in* $\mathbb{T}^n$. *Then, there exist a unique viscosity solution* $u \in C(\overline{Q}_T) \cap W^{1,1}_\infty(Q_T)$ *of* (1.5) *satisfying* $u(\cdot, 0) = g$ *in* $\mathbb{T}^n$, *a constant* $C > 0$, *independent of* $\varepsilon, \delta, \gamma \in (0, 1)$, *and a continuous function* $\omega^* : [0, \infty) \to [0, \infty)$ *such that* $\omega^*(0) = 0$ *and*

$$\|u^{\varepsilon,\delta,\gamma} - u\|_{L^\infty(Q_T)} \le C(\varepsilon + \delta) + \omega^*(\gamma).$$

3.2 Examples

We shall present examples for G and G^γ, to which Theorem 3.2 and Corollary 3.3 can be applied.

Example 3.4 The first example corresponds to second-order linear uniformly parabolic equations, i.e.,G and G^γ are defined by (2.27) and (2.29), where $A = (a^{ij}) : \overline{Q}_T \to S^n$, $b = (b^i) : \overline{Q}_T \to \mathbb{R}^n$ and $c, f : \overline{Q}_T \to \mathbb{R}$ are Lipschitz continuous functions such that

$$\langle A(x, t)\eta, \eta\rangle \ge \lambda|\eta|^2 \quad \text{for } \eta \in \mathbb{R}^n \text{ and } (x, t) \in \overline{Q}_T,$$

and there exists $\nu \in \mathbb{R}$ such that (2.28). Here, $A^\gamma, b^\gamma, c^\gamma, f^\gamma$ are smooth functions by mollifications of A, b, c, f, respectively.

Example 3.5 The next one corresponds to the mean curvature flow equation of graphs

$$G(\xi, X) = -\operatorname{tr}\left[\left(I - \frac{\xi \otimes \xi}{1 + |\xi|^2}\right)X\right], \quad \text{and} \quad G^\gamma \equiv G.$$

We refer to [10] for an equation by adding vanishing viscosity term $-\delta\Delta$, which enables us to have classical approximate solutions.

Acknowledgements S. K. is supported by JSPS KAKENHI Grant Number 20H01817. T. K. is supported by JSPS KAKENHI Grant Number JP18K13436 and Tottori University of Environmental Studies Grant-in-Aid for Special Research.

4 Appendix

4.1 Comparison Principle Under (1.13)

In the theory of viscosity solutions, the comparison principle for (4.1) means that if $u \in C(\overline{Q}_T)$ and $v \in C(\overline{Q}_T)$ are, respectively, a viscosity subsolution and a viscosity supersolution of a PDE in Q_T satisfying $u(\cdot, 0) \leq v(\cdot, 0)$ in $\mathbb{T}^n$, then it follows that $u \leq v$ in $\overline{Q}_T$. It is straightforward to verify that the uniqueness of viscosity solutions of the PDE under the initial condition if the comparison principle holds true. We refer to [3] for more general comparison principle to semicontinous viscosity solutions.

Remark 4.1 When $\nu < 0$ in (1.13) (ii), we need one more assumption on G. In other words, if $\nu \geq 0$ in (1.13) such as p-Laplace operators for $2 \leq p \leq \infty$, then it is not necessary to assume (4.2) in Sect. 5.2 below because we do not need to multiply $e^{-\nu t}$ to u in the proof of Proposition 4.5 below.

Now, we give the definition of viscosity solutions of

$$\partial_t u + G(x, t, u, Du, D^2 u) = 0 \quad \text{in } Q_T, \tag{4.1}$$

where $F : Q_T \times \mathbb{R} \times \mathbb{R} \times \mathbb{R}^n \times S^n \to \mathbb{R}$ is continuous.

We recall parabolic semi-jets $\mathcal{P}^{\pm}u(x, t)$ for $u : Q_T \to \mathbb{R}$ at $(x, t) \in Q_T$.

$$\mathcal{P}^{+}u(x,t) := \left\{ \begin{array}{c} (q, \xi, X) \\ \in \mathbb{R} \times \mathbb{R}^n \times S^n \end{array} \middle| \begin{array}{l} u(y,s) \leq u(x,t) + q(s-t) + \langle \xi, y-x \rangle \\ \qquad +\frac{1}{2}\langle X(y-x), y-x \rangle \\ \qquad +o(|y-x|^2 + |t-s|) \text{ for } (y,s) \in Q_T \end{array} \right\}$$

and $\mathcal{P}^{-}u(x,t) := -\mathcal{P}^{+}(-u)(x,t)$.

We also need closures of $\mathcal{P}^{\pm}u(x,t)$ for $u \in C(Q_T)$ at $(x,t) \in Q_T$:

$$\overline{\mathcal{P}}^{\pm}u(x,t) := \left\{ \begin{array}{c} (q, \xi, X) \\ \in \mathbb{R} \times \mathbb{R}^n \times S^n \end{array} \middle| \begin{array}{c} \text{There exist } (x_k, t_k) \in Q_T \text{ and} \\ (q_k, \xi_k, X_k) \in \mathcal{P}^{\pm}u(x_k, t_k) \text{ such that} \\ \lim_{k\to\infty} (x_k, t_k, q_k, \xi_k, X_k) = (x, t, q, \xi, X) \end{array} \right\}$$

Definition 4.2 We call $u \in C(Q_T)$ a viscosity subsolution (resp., supersolution) of (4.1) if

$$q + G(x, t, u(x,t), \xi, X) \leq 0 \quad (\text{resp., } q + G(x, t, u(x,t), \xi, X) \geq 0)$$

provided that $(x,t) \in Q_T$ and $(q,\xi,X) \in \overline{\mathcal{P}}^{+}u(x,t)$ (resp., $\overline{\mathcal{P}}^{-}u(x,t)$). We also call $u \in C(Q_T)$ a viscosity solution of (4.1) if it is both viscosity subsolution and supersolution of (4.1).

Remark 4.3 It is well-known that $u \in C(Q_T)$ is a viscosity subsolution (resp., supersolution) of (4.1) if and only if

$$\partial_t\eta(x,t) + G(x,t,u(x,t),D\eta(x,t),D^2\eta(x,t)) \le 0 \quad (\text{resp.,} \ \ge 0)$$

provided that for any $\eta \in C^{2,1}(Q_T)$, $u-\eta$ takes its local maximum (resp., minimum) at (x,t). We refer to [3, 12, 15] for equivalent definitions of viscosity solutions, which will be used to prove Propositions 4.4 and 4.5.

In order to involve the case when $\nu < 0$ in (1.13) (i), we will need to use a typical argument in parabolic PDEs. To this end, we prepare the following proposition with no proof since it is easy from the definition or the equivalent definition in Remark 4.3.

Proposition 4.4 *Assume that $G \in C(Q_T \times \mathbb{R} \times \mathbb{R}^n \times S^n)$. Let $u \in C(Q_T)$ be a viscosity subsolution (resp., supersolution) of* (4.1). *Setting $U(x,t) := e^{\nu t}u(x,t)$ for $\nu \in \mathbb{R}$, we see that U is a viscosity subsolution (resp., supersolution) of*

$$\partial_t u + \hat{G}(x,t,u,Du,D^2u) = 0 \quad \text{in } Q_T,$$

where

$$\hat{G}(x,t,r,\xi,X) := -\nu r + e^{\nu t}G(x,t,e^{-\nu t}re^{-\nu t}\xi, e^{-\nu t}X)$$

for $(x,t) \in Q_T$ and $(r,q,\xi,X) \in \mathbb{R} \times \mathbb{R} \times \mathbb{R}^n \times S^n$.

We suppose the following hypothesis, which is only mentioned in [15] to our knowledge.

$$\begin{cases} \text{There is } \hat{\omega} : [0,\infty) \to [0,\infty) \text{ such that } \lim_{s\to 0} \hat{\omega}(s) = 0, \\ \text{and } \dfrac{1}{\sigma}G(x,t,\sigma r,\sigma\xi,\sigma X) - \dfrac{1}{\tau}G(x,t,\tau r,\tau\xi,\tau X) \le \hat{\omega}(|\sigma-\tau|) \\ \text{for } \sigma,\tau > 1,\ (x,t) \in Q_T \text{ and } (r,\xi,X) \in \mathbb{R} \times \mathbb{R}^n \times S^n. \end{cases} \tag{4.2}$$

It is immediate to see that (4.2) is true if f is bounded when G is given in (2.27). We note again that we do not need the above (4.2) to deal with G which do not depend on the r variable such as p-Laplace operators ($2 \le p \le \infty$) since we can take $\nu = 0$ in (1.13) (i).

Proposition 4.5 *Assume* (1.13) *and* (4.2). *If u and $v \in C(\overline{Q}_T)$ are, respectively, a viscosity subsolution and supersolution of* (4.1) *satisfying $u(x,0) \le v(x,0)$ for $x \in \mathbb{T}^n$, then $u \le v$ in $\overline{Q}_T$.*

Because our proof below is standard, we only give a sketch of proof.

Proof If ν in (1.13) (i) is nonnegative, we may take $\nu = 0$ below; $U(x,t) = e^0 u(x,t) = u(x,t)$. Therefore, we shall suppose $\nu < 0$ in what follows.

Define $U(x,t) := e^{\nu t} u(x,t)$ and $V(x,t) := e^{\nu t} v(x,t)$ for $\overline{Q}_T$. Setting $\Theta_\mu := \sup_{(x,t)\in Q_T} \left(U(x,t) - V(x,t) - \frac{\mu}{T-t} \right)$ for $\mu \in (0,1)$, we define

$$2\theta := \liminf_{\mu\to 0} \Theta_\mu .$$

If $\theta \le 0$, then we have $U - V \le 0$ in Q_T, which concludes the proof. Thus, we may suppose

$$\Theta_\mu \ge \theta > 0 \quad \text{for small } \mu \in (0,1)$$

along a subsequence $\{\mu_k\}_{k=1}^\infty$ if necessary.

We shall adapt the doubling argument; considering

$$W(x,t,y,s) := U(x,t) - V(y,s) - \frac{\mu}{T-t} - \frac{\alpha}{2}(|x-y|^2 + (t-s)^2) \quad \text{for } \alpha > 1,$$

we let $(x_\alpha, t_\alpha), (y_\alpha, s_\alpha) \in \mathbb{T}^n \times [0,T)$ be such that

$$W(x_\alpha, t_\alpha, y_\alpha, s_\alpha) \ge \Theta_\mu > 0.$$

It is standard to show that

$$\lim_{\alpha\to\infty} \alpha(|x_\alpha - y_\alpha|^2 + (t_\alpha - s_\alpha)^2) = 0,$$

which implies that there is $(\hat{x}, \hat{t}) \in \mathbb{T}^n \times [0,T)$ such that

$$\lim_{\alpha\to\infty} (x_\alpha, t_\alpha) = \lim_{\alpha\to\infty} (y_\alpha, s_\alpha) = (\hat{x}, \hat{t}).$$

We also have $\Theta_\mu = W(\hat{x}, \hat{t}, \hat{x}, \hat{t}) > 0$, which yields $\hat{t} \in (0,T)$ since $W(\hat{x}, 0, \hat{x}, 0) \le 0$. Therefore, for large $\alpha > 1$, we have $t_\alpha \in (0,T)$.

Hence, thanks to the so-called Ishii's lemma in [3, 15], we have $X_\alpha, Y_\alpha \in S^n$ such that

$$\left(\alpha(t_\alpha - s_\alpha) + \frac{\mu}{(T-t_\alpha)^2}, \alpha(x_\alpha - y_\alpha), X_\alpha \right) \in \overline{\mathcal{P}}^+ U(x_\alpha, t_\alpha),$$

$$(\alpha(t_\alpha - s_\alpha), \alpha(x_\alpha - y_\alpha), -Y_\alpha) \in \overline{\mathcal{P}}^- V(y_\alpha, s_\alpha), \text{ and } \begin{pmatrix} X_\alpha & O \\ O & Y_\alpha \end{pmatrix} \le 3\alpha \begin{pmatrix} I & -I \\ -I & I \end{pmatrix}.$$

Therefore, setting $\xi_\alpha := \alpha(x_\alpha - y_\alpha)$, we have

$$\alpha(t_\alpha - s_\alpha) + \frac{\mu}{(T-t_\alpha)^2} - \nu U(x_\alpha, t_\alpha) + \hat{G}(x_\alpha, t_\alpha, U(x_\alpha, t_\alpha), \xi_\alpha, X_\alpha) \le 0 \quad (4.3)$$

and

$$\alpha(t_\alpha - s_\alpha) - \nu V(y_\alpha, s_\alpha) + \hat{G}(y_\alpha, s_\alpha, V(y_\alpha, s_\alpha), \xi_\alpha, -Y_\alpha) \geq 0, \tag{4.4}$$

where $\hat{G}$ is defined in Proposition 4.4. We shall drop α from $x_\alpha, y_\alpha, t_\alpha, s_\alpha, \xi_\alpha, X_\alpha, Y_\alpha$ in the proceeding calculations.

Noting (1.13) and letting $R := \max\{\|u\|_{L^\infty(Q_T)}, \|v\|_{L^\infty(Q_T)}\} > 0$, we have

$$\begin{aligned}
&\frac{\mu}{(T-t)^2} - \nu(U(x,t) - V(y,s)) \\
&\leq \hat{G}(y,s,V(y,s),\xi,-Y) - \hat{G}(x,t,U(x,t),\xi,X) \\
&\leq -\nu(U(x,t) - V(y,s)) + \hat{G}(y,s,U(x,t),\xi,-Y) - \hat{G}(x,t,U(x,t),\xi,X).
\end{aligned}$$

Hence, since we may suppose $t, s \in (0, T)$, by (4.2), remembering $\nu < 0$, we have

$$\begin{aligned}
&\frac{\mu}{(T-t)^2} \\
&\leq e^{\nu s} G(y,s,e^{\nu(t-s)}u(x,t), e^{-\nu s}\xi, -e^{-\nu s}Y) - e^{\nu t} G(x,t,u(x,t), e^{-\nu t}\xi, e^{-\nu t}X) \\
&\leq e^{\nu s}\{G(y,s,e^{\nu(t-s)}u(x,t), e^{-\nu s}\xi, e^{-\nu s}Y) - G(x,t,e^{\nu(t-s)}u(x,t), e^{-\nu s}\xi, e^{-\nu s}X)\} \\
&\quad + \hat{\omega}\left(\left|e^{-\nu t} - e^{-\nu s}\right|\right).
\end{aligned}$$

In view of (1.13) (*ii*), we have

$$\frac{\mu}{(T-t)^2} \leq e^{\nu s}\omega_{e^{-\nu T}R}\left(\alpha e^{-\nu s}|x-y|^2 + |x-y| + |t-s|\right) + \hat{\omega}\left(\left|e^{-\nu t} - e^{-\nu s}\right|\right),$$

which gives a contradiction by sending $\alpha \to \infty$. □

This argument works also for obstacle problems (1.5), and the associated penalty equations (1.10) when $0 \leq \delta < 1$.

Corollary 4.6 *Assume* (1.13) *and* (4.2). *If u and $v \in C(\overline{Q}_T)$ are, respectively, a viscosity subsolution and supersolution of* (1.5) *(resp.,* (1.10) *for* $0 \leq \delta < 1$*) satisfying* $u(x,0) \leq v(x,0)$ *for* $x \in \mathbb{T}^n$, *then* $u \leq v$ *in* $\overline{Q}_T$.

Proof (1.5) case: Using the same notations in the above proof, we verify that U (resp., V) is a viscosity subsolution (resp., supersolution) of

$$\min\{\max\{u_t + \hat{G}(x,t,u,Du,D^2u), u - \hat{\Phi}\}, u - \hat{\Psi}\} = 0 \quad \text{in } Q_T,$$

where $\hat{\Phi}(x,t) := e^{\nu t}\Phi(x,t)$, $\hat{\Psi}(x,t) := e^{\nu t}\Psi(x,t)$, and

$$\hat{G}(x,t,r,\xi,X) := -\nu r + e^{\nu t} G(x,t,e^{-\nu t}r, e^{-\nu t}\xi, e^{-\nu t}X)$$

for $(x,t) \in Q_T$ and $(r,\xi,X) \in \mathbb{R} \times \mathbb{R}^n \times S^n$.

Following the same argument in the above, we use the same notations (x_α, t_α), $(y_\alpha, s_\alpha) \in Q_T$. If $U \leq \hat{\Psi}$ at (x_α, t_α), then noting

$$\hat{\Psi}(y_\alpha, s_\alpha) \le V(y_\alpha, s_\alpha),$$

and sending $\alpha \to \infty$, we have $U(\hat{x}, \hat{t}) \le V(\hat{x}, \hat{t})$, which implies a contradiction. Hence, we may suppose $U(x_\alpha, t_\alpha) > \hat{\Psi}(x_\alpha, t_\alpha)$ for large $\alpha > 1$. If we moreover assume $V \ge \hat{\Phi}$ at (y_α, s_α), then noting $U \le \hat{\Phi}$ at (x_α, t_α), we have $U \le V$ at $(\hat{x}, \hat{t})$, which gives a contradiction again. Thus, we may suppose that $U > \hat{\Psi}$ at (x_α, t_α) and $V < \hat{\Phi}$ at (y_α, s_α) for large $\alpha > 1$. Therefore, we have (4.3) and (4.4), from which we derive a contradiction.

(1.10) case: It is easy to observe that the above argument works by exactly the same manner for penalized PDEs (1.10). Thus, we leave the details to the readers. □

4.2 *A Proof of* (2.30)

Although the proof below is exactly same as in [26], we repeat it since we allow A to depend on t-variable.

Proposition 4.7 (cf. Lemma 1.7.1 in [26]) *Let $A = (a^{ij}) : \mathbb{R}^n \times \mathbb{R} \to S^n$ satisfy $a^{ij} \in C^2(\mathbb{R}^n \times \mathbb{R})$ and $A \ge O$ in $\mathbb{R}^n \times \mathbb{R}$. Let $u \in C^2(\mathbb{R}^n \times \mathbb{R})$. Then, there exists a constant $\hat{C} = \hat{C}(\|A\|_{W^{2,\infty}(\mathbb{R}^n \times \mathbb{R})}) > 0$ such that*

$$\left(\sum_{i,j=1}^{n} a_z^{ij} u_{x_i x_j} \right)^2 \le \hat{C} \sum_{i,j,l=1}^{n} a^{ij} u_{x_i x_l} u_{x_j x_l} \quad \text{in } \mathbb{R}^n \times \mathbb{R} \tag{4.5}$$

for $z = x_1, \ldots, x_n, t$.

Following the argument in [26], we prepare the following lemma.

Lemma 4.8 *Let $f \in C^2(\mathbb{R})$ satisfy $\|f''\|_{L^\infty(\mathbb{R})} < \infty$, and $f \ge 0$ in $\mathbb{R}$. Then,*

$$f'(x)^2 \le 2f(x)\|f''\|_{L^\infty(\mathbb{R})} \quad \text{for } x \in \mathbb{R}.$$

Proof When $f(x) = 0$, then $f'(x) = 0$ is trivial. Thus, we may suppose $f(x) > 0$.
Suppose the contrary; there exists $x_0 \in \mathbb{R}$ such that

$$f'(x_0)^2 > 2f(x_0)\|f''\|_\infty.$$

Hence, $f'(x_0) \ne 0$. Again, we may assume $f(x_0) > 0$ since otherwise, we have $f'(x_0) = 0$.

We find $\hat{x}$ such that

$$\begin{aligned} f\left(x_0 - 2\frac{f(x_0)}{f'(x_0)}\right) &= f(x_0) - 2\frac{f(x_0)}{f'(x_0)}f'(x_0) + 2\frac{f(x_0)^2}{f'(x_0)^2}f''(\hat{x}) \\ &= -f(x_0)\left\{1 - \frac{2f''(\hat{x})f(x_0)}{f'(x_0)^2}\right\} < 0, \end{aligned}$$

which gives a contradiction. □

Proof of Proposition 4.7 Fix any $(x', t') \in \mathbb{R}^n \times \mathbb{R}$. There exists an orthogonal matrix T such that

$$\Lambda := \begin{pmatrix} \lambda_1 & 0 & \cdots & 0 \\ 0 & \lambda_2 & \cdots & \vdots \\ \vdots & \vdots & \ddots & 0 \\ 0 & 0 & \cdots & \lambda_n \end{pmatrix} = {}^tTA(x', t')T,$$

where ${}^tT$ denotes the transport of T. Setting $y = {}^tTx$, for any $u \in C^2(\mathbb{R}^n \times \mathbb{R})$, we define

$$\tilde{u}(y, t) := u(Ty, t) = u(x, t).$$

Notice that

$$D_y\tilde{u}(y, t) = {}^tTD_xu(x, t), \quad D_y^2\tilde{u}(y, t) = {}^tTD_x^2u(x, t)T.$$

Letting $y' = {}^tTx'$, we have

$$\begin{aligned} {}^tD_x^2u(x', t')A(x', t')D_x^2u(x', t') &= T{}^tD_y^2\tilde{u}(y', t'){}^tTA(x', t')TD_y^2\tilde{u}(y', t'){}^tT \\ &= T{}^tD_y^2\tilde{u}(y', t')\Lambda D_y^2\tilde{u}(y', t'){}^tT. \end{aligned}$$

Notice that

$$\begin{aligned} a^{ij}(x', t')u_{x_ix_l}(x', t')u_{x_jx_l}(x', t') &= \mathrm{tr}\left[{}^tD_x^2u(x', t')A(x', t')D_x^2u(x', t')\right] \\ &= \mathrm{tr}\left[T{}^tD_y^2\tilde{u}(y', t')\Lambda D_y^2\tilde{u}(y', t'){}^tT\right] \\ &= \mathrm{tr}\left[{}^tD_y^2\tilde{u}(y', t')\Lambda D_y^2\tilde{u}(y', t')\right] \\ &= \lambda_i\tilde{u}_{y_iy_j}(y', t')\tilde{u}_{y_iy_j}(y', t'). \end{aligned} \tag{4.6}$$

Setting $B(x, t) = (b^{ij}(x, t)) := {}^tTA(x, t)T \in S^n$, the definition of T yields

$$b^{ij}(x', t') = 0 \quad \text{provided } 1 \le i < j \le n \tag{4.7}$$

while we have

$$A_z(x,t) = TB_z(x,t)^tT \text{ and } A_{zz}(x,t) = TB_{zz}(x,t)^tT \quad \text{for } z = x_1, \dots, x_n, t.$$

Thus, we note that there is $C_0 = C_0(n) > 0$ such that

$$\|B_z\|_\infty \le C_0\|A_z\|_\infty \text{ and } \|B_{zz}\|_\infty \le C_0\|A_{zz}\|_\infty \quad \text{for } z = x_1, \dots, x_n, t$$

because T is an orthogonal matrix.

We first observe

$$a_z^{ij}(x',t')u_{x_ix_j}(x',t') = \mathrm{tr}\left[TB_z(x',t')D_y^2\tilde{u}(y',t')^tT\right] = b_z^{ij}(x',t')\tilde{u}_{y_iy_j}(y',t'). \qquad (4.8)$$

Recalling $B \ge 0$ (i.e. $b^{ii} \ge 0$), by Lemma 4.8, there exists $C_1 = C_1(\|A\|_{W^{2,\infty}(\mathbb{R}^n\times\mathbb{R})}) > 0$ such that for $z = x_1, \dots, x_n, t$,

$$(b_z^{ii})^2 \le C_1 b^{ii} \quad \text{in } \mathbb{R}^n \times \mathbb{R}. \qquad (4.9)$$

Furthermore, since $0 \le b^{ii} + 2b^{ij} + b^{jj}$ for $1 \le i < j \le n$, we also find $C_2 = C_2$ $(\|A\|_{W^{2,\infty}(\mathbb{R}^n\times\mathbb{R})})$ such that for $z = x_1, \dots, x_n, t$,

$$(b_z^{ii} + 2b_z^{ij} + b_z^{jj})^2 \le C_2(b^{ii} + 2b^{ij} + b^{jj}) \quad \text{in } \mathbb{R}^n \times \mathbb{R}. \qquad (4.10)$$

Since (4.10) yields

$$4(b_z^{ij})^2 \le C_2(b^{ii} + 2b^{ij} + b^{jj}) - 4b_z^{ii}b_z^{ij} - 4b_z^{ij}b_z^{jj} - 2b_z^{ii}b_z^{jj},$$

by (4.9), we find $C_3 = C_3(\|A|_{W_\infty^{2,1}(\mathbb{R}^n\times\mathbb{R})}) > 0$ such that

$$(b_z^{ij})^2 \le C_2(b^{ii} + 2b^{ij} + b^{jj}) + C_3(b^{ii} + b^{jj}) \quad \text{in } \mathbb{R}^n \times \mathbb{R}.$$

Therefore, recalling (4.7), we have

$$(b_z^{ij}(x',t'))^2 \le (C_2 + C_3)\{b^{ii}(x',t') + b^{jj}(x',t')\}.$$

Hence, in view of (4.8), there is $C_4 = C_4(n) > 0$ such that

$$\begin{aligned}\left(a_z^{ij}(x',t')u_{x_ix_j}(x',t')\right)^2 &= \left(b_z^{ij}(x',t')\tilde{u}_{y_iy_j}(y',t')\right)^2 \\ &\le C_4 b_z^{ij}(x',t')^2\tilde{u}_{y_iy_j}(y',t')^2 \\ &\le C_4(C_2 + C_3)\left\{(b^{ii}(x',t') + b^{jj}(x',t'))\tilde{u}_{y_iy_j}(y',t')^2\right\}.\end{aligned}$$

Therefore, by (4.6), there exists $\hat{C} = \hat{C}(\|A\|_{W^{2,\infty}(\mathbb{R}^n\times\mathbb{R})}) > 0$ such that (4.5) holds true. □

References

1. G. Barles, A weak Bernstein method for fully nonlinear elliptic equations. Differ. Integr. Equ. **4**(2), 241–262 (1991)
2. F. Cagnetti, D. Gomez, H.V. Tran, Adjoint methods for obstacle problems and weakly coupled systems of PDE. ESAIM Control Optim. Calc. Var. **19**(3), 754–779 (2013)
3. M.G. Crandall, H. Ishii, P.-L. Lions, User's guide to viscosity solutions of second order partial differential equations. Bull. Am. Math. Soc. **27**(1), 1–67 (1992)
4. M.G. Crandall, P.-L. Lions, Two approximations of solutions of Hamilton-Jacobi equations. Math. Comput. **43**(167), 1–19 (1984)
5. L.C. Evans, Classical solutions of fully nonlinear, convex, second-order elliptic equations. Commun. Pure Appl. Math. **35**(3), 333–363 (1982)
6. L.C. Evans, Adjoint and compensated compactness methods for Hamilton-Jacobi PDE. Arch. Ration. Mech. Anal. **197**(3), 1053–1088 (2010)
7. L.C. Evans, P.-L. Lions, Fully nonlinear second order elliptic equations with large zeroth order coefficient. Ann. Inst. Fourier (Grenoble) **31**(2), 175–191 (1981)
8. L.C. Evans, J.-L. Menaldi, Gradient bounds for solutions of degenerate variational inequalities. Appl. Math. Optim. **7**(3), 247–252 (1981)
9. L.C. Evans, J.-L. Menaldi, Correction: "Gradient bounds for solutions of degenerate variational inequalities. Appl. Math. Optim. **8**(2), 197–197 (1982)
10. L.C. Evans, J. Spruck, Motion of level sets by mean curvature I. J. Differ. Geom. **33**, 635–681 (1991)
11. A. Friedman, *Partial Differential Equations of Parabolic Type* (Prentice-Hall, Englewood Cliffs, 1964)
12. Y. Giga, *Surface Evolution Equations, A Level Set Approach, Monographs in Mathematics*, vol. 99 (Birkhäuser Verlag, Basel, 2006)
13. K. Ishii, H. Kamata and S. Koike, Remarks on viscosity solutions for mean curvature flow with obstacles, mathematics for nonlinear phenomena-analysis and computation, in *Springer Proceedings in Mathematics and Statistics*, vol. 215 (Springer, Cham, 2017), pp. 83–103
14. S. Koike, A Beginner's Guide to the Theory of Viscosity Solutions, MSJ memoir. Math. Soc. Japan **13** (2004)
15. S. Koike, Viscosity Solutions-Around Comparison Principle-(*Japanese*), Kyoritsu Koza, vol. 23, (Kyoritsu Shuppan, 2016)
16. S. Koike, Regularity of solutions of obstacle problems-old & new-nonlinear partial differential equations for future applications. Springer Proc. Math. Stat. **346**, 205–243 (2021)
17. S. Koike, T. Kosugi, M. Naito, On the rate of convergence of solutions in free boundary problems via penalization. J. Math. Anal. Appl. **457**(1), 436–460 (2018)
18. D. Kinderlehrer, G. Stampacchia, An Introduction to Variational Inequalities and Their Applications. *Pure and Applied Mathematics*, vol. 88 (Academic Press, 1980)
19. O.A. Ladyženskaja, V.A. Solonnikov, N.N. Ural'ceva, Linear and Quasilinear Equations of Parabolic Type. Transl. Math. Monogr. Am. Math. Soc. Providence **23** (1968)
20. G.M. Lieberman, *Second Order Parabolic Differential Equations* (World Scientific Publishing, River Edge, 1996)
21. P. Lindqvist, *Notes on the Infinity Laplace Equation, Springer Briefs in Mathematics* (Springer, 2016)
22. P. Lindqvist, *Notes on the Stationary p-Laplace Equation, Springer Briefs in Mathematics* (Springer, 2019)
23. P.-L. Lions, P.E. Souganidis, Fully nonlinear second-order degenerate elliptic equations with large zeroth-order coefficients. Indiana Univ. Math. J. **42**(4), 1525–1543 (1993)
24. G. Mercier, M. Novaga, Mean curvature flow with obstacles: existence, uniqueness and regularity of solutions. Interfaces Free Bound. **17**(3), 399–426 (2015)
25. H. Mitake, H.V. Tran, Large-time behavior for obstacle problems for degenerate viscous Hamilton-Jacobi equations. Calc. Var. Partial Differ. Equ. **54**(2), 2039–2058 (2015)

26. O.A. Oleĭnik, E.V. Radkevič, *Second Order Equations with Nonnegative Characteristic Form* (Plenum Press, New York-London, 1973)
27. H.V. Tran, Adjoint methods for static Hamilton-Jacobi equations. Calc. Var. Partial Differ. Equ. **41**(3–4), 301–319 (2011)

On a Compatibility Condition for the Navier-Stokes Solutions in Maximal L^p-regularity Class

Hideo Kozono and Senjo Shimizu

Dedicated to Professor Tohru Ozawa on the occasion of his sixtieth birthday

Abstract We consider a compatibility condition on the initial data a and the external force f for the initial-boundary value problem of the Navier-Stokes equations with no-slip condition on $\partial\Omega$ in a bounded domain $\Omega \subset \mathbb{R}^n$. Our class of solutions is based on that of maximal L^s-regularity as $W^{1,s}(0, T; \mathcal{D}(A_r^{\varepsilon})) \cap L^s(0, T; \mathcal{D}(A_r^{1+\varepsilon}))$, where A_r denotes the Stokes operator in $L^r_\sigma(\Omega)$. We show that if the solution belongs to such a class for $\varepsilon > 1/s + 1/2r$, then a and f necessarily satisfy

$$A_r a + P_r(a \cdot \nabla a) = P_r f(0) \quad \text{on } \partial\Omega,$$

where P_r denotes the L^r-Helmholtz projection in Ω. Simultaneously, we construct the solution in such a class with $0 \le \varepsilon < 1/2r$ for a and f without such a compatibility condition as above provided $2/s + n/r = 3 + 2\varepsilon$.

Keywords Compatibility conditions · Navier-Stokes equations · Maximal L^s-regularity

2020 Mathematics Subject Classification 35Q30 · 76D05

H. Kozono
Department of Mathematics, Waseda University, Tokyo 169-8555, Japan
e-mail: kozono@waseda.jp; hideo.kozono@tohoku.ac.jp

Mathematical Research Center for Co-creative Society, Tohoku University, Sendai 980-8578, Japan

S. Shimizu (✉)
Department of Mathematics, Faculty of Science, Kyoto University, Kyoto 606-8502, Japan
e-mail: shimizu.senjo.5s@kyoto-u.ac.jp

S. Machihara (ed.), *Mathematical Physics and Its Interactions*, Springer Proceedings in Mathematics & Statistics 451, https://doi.org/10.1007/978-981-97-0364-7_4

1 Introduction

In a bounded domain $\Omega \subset \mathbb{R}^n (n \geq 2)$ with the smooth boundary $\partial\Omega$ we consider the initial boundary value problem of the Navier-Stokes equations;

$$\begin{cases} \dfrac{\partial u}{\partial t} - \Delta u + u \cdot \nabla u + \nabla \pi = f & \text{in } \Omega \times (0, T), \\ \operatorname{div} u = 0 & \text{in } \Omega \times (0, T), \\ u = 0 & \text{on } \partial\Omega \times (0, T), \\ u|_{t=0} = a & \text{in } \Omega, \end{cases} \tag{NS}$$

where $u = u(x, t) = (u_1(x, t), \ldots, u_n(x, t))$ and $\pi = \pi(x, t)$ denote the unknown velocity and the unknown pressure, respectively, while $a = a(x) = (a_1(x), \ldots, a_n(x))$ and $f = f(x, t) = (f_1(x, t), \ldots, f_n(x, t))$ are the given initial velocity and the given external force, respectively. In this paper, we consider a compatibility condition on a and f if the solution u of (NS) has more regularity in $W^{1,s}(0, T; \mathcal{D}(A_r^{\varepsilon})) \cap L^s(0, T; \mathcal{D}(A_r^{1+\varepsilon}))$, where $A_r = -P_r\Delta$ is the Stokes operator in $L^r_\sigma(\Omega)$ with P_r denoting the L^r-Helmholtz projection on Ω. This problem was first proposed by Rautmann [21]. Indeed, he considered the three dimensional case in $L^2_\sigma(\Omega)$ and proved that if the solution u is in $C([0, T]; \mathcal{D}(A_2^{1+\varepsilon}))$ for $\varepsilon > 1/4$, then a and f necessarily satisfy

$$A_2 a + P_2(a \cdot \nabla a) = P_2 f(0) \quad \text{on } \partial\Omega. \tag{1.1}$$

He also showed that if $a \in \mathcal{D}(A_2^{1+\varepsilon})$ and $Pf \in C([0, T]; \mathcal{D}(A_2^{\varepsilon}))$ for $0 \leq \varepsilon < 1/4$, then without any compatibility condition as (1.1), the solution u necessarily fulfills $u(t) \to a$ in the topology of $\mathcal{D}(A_2^{1+\varepsilon})$ as $t \to +0$. For more general cases of $u \in C([0, T]; \mathcal{D}(A_r^{1+\varepsilon}))$ for $n \geq 2$ and for $1 < r < \infty$, von Wahl [26, 27] proved the corresponding results to the threshold $\varepsilon = 1/2r$.

In the present paper, we consider another class of solutions such as

$$W^{1,s}(0, T; \mathcal{D}(A_r^{\varepsilon})) \cap L^s(0, T; \mathcal{D}(A_r^{1+\varepsilon})) \tag{1.2}$$

which is based on maximal L^s-regularity of the Stokes operator A_r. It has been proved by Giga-Sohr [15] that for every $a \in (L^r_\sigma(\Omega), \mathcal{D}(A_r))_{1-1/s,s} \equiv \mathcal{D}(A_{r,s}^{1-1/s})$ and $f \in L^s(0, T; L^r_\sigma(\Omega))$ with $2/s + n/r = 3$, there exists $0 < T' \leq T$ and a solution u of (NS) on $(0, T')$ such that $u \in W^{1,s}(0, T'; L^r_\sigma(\Omega)) \cap L^s(0, T'; \mathcal{D}(A_r))$. By the trace theorem in the space and time, such a solution u satisfies $u(t) \to a$ as $t \to +0$ in the topology of $\mathcal{D}(A_{r,s}^{1-1/s})$. Similarly, if the solution u has more regularity as (1.2), then it necessarily holds that $u(t) \to a$ as $t \to +0$ in the topology of $\mathcal{D}(A_{r,s}^{1+\varepsilon-1/s})$. We shall prove that if the solution u belongs to the class (1.2) for $\varepsilon > 1/s + 1/2r$, then a and f need to satisfy the compatibility condition (1.1). It seems a natural question whether or not for $0 \leq \varepsilon \leq 1/s + 1/2r$ such a compatibility condition on a and f is redundant if the solution u belongs to the class (1.2). We shall give a partially positive answer to this question provided $0 \le \varepsilon < 1/2r$. Unfortunately, it is still open how the situation is for $1/2r \le \varepsilon \le 1/s + 1/2r$.

Before stating our main results, we introduce some notations. $L^r(\Omega)$ stands for an L^r-vector or scalar functions on Ω with the norm $\|\cdot\|_{L^r(\Omega)}$ for $1 < r < \infty$. Let $C^\infty_{0,\sigma}(\Omega) = \{\varphi \in C^\infty_0(\Omega);\, \mathrm{div}\,\varphi = 0\}$. $L^r_\sigma(\Omega)$ denotes the closure of $C^\infty_{0,\sigma}(\Omega)$ in $L^r(\Omega)$. We denote by P_r the Helmholtz projection from $L^r(\Omega)$ onto $L^r_\sigma(\Omega)$. See, e.g., Fujiwara-Morimoto [9], Solonnikov [24] and Simader-Sohr [23]. The Stokes operator $A_r = -P_r\Delta$ with the domain $\mathcal{D}(A_r) = \{u \in H^{2,r}(\Omega);\, u|_{\partial\Omega} = 0\} \cap L^r_\sigma(\Omega)$ is defined. It is known by Giga [10, 11] that $-A_r$ generates the uniformly bounded holomorphic semigroup $\{e^{-tA_r}\}_{t\ge 0}$ and the domain $\mathcal{D}(A^\alpha_r)$ of the fractional power A^α_r for $0 \le \alpha \le 1$ is characterized as $\mathcal{D}(A^\alpha_r) = [L^r_\sigma(\Omega), \mathcal{D}(A_r)]_\alpha$, where $[X, Y]_\alpha$ denotes the complex interpolation space of the couple of Banach spaces X and Y. Furthermore, A_r satisfies L^s-maximal regularity theorem in $L^r_\sigma(\Omega)$ for $1 < s, r < \infty$. See e.g., Giga-Sohr [15]. Finally, for $1 < s, r < \infty$, $0 \le \alpha_0 < \alpha_1 < \infty$ and $0 < \theta < 1$ we define $\mathcal{D}(A^\alpha_{r,s})$ with $\alpha = (1-\theta)\alpha_0 + \theta\alpha_1$ by

$$\mathcal{D}(A^\alpha_{r,s}) \equiv (\mathcal{D}(A^{\alpha_0}_r), \mathcal{D}(A^{\alpha_1}_r))_{\theta,s},$$

where $(X, Y)_{\theta,s}$ denotes the real interpolation space. Since A_r has bounded pure imaginary powers A^{it}_r in $L^r_\sigma(\Omega)$ for $t \in \mathbb{R}$ (see also Giga [11]), it follows from Amann [1, III 4.7.5 Theorem] that $\mathcal{D}(A^\alpha_{r,s})$ is a Banach space with the graph norm of A^α_r in $B^0_{r,s}(\Omega)$.

Then the original system (NS) of equations is rewritten to the abstract evolution equation:

$$\begin{cases} \frac{du}{dt} + Au + P(u\cdot\nabla u) = Pf \ \text{on } (0, T), \\ u(0) = a. \end{cases} \tag{1.3}$$

Our first result on a compatibility condition on a and f now reads:

Theorem 1.1 *Let $n/2 < r < \infty$ and $\varepsilon > 1/s + 1/2r$. Let $a \in \mathcal{D}(A^{1+\varepsilon-\frac{1}{s}}_{r,s})$, $Pf \in L^s(0, T; \mathcal{D}(A^\varepsilon_r)) \cap C([0, T]; \mathcal{D}(A^{\varepsilon-1/s}_{r,s}))$. If u is a solution of* (1.3) *satisfying*

$$u \in W^{1,s}(0, T; \mathcal{D}(A^\varepsilon_r)) \cap L^s(0, T; \mathcal{D}(A^{1+\varepsilon}_r)), \tag{1.4}$$

then we have

$$Aa + P(a\cdot\nabla a) = Pf(0) \ \text{ in } \ B^{2(\varepsilon-\frac{1}{s})-\frac{1}{r}}_{r,s}(\partial\Omega), \tag{1.5}$$

where $B^\alpha_{r,s}(\partial\Omega) \equiv (H^{\alpha_0,r}(\partial\Omega), H^{\alpha_1,r}(\partial\Omega))_{\theta,s}$ with $\alpha = (1-\theta)\alpha_0 + \theta\alpha_1$ for $0 \le \alpha_0 < \alpha_1 < \infty$ and $0 < \theta < 1$.

In addition, if $f \in L^s(0, T; H^{2\varepsilon,r}(\Omega)) \cap C([0, T]; H^{2(\varepsilon-1/s),r}(\Omega))$, then it holds that

$$-\Delta a + \nabla p(0) = f(0) \ \text{ in } B^{2(\varepsilon-\frac{1}{s})-\frac{1}{r}}_{r,s}(\partial\Omega) \tag{1.6}$$

for some $p \in L^s(0, T; H^{1+2\varepsilon,r}(\Omega)) \cap C([0, T]; H^{1+2(\varepsilon-1/s),r}(\Omega))$.

Our second result is unique existence of solutions u to (1.3) in the class (1.4) having the scaling invariant property provided $0 \le \varepsilon < 1/2r$. We also show gain of more regularity such as (1.4) if u is a solution of (1.3) in the usual scaling invariant class of maximal L^s-regularity. In both cases, compatibility conditions such as (1.5) and (1.6) are redundant.

Theorem 1.2 (i) *Let* $1 < r, s < \infty$ *and* $\varepsilon \ge 0$ *satisfy*

$$1 < r < n, \quad 0 \le \varepsilon < \frac{1}{2r}, \quad \frac{2}{s} + \frac{n}{r} = 3 + 2\varepsilon. \tag{1.7}$$

For every $a \in \mathcal{D}(A_{r,s}^{1+\varepsilon-\frac{1}{s}})$ *and every* $Pf \in L^s(0,T;\mathcal{D}(A_r^{\varepsilon})) \cap C([0,T);\mathcal{D}(A_{r,s}^{\varepsilon-\frac{1}{s}}))$ *there exist* $0 < T' \le T$ *and a unique solution* u *of (1.3) on* $(0,T')$ *such that*

$$u \in W^{1,s}(0,T';\mathcal{D}(A_r^{\varepsilon})) \cap L^s(0,T';\mathcal{D}(A_r^{1+\varepsilon})). \tag{1.8}$$

Moreover, such a solution u *satisfies*

$$u(t) \to a \quad \text{in the norm of } \mathcal{D}(A_{r,s}^{1+\varepsilon-\frac{1}{s}}) \text{ as } t \to +0. \tag{1.9}$$

(ii) *Let*

$$a \in \mathcal{D}(A_{r_*,s_*}^{1-\frac{1}{s_*}}), \quad Pf \in L^{s_*}(0,T;L_\sigma^{r_*}(\Omega)) \quad \text{for} \frac{2}{s_*} + \frac{n}{r_*} = 3 \text{ with } n/3 < r_* < n. \tag{1.10}$$

Suppose that u *is a solution of (1.3) satisfying*

$$u \in W^{1,s_*}(0,T;L_\sigma^{r_*}(\Omega)) \cap L^{s_*}(0,T;\mathcal{D}(A_{r_*})). \tag{1.11}$$

Assume that $1 < s < \infty$ *and* $1 < r < n$ *satisfy*

$$2 < \frac{2}{s} + \frac{n}{r}, \quad 2 - \frac{2}{s} < \frac{n}{r_*} < \frac{n}{r} + 1. \tag{1.12}$$

In addition to (1.10), if $a \in \mathcal{D}(A_{r,s}^{1+\varepsilon-\frac{1}{s}})$ *and* $Pf \in L^s(0,T;\mathcal{D}(A_r^{\varepsilon})) \cap C([0,T);\mathcal{D}(A_{r,s}^{\varepsilon-\frac{1}{s}}))$ *for* $0 \le \varepsilon < 1/2r$*, then we have (1.8) with* T' *replaced by* T*, and (1.9).*

Remark 1.3 (i) There is a gap of ε between $1/2r$ and $1/s + 1/2r$ in Theorems 1.1 and 1.2. This is closely related to the fact that the projection P_r breaks down the homogeneous boundary condition of functions on $\partial\Omega$. Roughly speaking, we do not know whether $P_r f|_{\partial\Omega} = 0$ even though $f \in H_0^{1,r}(\Omega)$. More precisely, if $0 \le \varepsilon < 1/2r$, then $\mathcal{D}(A_r^{\varepsilon})$ does not have any boundary condition on $\partial\Omega$ so that $A_r^{\varepsilon} Pf$ is well-defined for all $f \in \mathcal{D}(A_r^{\varepsilon})$. However, if $\varepsilon > 1/2r$, then $\mathcal{D}(A_r^{\varepsilon}) = \{f \in H^{2\varepsilon,r}(\Omega); f|_{\partial\Omega} = 0\} \cap L_\sigma^r(\Omega)$(see e.g., Fujiwara [8, Theorem 6.7] and Giga [11, Theorem 3]). Hence, in the case when $1/2r \le \varepsilon \le 1/s + 1/2r$, we do not know whether or not $P(u \cdot \nabla u)$ belongs to $\mathcal{D}(A_r^{\varepsilon})$ for $u \in \mathcal{D}(A_r^{1+\varepsilon})$, which cases difficulty to construct the solution u of (1.3) in the class (1.8).

(ii) Rautmann [21] considered the compatibility condition (1.5) for the solution of (1.3) in the class

$$u \in C([0,T); L^2_\sigma(\Omega)) \cap C((0,T); \mathcal{D}(A_2)) \cap C^1((0,T); L^2_\sigma(\Omega)). \tag{1.13}$$

He showed that if $0 \le \varepsilon < 1/4$ and if $a \in \mathcal{D}(A_2^{1+\varepsilon})$ and $Pf \in C^{\mu}([0,T); \mathcal{D}(A_2^{\varepsilon}))$ with $0 < \mu < 1$, then such a solution u behaves like

$$u(t) \to a \quad \text{in } \mathcal{D}(A_2^{1+\varepsilon}) \text{ as } t \to +0. \tag{1.14}$$

However, if $\varepsilon > 1/4$ and if (1.14) holds, then a and f necessarily satisfy (1.5) in $H^{2\varepsilon-1/2,2}(\partial\Omega)$. Later on, von Wahl [26, 27] generalized such a result in the L^r-setting and proved that $\varepsilon = 1/2r$ is the threshold exponent whether or not the behavior like $u(t) \to a$ in $\mathcal{D}(A_r^{1+\varepsilon})$ as $t \to +0$ implies the compatibility condition (1.5) in $H^{2\varepsilon-1/r,r}(\partial\Omega)$. The class (1.13) is based on the analyticity of the semi-group $\{e^{-tA_r}\}_{t\ge 0}$ in $L^r_\sigma(\Omega)$. On the other hand, our class (1.4) is related to maximal L^s-regularity of A_r in $L^r_\sigma(\Omega)$. Theorem 1.1 means if $\varepsilon > 1/s + 1/2r$, then the class (1.4) necessarily yields the compatibility condition (1.5). Unfortunately, Theorem 1.2 does not give any information on its converse direction in the case $1/2r \le \varepsilon \le 1/s + 1/2r$.

The proof of Theorems 1.1 and 1.2 are given in Sects. 2 and 3, respcctively.

2 Proof of Theorem 1.1

For the proof of Theorem 1.1, the following is a key lemma.

Lemma 2.1 *Let $n/2 < r < \infty$ and $\varepsilon > 1/s + 1/2r$. If $u, v \in \mathcal{D}(A_{r,s}^{1+\varepsilon-\frac{1}{s}})$, then*

$$P(u \cdot \nabla v) \in B^{2(\varepsilon-\frac{1}{s})}_{r,s}(\Omega)$$

with the estimate

$$\|P(u \cdot \nabla v)\|_{B^{2(\varepsilon-\frac{1}{s})}_{r,s}(\Omega)} \le C\|Au\|_{B^{2(\varepsilon-\frac{1}{s})}_{r,s}(\Omega)}\|Av\|_{B^{2(\varepsilon-\frac{1}{s})}_{r,s}(\Omega)}, \tag{2.1}$$

where $C = C(n,s,r,\varepsilon)$.

Proof First of all, we remark that for every $m \in \mathbb{N}$ the boundedness of Helmholtz projection

$$\|Pf\|_{H^{m,r}(\Omega)} \le C\|f\|_{H^{m,r}(\Omega)} \tag{2.2}$$

holds, where $C = C(n,r,m)$. Indeed, for every $f \in H^{m,r}(\Omega)$, Pf is expressed as $Pf = f - \nabla q$, where q is the solution of the following Neumann problem

$$\begin{cases} \Delta q = \operatorname{div} f & \text{in } \Omega, \\ \dfrac{\partial q}{\partial \nu} = f \cdot \nu & \text{on } \partial\Omega. \end{cases}$$

It is known that such a solution q is subject to the estimate

$$\|\nabla q\|_{H^{m,r}(\Omega)} \le C \left(\|\operatorname{div} f\|_{H^{m-1,r}(\Omega)} + \|f \cdot \nu\|_{H^{m-1/r,r}(\partial\Omega)} \right) \le C \|f\|_{H^{m,r}(\Omega)},$$

which yields (2.2).

Since $\varepsilon > 1/s + 1/2r$, we may assume that $\varepsilon < 1/s + 1/2$, which yields that $1/r < 2(\varepsilon - 1/s) < 1$. We shall prove (2.1) by an interpolation method in the cases of $n/2 < r \le n$ and $n < r < \infty$.

(1) The case $n < r < \infty$. In the first step, we interpolate $\|P(u \cdot \nabla v)\|_{L^r(\Omega)}$ and $\|P(u \cdot \nabla v)\|_{H^{1,r}(\Omega)}$. By (2.2) with $m = 1$, we have

$$\|P(u \cdot \nabla v)\|_{H^{1,r}(\Omega)} \le C \left(\|u \cdot \nabla v\|_{L^r(\Omega)} + \|Du \cdot Dv\|_{L^r(\Omega)} + \|u \cdot D^2 v\|_{L^r(\Omega)} \right).$$

Since $n < r$, it holds that $\mathcal{D}(A_r^{1/2}) \hookrightarrow H^{1,r}(\Omega) \hookrightarrow L^\infty(\Omega)$ and that $\mathcal{D}(A_r) \hookrightarrow H^{2,r}(\Omega) \hookrightarrow H^{1,\infty}(\Omega)$, so we have

$$\begin{aligned} \|u \cdot D^2 v\|_{L^r(\Omega)} &\le \|u\|_{L^\infty(\Omega)} \|D^2 v\|_{L^r(\Omega)} \le C \|A^{1/2} u\|_{L^r(\Omega)} \|Av\|_{L^r(\Omega)}, \\ \|u \cdot \nabla v\|_{L^r(\Omega)} &\le \|u\|_{L^r(\Omega)} \|\nabla v\|_{L^\infty(\Omega)} \le C \|u\|_{L^r(\Omega)} \|Av\|_{L^r(\Omega)}, \\ \|Du \cdot Dv\|_{L^r(\Omega)} &\le \|Du\|_{L^r(\Omega)} \|Dv\|_{L^\infty(\Omega)} \le C \|A^{1/2} u\|_{L^r(\Omega)} \|Av\|_{L^r(\Omega)}, \end{aligned}$$

which yields

$$\|P(u \cdot \nabla v)\|_{H^{1,r}(\Omega)} \le C \|A^{1/2} u\|_{L^r(\Omega)} \|Av\|_{L^r(\Omega)}. \tag{2.3}$$

Similarly, we have

$$\|P(u \cdot \nabla v)\|_{L^r(\Omega)} \le C \|u \cdot \nabla v\|_{L^r(\Omega)} \le C \|u\|_{L^r(\Omega)} \|\nabla v\|_{L^\infty(\Omega)} \le C \|u\|_{L^r(\Omega)} \|Av\|_{L^r(\Omega)}. \tag{2.4}$$

We obtain from (2.4) and (2.3) that for fixed $v \in \mathcal{D}(A_r)$

$$\begin{aligned} &L^r_\sigma(\Omega) \ni u \mapsto P(u \cdot \nabla v) \in L^r(\Omega) \quad \text{with the bound } \|Av\|_{L^r(\Omega)}, \\ &\mathcal{D}(A_r^{1/2}) \ni u \mapsto P(u \cdot \nabla v) \in H^{1,r}(\Omega) \quad \text{with the bound } \|Av\|_{L^r(\Omega)}. \end{aligned}$$

Hence, it follows from the real interpolation that for $0 < \theta < 1$ and $1 \le s \le \infty$

$$(L^r(\Omega), \mathcal{D}(A_r^{1/2}))_{\theta,s} \ni u \mapsto P(u \cdot \nabla v) \in (L^r(\Omega), H^{1,r}(\Omega))_{\theta,s}$$

with the estimate

$$\|P(u \cdot \nabla v)\|_{B^\theta_{r,s}(\Omega)} \le C \|A^{\theta/2} u\|_{B^0_{r,s}(\Omega)} \|Av\|_{L^r(\Omega)}. \tag{2.5}$$

In the next step, we interpolate $\|P(u \cdot \nabla v)\|_{L^r(\Omega)}$ and $\|P(u \cdot \nabla v)\|_{H^{2,r}(\Omega)}$. By (2.2) with $m = 2$, we have

$$\|P(u \cdot \nabla v)\|_{H^{2,r}(\Omega)} \le C(\|u \cdot \nabla v\|_{H^{1,r}(\Omega)} + \|2Du \cdot D^2 v + u \cdot D^3 v + D^2 u \cdot Dv\|_{L^r(\Omega)}).$$

Since $n < r < \infty$, it holds that $\mathcal{D}(A_r^{1+1/2}) \hookrightarrow H^{3,r}(\Omega) \hookrightarrow H^{2,\infty}(\Omega)$, which yields that

$$\|Du \cdot D^2 v\|_{L^r(\Omega)} \le \|Du\|_{L^r(\Omega)} \|D^2 v\|_{L^\infty(\Omega)} \le C\|A^{1/2}u\|_{L^r(\Omega)} \|A^{1+1/2} v\|_{L^r(\Omega)}.$$

Again by embedding $\mathcal{D}(A_r^{1/2}) \hookrightarrow H^{1,r}(\Omega) \hookrightarrow L^\infty(\Omega), \mathcal{D}(A_r) \hookrightarrow H^{2,r}(\Omega) \hookrightarrow H^{1,\infty}$ (Ω) for $n < r < \infty$, we have

$$\|u \cdot D^3 v\|_{L^r(\Omega)} \le \|u\|_{L^\infty(\Omega)} \|D^3 v\|_{L^r(\Omega)} \le C\|A^{1/2}u\|_{L^r(\Omega)} \|A^{1+1/2} v\|_{L^r(\Omega)},$$
$$\|D^2 u \cdot Dv\|_{L^r(\Omega)} \le \|D^2 u\|_{L^r(\Omega)} \|Dv\|_{L^\infty(\Omega)} \le C\|Au\|_{L^r(\Omega)} \|Av\|_{L^r(\Omega)}.$$

Combining the above estimates with (2.3), we obtain

$$\|P(u \cdot \nabla v)\|_{H^{2,r}(\Omega)} \le C\|Au\|_{L^r(\Omega)} \|A^{1+1/2} v\|_{L^r(\Omega)}. \tag{2.6}$$

Then it follows from (2.4) and (2.6) that for fixed $v \in \mathcal{D}(A_r^{1+1/2})$

$$L^r_\sigma(\Omega) \ni u \mapsto P(u \cdot \nabla v) \in L^r(\Omega) \qquad \text{with bound} \|Av\|_{L^r(\Omega)},$$
$$\mathcal{D}(A_r) \ni u \mapsto P(u \cdot \nabla v) \in H^{2,r}(\Omega) \qquad \text{with bound } \|A^{1+1/2} v\|_{L^r(\Omega)},$$

which yields by the real interpolation that for $0 < \eta < 1$ and $1 \le s \le \infty$

$$(L^r_\sigma(\Omega), \mathcal{D}(A_r))_{\eta,s} \ni u \mapsto P(u \cdot \nabla v) \in (L^r(\Omega), H^{2,r}(\Omega))_{\eta,s}$$

with the estimate

$$\begin{aligned}\|P(u \cdot \nabla v)\|_{B^{2\eta}_{r,s}(\Omega)} &\le C\|A^\eta u\|_{B^0_{r,s}(\Omega)} \|Av\|^{1-\eta}_{L^r(\Omega)} \|A^{1+1/2} v\|^\eta_{L^r(\Omega)} \\ &\le C\|A^\eta u\|_{B^0_{r,s}(\Omega)} \|A^{1+1/2} v\|_{L^r(\Omega)}.\end{aligned} \tag{2.7}$$

By (2.5) and (2.7) we see that

$$\mathcal{D}(A_r) \ni v \mapsto P(u \cdot \nabla v) \in B^\theta_{r,s}(\Omega) \qquad \text{with bound } \|A^{\frac{\theta}{2}} u\|_{B^0_{r,s}(\Omega)},$$
$$\mathcal{D}(A_r^{1+\frac{1}{2}}) \ni v \mapsto P(u \cdot \nabla v) \in B^{2\eta}_{r,s}(\Omega) \qquad \text{with bound } \|A^\eta u\|_{L^r(\Omega)},$$

which yields by the real interpolation that for $0 < \mu < 1$ and $1 \le s \le \infty$

$$(\mathcal{D}(A_r), \mathcal{D}(A_r^{1+\frac{1}{2}}))_{\mu,s} \ni v \mapsto P(u \cdot \nabla v) \in (B^\theta_{r,s}(\Omega), B^{2\eta}_{r,s}(\Omega))_{\mu,s}$$

with the estimate

$$\|P(u\cdot\nabla v)\|_{B^{(1-\mu)\theta+2\mu\eta}_{r,s}(\Omega)} \le C\|A^{\frac{\theta}{2}}u\|^{1-\mu}_{B^0_{r,s}(\Omega)}\|A^{\eta}u\|^{\mu}_{B^0_{r,s}(\Omega)}\|A^{1+\frac{\mu}{2}}v\|_{B^0_{r,s}(\Omega)}$$
$$\le C\|A^{\max\{\frac{\theta}{2},\eta\}}u\|_{B^0_{r,s}(\Omega)}\|A^{1+\frac{\mu}{2}}v\|_{B^0_{r,s}(\Omega)}.$$

Taking $\eta=\theta/2$ and $\mu=\theta=2(\varepsilon-1/s)\in(1/r,1)$ in the above estimate, we have

$$\|P(u\cdot\nabla v)\|_{B^{2(\varepsilon-\frac{1}{s})}_{r,s}(\Omega)} \le C\|u\|_{B^{2(\varepsilon-\frac{1}{s})}_{r,s}(\Omega)}\|Av\|_{B^{2(\varepsilon-\frac{1}{s})}_{r,s}(\Omega)},$$

which implies (2.1).

(2) The case $n/2<r\le n$. In the same way as in the case $n<r<\infty$, we first interpolate $\|P(u\cdot\nabla v)\|_{L^r(\Omega)}$ and $\|P(u\cdot\nabla v)\|_{H^{1,r}(\Omega)}$. Assume that $n/2<r<n$. We define $r<r_*<\infty$ and $0<\alpha_r<1/2$ by $1/r_*=1/r-1/n$ and $\alpha_r=\frac{n}{2}(\frac{1}{r}-\frac{1}{n})$, respectively. Since $n/2<r<n$, we have $\mathcal{D}(A_r)\hookrightarrow H^{2,r}(\Omega)\hookrightarrow H^{1,r_*}(\Omega)$. Moreover, it holds that $\mathcal{D}(A_r^{\alpha_r})\hookrightarrow H^{2\alpha_r,r}(\Omega)\hookrightarrow L^n(\Omega)$. Hence it follows from the Hölder inequality that

$$\|P(u\cdot\nabla v)\|_{L^r(\Omega)}\le C\|u\cdot\nabla v\|_{L^r(\Omega)}\le C\|u\|_{L^n(\Omega)}\|\nabla v\|_{L^{r_*}(\Omega)}\le C\|A^{\alpha_r}u\|_{L^r(\Omega)}\|Av\|_{L^r(\Omega)}. \tag{2.8}$$

In the case that $r=n$, we take some $n<q<\infty$ and define $0<\alpha<1/2$ and $n<p<\infty$ by $\alpha\equiv n/2q$ and $1/p\equiv 1/n-1/q$, respectively. Since $\mathcal{D}(A_n)\hookrightarrow H^{2,n}(\Omega)\hookrightarrow H^{1,q}(\Omega)$ and since $\mathcal{D}(A_n^{\alpha})\hookrightarrow H^{2\alpha,n}(\Omega)\hookrightarrow L^p(\Omega)$, it follows from the Hölder inequality that

$$\|P(u\cdot\nabla v)\|_{L^n(\Omega)}\le C\|u\|_{L^p(\Omega)}\|\nabla v\|_{L^q(\Omega)}\le C\|A^{\alpha}u\|_{L^n(\Omega)}\|Av\|_{L^n(\Omega)}. \tag{2.9}$$

By (2.8) and (2.9), we have

$$\|P(u\cdot\nabla v)\|_{L^r(\Omega)}\le C\begin{cases}\|A^{\alpha_r}u\|_{L^r(\Omega)}\|Av\|_{L^r(\Omega)}, & n/2<r<n,\\ \|A^{\frac{n}{2q}}u\|_{L^n(\Omega)}\|Av\|_{L^n(\Omega)}, & r=n.\end{cases} \tag{2.10}$$

Hereafter we simply write

$$\alpha_r=\begin{cases}\frac{n}{2}\left(\frac{1}{r}-\frac{1}{n}\right), & n/2<r<n,\\ \frac{n}{2q}, & r=n.\end{cases}$$

Notice that $0<\alpha_r<1/2$.

Next we see

$$\|P(u\cdot\nabla v)\|_{H^{1,r}(\Omega)}\le C\left(\|u\cdot\nabla v\|_{L^r(\Omega)}+\|Du\cdot Dv\|_{L^r(\Omega)}+\|u\cdot D^2v\|_{L^r(\Omega)}\right).$$

Since $n/2<r\le n$, it holds that $\mathcal{D}(A_r)\hookrightarrow H^{2,r}(\Omega)\hookrightarrow L^\infty(\Omega)$, which yields that

$$\|u \cdot D^2 v\|_{L^r(\Omega)} \leq \|u\|_{L^\infty(\Omega)} \|D^2 v\|_{L^r(\Omega)} \leq C\|Au\|_{L^r(\Omega)} \|Av\|_{L^r(\Omega)}.$$

Assume that $n/2 < r < n$. Define r_* by $1/r_* = 1/r - 1/n$ and we have $\mathcal{D}(A_r) \hookrightarrow H^{2,r}(\Omega) \hookrightarrow H^{1,r_*}(\Omega)$. Moreover, since $\mathcal{D}(A_r^{1/2+\alpha_r}) \hookrightarrow H^{1+2\alpha_r,r}(\Omega) \hookrightarrow H^{1,n}(\Omega)$, it follows from the Hölder inequality that

$$\|Du \cdot Dv\|_{L^r(\Omega)} \leq \|Du\|_{L^n(\Omega)} \|Dv\|_{L^{r_*}(\Omega)} \leq C\|A^{\frac{1}{2}+\alpha_r} u\|_{L^r(\Omega)} \|Av\|_{L^r(\Omega)}.$$

Similarly to (2.8), we also have

$$\|u \cdot \nabla v\|_{L^r(\Omega)} \leq \|A^{\alpha_r} u\|_{L^r(\Omega)} \|Av\|_{L^r(\Omega)}.$$

For $r = n$, we define $1/p = 1/n - 1/q$ and $\alpha = n/2q$ with some $n < q < \infty$. Since $\mathcal{D}(A_n^{1/2+\alpha}) \hookrightarrow H^{1+2\alpha,n}(\Omega) \hookrightarrow H^{1,p}(\Omega)$ and since $\mathcal{D}(A_n) \hookrightarrow H^{1,q}(\Omega)$, it follows from the Hölder inequality that

$$\begin{aligned}\|Du \cdot Dv\|_{L^n(\Omega)} &\leq \|Du\|_{L^p(\Omega)} \|Dv\|_{L^q(\Omega)} \leq C\|A^{\frac{1}{2}+\alpha} u\|_{L^n(\Omega)} \|Av\|_{L^n(\Omega)} \\ &\leq C\|Au\|_{L^n(\Omega)} \|Av\|_{L^n(\Omega)}.\end{aligned}$$

Therefore we have for $n/2 < r \leq n$,

$$\|P(u \cdot \nabla v)\|_{H^{1,r}(\Omega)} \leq C\|Au\|_{L^r(\Omega)} \|Av\|_{L^r(\Omega)}. \tag{2.11}$$

Then it follows from (2.10) and (2.11) that for fixed $v \in \mathcal{D}(A_r)$,

$$\begin{aligned}\mathcal{D}(A^{\alpha_r}) \ni u &\mapsto P(u \cdot \nabla v) \in L^r(\Omega) \qquad \text{with bound } \|Av\|_{L^r(\Omega)}, \\ \mathcal{D}(A_r) \ni u &\mapsto P(u \cdot \nabla v) \in H^{1,r}(\Omega) \qquad \text{with bound } \|Av\|_{L^r(\Omega)},\end{aligned}$$

which yields by the real interpolation that for $0 < \theta < 1$ and $1 \leq s \leq \infty$

$$(\mathcal{D}(A_r^{\alpha_r}), \mathcal{D}(A_r))_{\theta,s} \ni u \mapsto P(u \cdot \nabla v) \in (L^r(\Omega), H^{1,r}(\Omega))_{\theta,s}$$

with the estimate

$$\|P(u \cdot \nabla v)\|_{B^\theta_{r,s}(\Omega)} \leq C\|A^{(1-\theta)\alpha_r+\theta} u\|_{B^0_{r,s}(\Omega)} \|Av\|_{L^r(\Omega)}. \tag{2.12}$$

In the next step, we interpolate $\|P(u \cdot \nabla v)\|_{L^r(\Omega)}$ and $\|P(u \cdot \nabla v)\|_{H^{2,r}(\Omega)}$. In the case when $n/2 < r < n$, taking $1/r_* = 1/r - 1/n \in (0, 1)$ and using the Hölder inequality we have

$$\begin{aligned}\|D^2(u \cdot \nabla v)\|_{L^r(\Omega)} &\leq \|D^2 u \cdot Du\|_{L^r(\Omega)} + 2\|Du \cdot D^2 v\|_{L^r(\Omega)} + \|u \cdot D^3 v\|_{L^r(\Omega)} \\ &\leq \|D^2 u\|_{L^r(\Omega)} \|Dv\|_{L^\infty(\Omega)} + 2\|Du\|_{L^n(\Omega)} \|D^2 v\|_{L^{r_*}(\Omega)} \\ &\quad + \|u\|_{L^\infty(\Omega)} \|D^3 v\|_{L^r(\Omega)}.\end{aligned}$$

Since $\mathcal{D}(A_r) \hookrightarrow H^{2,r}(\Omega) \hookrightarrow H^{1,n}(\Omega) \cap L^\infty(\Omega)$ and since $\mathcal{D}(A_r^{1+1/2}) \hookrightarrow H^{3,r}(\Omega) \hookrightarrow H^{2,r_*}(\Omega) \hookrightarrow H^{1,\infty}(\Omega)$, implied by $n/2 < r < n < r_*$, we obtain from the above estimate that

$$\|D^2(u \cdot \nabla v)\|_{L^r(\Omega)} \le C\|Au\|_{L^r(\Omega)}\|A^{1+\frac{1}{2}}v\|_{L^r(\Omega)},$$

which yields with the aid of (2.2) and (2.11) that

$$\|P(u \cdot \nabla v)\|_{H^{2,r}(\Omega)} \le C\|Au\|_{L^r(\Omega)}\|A^{1+\frac{1}{2}}v\|_{L^r(\Omega)}. \tag{2.13}$$

In the case $r = n$, taking $n < p, q < \infty$ such as $1/n = 1/p + 1/q$, we have by the Hölder inequality that

$$\begin{aligned}\|D^2(u \cdot \nabla v)\|_{L^n(\Omega)} \le \|D^2u\|_{L^n(\Omega)}\|Dv\|_{L^\infty(\Omega)} + 2\|Du\|_{L^p(\Omega)}\|D^2v\|_{L^q(\Omega)}\\ + \|u\|_{L^\infty(\Omega)}\|D^3v\|_{L^n(\Omega)}.\end{aligned}$$

Since $D(A_n) \hookrightarrow H^{2,n}(\Omega) \hookrightarrow H^{1,p}(\Omega) \cap L^\infty(\Omega)$ and since $D(A_n^{1+\frac{1}{2}}) \hookrightarrow H^{3,n}(\Omega) \hookrightarrow H^{2,q}(\Omega) \cap H^{1,\infty}(\Omega)$, we obtain from the above estimate that

$$\|D^2(u \cdot \nabla v)\|_{L^n(\Omega)} \le C\|Au\|_{L^n(\Omega)}\|A^{1+\frac{1}{2}}v\|_{L^n(\Omega)},$$

which yields with the aid of (2.11) that

$$\|P(u \cdot \nabla v)\|_{H^{2,n}(\Omega)} \le C\|Au\|_{L^n(\Omega)}\|A^{1+\frac{1}{2}}v\|_{L^n(\Omega)}. \tag{2.14}$$

Thus by (2.13) and (2.14), we have for $n/2 < r \le n$

$$\|P(u \cdot \nabla v)\|_{H^{2,r}(\Omega)} \le C\|Au\|_{L^r(\Omega)}\|A^{1+\frac{1}{2}}v\|_{L^r(\Omega)}. \tag{2.15}$$

Now it follows from (2.10) and (2.15) that for fixed $v \in \mathcal{D}(A_r^{1+1/2})$

$$\begin{aligned}\mathcal{D}(A_r^{\alpha_r}) \ni u \mapsto P(u \cdot \nabla v) \in L^r(\Omega) \quad &\text{with bound } C\|Av\|_{L^r(\Omega)},\\ \mathcal{D}(A_r) \ni u \mapsto P(u \cdot \nabla v) \in H^{2,r}(\Omega) \quad &\text{with bound } C\|A^{1+1/2}v\|_{L^r(\Omega)},\end{aligned}$$

it holds that for $0 < \eta < 1$ and $1 \le s \le \infty$

$$(\mathcal{D}(A_r^{\alpha_r}), \mathcal{D}(A_r))_{\eta,s} \ni u \mapsto P(u \cdot \nabla v) \in (L^r(\Omega), H^{2,r}(\Omega))_{\eta,s}$$

with the estimate

$$\begin{aligned}\|P(u \cdot \nabla v)\|_{B^{2\eta}_{r,s}(\Omega)} &\le C\|A^{(1-\eta)\alpha_r+\eta}u\|_{B^0_{r,s}(\Omega)}\|Av\|^{1-\eta}_{L^r(\Omega)}\|A^{1+1/2}v\|^{\eta}_{L^r(\Omega)}\\ &\le C\|A^{(1-\eta)\alpha_r+\eta}u\|_{B^0_{r,s}(\Omega)}\|A^{1+1/2}v\|_{L^r(\Omega)}.\end{aligned} \tag{2.16}$$

Finally, we interpolate (2.12) and (2.16). Indeed, we see by (2.12) that for fixed $u \in \mathcal{D}(A_{r,s}^{(1-\theta)\alpha_r+\theta})$

$$\mathcal{D}(A_r) \ni v \mapsto P(u \cdot \nabla v) \in B^{\theta}_{r,s}(\Omega) \quad \text{with bound } C\|A^{(1-\theta)\alpha_r+\theta}u\|_{B^0_{r,s}(\Omega)}.$$

Similarly, we see by (2.16) that for fixed $u \in \mathcal{D}(A_{r,s}^{(1-\eta)\alpha_r+\eta})$

$$\mathcal{D}(A_r^{1+1/2}) \ni v \mapsto P(u \cdot \nabla v) \in B^{2\eta}_{r,s}(\Omega) \quad \text{with bound } C\|A^{(1-\eta)\alpha_r+\eta}u\|_{B^0_{r,s}(\Omega)}.$$

Hence it follows from the real interpolation that for $0 < \mu < 1$ and $1 \le s \le \infty$

$$(\mathcal{D}(A_r), \mathcal{D}(A_r^{1+1/2}))_{\mu,s} \ni v \mapsto P(u \cdot \nabla v) \in (B^{\theta}_{r,s}(\Omega), B^{2\eta}_{r,s}(\Omega))_{\mu,s}$$

with the estimate

$$\begin{aligned}\|P(u \cdot \nabla v)\|_{B^{(1-\mu)\theta+2\eta\mu}_{r,s}(\Omega)} &\le C\|A^{(1-\theta)\alpha_r+\theta}u\|^{1-\mu}_{B^0_{r,s}(\Omega)}\|A^{(1-\eta)\alpha_r+\eta}u\|^{\mu}_{B^0_{r,s}(\Omega)}\|A^{1+\frac{\mu}{2}}v\|_{B^0_{r,s}(\Omega)}\\ &\le C\|A^{\max\{(1-\theta)\alpha_r+\theta,(1-\eta)\alpha_r+\eta\}}u\|_{B^0_{r,s}(\Omega)}\|A^{1+\frac{\mu}{2}}v\|_{B^0_{r,s}(\Omega)}.\end{aligned}$$

If we take $\eta = \theta/2$ and $\mu = \theta = 2(\varepsilon - 1/s) \in (1/r, 1)$, then it holds that

$$\begin{aligned}\max\{(1-\theta)\alpha_r + \theta, (1-\eta)\alpha_r + \eta\} &= (1-\theta)\alpha_r + \theta\\ &= \alpha_r + 2(\varepsilon - \frac{1}{s})(1-\alpha_r) < 1\end{aligned}$$

and therefore we obtain (2.1). This completes the proof of Lemma 2.1. □

Proof of Theorem 1.1. By (1.4), it holds that

$$u \in BUC\big([0,T); (\mathcal{D}(A_r^{\varepsilon}), \mathcal{D}(A_r^{1+\varepsilon}))_{1-\frac{1}{s},s}\big) \subset BUC\big([0,T); B^{2+2(\varepsilon-1/s)}_{r,s}(\Omega)\big). \tag{2.17}$$

Since u is a solution of (1.3), we have by (2.17) and Lemma 2.1 that

$$\partial_t u = -Au - P(u \cdot \nabla u) + Pf \in BUC([0,T); B^{2(\varepsilon-1/s)}_{r,s}(\Omega)). \tag{2.18}$$

Since $2(\varepsilon - 1/s) - 1/r > 0$, the trace operator $\gamma g \equiv g|_{\partial\Omega}$ extends the continuous mapping

$$\gamma : B^{2(\varepsilon-1/s)}_{r,s}(\Omega) \to B^{2(\varepsilon-1/s)-1/r}_{r,s}(\partial\Omega),$$

from which and (2.18) it follows that we have

$$\gamma(\partial_t u) = \gamma(-Au - P(u \cdot \nabla u) + Pf) \in BUC([0,T); B^{2(\varepsilon-1/s)-1/r}_{r,s}(\partial\Omega)). \tag{2.19}$$

Since $\gamma u(t) = 0$, we have $\gamma(\partial_t u(t)) = 0$ for $t \in (0,T)$, and hence it holds by (2.19) that

$$\gamma(-Au(t) - P(u \cdot \nabla u)(t) + Pf(t)) = 0 \quad \text{in } B^{2(\varepsilon-1/s)-1/r}_{r,s}(\partial\Omega) \text{ for all } t \in (0, T). \tag{2.20}$$

Now, letting $t \to +0$ in (2.20), again from (2.19) we obtain the desired compatibility condition (1.5). Furthermore, if $f \in L^s(0, T; H^{2\varepsilon,r}(\Omega)) \cap C([0, T]; H^{2(\varepsilon-1/s),r}(\Omega))$, then it follows from (1.3) and (2.2) that there exists some $p \in L^s(0, T; H^{1+2\varepsilon,r}(\Omega)) \cap C([0, T]; H^{1+2(\varepsilon-1/s),r}(\Omega))$ such that

$$\partial_t u(t) - \Delta u(t) + u \cdot \nabla u(t) + \nabla p(t) = f(t) \quad \text{for all } t \in (0, T). \tag{2.21}$$

Since $\gamma(\partial_t u(t)) = \gamma(u \cdot \nabla u)(t) = 0$ for $t \in (0, T)$, we have by (2.21) that

$$\gamma(-\Delta u(t) + \nabla p(t)) = \gamma f(t) \quad \text{for all } t \in (0, T).$$

Letting $t \to +0$ in the above identity, we have the desired compatibility condition (1.6). This completes the proof of Theorem 1.1. □

3 Proof of Theorem 1.2

We first show L^s-maximal regularity of A_r in the Banach space $\mathcal{D}(A_r^\varepsilon)$ with the norm $\|u\|_{\mathcal{D}(A_r^\varepsilon)} = \|A_r^\varepsilon u\|_{L^r(\Omega)}$. Notice that $0 \in \rho(A_r)$ with $\rho(A_r)$ denoting the resolvent set of A_r in $L^r_\sigma(\Omega)$.

Lemma 3.1 *Let* $1 < s,\ r < \infty$, $0 < \varepsilon < 1$ *and* $0 < T \le \infty$. *For every* $a \in \mathcal{D}(A^{1+\varepsilon-1/s}_{r,s})$ *and* $Pf \in L^s(0, T; \mathcal{D}(A_r^\varepsilon))$, *there exists a unique solution*

$$u \in W^{1,s}(0, T; \mathcal{D}(A_r^\varepsilon)) \cap L^s(0, T; \mathcal{D}(A_r^{1+\varepsilon}))$$

to

$$\begin{cases} \dfrac{du}{dt} + Au = Pf & \text{on } (0, T), \\ u(0) = a \end{cases}$$

with the estimate

$$\|\partial_t u\|_{L^s(0,T;\mathcal{D}(A_r^\varepsilon))} + \|Au\|_{L^s(0,T;\mathcal{D}(A_r^\varepsilon))} \le C\left(\|a\|_{\mathcal{D}(A_{r,s}^{1+\varepsilon-\frac{1}{s}})} + \|Pf\|_{L^s(0,T;\mathcal{D}(A_r^\varepsilon))}\right) \tag{3.1}$$

with the constant $C = C(n, r, s, \varepsilon)$.

Proof The assertion of this lemma implies that A_r has maximal L^s-regularity in the Banach space $\mathcal{D}(A_r^\varepsilon)$, and hence it suffices to show that $\lambda(\lambda + A_r)^{-1}$ is $\mathcal{R}$-bounded in $\mathcal{D}(A_r^\varepsilon)$ for $\lambda \in \Sigma_\omega$, where $\Sigma_\omega \equiv \{z \in \mathbb{C};\ |\arg z| < \pi - \omega,\ z \ne 0\}$ for some $\omega \in (0, \pi/2)$. Since $\lambda(\lambda + A_r)^{-1}$ is $\mathcal{R}$-bounded in $L^r_\sigma(\Omega)$ for $\lambda \in \Sigma_\omega$, we may show that

$$(\lambda + A_r)^{-1} A_r^\varepsilon f = A_r^\varepsilon (\lambda + A_r)^{-1} f \tag{3.2}$$

for all $f \in \mathcal{D}(A_r^{\varepsilon})$ and all $\lambda \in \Sigma_\omega$. Since Ω is a bounded domain, A_r has the bounded inverse in $L^r_\sigma(\Omega)$ and hence $A_r^{-\varepsilon}$ has an expression

$$A_r^{-\varepsilon} = \frac{\sin \pi\varepsilon}{\pi} \int_0^\infty \mu^{-\varepsilon}(\mu + A_r)^{-1}\, d\mu.$$

Since $(\mu + A_r)^{-1}$ with $\mu \in (0, \infty)$ commutes $(\lambda + A_r)^{-1}$ for all $\lambda \in \Sigma_\omega$, it holds that

$$A_r^{-\varepsilon}(\lambda + A_r)^{-1}A_r^{\varepsilon}f = (\lambda + A_r)^{-1}A_r^{-\varepsilon}A_r^{\varepsilon}f = (\lambda + A_r)^{-1}f, \quad f \in \mathcal{D}(A_r^{\varepsilon}), \ \lambda \in \Sigma_\omega,$$

which yields (3.2). This proves Lemma 3.1. □

In what follows, for simplicity of notation, we shall abbreviate the basic norm of maximal regularity as

$$\|\partial_t u, Au\|_{L^{r,\alpha}(0,T;X)} \equiv \|\partial_t u\|_{L^{r,\alpha}(0,T;X)} + \|Au\|_{L^{r,\alpha}(0,T;X)}.$$

We next establish the following bilinear estimate for the nonlinear term of (3.38). □

Lemma 3.2 *Let* $1 < s,\ r < \infty$ *and* $\varepsilon > 0$ *satisfy*

$$1 < r < n, \quad 0 < \varepsilon < \frac{1}{2r}, \quad \frac{2}{s} + \frac{n}{r} = 3 + 2\varepsilon. \tag{3.3}$$

For every u *and* v *such that* $\partial_t u, Au, \partial_t v, Av \in L^s(0, T; \mathcal{D}(A_r^{\varepsilon}))$, $0 < T \le \infty$ *with* $u(0), v(0) \in \mathcal{D}(A_{r,s}^{1+\varepsilon-1/s})$, *we have* $P(u \cdot \nabla v) \in L^s(0, T; \mathcal{D}(A_r^{\varepsilon}))$ *with the estimate*

$$\begin{aligned} &\|A^{\varepsilon}P(u \cdot \nabla v)\|_{L^s(0,T;L^r(\Omega))} \\ &\le C(\|e^{-tA}u(0)\|_{L^{s_0}(0,T;L^{r_0}(\Omega))} + \|\nabla e^{-tA}u(0)\|_{L^{s_0}(0,T;L^{\tilde r_0}(\Omega))} + \|\partial_t u, Au\|_{L^s(0,T;\mathcal{D}(A_r^{\varepsilon}))}) \\ &\quad \times (\|e^{-tA}v(0)\|_{L^{s_1}(0,T;H^{1+2\varepsilon,r_1}(\Omega))} + \|\partial_t v, Av\|_{L^s(0,T;\mathcal{D}(A_r^{\varepsilon}))}) \end{aligned} \tag{3.4}$$

and

$$\|e^{-tA}u(0)\|_{L^{s_0}(0,T;L^{r_0}(\Omega))} + \|\nabla e^{-tA}u(0)\|_{L^{s_0}(0,T;L^{\tilde r_0}(\Omega))} \le C\|u(0)\|_{\mathcal{D}(A_{r,s}^{1+\varepsilon-\frac{1}{s}})}, \tag{3.5}$$

$$\|e^{-tA}v(0)\|_{L^{s_1}(0,T;H^{1+2\varepsilon,r_1}(\Omega))} \le C\|v(0)\|_{\mathcal{D}(A_{r,s}^{1+\varepsilon-\frac{1}{s}})} \tag{3.6}$$

for some $r_0, \tilde r_0, r_1 > r$ *and* $s_0, s_1 > s$ *satisfying* $2/s_0 + n/r_0 = 1$, $2/s_0 + n/\tilde r_0 = 2$ *and* $2/s_1 + n/r_1 = 2 + 2\varepsilon$. *Here* $C = C(n, s, r, \varepsilon, r_0)$ *is a constant independent of* T.

In order to prove Lemma 3.2 we use the following proposition.

Proposition 3.3 *Let* $1 < s,\ r < \infty$ *and* $0 \le \varepsilon < 1/2r$.
(i) *For* $a \in \mathcal{D}(A_{r,s}^{1+\varepsilon-1/s})$, *it holds that*

$$e^{-tA}a \in L^{s_0,s}(0,\infty; H^{\beta_0,r_0}(\Omega)) \tag{3.7}$$

for

$$\frac{2}{s_0} + \frac{n}{r_0} - \beta_0 = \frac{2}{s} + \frac{n}{r} - 2 - 2\varepsilon \tag{3.8}$$

with $s \le s_0$, $r < r_0$ *and* $0 \le \beta_0$ *with the estimate*

$$\|e^{-tA}a\|_{L^{s_0,s}(0,\infty;H^{\beta_0,r_0}(\Omega))} \le C\|a\|_{\mathcal{D}(A_{r,s}^{1+\varepsilon-\frac{1}{s}})}, \tag{3.9}$$

where $C = C(n, r, s, r_0, s_0, \beta_0, \varepsilon)$.

(ii) *Let* $\partial_t u$, $Au \in L^s(0,T;\mathcal{D}(A_r^\varepsilon))$. *Then we have* $u \in L^{s_0}(0,T;H^{\beta_0,r_0}(\Omega))$ *for* $s < s_0$, $r < r_0$ *and* $0 \le \beta_0$ *satisfying (3.8) with the estimate*

$$\|u\|_{L^{s_0}(0,T;H^{\beta_0,r_0}(\Omega))} \le C\left(\|e^{-tA}u(0)\|_{L^{s_0}(0,\infty;H^{\beta_0,r_0}(\Omega))} + \|\partial_t u,\ Au\|_{L^s(0,T;\mathcal{D}(A_r^\varepsilon))}\right), \tag{3.10}$$

where $C = C(n, r, s, r_0, s_0, \beta_0, \varepsilon)$.

Proof (i) If $\beta_0 \le \beta$ and $r \le r_0$ satisfy

$$\frac{1}{r_0} - \frac{\beta_0}{n} = \frac{1}{r} - \frac{\beta}{n}, \tag{3.11}$$

then it holds that $\mathcal{D}(A_r^{\beta/2}) \subset H^{\beta,r}(\Omega) \subset H^{\beta_0,r_0}(\Omega)$. For sake of simplicity, we set $\alpha = 1 + \varepsilon - 1/s$ and $\beta = n(1/r - 1/r_0) + \beta_0$. We take α_i $(i = 0, 1)$ such as $0 < \alpha_0 < \alpha < \alpha_1 < \beta/2$. Let $a \in \mathcal{D}(A_r^{\alpha_i})$. Since $\mathcal{D}(A_r^{\beta/2}) \subset H^{\beta_0,r_0}(\Omega)$, we have by the L^p-L^q estimate of $\{e^{-tA}\}_{t\ge 0}$ that

$$\begin{aligned}\|e^{-tA}a\|_{H^{\beta_0,r_0}(\Omega)} &\le C\|A^{\beta/2}e^{-tA}a\|_{L^r(\Omega)}\\ &\le C\|A^{\beta/2-\alpha_i}e^{-tA}A^{\alpha_i}a\|_{L^r(\Omega)}\\ &\le Ct^{\alpha_i-\beta/2}\|A^{\alpha_i}a\|_{L^r(\Omega)}.\end{aligned}$$

This means the mapping

$$\mathcal{D}(A_r^{\alpha_i}) \ni a \mapsto \|e^{-tA}a\|_{H^{\beta_0,r_0}(\Omega)} \in L^{q_i,\infty}(0,\infty), \quad i = 0, 1$$

with

$$\frac{1}{q_i} = \frac{\beta}{2} - \alpha_i, \quad i = 0, 1 \tag{3.12}$$

is a bounded subadditive operator. Hence, we have by the real interpolation that

$$\left(\mathcal{D}(A_r^{\alpha_0}), \mathcal{D}(A_r^{\alpha_1})\right)_{\theta,s} \ni a \mapsto \|e^{-tA}a\|_{H^{\beta_0,r_0}(\Omega)} \in \left(L^{q_0,\infty}(0,\infty), (L^{q_1,\infty}(0,\infty)\right)_{\theta,s}$$

with $0 < \theta < 1$ and $1 \le s \le \infty$, which yields that

$$\mathcal{D}(A^{\alpha}_{r,s}) \ni a \mapsto \|e^{-tA}a\|_{H^{\beta_0,r_0}(\Omega)} \in L^{q,s}(0,\infty) \tag{3.13}$$

with $\alpha = (1-\theta)\alpha_0 + \theta\alpha_1$ and $1/q = (1-\theta)/q_0 + \theta/q_1$. By (3.11) and (3.12) it holds that

$$\frac{1}{q} = \frac{\beta}{2} - \alpha = \frac{n}{2}\Big(\frac{1}{r} - \frac{1}{r_0}\Big) + \frac{\beta_0}{2} - \alpha = \frac{1}{s_0}, \tag{3.14}$$

which shows (3.9).

(ii) Let $f \in L^s(0,T;\mathcal{D}(A_r^{\varepsilon}))$ and let v be a solution of

$$\begin{cases} \partial_t v + Av = f, \\ \quad v(0) = 0. \end{cases}$$

By Lemma 3.1 we have that the solution v belongs to $L^s(0,T;\mathcal{D}(A_r^{1+\varepsilon}))$. Since $s < s_0$, we have by (3.8) that

$$\frac{n}{r_0} - \beta_0 > \frac{n}{r} - 2 - 2\varepsilon, \tag{3.15}$$

which yields the embedding relation $\mathcal{D}(A_r^{1+\varepsilon}) \subset H^{\beta_0,r_0}(\Omega)$. Define $\gamma > 0$ by

$$\gamma = \frac{n}{2}\big(\frac{1}{r} - \frac{1}{r_0}\big), \tag{3.16}$$

and we have the embedding $\mathcal{D}(A_r^{\gamma}) \hookrightarrow L^{r_0}(\Omega)$. Since v has an expression $v(t) = \int_0^t e^{-(t-\tau)A}f(\tau)\,d\tau$, we have

$$\begin{aligned}
\|v(t)\|_{H^{\beta_0,r_0}(\Omega)} &\le C\|A^{\frac{\beta_0}{2}}v(t)\|_{L^{r_0}(\Omega)} \\
&= C\left\|\int_0^t A^{\frac{\beta_0}{2}}e^{-(t-\tau)A}f(\tau)\,d\tau\right\|_{L^{r_0}(\Omega)} \\
&\le C\int_0^t \|A^{\frac{\beta_0}{2}-\varepsilon}e^{-(t-\tau)A}A^{\varepsilon}f(\tau)\|_{L^{r_0}(\Omega)}\,d\tau \\
&\le C\int_0^t \|A^{\gamma+\frac{\beta_0}{2}-\varepsilon}e^{-(t-\tau)A}A^{\varepsilon}f(\tau)\|_{L^{r}(\Omega)}\,d\tau \\
&\le C\int_0^t (t-\tau)^{-\gamma-\frac{\beta_0}{2}+\varepsilon}\|A^{\varepsilon}f(\tau)\|_{L^{r}(\Omega)}\,d\tau \\
&\le C\int_0^t (t-\tau)^{-\frac{n}{2}(\frac{1}{r}-\frac{1}{r_0})-\frac{\beta_0}{2}+\varepsilon}\|A^{\varepsilon}f(\tau)\|_{L^{r}(\Omega)}\,d\tau \\
&\le C\int_0^t (t-\tau)^{\mu-1}\|A^{\varepsilon}f(\tau)\|_{L^{r}(\Omega)}\,d\tau,
\end{aligned} \tag{3.17}$$

where $\mu = 1 - \frac{n}{2}(\frac{1}{r} - \frac{1}{r_0}) - \frac{\beta_0}{2} + \varepsilon = \frac{1}{s} - \frac{1}{s_0}$, implied by (3.8). Since $s < s_0$, it holds $0 < \mu < 1$. Hence, it follows from the Hardy-Littlewood-Sobolev inequality that

$$\|v\|_{L^{s_0}(0,T;H^{\beta_0,r_0}(\Omega))} \le C\|f\|_{L^s(0,T;\mathcal{D}(A_r^\varepsilon))} \le C\|\partial_t v,\ Av\|_{L^s(0,T;\mathcal{D}(A_r^\varepsilon))},$$

which implies (3.10) for $u(0) = 0$. For general u, we have an expression $u(t) = e^{-tA}u(0) + v(t)$, so the desired estimate (3.10) is an immediate consequence of the above estimate. This completes the proof of Proposition 3.3. □

It should be noted that for the estimate (3.9) in (i), we may assume $s \le s_0$. However, for the estimate (3.10) in (ii), we impose $s < s_0$ to apply the Hardy-Littlewood-Sobolev inequality.

Proof of Lemma 3.2. Since $2/s + n/r = 3 + 2\varepsilon$, we may take r_0 in such a way that

$$\frac{1}{r} - \frac{2+2\varepsilon}{n} < \frac{1}{r_0} < \min\left\{\frac{1}{n}, \frac{1}{r} - \frac{1}{n}\right\}. \tag{3.18}$$

Next, we define $1 < r_1, \tilde{r}_0, \tilde{r}_1, s_0, s_1 < \infty$ by

$$\frac{1}{r_1} = \frac{1}{r} - \frac{1}{r_0}, \quad \frac{1}{\tilde{r}_0} = \frac{1}{r_0} + \frac{1}{n}, \quad \frac{1}{\tilde{r}_1} = \frac{1}{r} - \frac{1}{r_0} - \frac{1}{n}, \tag{3.19}$$

$$\frac{1}{s_0} = \frac{1}{2}\left(1 - \frac{n}{r_0}\right), \quad \frac{1}{s_1} = \frac{1}{s} - \frac{1}{2}\left(1 - \frac{n}{r_0}\right). \tag{3.20}$$

By (3.18) and (3.19) we have $1 < r_1 < n$ so that the embedding $H^{2,r_1}(\Omega) \subset H^{1,\tilde{r}_1}(\Omega)$ holds. By the Hölder inequality, it holds that

$$\|u \cdot \nabla v\|_{L^r(\Omega)} \le \|u\|_{L^{r_0}(\Omega)}\|\nabla v\|_{L^{r_1}(\Omega)} \le C\|u\|_{L^{r_0}(\Omega)}\|A^{\frac{1}{2}}v\|_{L^{r_1}(\Omega)}, \tag{3.21}$$

$$\begin{aligned}\|\nabla(u \cdot \nabla v)\|_{L^r(\Omega)} &= \|Du \cdot Dv\|_{L^r(\Omega)} + \|u \cdot D^2 v\|_{L^r(\Omega)}\\ &\le \|Du\|_{L^{\tilde{r}_0}(\Omega)}\|Dv\|_{L^{\tilde{r}_1}(\Omega)} + \|u\|_{L^{r_0}(\Omega)}\|D^2 v\|_{L^{r_1}(\Omega)}\\ &\le C\left(\|Du\|_{L^{\tilde{r}_0}(\Omega)}\|Dv\|_{L^{\tilde{r}_1}(\Omega)} + \|u\|_{L^{r_0}(\Omega)}\|v\|_{H^{2,r_1}(\Omega)}\right)\\ &\le C\left(\|Du\|_{L^{\tilde{r}_0}(\Omega)} + \|u\|_{L^{r_0}(\Omega)}\right)\|v\|_{H^{2,r_1}(\Omega)}.\end{aligned} \tag{3.22}$$

From (3.21) we obtain that for $u \in L^{r_0}(\Omega)$

$$\mathcal{D}(A_{r_1}^{1/2}) \ni v \mapsto u \cdot \nabla v \in L^r(\Omega) \text{ with the bound } C\|u\|_{L^{r_0}(\Omega)}. \tag{3.23}$$

From (3.21) and (3.22) we obtain that for $u \in H^{1,\tilde{r}_0}(\Omega) \cap L^{r_0}(\Omega)$

$$\mathcal{D}(A_{r_1}) \ni v \mapsto u \cdot \nabla v \in H^{1,r}(\Omega) \text{ with the bound } C(\|Du\|_{L^{\tilde{r}_0}(\Omega)} + \|u\|_{L^{r_0}(\Omega)}). \tag{3.24}$$

It follows from (3.23), (3.24) and the interpolation theorem that

$$\begin{aligned} &\mathcal{D}(A_{r_1}^{\frac{1}{2}+\varepsilon}) \ni v \mapsto u \cdot \nabla v \in H^{2\varepsilon,r}(\Omega) \text{ with the bound} \\ &C\|u\|_{L^{r_0}(\Omega)}^{1-2\varepsilon}(\|Du\|_{L^{\tilde{r}_0}(\Omega)} + \|u\|_{L^{r_0}(\Omega)})^{2\varepsilon}. \end{aligned} \tag{3.25}$$

Since $2\varepsilon < \frac{1}{r}$, we have by (3.25) that $P(u \cdot \nabla v) \in \mathcal{D}(A_r^{\varepsilon})$ for $u \in H^{1,\tilde{r}_0}(\Omega) \cap L^{r_0}(\Omega)$ and $v \in \mathcal{D}(A_{r_1}^{\frac{1}{2}+\varepsilon})$ with the estimate

$$\begin{aligned} \|A^{\varepsilon}P(u \cdot \nabla v)\|_{L^r(\Omega)} &\le C\|P(u \cdot \nabla v)\|_{H^{2\varepsilon,r}(\Omega)} \le C\|u \cdot \nabla v\|_{H^{2\varepsilon,r}(\Omega)} \\ &\le C\|u\|_{L^{r_0}(\Omega)}^{1-2\varepsilon}(\|Du\|_{L^{\tilde{r}_0}(\Omega)} + \|u\|_{L^{r_0}(\Omega)})^{2\varepsilon}\|v\|_{H^{1+2\varepsilon,r_1}(\Omega)}. \end{aligned}$$

Since $1/s = 1/s_0 + 1/s_1$, we have by the Hölder inequality

$$\begin{aligned} &\|A^{\varepsilon}P(u \cdot \nabla v)\|_{L^s(0,T;L^r(\Omega))} \\ \le &C\|u\|_{L^{s_0}(0,T;L^{r_0}(\Omega))}^{1-2\varepsilon}\Big(\|Du\|_{L^{s_0}(0,T;L^{\tilde{r}_0}(\Omega))} + \|u\|_{L^{s_0}(0,T;L^{r_0}(\Omega))}\Big)^{2\varepsilon}\|v\|_{L^{s_1}(0,T;H^{1+2\varepsilon,r_1}(\Omega))} \\ \le &C\Big(\|u\|_{L^{s_0}(0,T;L^{r_0}(\Omega))} + \|Du\|_{L^{s_0}(0,T;L^{\tilde{r}_0}(\Omega))}\Big)\|v\|_{L^{s_1}(0,T;H^{1+2\varepsilon,r_1}(\Omega))}. \end{aligned}$$

By (3.19) and (3.20) we have $r_0, \tilde{r}_0, r_1 > r$ and $s_0, s_1 > s$ with

$$1 = \frac{2}{s_0} + \frac{n}{r_0} = \frac{2}{s_0} + \frac{n}{\tilde{r}_0} - 1 = \frac{2}{s_1} + \frac{n}{r_1} - (1 + 2\varepsilon) = \frac{2}{s} + \frac{n}{r} - 2 - 2\varepsilon.$$

Hence it follows from Proposition 3.3 that

$$\begin{aligned} &\|u\|_{L^{s_0}(0,T;L^{r_0}(\Omega))} \\ &\le C(\|e^{-tA}u(0)\|_{L^{s_0}(0,T;L^{r_0}(\Omega))} + \|\partial_t u, Au\|_{L^s(0,T;\mathcal{D}(A_r^{\varepsilon}))}) \\ &\le C(\|u(0)\|_{\mathcal{D}(A_{r,s}^{1+\varepsilon-\frac{1}{s}})} + \|\partial_t u, Au\|_{L^s(0,T;\mathcal{D}(A_r^{\varepsilon}))}), \end{aligned} \tag{3.26}$$

$$\begin{aligned} &\|\nabla u\|_{L^{s_0}(0,T;L^{\tilde{r}_0}(\Omega))} \\ &\le C(\|\nabla e^{-tA}u(0)\|_{L^{s_0}(0,T;L^{\tilde{r}_0}(\Omega))} + \|\partial_t u, Au\|_{L^s(0,T;\mathcal{D}(A_r^{\varepsilon}))}) \\ &\le C(\|u(0)\|_{\mathcal{D}(A_{r,s}^{1+\varepsilon-\frac{1}{s}})} + \|\partial_t u, Au\|_{L^s(0,T;\mathcal{D}(A_r^{\varepsilon}))}) \end{aligned} \tag{3.27}$$

$$\begin{aligned}&\|v\|_{L^{s_1}(0,T;H^{1+2\varepsilon,r_1}(\Omega))}\\&\le C(\|e^{-tA}v(0)\|_{L^{s_1}(0,T;H^{1+2\varepsilon,r_1}(\Omega))}+\|\partial_t v, Av\|_{L^s(0,T;\mathcal{D}(A_r^{\varepsilon}))})\\&\le C(\|v(0)\|_{\mathcal{D}(A_{r,s}^{1+\varepsilon-\frac{1}{s}})}+\|\partial_t v, Av\|_{L^s(0,T;\mathcal{D}(A_r^{\varepsilon}))}).\end{aligned}\tag{3.28}$$

Now, the desired estimates (3.4), (3.5) and (3.6) are consequences of (3.26)–(3.28). This proves Lemma 3.2. □

We next consider another bilinear estimate in which one of two functions is in the scaling invariant class.

Lemma 3.4 *Let* $1 < s_* < \infty$ *and* $n/3 < r_* < n$ *satisfy*

$$2/s_* + n/r_* = 3.$$

Assume that $1 < s < \infty$ *and* $1 < r < n$ *satisfy the condition (1.12). For every* $u \in W^{1,s_*}(0,T; L^{r_*}_\sigma(\Omega)) \cap L^{s_*}(0,T; \mathcal{D}(A_{r_*}))$ *and* $v \in W^{1,s}(0,T; D(A_r^{\varepsilon})) \cap L^s(0,T; \mathcal{D}(A_r^{1+\varepsilon}))$ *with* $0 \le \varepsilon < 1/2r$ *we have* $P(u \cdot \nabla v) \in L^s(0,T; \mathcal{D}(A_r^{\varepsilon}))$ *with the estimate*

$$\begin{aligned}&\|A^{\varepsilon}P(u\cdot\nabla v)\|_{L^s(0,T;L^r(\Omega))}\\&\le C(\|e^{-tA}u(0)\|_{L^{s_0}(0,T;L^{r_0}(\Omega))}+\|\nabla e^{-tA}u(0)\|_{L^{s_0}(0,T;L^{\tilde r_0}(\Omega))}+\|\partial_t u, Au\|_{L^{s_*}(0,T;L^{r_*}(\Omega))})\\&\quad\times(\|e^{-tA}v(0)\|_{L^{s_1}(0,T;H^{1+2\varepsilon,r_1}(\Omega))}+\|\partial_t v, Av\|_{L^s(0,T;\mathcal{D}(A_r^{\varepsilon}))})\end{aligned}\tag{3.29}$$

and

$$\|e^{-tA}u(0)\|_{L^{s_0}(0,T;L^{r_0}(\Omega))}+\|\nabla e^{-tA}u(0)\|_{L^{s_0}(0,T;L^{\tilde r_0}(\Omega))}\le C\|u(0)\|_{D(A_{r_*,s_*}^{1-\frac{1}{s_*}})},\tag{3.30}$$

$$\|e^{-tA}v(0)\|_{L^{s_1}(0,T;H^{1+2\varepsilon,r_1}(\Omega))}\le C\|v(0)\|_{D(A_{r,s}^{1+\varepsilon-\frac{1}{s}})}\tag{3.31}$$

for some $r_0, \tilde r_0 > r_*$ *and* $s_0 > s_*$ *satisfying* $2/s_0 + n/r_0 = 1$, $2/s_0 + n/\tilde r_0 = 2$, *and for some* $r_1 > r$ *and* $s_1 > s$ *satisfying* $2/s_1 + n/r_1 = 2/s + n/r - 1$. *Here* $C = C(n, s, r, s_*, r_*)$ *is a constant independent of* T.

Proof By (1.12) we may take r_0 in such a way that

$$\max\left\{\frac{1}{n}\left(1-\frac{2}{s}\right), \frac{1}{r_*}-\frac{2}{n}\right\} < \frac{1}{r_0} < \min\left\{\frac{1}{n}, \frac{1}{r}-\frac{1}{n}, \frac{1}{r_*}-\frac{1}{n}\right\}.\tag{3.32}$$

Next, we define $1 < r_1, \tilde r_0, \tilde r_1, s_0, s_1 < \infty$ in the same manner as in (3.19) and (3.20). Since $0 < \varepsilon < 1/2r$, we have similarly to the proof of Lema 3.2 that

$$\|A^{\varepsilon}P(u\cdot\nabla v)\|_{L^s(0,T;L^r(\Omega))}\le C(\|u\|_{L^{s_0}(0,T;L^{r_0}(\Omega))}+\|\nabla u\|_{L^{s_0}(0,T;L^{\tilde r_0}(\Omega))})\|v\|_{L^{s_1}(0,T;H^{1+2\varepsilon,r_1}(\Omega))}.\tag{3.33}$$

Since $2/s_* + n/r_* = 3$, we have by (3.19), (3.20) and (3.32) that

$$s_* < s_0,\quad r_* < r_0,\quad 2/s_0 + n/r_0 = 1 = 2/s_* + n/r_* - 2,$$

and hence it follows from Proposition 3.3 with $\beta_0 = \varepsilon = 0$, $s = s_*$, $r = r_*$ in (3.8) that

$$\begin{aligned} &\|u\|_{L^{s_0}(0,T;L^{r_0}(\Omega))} \\ &\le C(\|e^{-tA}u(0)\|_{L^{s_0}(0,T;L^{r_0}(\Omega))} + \|\partial_t u, Au\|_{L^{s_*}(0,T;L^{r_*}(\Omega))}) \\ &\le C(\|u(0)\|_{\mathcal{D}(A_{r_*,s_*}^{1-\frac{1}{s_*}})} + \|\partial_t u, Au\|_{L^{s_*}(0,T;L^{r_*}(\Omega))}). \end{aligned} \tag{3.34}$$

Similarly, by (3.19) and (3.20) we have

$$r_* < \tilde{r}_0, \quad 2/s_0 + n/\tilde{r}_0 - 1 = 1 = 2/s_* + n/r_* - 2,$$

and it follows from Proposition 3.3 with $r_0 = \tilde{r}_0$, $\beta_0 = 1$, $\varepsilon = 0$, $s = s_*$, $r = r_*$ in (3.8) that

$$\begin{aligned} &\|\nabla u\|_{L^{s_0}(0,T;L^{\tilde{r}_0}(\Omega))} \\ &\le C(\|\nabla e^{-tA}u(0)\|_{L^{s_0}(0,T;L^{\tilde{r}_0}(\Omega))} + \|\partial_t u, Au\|_{L^{s_*}(0,T;L^{r_*}(\Omega))}) \\ &\le C(\|u(0)\|_{\mathcal{D}(A_{r_*,s_*}^{1-\frac{1}{s_*}})} + \|\partial_t u, Au\|_{L^{s_*}(0,T;L^{r_*}(\Omega))}). \end{aligned} \tag{3.35}$$

Again by (3.19) and (3.20) we have

$$s < s_1, \quad r < r_1, \quad 2/s_1 + n/r_1 - (1 + 2\varepsilon) = 2/s + n/r - 2 - 2\varepsilon,$$

and it follows from Proposition 3.3 with $s_0 = s_1$, $r_0 = r_1$, $\beta_0 = 1 + 2\varepsilon$ in (3.8) that

$$\begin{aligned} &\|v\|_{L^{s_1}(0,T;H^{1+2\varepsilon,r_1}(\Omega))} \\ &\le C(\|e^{-tA}v(0)\|_{L^{s_1}(0,T;H^{1+2\varepsilon,r_1}(\Omega))} + \|\partial_t v, Av\|_{L^{s}(0,T;\mathcal{D}(A_r^{\varepsilon}))}) \\ &\le C(\|v(0)\|_{\mathcal{D}(A_{r,s}^{1+\varepsilon-\frac{1}{s}})} + \|\partial_t v, Av\|_{L^{s}(0,T;\mathcal{D}(A_r^{\varepsilon}))}). \end{aligned} \tag{3.36}$$

The desired estimates (3.29), (3.30) and (3.31) are consequences of (3.33)–(3.36). This proves Lemma 3.4. □

Proof of Theorem 1.2. (i) We find the solution u of (1.3) in the form

$$u(t) = u_0(t) + v(t)$$

where u_0 and v satisfy

$$\begin{cases} \frac{du_0}{dt} + Au_0 = Pf(t) \text{ on } (0, T), \\ u_0(0) = a, \end{cases} \tag{3.37}$$

and

$$\begin{cases} \frac{dv}{dt} + Av = -P(u_0 \cdot \nabla v + v \cdot \nabla u_0 + v \cdot \nabla v + u_0 \cdot \nabla u_0) \text{ on } (0, T), \\ v(0) = 0. \end{cases} \tag{3.38}$$

Let us first construct the solution u_0 of (3.37). Since $a \in \mathcal{D}(A_{r,s}^{1+\varepsilon-\frac{1}{s}})$ and $Pf \in L^s(0, T; \mathcal{D}(A_r^\varepsilon)) \cap C([0, T); \mathcal{D}(A_{r,s}^{\varepsilon-\frac{1}{s}}))$, by Lemma 3.1, there exists a unique solution of (3.37) such that

$$u_0 \in W^{1,s}(0, T; \mathcal{D}(A_r^\varepsilon)) \cap L^s(0, T; \mathcal{D}(A_r^{1+\varepsilon})). \tag{3.39}$$

Moreover, such u_0 is subject to the estimate

$$\|\partial_t u_0\|_{L^s(0,T;\mathcal{D}(A_r^\varepsilon))} + \|Au_0\|_{L^s(0,T;\mathcal{D}(A_r^\varepsilon))} \le C\left(\|a\|_{\mathcal{D}(A_{r,s}^{1+\varepsilon-\frac{1}{s}})} + \|Pf\|_{L^s(0,T;\mathcal{D}(A_r^\varepsilon))}\right) \tag{3.40}$$

with the constant $C = C(n, r, s, \varepsilon)$ independent of T.

We next consider (3.38). Let us define the approximating solution $\{v_j\}_{j=0}^m$ with $v_0(t) \equiv 0$ by

$$\begin{cases} \dfrac{dv_{j+1}}{dt} + Av_{j+1} = -P(u_0 \cdot \nabla v_j + v_j \cdot \nabla u_0 + v_j \cdot \nabla v_j + u_0 \cdot \nabla u_0) \quad \text{on } (0, T), \\ v_{j+1}(0) = 0. \end{cases} \tag{3.41}$$

We introduce the function space X_T by

$$X_T \equiv \{v \in C([0, T]; \mathcal{D}(A^{1+\varepsilon-\frac{1}{s}})); \partial_t v, Av \in L^s(0, T; \mathcal{D}(A_r^\varepsilon)), v(0) = 0\}.$$

Equipped with the norm

$$\|v\|_{X_T} \equiv \|\partial_t v\|_{L^s(0,T;\mathcal{D}(A_r^\varepsilon))} + \|Av\|_{L^s(0,T;\mathcal{D}(A_r^\varepsilon))},$$

X_T is the Banach space. Obviously, we have $v_0 \in X_T$. Assume that $v_j \in X_T$. Then it follows from Lemma 3.2 and Proposition 3.3 that $P(u_0 \cdot \nabla v_j + v_j \cdot \nabla u_0 + v_j \cdot \nabla v_j + u_0 \cdot \nabla u_0) \in L^s(0, T; \mathcal{D}(A_r^\varepsilon))$ with the estimate

$$\begin{aligned} &\|P(u_0 \cdot \nabla v_j + v_j \cdot \nabla u_0 + v_j \cdot \nabla v_j + u_0 \cdot \nabla u_0)\|_{L^s(0,T;\mathcal{D}(A_r^\varepsilon))} \\ &\le C\left(2D_T(a,f)\|v_j\|_{X_T} + \|v_j\|_{X_T}^2 + D_T(a,f)^2\right), \end{aligned} \tag{3.42}$$

with $C = C(n, r, s, \varepsilon, r_0, s_0, \varepsilon)$ independent of T and j, where

$$\begin{aligned} &D_T(a,f) \\ &\equiv \|e^{-tA}a\|_{L^{s_0}(0,T;L^{r_0}(\Omega))} + \|\nabla e^{-tA}a\|_{L^{s_0}(0,T;L^{\tilde{r}_0}(\Omega))} + \|e^{-tA}a\|_{L^{s_1}(0,T;H^{1+2\varepsilon,r_1}(\Omega))} + \|u_0\|_{X_T}. \end{aligned} \tag{3.43}$$

Hence it follows from (3.42) and Lemma 3.1 that there exists a unique solution $v_{j+1} \in X_T$ of (3.41). Moreover, by (3.1) and (3.42), we have that

$$\|v_{j+1}\|_{X_T} \leq C\left(2D_T(a,f)\|v_j\|_{X_T} + \|v_j\|_{X_T}^2 + D_T(a,f)^2\right). \tag{3.44}$$

By induction, it holds that $v_j \in X_T$ for all $j = 0, 1, 2, \ldots$. If there exists $0 < T' \leq T$ such that

$$D_{T'}(a,f) < \frac{1}{4C}, \tag{3.45}$$

then we obtain from (3.44) that

$$\|v_j\|_{X_{T'}} \leq \frac{1}{2C}\left(1 - 2CD_{T'}(a,f) - \sqrt{1 - 4CD_{T'}(a,f)}\right) \equiv K_{T'} \tag{3.46}$$

for all $j = 1, 2, \ldots$. Furthermore, taking $w_j \equiv v_j - v_{j-1}$, we have by the similar argument to the above that

$$\|w_{j+1}\|_{X_{T'}} \leq \{2C\,(D_{T'}(a,f) + K_{T'})\}^j \|v_1\|_{X_{T'}}, \quad j = 1, 2, \ldots. \tag{3.47}$$

For more details, see for instance [16, Sect. 4]. Since we have by (3.46)

$$2C\,(D_{T'}(a,f) + K_{T'}) = 1 - \sqrt{1 - 4CD_{T'}(a,f)} < 1,$$

it follows from (3.47) that

$$\sum_{j=1}^{\infty} \|w_j\|_{X_{T'}} < \infty,$$

which yields the limit $v \in X_{T'}$ of $\{v_j\}_{j=0}^{\infty}$. Letting $j \to \infty$ in (3.41), we see by Lemma 3.2 that v is the solution of (3.38) on $(0, T')$ provided the condition (3.45) is fulfilled. The validity of (3.45) is a consequence of (3.5), (3.6) and (3.40), so we obtain the solution u of (1.3) on $(0, T')$ satisfying (1.8) and (1.9). The proof for uniqueness is rather standard, so we may omit it.

(ii) Let u be a solution of (1.3) satisfying (1.11). We consider the following Cauchy problem on linear evolution equation

$$\begin{cases} \dfrac{dw}{dt} + Aw + P(u \cdot \nabla w) = Pf \quad \text{on } (0, T), \\ w(0) = a. \end{cases} \tag{3.48}$$

Based on Lemma 3.4, we see by the similar argument to the above (i) that there exists some positive constant δ such that if $0 < T' \leq T$ satisfies

$$\|e^{-tA}a\|_{L^{s_0}(0,T';L^{r_0}(\Omega))} + \|\nabla e^{-tA}a\|_{L^{s_0}(0,T';L^{\tilde{r}_0}(\Omega))} + \|\partial_t u, Au\|_{L^{s_*}(0,T';L^{r_*}(\Omega))} \leq \delta, \tag{3.49}$$

then there exists a unique solution w of (3.48) on $(0, T')$ satisfying

$$w \in W^{1,s}(0, T'; \mathcal{D}(A_r^{\varepsilon})) \cap L^s(0, T'; \mathcal{D}(A_r^{1+\varepsilon})). \tag{3.50}$$

By (1.11) and (3.30), there exists $\tau > 0$ such that

$$\|e^{-tA}a\|_{L^{s_0}(t,t+\tau;L^{r_0}(\Omega))} + \|\nabla e^{-tA}a\|_{L^{s_0}(t,t+\tau;L^{\tilde{r}_0}(\Omega))} + \|\partial_t u, Au\|_{L^{s_*}(t,t+\tau;L^{r_*}(\Omega))} \le \delta, \tag{3.51}$$

for all $0 < t < T - \tau$. It should be noted for validity of (3.51) a and Pf themselves may be large. Hence, in the first step we may take $T' = \tau$ so that the condition (3.49) is fulfilled, which yields the solution w of (3.48) on $[0, \tau]$ in the class of (3.50) with $T' = \tau$. In the next step, starting from $t = \tau$, we construct the solution w of (3.48) on $[\tau, 2\tau]$, which yields the solution on $[0, 2\tau]$. Proceeding this procedure beyond $t = 2\tau$, within finitely many steps, we obtain the solution w of (3.48) on the whole interval $(0, T)$ satisfying

$$w \in W^{1,s}(0, T; \mathcal{D}(A_r^{\varepsilon})) \cap L^s(0, T; \mathcal{D}(A_r^{1+\varepsilon})). \tag{3.52}$$

Furthermore, since a and Pf satisfy (1.10), it is easy to show that

$$w \in W^{1,s_*}(0, T; L^{r_*}_{\sigma}(\Omega)) \cap L^{s_*}(0, T; \mathcal{D}(A_{r_*})). \tag{3.53}$$

Since both u and w are in the scaling invariant class as in (1.11) and (3.53), it follows from uniqueness of solutions to (1.3) that

$$u = w \quad \text{on } [0, T),$$

which yields the desired regularity of u. This completes the proof of Theorem 1.2. □

Acknowledgements The research of the project was partially supported by JSPS Fostering Joint Research Program (B)-18KK0072. The research of H. K. was partially supported by JSPS Grant-in-Aid for Scientific Research (S)-16H06339. The research of S. S. was partially supported by JSPS Grant-in-Aid for Scientific Research (B)-16H03945 and (B)-21H00992. The authors declare that there is no conflict of interest. Finally, the authors would like to express their sincere thanks to referees for their valuable comments.

References

1. H. Amann, *Linear and Quasilinear Parabolic Problems, Volume I: Abstract Linear Theory* (Birkhäuser, Basel, 1995)
2. J. Bergh, J. Löfström, *Interpolation Spaces An Introduction*. Grundlehren der mathematische Wisssenshaften, vol. 223 (Springer, Berlin-Heidelberg-New York, 1976)
3. R. Denk, An introduction to maximal regularity for parabolic evolution equations, in *Nonlinear Partial Differential Equations for Future Applications*, vol. 346 (Springer Proceedings in Mathematics & Statistics, 2021), pp. 1–70
4. W. Desch, M. Hieber, J. Prüss, L^p-Theory of the Stokes equations in a half space. J. Evol. Equ. **1**, 115–142 (2001)

5. R. Farwig, H. Sohr, Optimal initial value conditions for the existence of local strong solutions of the Navier-Stokes equations. Math. Ann. **345**, 631–642 (2009)
6. R. Farwig, H. Sohr, W. Varnhorn, Optimal initial value conditions for local strong solutions of the Navier-Stokes equations. Ann. Univ. Ferrara **55**, 89–110 (2009)
7. H. Fujita, T. Kato, On the nonstationary Navier-Stokes system. Arch. Rational Mech. Anal. **16**, 269–315 (1964)
8. D. Fujiwara, On the asymptotic behaviour of the Green operators for elliptic boundary problems and the pure imaginary powers of some second order operators. J. Math. Soc. Jpn. **21**, 481–522 (1969)
9. D. Fujiwara, H. Morimoto, An L_r theorem of the Helmholtz decomposition of vector fields. J. Fac. Sci. Univ. Tokyo, Sec. IA **24**, 685–700 (1977)
10. Y. Giga, Analyticity of the semigroups generated by the Stokes operator in L_r spaces. Math. Z. **178**, 297–329 (1981)
11. Y. Giga, Domains of fractional powers of the Stokes operator in L^r spaces. Arch. Rat. Mech. Anal. **89**, 251–265 (1985)
12. Y. Giga, Solutions for semilinear parabolic equations in L^p and regularity of weak solutions of the Navier-Stokes system. J. Diff. Equ. **61**, 186–212 (1986)
13. Y. Giga, T. Miyakawa, Solutions in L^r of the Navier-Stokes initial value problem. Arch. Rat. Mech. Anal. **89**, 267–281 (1985)
14. Y. Giga, H. Sohr, On the Stokes operator in exterior domains. J. Fac. Sci. Univ. Tokyo Sect. IA Math. **36**, 103–130 (1989)
15. Y. Giga, H. Sohr, Abstract L^p estimates for the Cauchy problem with applications to the Navier-Stokes equations in exterior domains. J. Funct. Anal. **102**, 72–94 (1991)
16. H. Kozono, S. Shimizu, Strong solutions of the Navier-Stokes equations based on the maximal Lorentz regularity theorem in Besov spaces. J. Funct. Anal. **276**, 896–931 (2019)
17. A. Lunardi, Interpolation theory, Appunti. Scuola Normale Superiore di Pisa, vol. 16, Edizioni della Normale, Pisa (2018)
18. T. Miyakawa, On nonstationary solutions of the Navier-Stokes equations in an exterior domain. Hiroshima Math. J. **12**, 115–140 (1982)
19. J. Prüss, Maximal regularity for evolution equations in L_p-spaces. Conf. Semin. Mat. Univ. Bari **285**, 1–39 (2003)
20. J. Prüss, G. Simonett, Moving interfaces and quasi-linear parabolic differential equations. *Monographs in Mathematics*, vol. 105 (Birkhäuser, Basel, 2016)
21. R. Rautmann, On optimum regularity of Navier-Stokes solutions at time $t = 0$. Math. Z. **184**, 141–149 (1983)
22. M. Reed, B. Simon, *Method of Modern Mathematical Physics II: Fourier Analysis, Self-Adjointness* (Academic Press, San Diego-New York-Berkeley-Boston-London-Sydney-Tokyo-Toronto, 1975)
23. C.G. Simader, H. Sohr, A new approach to the Helmholtz decomposition and the Neumann problem in L^q-spaces for bounded and exterior domains. Mathematical problems relating to the Navier-Stokes equation, in *Series Advances in Mathematics: Scientific*, vol. 11 (World Scientific Publishing, River Edge, NJ, 1992), pp. 1–35
24. V.A. Solonnikov, Estimates for solutions of nonstationary Navier-Stokes equations. J. Soviet Math. **8**, 467–529 (1977)
25. H. Triebel, *Interpolation Theory, Function Spaces, Differential Operators*, North-Holland, Amsterdam-New York-Oxford (1978)
26. W. von Wahl, Über das Verhalten für $t \to 0$ der Lösungen nichtlinearer parabolischer Gleichungen, insbesondere der Gleichungen von Navier-Stokes. Bayreuth. Math. Schr. No. **16**, 151–277 (1984)
27. W. von Wahl, The equations of Navier-Stokes and abstract parabolic equations. *Aspects of Mathematics*, E8. (Friedr. Vieweg & Sohn, Braunschweig, 1985)

Asymptotic Behavior in Time of Solution to System of Cubic Nonlinear Schrödinger Equations in One Space Dimension

Satoshi Masaki, Jun-Ichi Segata, and Kota Uriya

Abstract In this paper, we consider the large time asymptotic behavior of solutions to systems of two cubic nonlinear Schrödinger equations in one space dimension. It turns out that for a system there exists a small solution of which asymptotic profile is a sum of two parts oscillating in a different way. This kind of behavior seems new. Further, several examples of systems which admit solution with several types of behavior such as modified scattering, nonlinear amplification, and nonlinear dissipation, are given. We also extend our previous classification result of nonlinear cubic systems.

Keywords Nonlinear Schrödinger equation · Asymptotic behavior of solutions · Long-range scattering · Normalization of systems · Matrix representation

2020 Mathematical Subject Classification: Primary 35Q55 · Secondary 35A22, 35B40

Dedicated to Professor Tohru Ozawa on the occasion of his sixtieth birthday.

S. Masaki
Department of Mathematics, Faculty of Science, Hokkaido University, Kita 10, Nishi 8, Kita-Ku, Sapporo, Hokkaido, Osaka 060-0810, Japan
e-mail: masaki@math.sci.hokudai.ac.jp

J.-I. Segata (✉)
Faculty of Mathematics, Kyushu University, Fukuoka 819-0395, Japan
e-mail: segata@math.kyushu-u.ac.jp

K. Uriya
Department of Applied Mathematics, Faculty of Science, Okayama University of Science, Okayama 700-0005, Japan
e-mail: uriya@xmath.ous.ac.jp

S. Machihara (ed.), *Mathematical Physics and Its Interactions*, Springer Proceedings in Mathematics & Statistics 451, https://doi.org/10.1007/978-981-97-0364-7_5

1 Introduction

The paper is devoted to the study of the asymptotic behavior in time of solutions to the Cauchy problem of the following two systems of cubic nonlinear Schrödinger (NLS) equations in one space dimension: The first one is

$$\begin{cases} i\partial_t u_1 + \frac{1}{2}\partial_x^2 u_1 = 3\lambda_1 |u_1|^2 u_1, & t \in \mathbb{R},\ x \in \mathbb{R}, \\ i\partial_t u_2 + \frac{1}{2}\partial_x^2 u_2 = \lambda_6 (2|u_1|^2 u_2 + u_1^2 \overline{u_2}), & t \in \mathbb{R},\ x \in \mathbb{R}, \\ u_1(0,x) = u_{1,0}(x), \quad u_2(0,x) = u_{2,0}(x), & x \in \mathbb{R} \end{cases} \tag{1.1}$$

and

$$\begin{cases} i\partial_t u_1 + \frac{1}{2}\partial_x^2 u_1 = 0, & t \in \mathbb{R},\ x \in \mathbb{R}, \\ i\partial_t u_2 + \frac{1}{2}\partial_x^2 u_2 = 3|u_1|^2 u_1, & t \in \mathbb{R},\ x \in \mathbb{R}, \\ u_1(0,x) = u_{1,0}(x), \quad u_2(0,x) = u_{2,0}(x), & x \in \mathbb{R} \end{cases} \tag{1.2}$$

is the second, where $u_j : \mathbb{R} \times \mathbb{R} \to \mathbb{C}$ $(j = 1, 2)$ are unknown functions, $u_{j,0} : \mathbb{R} \to \mathbb{C}$ $(j = 1, 2)$ are given functions, and λ_1 and λ_6 are real constants satisfying $(\lambda_1, \lambda_6) \neq (0, 0)$ and $(\lambda_6 - \lambda_1)(\lambda_6 - 3\lambda_1) \geqslant 0$.

The systems (1.1) and (1.2) are particular cases of

$$\begin{cases} i\partial_t u_1 + \frac{1}{2}\partial_x^2 u_1 = 3\lambda_1 |u_1|^2 u_1 + \lambda_2 (2|u_1|^2 u_2 + u_1^2 \overline{u_2}) + \lambda_3 (2u_1 |u_2|^2 + \overline{u_1} u_2^2) + 3\lambda_4 |u_2|^2 u_2, \\ i\partial_t u_2 + \frac{1}{2}\partial_x^2 u_2 = 3\lambda_5 |u_1|^2 u_1 + \lambda_6 (2|u_1|^2 u_2 + u_1^2 \overline{u_2}) + \lambda_7 (2u_1 |u_2|^2 + \overline{u_1} u_2^2) + 3\lambda_8 |u_2|^2 u_2, \end{cases} \tag{1.3}$$

where $t \in \mathbb{R}$, $x \in \mathbb{R}$ and λ_j $(j = 1, \ldots, 8)$ are real constants. The system (1.3) includes several important physical models such as Manakov system [21] or a system describing spinor Bose-Einstein condensate [10]. Due to a classification result in our previous study [22], systems of the form (1.3) are classified according to the number of mass-like conserved quantities, which is connected to the complexity of the behavior of solution. In view of the classification theory, the study of the peculiar systems (1.1) and (1.2) have an importance. We discuss it in detail below.

It is well known that the cubic nonlinearity is critical in one dimension from the view point of the asymptotic behavior of solutions to the NLS equations and systems. As for the single equation

$$i\partial_t u + \frac{1}{2}\Delta u = \lambda |u|^{p-1} u, \qquad t \in \mathbb{R},\ x \in \mathbb{R}^d, \tag{1.4}$$

where $\lambda \in \mathbb{R}$, $p = 1 + 2/d$ is known to be the critical exponent (see [1, 29, 32]). In the critical case $p = 1 + 2/d$, the long-range scattering occurs, namely, a class of solutions to (1.4) satisfies

$$u(t) \to t^{-\frac{d}{2}} W\left(\frac{\cdot}{t}\right) e^{\frac{i|\cdot|^2}{2t} - i\lambda |W(\frac{\cdot}{t})|^{\frac{2}{d}} \log t - i\frac{\pi}{4}d} \quad \text{as } t \to \infty, \tag{1.5}$$

for some function $W \in L^\infty$ in a suitable topology. This kind of asymptotic behavior is also called the modified scattering because it involves a phase correction (see Ozawa [25] and Ginibre–Ozawa [3] for the final value problem and Hayashi–Naumkin [8] for the initial value problem).

The system (1.3) is cubic and the asymptotic behavior of solutions depends on the coefficients of the nonlinearities. Our underlying motivation of the study in the present paper is to find the all possible behavior to the system of the form (1.3) and more general system

$$\begin{cases} i\partial_t u_1 + \frac{1}{2}\partial_x^2 u_1 = c_1|u_1|^2 u_1 + c_2|u_1|^2 u_2 + c_3 u_1^2 \overline{u_2} \\ \qquad\qquad\qquad + c_4 u_1 |u_2|^2 + c_5 \overline{u_1} u_2^2 + c_6 |u_2|^2 u_2, \\ i\partial_t u_2 + \frac{1}{2}\partial_x^2 u_2 = c_7|u_1|^2 u_1 + c_8|u_1|^2 u_2 + c_9 u_1^2 \overline{u_2} \\ \qquad\qquad\qquad + c_{10} u_1 |u_2|^2 + c_{11} \overline{u_1} u_2^2 + c_{12} |u_2|^2 u_2, \end{cases} \tag{1.6}$$

where c_j $(j = 1, \ldots, 12)$ are real constants. Even though the systems of the form (1.3) or (1.6) are somewhat restricted in the sense that the nonlinearities do not contain derivatives of unknowns and that the coefficients are real, the variety of behavior of solutions to these systems is still richer than the single equations has.

Now, let us recall previous results on the large time behavior of solutions to systems of the NLS equations. As for the cubic system in one dimension with nonlinearities with/without derivatives, the *null condition*, a sufficient condition on nonlinearity for the existence of a non-trivial solution which asymptotically behaves like a free solution, is obtained in [13] (see [31] for the single equation). In [17], the long-range scattering is obtained for a matrix-valued equation. As mentioned above, a quadratic nonlinearity is critical in two dimensions and the asymptotic behavior of solutions to the two dimensional quadratic systems is also extensively studied. In this case, the ratio of the masses of two components matters. This phenomenon is called mass-resonance (see [4–6, 11] for systems with non-derivative nonlinearities and [12, 26] for those with derivative nonlinearities). The phenomenon is studied also for the one dimensional cubic systems (see [23, 33]).

In this paper, we restrict ourselves to the case where the coefficients of the nonlinearities are real numbers, as mentioned above. It is known that the NLS equations/systems with imaginary coefficients admit solution with different kinds of behavior. A typical example is the single NLS equation (1.4). If the coefficient λ is an imaginary number, a nonlinear amplification/dissipation phenomenon takes place. More precisely, if $\lambda \in \mathbb{C} \setminus \mathbb{R}$ then (1.4) is dissipative in one time direction and amplifying in the other direction. The sign of $\operatorname{Im} \lambda$ decides which direction is the dissipative direction. In the dissipative direction, it has shown that the small solution decays faster than the free Schrödinger evolution (see [7, 9, 16, 24, 27, 28]). On the other hand, Kita [15] showed that, in the amplifying direction, there exists an arbitrarily small data which gives a blowing-up solution. Systems with the dissipative structure is also intensively studied. See [14] for systems with non-derivative nonlinearities and [20] for those with derivative nonlinearities. Recently, new type of behavior of solution is found in a certain system of the cubic NLS equations in one dimension in [18, 19].

It turns out that there exists a system of the form (1.6) such that the amplification/dissipation phenomenon takes place, although the coefficients of the nonlinearities are real. Further, the system admits the following three types of solutions; (i) blowup forward in time and dissipative decay backward in time; (ii) blowup backward in time and dissipative decay forward in time; (iii) blowup for both time directions. This system is an evidence for the richness of the variety of behaviors for systems. We discuss this system in Appendix B.

In [22], the authors considered the system of cubic nonlinear Klein–Gordon equations in one space dimension:

$$\begin{cases} (\Box + 1)u_1 = \lambda_1 u_1^3 + \lambda_2 u_1^2 u_2 + \lambda_3 u_1 u_2^2 + \lambda_4 u_2^3, & t \in \mathbb{R},\ x \in \mathbb{R}, \\ (\Box + 1)u_2 = \lambda_5 u_1^3 + \lambda_6 u_1^2 u_2 + \lambda_7 u_1 u_2^2 + \lambda_8 u_2^3, & t \in \mathbb{R},\ x \in \mathbb{R}, \end{cases} \tag{1.7}$$

where $u_j : \mathbb{R} \times \mathbb{R} \to \mathbb{R}$ $(j = 1, 2)$ are real-valued unknowns, $\Box = \partial_t^2 - \partial_x^2$ is the d'Alembertian, and $\lambda_1, \dots, \lambda_8$ are real constants. This system is closed under the linear transformation of unknowns. An equivalence relation between two systems is then naturally introduced by the linear transformation of unknowns. They give a classification result by considering a quotient set of the systems with respect to the equivalence relation. The classification result is applicable to (1.3). This is because the change of coefficients caused by the linear transformation of unknown is identical to that for (1.7). This agrees with the fact that the asymptotic profile for a solution of (1.7) and that for a solution of (1.3) are, at least formally, described by the same ODE system

$$\begin{cases} i\partial_t \alpha_1 = 3\lambda_1 |\alpha_1|^2 \alpha_1 + \lambda_2 (2|\alpha_1|^2 \alpha_2 + \alpha_1^2 \overline{\alpha_2}) + \lambda_3 (2\alpha_1 |\alpha_2|^2 + \overline{\alpha_1} \alpha_2^2) + 3\lambda_4 |\alpha_2|^2 \alpha_2, \\ i\partial_t \alpha_2 = 3\lambda_5 |\alpha_1|^2 \alpha_1 + \lambda_6 (2|\alpha_1|^2 \alpha_2 + \alpha_1^2 \overline{\alpha_2}) + \lambda_7 (2\alpha_1 |\alpha_2|^2 + \overline{\alpha_1} \alpha_2^2) + 3\lambda_8 |\alpha_2|^2 \alpha_2. \end{cases} \tag{1.8}$$

Hereinafter, we refer to the system as *limit ODE system*. The classification result is also applicable to the system (1.8).

Let us quickly review the classification result in [22] by taking (1.3) as an example. The key ingredient is the introduction of a matrix representation of a system: A system (1.3) is identified with a matrix

$$A = \begin{pmatrix} \lambda_2 & -3\lambda_1 + \lambda_6 & -3\lambda_5 \\ \lambda_3 & -\lambda_2 + \lambda_7 & -\lambda_6 \\ 3\lambda_4 & 3\lambda_8 - \lambda_3 & -\lambda_7 \end{pmatrix}. \tag{1.9}$$

It then turns out that the change caused by the linear transformation of unknowns is clearly formulated as a matrix manipulation and, moreover, the characteristic properties such as conservation laws are well described by the matrix (see Section A.4). In particular, rank A is an invariant quantity which indicates the number of mass-like conserved quantities. Roughly speaking, the behavior of solution becomes complicated as rank A increases. In [22], the authors classify two subsets of systems. One is the set of systems such that rank $A = 1$ and the other is the set of systems such that $B = O$, where B is the matrix defined from the coefficients of the system as

$$B := \begin{pmatrix} -12\lambda_5 & 3(\lambda_1 - \lambda_6) & 2(\lambda_2 - \lambda_7) \\ 3(\lambda_1 - \lambda_6) & 2(\lambda_2 - \lambda_7) & 3(\lambda_3 - \lambda_8) \\ 2(\lambda_2 - \lambda_7) & 3(\lambda_3 - \lambda_8) & 12\lambda_4 \end{pmatrix} \tag{1.10}$$

and $O \in M_3(\mathbb{R})$ is the zero matrix. The quotient set of the first subset contains 9 equivalent classes. And the quotient set of the second does 5 equivalent classes.

In Appendix A, we review the classification result in more detail and extend it to the set of systems of the form (1.6). Note that the system of the form (1.6) is identified with a vector $(c_j)_{1\leqslant j\leqslant 12} \in \mathbb{R}^{12}$. Thus, it is difficult to correspond it with a 3×3 matrix like (1.9). Hence, we introduce a new way to represent a system. This is an extension of the matrix representation of (1.3). This enables us to formulate the equivalence relation in a clear way and describe the validity of conservation laws for the generalized system (1.6). As an application, a global existence result for (1.6) is shown in Proposition A.8. Note that the local well-posedness of the Cauchy problem of the system (1.6) is obtained by a standard theory in several function spaces such as the Lebesgue space $L^2(\mathbb{R})$, the Sobolev space $H^1(\mathbb{R})$ and so on, see [2] for instance.

Let us now discuss systems (1.1) and (1.2) in view of the classification result. As for the systems (1.7) and (1.3), the behavior of solutions is previously studied in the case that rank $A \leqslant 1$ or $B = O$. The conditions are connected to the existence of conserved quantities for theses systems and also for the corresponding limit ODE system (1.8). Notice that the system (1.2) satisfies rank $A = 1$. The end-point case $\lambda_6 = 3\lambda_1$ of (1.1) also corresponds to the case rank $A = 1$. In the other end-point case $\lambda_6 = \lambda_1$, one has $B = O$. We here remark that these cases are previously studied for the corresponding Klein–Gordon system (1.2) in [22, 30]. When $(\lambda_6 - \lambda_1)(\lambda_6 - 3\lambda_1) > 0$, we have rank $A = 2$ and $B \neq O$. As far as the authors know, this case is new. It is revealed that the asymptotic profile for the second component involves two parts which oscillate in a different way. This is the main result of this paper.

1.1 Main Results

To state main results, we define the weighted L^2 space $H^{0,1}(\mathbb{R})$ by

$$H^{0,1}(\mathbb{R}) = \left\{ f \in L^2(\mathbb{R}) \mid \|f\|_{H^{0,1}} = \|\langle x\rangle f\|_{L^2} < \infty \right\},$$

where $\langle x\rangle = \sqrt{1 + |x|^2}$.

Let us first consider the case $(\lambda_6 - \lambda_1)(\lambda_6 - 3\lambda_1) > 0$ of (1.1). In this case, one has rank $A = 2$ and $B \neq O$, where A and B are defined in (1.9) and (1.10), respectively. We have the following result on asymptotic behavior of solution to (1.1).

Theorem 1.1 *Suppose* $(\lambda_6 - \lambda_1)(\lambda_6 - 3\lambda_1) > 0$. *Let* $0 < \gamma < \delta < 1/100$. *Then there exists* $\varepsilon_0 > 0$ *such that for any* $u_{j,0} \in H^1(\mathbb{R}) \cap H^{0,1}(\mathbb{R})$ *satisfying* $\varepsilon := \sum_{j=1}^{2} (\|u_{j,0}\|_{H^1} + \|u_{j,0}\|_{H^{0,1}}) \leqslant \varepsilon_0$, *there exists a unique global solution* $u_j \in C(\mathbb{R}, H^1(\mathbb{R}) \cap H^{0,1}(\mathbb{R}))$ *of (1.1) satisfying*

$$\|u_1(t)\|_{H_x^{0,1}} \lesssim \varepsilon\langle t\rangle^{\gamma}, \quad \|u_1(t)\|_{L_x^\infty} \lesssim \varepsilon\langle t\rangle^{-\frac{1}{2}},$$
$$\|u_2(t)\|_{H_x^{0,1}} \lesssim \varepsilon\langle t\rangle^{\delta}, \quad \|u_2(t)\|_{L_x^\infty} \lesssim \varepsilon\langle t\rangle^{-\frac{1}{2}}$$

for any $t \in \mathbb{R}$. *Furthermore, there exist two functions* $W_1, W_2 \in L^\infty$ *such that*

$$u_1(t) = t^{-\frac{1}{2}} W_1\left(\frac{x}{t}\right) e^{\frac{ix^2}{2t} - i3\lambda_1 |W_1(\frac{x}{t})|^2 \log t - i\frac{\pi}{4}} + O(t^{-\frac{3}{4}+\gamma}), \tag{1.11}$$

$$\begin{aligned} u_2(t) = t^{-\frac{1}{2}} \mathbf{1}_{\{W_1 \neq 0\}}\left(\frac{x}{t}\right) \Bigg[&\lambda_6 \left(\frac{W_1^2}{|W_1|^2} W_2\right)\left(\frac{x}{t}\right) e^{i(-3\lambda_1 - \lambda_c)|W_1(\frac{x}{t})|^2 \log t} \\ &+ (3\lambda_1 - 2\lambda_6 + \lambda_c)\overline{W_2}\left(\frac{x}{t}\right) e^{i(-3\lambda_1 + \lambda_c)|W_1(\frac{x}{t})|^2 \log t} \Bigg] e^{\frac{ix^2}{2t} - i\frac{\pi}{4}} \\ &+ O(t^{-\frac{3}{4}+\delta}), \end{aligned} \tag{1.12}$$

in $L^\infty(\mathbb{R})$ *as* $t \to \infty$, *where* $\lambda_c = \sqrt{3(\lambda_6 - \lambda_1)(\lambda_6 - 3\lambda_1)}$. *Similar asymptotic formulas for* u_j *hold for* $t \to -\infty$.

Remark 1.2 If we assume the opposite inequality $(\lambda_6 - \lambda_1)(\lambda_6 - 3\lambda_1) < 0$, we formally obtain the same asymptotic profile of the solution. However, since λ_c is an imaginary number in this case, we see that one part of the asymptotic profile grows and the other part decays, which is problematic in obtaining rigorous asymptotics.

Let us move to the limiting case, $\lambda_6 = \lambda_1$ or $\lambda_6 = 3\lambda_1$, of (1.1). Remark that rank $A = 1$ if $\lambda_6 = 3\lambda_1$ and $B = O$ if $\lambda_6 = \lambda_1$. In this case, we have the following result.

Theorem 1.3 *Suppose* $\lambda_6 = 3\lambda_1 \neq 0$ *or* $\lambda_6 = \lambda_1 \neq 0$. *Let* $0 < \gamma < \delta < 1/100$. *Then there exists* $\varepsilon_0 > 0$ *such that for any* $u_{j,0} \in H^1(\mathbb{R}) \cap H^{0,1}(\mathbb{R})$ *satisfying* $\varepsilon := \sum_{j=1}^{2}(\|u_{j,0}\|_{H^1} + \|u_{j,0}\|_{H^{0,1}}) \leqslant \varepsilon_0$, *there exists a unique global solution* $u_j \in C(\mathbb{R}, H^1(\mathbb{R}) \cap H^{0,1}(\mathbb{R}))$ *of (1.1) satisfying*

$$\|u_1(t)\|_{H_x^{0,1}} \lesssim \varepsilon\langle t\rangle^{\gamma}, \quad \|u_1(t)\|_{L_x^\infty} \lesssim \varepsilon\langle t\rangle^{-\frac{1}{2}},$$
$$\|u_2(t)\|_{H_x^{0,1}} \lesssim \varepsilon\langle t\rangle^{\delta}, \quad \|u_2(t)\|_{L_x^\infty} \lesssim \varepsilon\langle t\rangle^{-\frac{1}{2}} \log\langle t\rangle$$

for any $t \in \mathbb{R}$. *Furthermore, there exist two functions* $W_j \in L^\infty$ *such that*

$$u_1(t) = t^{-\frac{1}{2}} W_1\left(\frac{x}{t}\right) e^{\frac{ix^2}{2t} - i3\lambda_1 |W_1(\frac{x}{t})|^2 \log t - i\frac{\pi}{4}} + O(t^{-\frac{3}{4}+\gamma}), \tag{1.13}$$

$$u_2(t) = t^{-\frac{1}{2}} \left\{W\left(\frac{x}{t}\right) \log t + W_2\left(\frac{x}{t}\right)\right\} e^{\frac{ix^2}{2t} - i3\lambda_1 |W_1(\frac{x}{t})|^2 \log t - i\frac{\pi}{4}} + O(t^{-\frac{3}{4}+\delta}) \tag{1.14}$$

in $L^\infty(\mathbb{R})$ *as* $t \to \infty$, *where* W *is given by*

$$W(y) = \begin{cases} -6i\lambda_1 W_1(y) \operatorname{Re}\left[W_1(y)\overline{W_2(y)}\right] & \text{if } \lambda_6 = 3\lambda_1, \\ 2\lambda_1 W_1(y) \operatorname{Im}\left[W_1(y)\overline{W_2(y)}\right] & \text{if } \lambda_6 = \lambda_1. \end{cases}$$

Similar asymptotic formulas for u_j *hold for* $t \to -\infty$.

Remark 1.4 An intuitive summary of Theorems 1.1 and 1.3 is as follows. In the non-limiting case $(\lambda_6 - \lambda_1)(\lambda_6 - 3\lambda_1) > 0$, we have $\lambda_c > 0$ and hence the asymptotic profile of the second component u_2 has two parts which have different phase modifications, i.e., which oscillate in a different way. However, in the limiting case $(\lambda_6 - \lambda_1)(\lambda_6 - 3\lambda_1) = 0$, we have $\lambda_c = 0$ and the oscillation of the two parts become the same. The coincidence causes a logarithmic amplitude correction together with the fact that the coefficients of the two parts becomes the same when $\lambda_6 = \lambda_1$ and the same modulus with the different signs when $\lambda_6 = 3\lambda_1$.

Remark 1.5 In the same way as in the proof of Theorem 1.3, it is not hard to see that

$$\|u_2(t)\|_{L^2_x} \lesssim \varepsilon \log t$$

for any $t \geqslant 2$. Furthermore, we see that $W_j \in L^2 \cap L^\infty$ and

$$u_2(t) = t^{-\frac{1}{2}} \left\{W\left(\frac{x}{t}\right) \log t + W_2\left(\frac{x}{t}\right)\right\} e^{\frac{ix^2}{2t} - i3\lambda_1 |W_1(\frac{x}{t})|^2 \log t - i\frac{\pi}{4}} + O(t^{-\frac{1}{4}+\delta})$$

in $L^2(\mathbb{R})$ as $t \to \infty$, where W is given in Theorem 1.3. From this asymptotic formula, we have the lower bound of L^2 norm of u_2:

$$\|u_2(t)\|_{L^2_x} \geqslant \|W\|_{L^2_x} \log t - C\varepsilon.$$

for $t \geqslant 2$.

We turn to (1.2). The asymptotic behavior of solutions to the corresponding system of nonlinear Klein–Gordon equations is studied in Sunagawa [30]. As for (1.2), merely the growth of the L^2-norm of the solution in a logarithmic rate is given (see [13]). Here, we obtain the explicit asymptotic formula of the solution.

Theorem 1.6 *Let* $0 < \gamma < 1/100$. *Then there exists* $\varepsilon_0 > 0$ *such that for any* $u_{j,0} \in H^1(\mathbb{R}) \cap H^{0,1}(\mathbb{R})$ *satisfying* $\varepsilon := \sum_{j=1}^2 (\|u_{j,0}\|_{H^1} + \|u_{j,0}\|_{H^{0,1}}) \leqslant \varepsilon_0$, *there exists a unique global solution* $u_j \in C(\mathbb{R}, H^1(\mathbb{R}) \cap H^{0,1}(\mathbb{R}))$ *of (1.2) satisfying*

$$\|u_1(t)\|_{H_x^{0,1}} \lesssim \varepsilon, \quad \|u_1(t)\|_{L_x^\infty} \gtrsim \varepsilon\langle t\rangle^{-\frac{1}{2}},$$
$$\|u_2(t)\|_{H_x^{0,1}} \lesssim \varepsilon\langle t\rangle^{\gamma}, \quad \|u_2(t)\|_{L_x^\infty} \lesssim \varepsilon\langle t\rangle^{-\frac{1}{2}} \log\langle t\rangle$$

for any $t \in \mathbb{R}$. Furthermore, let $W_1 := \hat{u}_{1,0}$. Then there exists a function $W_2 \in L^\infty$ such that

$$u_2(t) = -it^{-\frac{1}{2}} \left\{ \left| W_1\left(\frac{x}{t}\right)\right|^2 W_1\left(\frac{x}{t}\right) \log t + W_2\left(\frac{x}{t}\right)\right\} e^{\frac{ix^2}{2t} - i\frac{\pi}{4}} + O(t^{-\frac{3}{4}+\delta}) \tag{1.15}$$

in $L^\infty(\mathbb{R})$ as $t \to \infty$. Similar asymptotic formula for u_2 holds for $t \to -\infty$.

Remark 1.7 Our classification argument suggests that the system (1.2) can be regarded as a limiting case of (1.1). One sees that the behavior of solution is similar in these two cases. Especially, it is common that the second component has a *logarithmic amplitude correction*. The difference is as follows: The behavior of solutions to (1.1) involves a logarithmic phase correction term. Further, the logarithmic amplitude correction depends not only on W_1 but also on W_2. This reflects the difference of the mechanism of appearance of logarithmic amplitude correction, which is, at least formally, easily verified by the analysis of the corresponding limit ODE systems.

Remark 1.8 Remark that the above theorems do not follow from the argument of Katayama–Sakoda [12] since (1.1) and (1.2) do not satisfy their assumption.

We now give the outline of the proofs of Theorems 1.1 and 1.3. Let $w_j := \mathcal{F}U(-t)u_j$, $j = 1, 2$, where $\mathcal{F}$ is the Fourier transform in space variable and $U(t)$ is the free Schrödinger group (see (2.2) for the definition). We shall obtain the large time asymptotics of u_j by deriving the large time asymptotic formula for w_j. Applying $\mathcal{F}U(-t)$ to (1.1), we obtain

$$\begin{cases} i\partial_t w_1 = 3\lambda_1 t^{-1}|w_1|^2 w_1 + o(t^{-1}), \\ i\partial_t w_2 = \lambda_6 t^{-1}(2|w_1|^2 w_2 + w_1^2 \overline{w_2}) + o(t^{-1}), \end{cases}$$

as $t \to \infty$. Hence we formally see that the large time behavior of (w_1, w_2) is given by system of ODE

$$\begin{cases} i\partial_t \alpha_1 = 3\lambda_1 t^{-1}|\alpha_1|^2 \alpha_1, \\ i\partial_t \alpha_2 = \lambda_6 t^{-1}(2|\alpha_1|^2 \alpha_2 + \alpha_1^2 \overline{\alpha_2}), \end{cases} \tag{1.16}$$

where $\alpha_j = \alpha_j(t, \xi)$. From the first equation of (1.16), we easily obtain

$$\alpha_1(t, \xi) = \alpha_1(0, \xi) e^{-3i\lambda_1 |\alpha_1(0,\xi)|^2 \log t}. \tag{1.17}$$

The key point to prove Theorems 1.1 and 1.3 is how to find a explicit form of the second component α_2. For the non-limiting case $(\lambda_6 - \lambda_1)(\lambda_6 - 3\lambda_1) > 0$, substituting (1.17) into (1.16), and letting $\beta(t,\xi) := \alpha_2(t,\xi)\exp\left(3i\lambda_1|\alpha_1(0,\xi)|^2\log t\right)$, we have

$$i\partial_t\beta(t,\xi) = (-3\lambda_1 + 2\lambda_6)t^{-1}|\alpha_1(0,\xi)|^2\beta(t,\xi) + \lambda_6 t^{-1}\alpha_1(0,\xi)^2\overline{\beta(t,\xi)}.$$

By diagonalizing the system for $(\beta, \overline{\beta})$, we have the explicit form of β (hence α_2). Note that α_2 has two parts which have different phase modifications. For the limiting case $(\lambda_6 - \lambda_1)(\lambda_6 - 3\lambda_1) = 0$, we use the conservation laws:

$$\mathrm{Re}\,[\alpha_1(t,\xi)\overline{\alpha}_2(t,\xi)] = \mathrm{Re}\,[\alpha_1(0,\xi)\overline{\alpha}_2(0,\xi)] \quad \text{for } \lambda_6 = 3\lambda_1, \tag{1.18}$$

$$\mathrm{Im}\,[\alpha_1(t,\xi)\overline{\alpha}_2(t,\xi)] = \mathrm{Im}\,[\alpha_1(0,\xi)\overline{\alpha}_2(0,\xi)] \quad \text{for } \lambda_6 = \lambda_1. \tag{1.19}$$

For the case $\lambda_6 = 3\lambda_1$, substituting (1.17) and (1.18) into the second equation of (1.16), we have the first order linear ODE with external force:

$$i\partial_t\alpha_2(t,\xi) - 3\lambda_1|\alpha_1(0,\xi)|^2\alpha_2(t,\xi) = \lambda_6 t^{-1}\,\mathrm{Re}\,[\alpha_1(0,\xi)\overline{\alpha}_2(0,\xi)]\,\alpha_1(0)e^{-3i\lambda_1|\alpha_1(0,\xi)|^2\log t}.$$

Hence we have the explicit form of α_2. Note that the external force "resonate" with the linear term, so α_2 have logarithmic factor in the amplitude part. In a similar way, we also obtain the explicit form of α_2 for the case $\lambda_6 = \lambda_1$. From the exact forms of α_j, we obtain asymptotics of w_j (hence u_j). We shall justify those formal arguments in Sects. 2 and 3.

The rest of the paper is organized as follows. In Sect. 2, we first prove our main result (Theorem 1.1). Then, we turn to the proofs of Theorems 1.3 and 1.6 in Sect. 3. Appendix A is devoted to the classification result of (1.6). A global well-posdeness result for (1.6) is given as an application in Proposition A.8. Finally, we exhibit an interesting example of system in Appendix B.

2 Proof of Theorem 1.1

In this section, we prove Theorem 1.1. Let us recall that (u_1, u_2) satisfies

$$\begin{cases} i\partial_t u_1 + \dfrac{1}{2}\partial_x^2 u_1 = 3\lambda_1|u_1|^2u_1, & t\in\mathbb{R},\ x\in\mathbb{R},\\ i\partial_t u_2 + \dfrac{1}{2}\partial_x^2 u_2 = \lambda_6(2|u_1|^2u_2 + u_1^2\overline{u_2}), & t\in\mathbb{R},\ x\in\mathbb{R},\\ u_1(0,x) = u_{1,0}(x), \qquad u_2(0,x) = u_{2,0}(x), & x\in\mathbb{R}, \end{cases} \tag{2.1}$$

where λ_1 and λ_6 satisfy $(\lambda_6 - \lambda_1)(\lambda_6 - 3\lambda_1) > 0$.

To analyze the solution to (2.1), we introduce several linear operators. Let $\{U(t)\}_{t\in\mathbb{R}}$ be a unitary group generated by $i\partial_x^2/2$, i.e.,

$$U(t) := \mathcal{F}^{-1} e^{-\frac{it\xi^2}{2}} \mathcal{F}, \tag{2.2}$$

where $\mathcal{F}$ and $\mathcal{F}^{-1}$ are usual Fourier transform and its inverse transform.

We define multiplication operator $M(t)$ and dilation operator $D(t)$ by

$$(M(t)f)(x) = e^{\frac{ix^2}{2t}} f(x), \quad (D(t)f)(x) = t^{-\frac{1}{2}} f\left(\frac{x}{t}\right) e^{-\frac{i\pi}{4}}, \quad t \in \mathbb{R}\backslash\{0\}.$$

Then we have well-known Dollard decomposition for free Schrödinger group:

$$U(t) = M(t)D(t)\mathcal{F}M(t). \tag{2.3}$$

Let $w_j := \mathcal{F}U(-t)u_j$, $j = 1, 2$. Then, applying $\mathcal{F}U(-t)$ to (2.1), we obtain

$$i\partial_t w_1 = 3\lambda_1 \mathcal{F}U(-t)|U(t)\mathcal{F}^{-1}w_1|^2 U(t)\mathcal{F}^{-1}w_1. \tag{2.4}$$

By using the Dollard decomposition (2.3), we easily see that

$$\mathcal{F}U(-t) = U(1/t)D^{-1}(t)M^{-1}(t), \tag{2.5}$$

$$U(t)\mathcal{F}^{-1} = M(t)D(t)U(-1/t). \tag{2.6}$$

We summarize several estimates for the operator $U(\pm 1/t)$.

Lemma 2.1 *(i) There exists a positive constant C such that for any* $0 < \alpha < 1/4$ *and* $\varphi \in H^1_\xi(\mathbb{R})$*, we have*

$$\|U(\pm 1/t)\varphi - \varphi\|_{L^\infty_\xi} \lesssim t^{-\alpha}\|\varphi\|_{H^1_\xi}. \tag{2.7}$$

(ii) There exists a positive constant C such that for any $\varphi \in H^1_\xi(\mathbb{R})$*, we have*

$$\|U(\pm 1/t)\varphi\|_{H^1_\xi} \lesssim \|\varphi\|_{H^1_\xi}. \tag{2.8}$$

Proof of Lemma 2.1 The proof easily follows from the explicit representation (2.3) of the unitary group $U(t)$. □

By (2.5) and (2.6), Eq. (2.4) can be rewritten as

$$\begin{aligned} i\partial_t w_1 &= 3\lambda_1 U(1/t)D^{-1}(t)M^{-1}(t)|M(t)D(t)U(-1/t)w_1|^2 M(t)D(t)U(-1/t)w_1 \\ &= 3\lambda_1 t^{-1} U(1/t)|U(-1/t)w_1|^2 U(-1/t)w_1. \end{aligned} \tag{2.9}$$

In a similar way, we have

$$i\partial_t w_2 = \lambda_6 t^{-1} U(1/t)\left\{2|U(-1/t)w_1|^2 U(-1/t)w_2 + (U(-1/t)w_1)^2 \overline{U(-1/t)w_2}\right\}. \tag{2.10}$$

We first obtain short time bounds of w_j.

Lemma 2.2 (Short time bounds) *There exists $\varepsilon_0 > 0$ such that for any $u_{j,0} \in H^1(\mathbb{R}) \cap H^{0,1}(\mathbb{R})$ satisfying $\varepsilon := \sum_{j=1}^{2}(\|u_{j,0}\|_{H^1} + \|u_{j,0}\|_{H^{0,1}}) \leqslant \varepsilon_0$, there exists a unique solution $u_j \in C([0,1], H^1(\mathbb{R}) \cap H^{0,1}(\mathbb{R}))$ of (2.1) satisfying*

$$\sup_{t\in[0,1]} \sum_{j=1}^{2} \|w_j(t)\|_{H^1_\xi} \lesssim \varepsilon.$$

Proof of Lemma 2.2 The proof follows from a standard well-posedness theory, see [2] for instance. Hence we omit the proof. □

Next we derive a long time bounds of w_j. We fix $0 < \gamma < \delta < 1/100$ and introduce

$$\begin{aligned}\|(w_1, w_2)\|_{X_T} := \sup_{t\in[1,T]} \Big\{ &\|w_1(t)\|_{L^\infty_\xi} + \langle t\rangle^{-\gamma}\|w_1(t)\|_{H^1_\xi} \\ &+ (\log\langle t\rangle)^{-1}\|w_2(t)\|_{L^\infty_\xi} + \langle t\rangle^{-\delta}\|w_2(t)\|_{H^1_\xi}\Big\}.\end{aligned}$$

Lemma 2.3 (Long time bounds) *There exists $\varepsilon_0 > 0$ such that for any $u_{j,0} \in H^1(\mathbb{R}) \cap H^{0,1}(\mathbb{R})$ satisfying $\varepsilon := \sum_{j=1}^{2}(\|u_{j,0}\|_{H^1} + \|u_{j,0}\|_{H^{0,1}}) \leqslant \varepsilon_0$, there exists a unique global solution $u_j \in C([0,\infty), H^1(\mathbb{R}) \cap H^{0,1}(\mathbb{R}))$ of (2.1) satisfying*

$$\|(w_1, w_2)\|_{X_\infty} \lesssim \varepsilon. \tag{2.11}$$

Remark 2.4 If (w_1, w_2) satisfy (2.11), then we obtain L^∞ decay estimates for solution u_1 of (2.1). Indeed, by Lemma 2.1, and (2.11), we see,

$$\begin{aligned}\|u_1(t)\|_{L^\infty_x} &= \|U(t)\mathcal{F}^{-1}w_1(t)\|_{L^\infty_x} \\ &= \|M(t)D(t)U(-1/t)w_1(t)\|_{L^\infty_\xi} \\ &\lesssim t^{-\frac{1}{2}}\|U(-1/t)w_1(t)\|_{L^\infty_\xi} \\ &\lesssim t^{-\frac{1}{2}}(\|w_1(t)\|_{L^\infty_\xi} + t^{-\alpha}\|w_1(t)\|_{H^1_\xi}) \\ &\lesssim \varepsilon(t^{-\frac{1}{2}} + t^{-\frac{1}{2}-\alpha+\gamma}),\end{aligned}$$

for any $t \geqslant 1$, where $0 < \alpha < 1/4$. Hence choosing α so that $0 < \gamma < \alpha$, we obtain

$$\|u_1(t)\|_{L^\infty_x} \lesssim \varepsilon t^{-\frac{1}{2}}.$$

In a similar way, we have (non-optimal) decay estimate for u_2:

$$\|u_2(t)\|_{L^\infty_x} \lesssim \varepsilon t^{-\frac{1}{2}} \log t$$

for any $t \geqslant 1$. We will show later that $u_2(t) = O(t^{-1/2})$ by analyzing the large time asymptotics of u_2.

Proof of Lemma 2.3 We first evaluate H^1 norm of w_1. By (2.9), we have

$$w_1(t) = w_1(1) - 3i\lambda_1 \int_1^t \tau^{-1} U(1/\tau)\left[|U(-1/\tau)w_1|^2 U(-1/\tau)w_1\right](\tau)d\tau.$$

Then, we see

$$\|w_1(t)\|_{H^1_\xi} \lesssim \|w_1(1)\|_{H^1_\xi} + \int_1^t \tau^{-1} \|U(1/\tau)\left[|U(-1/\tau)w_1|^2 U(-1/\tau)w_1\right]\|_{H^1_\xi} d\tau.$$

By Lemma 2.1,

$$\begin{aligned}
&\|U(1/\tau)\left[|U(-1/\tau)w_1|^2 U(-1/\tau)w_1\right]\|_{H^1_\xi} \\
&\lesssim \||U(-1/\tau)w_1|^2 U(-1/\tau)w_1\|_{H^1_\xi} \\
&\lesssim \|U(-1/\tau)w_1\|^2_{L^\infty_\xi} \|U(-1/\tau)w_1\|_{H^1_\xi} \\
&\lesssim (\|w_1\|_{L^\infty_\xi} + \tau^{-\alpha}\|w_1\|_{H^1_\xi})^2 \|w_1\|_{H^1_\xi} \\
&\lesssim (1 + \tau^{-\alpha+\gamma})^2 \tau^\gamma \|(w_1, w_2)\|^3_{X_T},
\end{aligned}$$

where $0 < \alpha < 1/4$. Choosing α so that $\gamma < \alpha < 1/4$ and using Lemma 2.2, we find

$$\begin{aligned}
\langle t\rangle^{-\gamma}\|w_1(t)\|_{H^1_\xi} &\lesssim \varepsilon + \langle t\rangle^{-\gamma}\|(w_1, w_2)\|^3_{X_T} \int_1^t \tau^{-1+\gamma} d\tau \\
&\lesssim \varepsilon + \frac{1}{\gamma}\|(w_1, w_2)\|^3_{X_T}.
\end{aligned} \tag{2.12}$$

Next we evaluate H^1 norm of w_2. By (2.10), we have

$$\begin{aligned}
w_2(t) &= w_2(1) \\
&- i\lambda_6 \int_1^t \tau^{-1} U(1/\tau)\left[2|U(-1/\tau)w_1|^2 U(-1/\tau)w_2 + (U(-1/\tau)w_1)^2 \overline{U(-1/\tau)w_2}\right](\tau)d\tau.
\end{aligned}$$

Then, we see

$$\begin{aligned}
&\|w_2(t)\|_{H^1_\xi} \lesssim \|w_2(1)\|_{H^1_\xi} \\
&+ \int_1^t \tau^{-1} \left\|U(1/\tau)\left[2|U(-1/\tau)w_1|^2 U(-1/\tau)w_2 + (U(-1/\tau)w_1)^2 \overline{U(-1/\tau)w_2}\right]\right\|_{H^1_\xi} d\tau.
\end{aligned}$$

By Lemma 2.1,

$$
\begin{aligned}
&\left\| U(1/\tau)\left[2|U(-1/\tau)w_1|^2U(-1/\tau)w_2 + (U(-1/\tau)w_1)^2\overline{U(-1/\tau)w_2}\right]\right\|_{H^1_\xi} \\
&\lesssim \||U(-1/\tau)w_1|^2U(-1/\tau)w_2\|_{H^1_\xi} \\
&\lesssim \|U(-1/\tau)w_1\|_{L^\infty_\xi}\|U(-1/\tau)w_1\|_{H^1_\xi}\|U(-1/\tau)w_2\|_{L^\infty_\xi} \\
&\quad + \|U(-1/\tau)w_1\|^2_{L^\infty_\xi}\|U(-1/\tau)w_2\|_{H^1_\xi} \\
&\lesssim (\|w_1\|_{L^\infty_\xi} + \tau^{-\alpha}\|w_1\|_{H^1_\xi})\|w_1\|_{H^1_\xi}(\|w_2\|_{L^\infty_\xi} + \tau^{-\alpha}\|w_2\|_{H^1_\xi}) \\
&\quad + (\|w_1\|_{L^\infty_\xi} + \tau^{-\alpha}\|w_1\|_{H^1_\xi})^2\|w_2\|_{H^1_\xi} \\
&\lesssim (1+\tau^{-\alpha+\gamma})\tau^\gamma(\log\tau + \tau^{-\alpha+\delta})\|(w_1, w_2)\|^3_{X_T} \\
&\quad + (1+\tau^{-\alpha+\gamma})^2\tau^\delta\|(w_1, w_2)\|^3_{X_T},
\end{aligned}
$$

where $0 < \alpha < 1/4$. Choosing α so that $\delta < \alpha < 1/4$ and using Lemma 2.2, we find

$$
\begin{aligned}
\langle t\rangle^{-\delta}\|w_2(t)\|_{H^1_\xi} &\lesssim \varepsilon + \langle t\rangle^{-\delta}\|(w_1, w_2)\|^3_{X_T}\int_1^t \tau^{-1+\delta}d\tau \\
&\lesssim \varepsilon + \frac{1}{\delta}\|(w_1, w_2)\|^3_{X_T}.
\end{aligned}
\tag{2.13}
$$

Next we derive L^∞ estimates for w_j. From viewpoint of the asymptotic formulas for $U(\pm 1/t)$ (Lemma 2.1), we decompose the nonlinear term as follows:

$$
i\partial_t w_1 = 3\lambda_1 t^{-1}|w_1|^2 w_1 + R_1, \tag{2.14}
$$

$$
i\partial_t w_2 = \lambda_6 t^{-1}(2|w_1|^2 w_2 + w_1^2\overline{w_2}) + R_2, \tag{2.15}
$$

where R_1 and R_2 are given by

$$
\begin{aligned}
R_1 &= 3\lambda_1 t^{-1}[U(1/t)|U(-1/t)w_1|^2U(-1/t)w_1 - |w_1|^2 w_1], \\
R_2 &= \lambda_6 t^{-1}[U(1/t)\{2|U(-1/t)w_1|^2U(-1/t)w_2 + (U(-1/t)w_1)^2\overline{U(-1/t)w_2}\} \\
&\qquad - (2|w_1|^2 w_2 + w_1^2\overline{w_2})].
\end{aligned}
$$

Since

$$
\begin{aligned}
R_1 &= 3\lambda_1 t^{-1}[|U(-1/t)w_1|^2U(-1/t)w_1 - |w_1|^2 w_1] \\
&\quad + 3\lambda_1 t^{-1}(U(1/t) - 1)|U(-1/t)w_1|^2U(-1/t)w_1,
\end{aligned}
$$

by Lemma 2.1, we have

$$\begin{aligned}
&\|R_1\|_{L^\infty_\xi} \qquad (2.16)\\
&\lesssim t^{-1}(\|U(-1/t)w_1\|_{L^\infty_\xi} + \|w_1\|_{L^\infty_\xi})^2\|U(-1/t)w_1 - w_1\|_{L^\infty_\xi}\\
&\quad + t^{-1-\alpha}\||U(-1/t)w_1|^2U(-1/t)w_1\|_{H^1_\xi}\\
&\lesssim t^{-1-\alpha}(\|w_1\|_{L^\infty_\xi} + t^{-\alpha}\|w_1\|_{H^1_\xi})^2\|w_1\|_{H^1_\xi}\\
&\quad + t^{-1-\alpha}\|U(-1/t)w_1\|^2_{L^\infty_\xi}\|U(-1/t)w_1\|_{H^1_\xi}\\
&\lesssim t^{-1-\alpha}(\|w_1\|_{L^\infty_\xi} + t^{-\alpha}\|w_1\|_{H^1_\xi})^2\|w_1\|_{H^1_\xi}\\
&\lesssim t^{-1-\alpha}(1 + t^{-\alpha+\gamma})^2t^\gamma\|(w_1, w_2)\|^3_{X_T}\\
&\lesssim t^{-1-\alpha+\gamma}\|(w_1, w_2)\|^3_{X_T}.
\end{aligned}$$

Since

$$\begin{aligned}
R_2 = {}& 3\lambda_1 t^{-1}[\{2|U(-1/t)w_1|^2U(-1/t)w_2 + (U(-1/t)w_1)^2\overline{U(-1/t)w_2}\}\\
&\qquad - (2|w_1|^2w_2 + w_1^2\overline{w_2})]\\
&+ 3\lambda_1 t^{-1}(U(1/t) - 1)\{2|U(-1/t)w_1|^2U(-1/t)w_2 + (U(-1/t)w_1)^2\overline{U(-1/t)w_2}\}
\end{aligned}$$

by Lemma 2.1, we have

$$\begin{aligned}
&\|R_2\|_{L^\infty_\xi} \qquad (2.17)\\
&\lesssim t^{-1}(\|U(-1/t)w_1\|_{L^\infty_\xi} + \|w_1\|_{L^\infty_\xi})\|U(-1/t)w_1 - w_1\|_{L^\infty_\xi}\|U(-1/t)w_2\|_{L^\infty_\xi}\\
&\quad + t^{-1}\|w_1\|^2_{L^\infty_\xi}\|U(-1/t)w_2 - w_2\|_{L^\infty_\xi}\\
&\quad + t^{-1-\alpha}\||U(-1/t)w_1|^2U(-1/t)w_2\|_{H^1_\xi}\\
&\lesssim t^{-1-\alpha}(\|w_1\|_{L^\infty_\xi} + t^{-\alpha}\|w_1\|_{H^1_\xi})\|w_1\|_{H^1_\xi}(\|w_2\|_{L^\infty_\xi} + t^{-\alpha}\|w_2\|_{H^1_\xi})\\
&\quad + t^{-1-\alpha}\|w_1\|^2_{L^\infty_\xi}\|w_2\|_{H^1_\xi}\\
&\quad + t^{-1-\alpha}\|U(-1/t)w_1\|^2_{L^\infty_\xi}\|U(-1/t)w_1\|_{H^1_\xi}\\
&\lesssim t^{-1-\alpha}(\|w_1\|_{L^\infty_\xi} + t^{-\alpha}\|w_1\|_{H^1_\xi})\|w_1\|_{H^1_\xi}(\|w_2\|_{L^\infty_\xi} + t^{-\alpha}\|w_2\|_{H^1_\xi})\\
&\quad + t^{-1-\alpha}(\|w_1\|_{L^\infty_\xi} + t^{-\alpha}\|w_1\|_{H^1_\xi})^2\|w_2\|_{H^1_\xi}\\
&\lesssim t^{-1-\alpha+\gamma}(1 + t^{-\alpha+\gamma})(\log t + t^{-\alpha+\delta})\|(w_1, w_2)\|^3_{X_T}\\
&\quad + t^{-1-\alpha+\delta}(1 + t^{-\alpha+\gamma})^2\|(w_1, w_2)\|^3_{X_T}\\
&\lesssim t^{-1-\alpha+\delta}\|(w_1, w_2)\|^3_{X_T}.
\end{aligned}$$

By (2.14),

$$\partial_t|w_1|^2 = 2\,\mathrm{Im}(R_1\overline{w}_1) \lesssim \|R_1(t)\|_{L^\infty_\xi}|w_1|.$$

Hence (2.16) yields

$$\partial_t |w_1| \lesssim \|R_1(t)\|_{L^\infty_\xi} \lesssim t^{-1-\alpha+\gamma} \|(w_1, w_2)\|^3_{X_T}. \tag{2.18}$$

Therefore

$$|w_1(t, \xi)| \lesssim \varepsilon + \|(w_1, w_2)\|^3_{X_T}. \tag{2.19}$$

By (2.14) and (2.15),

$$i\partial_t w_1 \overline{w_2} = 3\lambda_1 t^{-1} |w_1|^2 w_1 \overline{w_2} + R_1 \overline{w_2}, \tag{2.20}$$
$$i\overline{w_1} \partial_t w_2 = \lambda_6 t^{-1} (2|w_1|^2 \overline{w_1} w_2 + |w_1|^2 w_1 \overline{w_2}) + R_2 \overline{w_1}. \tag{2.21}$$

From (2.20) and (2.21), we find

$$\begin{aligned}
&\partial_t \{w_1 \overline{w_2} + \tilde{\lambda}_\pm \overline{w_1} w_2\} e^{\pm\lambda_c |w_1|^2 \log t} \\
&= \left(-i R_1 \overline{w_2} + i \overline{R_2} w_1 \pm 2i\lambda_c \log t w_1 \overline{w_2} \operatorname{Im}(R_1 \overline{w_1})\right) e^{\pm\lambda_c |w_1|^2 \log t} \\
&\quad + \tilde{\lambda}_\pm \left(i \overline{R_1} w_2 - i R_2 \overline{w_1} \pm 2i\lambda_c \log t \overline{w_1} w_2 \operatorname{Im}(R_1 \overline{w_1})\right) e^{\pm\lambda_c |w_1|^2 \log t} \\
&\lesssim |R_1| (1 + |w_1|^2 \log t) |w_2| + |R_2| |w_1|,
\end{aligned} \tag{2.22}$$

where

$$\tilde{\lambda}_\pm = \frac{-3\lambda_1 + 2\lambda_6 \pm \lambda_c}{\lambda_6}, \quad \lambda_c = \sqrt{3(\lambda_6 - \lambda_1)(\lambda_6 - 3\lambda_1)}. \tag{2.23}$$

Hence

$$|w_1 \overline{w_2}| \lesssim \varepsilon^2 + t^{-\alpha+\delta} (\|(w_1, w_2)\|^4_{X_T} + \|(w_1, w_2)\|^6_{X_T}).$$

Therefore (2.15) and (2.17) imply

$$\begin{aligned}
\partial_t |w_2| &\lesssim t^{-1} |w_1 \overline{w_2}| |w_1| + |R_2| \\
&\lesssim \varepsilon^2 t^{-1} \|(w_1, w_2)\|_{X_T} + t^{-1} \|(w_1, w_2)\|^5_{X_T} + t^{-1} \|(w_1, w_2)\|^7_{X_T} \\
&\quad + t^{-1-\alpha+\delta} \|(w_1, w_2)\|^3_{X_T}.
\end{aligned}$$

Integrating this in t, we have

$$\begin{aligned}
|w_2(t, \xi)| \lesssim \varepsilon &+ \varepsilon^2 \log t \|(w_1, w_2)\|_{X_T} + \log t \|(w_1, w_2)\|^5_{X_T} \\
&+ \log t \|(w_1, w_2)\|^7_{X_T} + \|(w_1, w_2)\|^3_{X_T}.
\end{aligned} \tag{2.24}$$

Collecting (2.12), (2.13), (2.19) and (2.24), we obtain

$$\|(w_1, w_2)\|_{X_T} \lesssim \varepsilon + \varepsilon^2 \|(w_1, w_2)\|_{X_T} + \|(w_1, w_2)\|^3_{X_T} + \|(w_1, w_2)\|^7_{X_T}.$$

The the standard continuity argument yields that if $\varepsilon = \varepsilon(\gamma, \delta)$ is sufficiently small, then we have

$$\|(w_1, w_2)\|_{X_T} \lesssim \varepsilon$$

for any $T \geqslant 1$. This completes the proof of Lemma 2.3. □

Remark 2.5 One sees from (2.14) and (2.15) that (w_1, w_2) solves (1.8) with an error (by regarding $\log t$ as a "time" variable).

Proof of Theorem 1.1 The global existence of solution to (2.1) and decay estimate for u_1 follow from Lemma 2.3 and Remark 2.4. We now derive the asymptotic formulas (1.11) and (1.12) for solution (u_1, u_2) to (2.1) as $t \to \infty$. As byproduct of the asymptotic formula (1.12), we obtain decay estimate for u_2.

By (2.18), we see that there exists a real-valued positive function $\widetilde{W}_1 \in L^\infty$ such that

$$\||w_1| - \widetilde{W}_1\|_{L^\infty_\xi} \lesssim \varepsilon^3 t^{-\alpha+\gamma}, \tag{2.25}$$

where $\gamma < \alpha < 1/4$. Substituting this into (2.14), we have

$$i\partial_t w_1 = 3\lambda_1 t^{-1} \widetilde{W}_1^2 w_1 + R_3, \tag{2.26}$$

where R_3 is given by

$$R_3 = 3\lambda_1 t^{-1}(|w_1|^2 - \widetilde{W}_1^2) w_1 + R_1.$$

Hence by (2.16) and (2.25),

$$\|R_3(t)\|_{L^\infty_\xi} \lesssim \varepsilon^3 t^{-1-\alpha+\gamma}.$$

Multiplying (2.26) by $\exp(3i\lambda_1 \widetilde{W}_1^2 \log t)$, we have

$$i\partial_t (w_1 e^{3i\lambda_1 \widetilde{W}_1^2 \log t}) = R_3 e^{3i\lambda_1 \widetilde{W}_1^2 \log t}.$$

This implies that there exists $W_1 \in L^\infty$ such that

$$\|w_1 e^{3i\lambda_1 \widetilde{W}_1^2 \log t} - W_1\|_{L^\infty_\xi} \lesssim \varepsilon^3 t^{-\alpha+\gamma}.$$

We easily see that $|W_1| = \widetilde{W}_1$ and obtain

$$\|w_1 - W_1 e^{-3i\lambda_1 |W_1|^2 \log t}\|_{L^\infty_\xi} \lesssim \varepsilon^3 t^{-\alpha+\gamma}. \tag{2.27}$$

To derive the asymptotic behavior of w_2, we introduce a new unknown function

$$\beta(t,\xi) = w_2(t,\xi)e^{3i\lambda_1|W_1(\xi)|^2\log t}.$$

Then by (2.15), we see

$$\begin{aligned} i\partial_t\beta &= \left\{\frac{\lambda_6}{t}(2|w_1|^2 w_2 + w_1^2\overline{w_2}) + R_2\right\} e^{3i\lambda_1|W_1|^2\log t} - \frac{3\lambda_1}{t}|W_1|^2 w_2 e^{3i\lambda_1|W_1|^2\log t} \\ &= \frac{1}{t}\left((-3\lambda_1 + 2\lambda_6)|W_1|^2\beta + \lambda_6 W_1^2\overline{\beta}\right) + R_4, \end{aligned} \tag{2.28}$$

where

$$\begin{aligned} R_4 = &\frac{2\lambda_6}{t}(|w_1|^2 - |W_1|^2)\beta + \frac{\lambda_6}{t}\left\{\left(w_1 e^{3i\lambda_1|W_1|^2\log t}\right)^2 - W_1^2\right\}\overline{\beta} \\ &+ R_2 e^{3i\lambda_1|W_1(\xi)|^2\log t}. \end{aligned}$$

It holds that

$$\begin{aligned} \|R_4\|_{L^\infty} &\lesssim t^{-1}(\|w_1\|_{L^\infty} + \|W_1\|_{L^\infty}) \left\| w_1 - W_1 e^{-3i\lambda_1|W_1(\xi)|^2\log t} \right\|_{L^\infty} \|w_2\|_{L^\infty} \\ &\quad + \|R_2\|_{L^\infty} \\ &\lesssim \varepsilon^3 t^{-1-\alpha+\delta}. \end{aligned}$$

From (2.28), we have

$$i\partial_t \begin{pmatrix} \beta \\ \overline{\beta} \end{pmatrix} = \frac{1}{t}\begin{pmatrix} (-3\lambda_1 + 2\lambda_6)|W_1|^2 & \lambda_6 W_1^2 \\ -\lambda_6 \overline{W_1}^2 & (3\lambda_1 - 2\lambda_6)|W_1|^2 \end{pmatrix} \begin{pmatrix} \beta \\ \overline{\beta} \end{pmatrix} + \begin{pmatrix} R_4 \\ -\overline{R_4} \end{pmatrix}. \tag{2.29}$$

Let $N := \{\xi \in \mathbb{R}\ ;\ W_1(\xi) = 0\}$. Since $i\partial_t\beta = R_4$ on N, we easily see that $\beta = O(t^{-\alpha+\gamma})$. Therefore we concentrate our attention to the case $\xi \in N^c$. If $\xi \in N^c$, then we see that the matrix

$$A = \begin{pmatrix} (-3\lambda_1 + 2\lambda_6)|W_1|^2 & \lambda_6 W_1^2 \\ -\lambda_6 \overline{W_1}^2 & (3\lambda_1 - 2\lambda_6)|W_1|^2 \end{pmatrix}$$

can be diagonalized by the matrix

$$P = \begin{pmatrix} \lambda_6 \frac{W_1^2}{|W_1|^2} & 3\lambda_1 - 2\lambda_6 + \lambda_c \\ 3\lambda_1 - 2\lambda_6 + \lambda_c & \lambda_6 \frac{\overline{W_1}^2}{|W_1|^2} \end{pmatrix},$$

where λ is given by (2.23). Diagonalizing the Eq. (2.29) by the matrix P, we have

$$i\partial_t \begin{pmatrix} \gamma \\ \overline{\gamma} \end{pmatrix} = \frac{1}{t} \begin{pmatrix} \lambda_c |W_1|^2 & 0 \\ 0 & -\lambda_c |W_1|^2 \end{pmatrix} \begin{pmatrix} \gamma \\ \overline{\gamma} \end{pmatrix} + P^{-1} \begin{pmatrix} R_4 \\ -\overline{R_4} \end{pmatrix},$$

where γ is given by

$$\begin{pmatrix} \gamma \\ \overline{\gamma} \end{pmatrix} = P^{-1} \begin{pmatrix} \beta \\ \overline{\beta} \end{pmatrix}.$$

For the first component, we have

$$i\partial_t \gamma = \frac{1}{t} \lambda_c |W_1|^2 \gamma - \frac{\lambda_c \overline{W}_1^2 R_4 + \tilde{\lambda}_- |W_1|^2 \overline{R_4}}{2\lambda_c \tilde{\lambda} |W_1|^2},$$

where $\tilde{\lambda}_-$ is given by (2.23). This implies that

$$i\partial_t \left(e^{i\lambda_c |W_1|^2 \log t} \gamma_1 \right) = -e^{i\lambda_c |W_1|^2 \log t} \frac{\lambda_c \overline{W}_1^2 R_4 + \tilde{\lambda}_- |W_1|^2 \overline{R_4}}{2\lambda_c \tilde{\lambda} |W_1|^2}.$$

By using the estimates for R_4, we see that there exists a function $W_2 \in L^\infty$ such that

$$\left\| e^{i\lambda_c |W_1|^2 \log t} \gamma_1 - W_2 \right\|_{L^\infty} \lesssim \varepsilon t^{-\alpha+\gamma}.$$

Since

$$\begin{pmatrix} \beta \\ \overline{\beta} \end{pmatrix} = P \begin{pmatrix} \gamma \\ \overline{\gamma} \end{pmatrix} = \begin{pmatrix} \lambda_6 \frac{W_1^2}{|W_1|^2} \gamma + (3\lambda_1 - 2\lambda_6 + \lambda_c) \overline{\gamma} \\ (3\lambda_1 - 2\lambda_6 + \lambda_c) \gamma + \lambda_6 \frac{\overline{W}_1^2}{|W_1|^2} \overline{\gamma} \end{pmatrix},$$

it follows that

$$\beta(t) = \lambda_6 \frac{W_1^2}{|W_1|^2} W_2 e^{-i\lambda_c |W_1|^2 \log t} + (3\lambda_1 - 2\lambda_6 + \lambda_c) \overline{W}_2 e^{i\lambda_c |W_1|^2 \log t} + O(t^{-\alpha+\gamma}).$$

Hence we see that

$$\begin{aligned} w_2(t,\xi) = {} & \lambda_6 \frac{W_1^2}{|W_1|^2} W_2 e^{i(-3\lambda_1 - \lambda_c)|W_1|^2 \log t} \\ & + (3\lambda_1 - 2\lambda_6 + \lambda_c) \overline{W}_2 e^{i(-3\lambda_1 + \lambda_c) \log t} + O(t^{-\alpha+\gamma}). \end{aligned}$$

By Lemma 2.1 and the asymptotic formula (2.27) for w_1, we have

$$\begin{aligned} u_1(t) &= U(t)\mathcal{F}^{-1}w_1 \\ &= M(t)D(t)U(-1/t)w_1 \\ &= M(t)D(t)w_1 + O(t^{-\frac{1}{2}-\alpha}) \\ &= t^{-\frac{1}{2}} W_1\left(\frac{x}{t}\right) e^{\frac{ix^2}{2t}-3i\lambda_1|W_1(\frac{x}{t})|^2\log t - i\frac{\pi}{4}} + O(t^{-\frac{1}{2}-\alpha+\gamma}), \end{aligned} \tag{2.30}$$

in L_x^∞ as $t \to \infty$. Hence we have (1.11). In a similar way, we obtain (1.12). This completes the proof. □

3 Proofs of Theorems 1.3 and 1.6

In this section, we prove Theorem 1.3. Since the proof of Theorem 1.6 is similar and simpler, we omit the proof of Theorem 1.6.

We first consider the case $\lambda_6 = 3\lambda_1$, i.e.,

$$\begin{cases} i\partial_t u_1 + \frac{1}{2}\partial_x^2 u_1 = 3\lambda_1|u_1|^2 u_1, & t \in \mathbb{R},\ x \in \mathbb{R}, \\ i\partial_t u_2 + \frac{1}{2}\partial_x^2 u_2 = 3\lambda_1(2|u_1|^2 u_2 + u_1^2\overline{u_2}), & t \in \mathbb{R},\ x \in \mathbb{R}, \\ u_1(0,x) = u_{1,0}(x), \qquad u_2(0,x) = u_{2,0}(x), & x \in \mathbb{R}. \end{cases} \tag{3.1}$$

Let $w_j := \mathcal{F}U(-t)u_j$, $j = 1, 2$. As in the proof of Theorem 1.1, by applying $\mathcal{F}U(-t)$ to (3.1), we obtain

$$i\partial_t w_1 = 3\lambda_1 t^{-1}U(1/t)|U(-1/t)w_1|^2 U(-1/t)w_1, \tag{3.2}$$

$$i\partial_t w_2 = 3\lambda_1 t^{-1}U(1/t)\left\{2|U(-1/t)w_1|^2 U(-1/t)w_2 + (U(-1/t)w_1)^2\overline{U(-1/t)w_2}\right\}. \tag{3.3}$$

We first obtain short time bounds of w_j.

Lemma 3.1 (Short time bounds) *There exists $\varepsilon_0 > 0$ such that for any $u_{j,0} \in H^1(\mathbb{R}) \cap H^{0,1}(\mathbb{R})$ satisfying $\varepsilon := \sum_{j=1}^2(\|u_{j,0}\|_{H^1} + \|u_{j,0}\|_{H^{0,1}}) \leqslant \varepsilon_0$, there exists a unique solution $u_j \in C([0,1], H^1(\mathbb{R}) \cap H^{0,1}(\mathbb{R}))$ of (3.1) satisfying*

$$\sup_{t\in[0,1]} \sum_{j=1}^2 \|w_j(t)\|_{H_\xi^1} \lesssim \varepsilon.$$

***Proof of Lemma* 3.1** The proof follows from a standard well-posedness theory, see [2] for instance. Hence we omit the proof. □

Next we derive a long time bounds of w_j. We fix $0 < \gamma < \delta < 1/100$ and introduce

$$\|(w_1, w_2)\|_{X_T} := \sup_{t\in[1,T]} \Big\{ \|w_1(t)\|_{L^\infty_\xi} + \langle t\rangle^{-\gamma}\|w_1(t)\|_{H^1_\xi} + (\log\langle t\rangle)^{-1}\|w_2(t)\|_{L^\infty_\xi} + \langle t\rangle^{-\delta}\|w_2(t)\|_{H^1_\xi} \Big\}.$$

Lemma 3.2 (Long time bounds) *There exists $\varepsilon_0 > 0$ such that for any $u_{j,0} \in H^1(\mathbb{R}) \cap H^{0,1}(\mathbb{R})$ satisfying $\varepsilon := \sum_{j=1}^{2}(\|u_{j,0}\|_{H^1} + \|u_{j,0}\|_{H^{0,1}}) \leqslant \varepsilon_0$, there exists a unique global solution $u_j \in C([0,\infty), H^1(\mathbb{R}) \cap H^{0,1}(\mathbb{R}))$ of (3.1) satisfying*

$$\|(w_1, w_2)\|_{X_\infty} \lesssim \varepsilon. \tag{3.4}$$

Remark 3.3 As in Remark 2.4, we see that if (w_1, w_2) satisfy (3.4), then we obtain

$$\|u_1(t)\|_{L^\infty_x} \lesssim \varepsilon t^{-\frac{1}{2}}, \quad \|u_2(t)\|_{L^\infty_x} \lesssim \varepsilon t^{-\frac{1}{2}} \log t$$

for any $t \geqslant 1$.

***Proof of Lemma* 3.2** As in the proof of Lemma 2.3, we have

$$\langle t\rangle^{-\gamma}\|w_1(t)\|_{H^1_\xi} \lesssim \varepsilon + \frac{1}{\gamma}\|(w_1, w_2)\|^3_{X_T}, \tag{3.5}$$

$$\langle t\rangle^{-\delta}\|w_2(t)\|_{H^1_\xi} \lesssim \varepsilon + \frac{1}{\delta}\|(w_1, w_2)\|^3_{X_T}. \tag{3.6}$$

Next we derive L^∞ estimates for w_j. Form viewpoint of the asymptotic formulas for $U(\pm 1/t)$ (Lemma 2.1), we decompose the nonlinear term as follows:

$$i\partial_t w_1 = 3\lambda_1 t^{-1}|w_1|^2 w_1 + R_1, \tag{3.7}$$

$$i\partial_t w_2 = 3\lambda_1 t^{-1}(2|w_1|^2 w_2 + w_1^2\overline{w_2}) + R_2, \tag{3.8}$$

where R_1 and R_2 are given by

$$\begin{aligned} R_1 &= 3\lambda_1 t^{-1}[U(1/t)|U(-1/t)w_1|^2 U(-1/t)w_1 - |w_1|^2 w_1], \\ R_2 &= 3\lambda_1 t^{-1}[U(1/t)\{2|U(-1/t)w_1|^2 U(-1/t)w_2 + (U(-1/t)w_1)^2\overline{U(-1/t)w_2}\} \\ &\quad - (2|w_1|^2 w_2 + w_1^2\overline{w_2})]. \end{aligned}$$

As in the proof of Theorem 1.1, Lemma 2.1 yields

$$\|R_1\|_{L^\infty_\xi} \lesssim t^{-1-\alpha+\gamma}\|(w_1, w_2)\|^3_{X_T}, \tag{3.9}$$

$$\|R_2\|_{L^\infty_\xi} \lesssim t^{-1-\alpha+\delta}\|(w_1, w_2)\|^3_{X_T}. \tag{3.10}$$

By (3.7),

$$\partial_t|w_1|^2 = 2\,\mathrm{Im}(R_1\overline{w_1}) \lesssim \|R_1(t)\|_{L^\infty_\xi}|w_1|.$$

Hence (3.9) yields

$$\partial_t|w_1| \lesssim \|R_1(t)\|_{L^\infty_\xi} \lesssim t^{-1-\alpha+\gamma}\|(w_1, w_2)\|^3_{X_T}. \tag{3.11}$$

Therefore

$$|w_1(t,\xi)| \lesssim \varepsilon + \|(w_1, w_2)\|^3_{X_T}. \tag{3.12}$$

By (3.7), (3.8), (3.9) and (3.10),

$$\begin{aligned}\partial_t\,\mathrm{Re}(w_1\overline{w_2}) &= \mathrm{Im}(R_1\overline{w_2}) + \mathrm{Im}(R_2\overline{w_1}) \\ &\lesssim \|R_1\|_{L^\infty_\xi}\|w_2\|_{L^\infty_\xi} + \|R_2\|_{L^\infty_\xi}\|w_1\|_{L^\infty_\xi} \\ &\lesssim (t^{-1-\alpha+\gamma}\log t + t^{-1-\alpha+\delta})\|(w_1, w_2)\|^4_{X_T} \\ &\lesssim t^{-1-\alpha+\delta}\|(w_1, w_2)\|^4_{X_T}.\end{aligned} \tag{3.13}$$

Integrating this in t, we obtain

$$|\,\mathrm{Re}(w_1\overline{w_2})| \lesssim \varepsilon^2 + \|(w_1, w_2)\|^4_{X_T}. \tag{3.14}$$

On the other hand, by (3.8),

$$i\partial_t w_2 = 3\lambda_1 t^{-1}|w_1|^2 w_2 + 6\lambda_1 t^{-1} w_1\,\mathrm{Re}(w_1\overline{w_2}) + R_2.$$

Hence,

$$\begin{aligned}\partial_t|w_2|^2 &= 12\lambda_1 t^{-1}\,\mathrm{Re}(w_1\overline{w_2})\,\mathrm{Im}(w_1\overline{w_2}) + 2\,\mathrm{Im}(R_2\overline{w_2}) \\ &\lesssim t^{-1}|\,\mathrm{Re}(w_1\overline{w_2})||w_1||w_2| + |R_2||w_2|.\end{aligned}$$

Therefore (3.10) and (3.14) imply

$$\begin{aligned}\partial_t|w_2| &\lesssim t^{-1}|\,\mathrm{Re}(w_1\overline{w_2})||w_1| + |R_2| \\ &\lesssim \varepsilon^2 t^{-1}\|(w_1, w_2)\|_{X_T} + t^{-1}\|(w_1, w_2)\|^5_{X_T} + t^{-1-\alpha+\delta}\|(w_1, w_2)\|^3_{X_T}.\end{aligned}$$

Integrating this in t, we have

$$\begin{aligned}|w_2(t,\xi)| \lesssim\ &\varepsilon + \varepsilon^2\log t\|(w_1, w_2)\|_{X_T} \\ &+ \log t\|(w_1, w_2)\|^5_{X_T} + \|(w_1, w_2)\|^3_{X_T}.\end{aligned} \tag{3.15}$$

Collecting (3.5), (3.6), (3.12) and (3.15), we obtain

$$\|(w_1, w_2)\|_{X_T} \lesssim \varepsilon + \varepsilon^2 \|(w_1, w_2)\|_{X_T} + \|(w_1, w_2)\|_{X_T}^3 + \|(w_1, w_2)\|_{X_T}^5.$$

The the standard continuity argument yields that if $\varepsilon = \varepsilon(\gamma, \delta)$ is sufficiently small, then we have

$$\|(w_1, w_2)\|_{X_T} \lesssim \varepsilon$$

for any $T \geqslant 1$. This completes the proof of Lemma 3.2. □

Proof of Theorem 1.3 The global existence and decay estimates for solution to (3.1) follow from Lemma 3.2 and Remark 2.4. We now derive the asymptotic formulas (1.13) and (1.14) for solution (u_1, u_2) to (3.1) as $t \to \infty$.

As in the proof of Theorem 1.1, we find that there exists $W_1 \in L^\infty$ such that

$$\|w_1 e^{3i\lambda_1 |W_1|^2 \log t} - W_1\|_{L^\infty_\xi} \lesssim \varepsilon^3 t^{-\alpha+\gamma}. \tag{3.16}$$

Furthermore, by (3.13), there exists a real-valued function $\widetilde{W} \in L^\infty$ such that

$$\| \operatorname{Re}(w_1 \overline{w_2}) - \widetilde{W} \|_{L^\infty_\xi} \lesssim \varepsilon^4 t^{-\alpha+\delta}. \tag{3.17}$$

Substituting this into (3.8), we have

$$i\partial_t w_2 = 3\lambda_1 t^{-1} |W_1|^2 w_2 + 6\lambda_1 t^{-1} W_1 \widetilde{W} e^{-3i\lambda_1 |W_1|^2 \log t} + R_4, \tag{3.18}$$

where

$$R_4 = 3\lambda_1 t^{-1} (|w_1|^2 - |W_1|^2) w_2 + 6\lambda_1 t^{-1} (w_1 \operatorname{Re}(w_1 \overline{w_2}) - W_1 \widetilde{W} e^{-3i\lambda_1 |W_1|^2 \log t}) \\ + R_2.$$

Hence by (3.10), (3.16) and (3.17),

$$\|R_4(t)\|_{L^\infty_\xi} \lesssim \varepsilon^5 t^{-1-\alpha+\delta}. \tag{3.19}$$

Multiplying (3.18) by $\exp(3i\lambda_1 |W_1|^2 \log t)$, we find

$$i\partial_t \left\{ w_2 e^{3i\lambda_1 |W_1|^2 \log t} \right\} = 6\lambda_1 t^{-1} W_1 \widetilde{W} + R_4 e^{3i\lambda_1 |W_1|^2 \log t}.$$

Since W_1 and $\widetilde{W}$ are independent of t, we see

$$i\partial_t \left\{ w_2 e^{3i\lambda_1 |W_1|^2 \log t} + 6i\lambda_1 \log t \, W_1 \widetilde{W} \right\} = R_4 e^{3i\lambda_1 |W_1|^2 \log t}.$$

Hence, by (3.19) we find that there exists $W_2 \in L^\infty$ such that

$$\|w_2 e^{3i\lambda_1|W_1|^2\log t} + 6i\lambda_1 \log t W_1 \widetilde{W} - W_2\|_{L^\infty_\xi} \lesssim \varepsilon^5 t^{-\alpha+\delta}. \tag{3.20}$$

Especially, we have $\widetilde{W} = \mathrm{Re}(W_1\overline{W_2})$.

In the same way as in the proof of (2.30), from the asymptotic formulas (3.16) for w_1 and (3.20) for w_2, we obtain (1.13) and (1.14).

For the case $\lambda_6 = \lambda_1$, it suffices to replace $\mathrm{Re}(w_1\overline{w_2})$ by $\mathrm{Im}(w_1\overline{w_2})$ in the proof for the case $\lambda_6 = 3\lambda_1$. This completes the proof of Theorem 1.3. □

Acknowledgements The authors express their gratitude to Professor Kenji Nakanishi for valuable comments on a preliminary version of the manuscript. The authors also thank the referee for careful reading of the manuscript and for fruitful suggestions. S.M. was supported by JSPS KAKENHI Grant Numbers JP17K14219, JP17H02854, JP17H02851, and JP18KK0386. J.S was supported by JSPS KAKENHI Grant Numbers JP17H02851, JP19H05597, JP20H00118, JP21H00993, and JP21K18588. K.U was supported by JSPS KAKENHI Grant Number JP19K14578.

Appendix A Classification for System of Cubic Nonlinear System

In this appendix, we consider classification of abstract cubic nonlinear systems of the form

$$\begin{cases} \mathcal{L}u_1 = c_1|u_1|^2u_1 + c_2|u_1|^2u_2 + c_3u_1^2\overline{u_2} + c_4u_1|u_2|^2 + c_5\overline{u_1}u_2^2 + c_6|u_2|^2u_2, \\ \mathcal{L}u_2 = c_7|u_1|^2u_1 + c_8|u_1|^2u_2 + c_9u_1^2\overline{u_2} + c_{10}u_1|u_2|^2 + c_{11}\overline{u_1}u_2^2 + c_{12}|u_2|^2u_2, \end{cases} \tag{A.1}$$

where $u_j : \Omega \to \mathbb{C}$ $(j = 1, 2)$ are complex-valued unknown functions defined on a set Ω, $\mathcal{L}$ is a $\mathbb{R}$-linear operator, and c_j $(j = 1, \ldots, 12)$ are real constants. Obviously, our system (1.6) is an example of (A.1). Another example of a system of the form (A.1) in our mind is the limit ODE system:

$$\begin{cases} i\partial_t\alpha_1 = c_1|\alpha_1|^2\alpha_1 + c_2|\alpha_1|^2\alpha_2 + c_3\alpha_1^2\overline{\alpha_2} + c_4\alpha_1|\alpha_2|^2 + c_5\overline{\alpha_1}\alpha_2^2 + c_6|\alpha_2|^2\alpha_2, \\ i\partial_t\alpha_2 = c_7|\alpha_1|^2\alpha_1 + c_8|\alpha_1|^2\alpha_2 + c_9\alpha_1^2\overline{\alpha_2} + c_{10}\alpha_1|\alpha_2|^2 + c_{11}\overline{\alpha_1}\alpha_2^2 + c_{12}|\alpha_2|^2\alpha_2, \end{cases} \tag{A.2}$$

which is a generalized version of (1.8). The system arises in the study of asymptotic behavior of solutions to (1.6). Indeed, the main part of the systems (2.14)–(2.15), and (3.7)–(3.8) are examples of (A.2). Although we discuss in a quite abstract setting, our main targets are (1.6) and (A.2).

Remark A.1 The complex conjugation in the nonlinearity can be also replaced by another $\mathbb{R}$-linear operation. If we take a derivative operator $\mathcal{D}$, i.e., an operation satisfying $\mathcal{D}(fg) = (\mathcal{D}f)g + f(\mathcal{D}g)$, then (A.1) takes the form

$$\begin{cases} \mathcal{L}u_1 = c_1 u_1^2 \mathcal{D}u_1 + c_2 u_1 u_2 \mathcal{D}u_1 + c_3 u_1^2 \mathcal{D}u_2 + c_4 u_1 u_2 \mathcal{D}u_2 + c_5 u_2^2 \mathcal{D}u_1 + c_6 u_2^2 \mathcal{D}u_2, \\ \mathcal{L}u_2 = c_7 u_1^2 \mathcal{D}u_1 + c_8 u_1 u_2 \mathcal{D}u_1 + c_9 u_1^2 \mathcal{D}u_2 + c_{10} u_1 u_2 \mathcal{D}u_2 + c_{11} u_2^2 \mathcal{D}u_1 + c_{12} u_2^2 \mathcal{D}u_2. \end{cases} \tag{A.3}$$

Our argument can be applied to a cubic system with a derivative type nonlinearity.

Our classification is done by (1) regarding two systems being equivalent if there exists a linear transformation of unknown between them and by (2) specifying the quotient set with respect to the equivalence relation. In other words, we find a collection of standard forms (or canonical forms) of the system of the form (A.1). The key ingredient of our argument is the matrix-kernel representation of a system. This representation has two good properties. One is that the change of the coefficients of the system is clearly described by means of a matrix manipulation. The other is that the matrix describes the characteristic quadratic quantities of the system, which correspond to the conserved quantities in the case of (1.6) and (A.2). The representation is an extension of a matrix representation of a subclass of systems used in [22].

The rest of Appendix A is organized as follows. In Sect. A.1, we introduce two sets of systems. One is the set of the form (A.1) and the other is its subset treated in the previous study. We then introduce the matrix-kernel representation of the system of the form (A.1) in Sect. A.2, The difference of the two classes is also discussed. Section A.3 is devoted to the first benefit of the matrix-kernel representation, i.e., the equivalence relation is clearly described in terms of matrix manipulations (see Theorem A.5). We next see how the matrix-kernel representation reflects the characteristic quantities of (1.6) and (A.2) in Sect. A.4. Then, we turn to the main subject, the classification of the systems. Our main results are Theorems A.12 and A.13 in Sect. A.5. We prove these theorems in Sects. A.6 and A.7.

A.1 Set of Cubic Nonlinear Systems

For each fixed Ω and $\mathcal{L}$, a system of the form (A.1) is naturally regarded as a vector $(c_j)_{1\leqslant j\leqslant 12} \in \mathbb{R}^{12}$ by identifying the system with a combination of constants of the nonlinearity. Let $\mathrm{CNS} = \mathrm{CNS}(\Omega, \mathcal{L})$ be a set of cubic nonlinear systems of the form (A.1).

We introduce a subclass of systems of the form

$$\begin{cases} \mathcal{L}u_1 = 3\lambda_1|u_1|^2u_1 + \lambda_2(2|u_1|^2u_2 + u_1^2\overline{u_2}) + \lambda_3(2u_1|u_2|^2 + \overline{u_1}u_2^2) + 3\lambda_4|u_2|^2u_2, \\ \mathcal{L}u_2 = 3\lambda_5|u_1|^2u_1 + \lambda_6(2|u_1|^2u_2 + u_1^2\overline{u_2}) + \lambda_7(2u_1|u_2|^2 + \overline{u_1}u_2^2) + 3\lambda_8|u_2|^2u_2, \end{cases} \tag{A.4}$$

where λ_j $(j = 1, \ldots, 8)$ are real constants. It is a particular case

$$\begin{aligned} &c_1 = 3\lambda_1, && c_2 = 2c_3 = 2\lambda_2, && c_4 = 2c_5 = 2\lambda_3, && c_6 = 3\lambda_4, \\ &c_7 = 3\lambda_5, && c_8 = 2c_9 = 2\lambda_6, && c_{10} = 2c_{11} = 2\lambda_7, && c_{12} = 3\lambda_8 \end{aligned}$$

of (A.1). Let $\mathrm{CNS}_A \subset \mathrm{CNS}$ be the set of systems of the form (A.4). Notice that if we take (1.6) as an example of CNS then CNS_A is the set of systems of the form (1.3). If (A.3) is taken as CNS then CNS_A is the set of systems with nonlinearities in the divergent form:

$$\begin{cases} \mathcal{L}u_1 = \lambda_1 \mathcal{D}(u_1^3) + \lambda_2 \mathcal{D}(u_1^2 u_2) + \lambda_3 \mathcal{D}(u_1 u_2^2) + \lambda_4 \mathcal{D}(u_2^3), \\ \mathcal{L}u_2 = \lambda_5 \mathcal{D}(u_1^3) + \lambda_6 \mathcal{D}(u_1^2 u_2) + \lambda_7 \mathcal{D}(u_1 u_2^2) + \lambda_8 \mathcal{D}(u_2^3). \end{cases}$$

The system (A.1) is closed under the change of variables of the form

$$\begin{pmatrix} v_1 \\ v_2 \end{pmatrix} = M \begin{pmatrix} u_1 \\ u_2 \end{pmatrix}, \quad M \in GL_2(\mathbb{R}). \tag{A.5}$$

Namely, if we define a new unknown (v_1, v_2) by the above transform then (v_1, v_2) solves a system of the same form, with a possibly different combination of c_j. Then, a natural equivalence relation $\sim$ on CNS is induced by the change of variables of this form. Our interest is in the study on the quotient set $\mathrm{CNS}/\sim$.

A.2 *Matrix Representation of Systems*

In the previous study in [22], we introduced a matrix representation of systems in CNS_A. The representation enables us to describe the equivalence relation by matrix multiplications. We briefly recall the representation. Then, we consider a similar representation of systems in CNS, which is our first main result in this appendix. It turns out that our representation is a natural extension of that for CNS_A.

A.2.1 Representation of Systems in CNS_A

To begin with let us recall the matrix representation for CNS_A. Let

$$\mathcal{Z} := \{A \in M_3(\mathbb{R}) \mid \operatorname{tr} A = 0\}.$$

A system $\sigma = (\lambda_j)_{1 \leqslant j \leqslant 8} \in \mathrm{CNS}_A$ parameterized as in (A.4) is identified with a matrix $A = A(\sigma) \in \mathcal{Z}$ defined by

$$A = \begin{pmatrix} \lambda_2 & -3\lambda_1 + \lambda_6 & -3\lambda_5 \\ \lambda_3 & -\lambda_2 + \lambda_7 & -\lambda_6 \\ 3\lambda_4 & 3\lambda_8 - \lambda_3 & -\lambda_7 \end{pmatrix}. \tag{1.9}$$

Now, we introduce a notation. For a given matrix $A = (a_{ij})_{1 \leqslant i,j \leqslant 3} \in \mathcal{Z}$, we denote the corresponding system $\sigma \in \mathrm{CNS}_A$ by

$$\begin{cases} \mathcal{L}u_1 = Q_{1,A}(u_1, u_2), \\ \mathcal{L}u_2 = Q_{2,A}(u_1, u_2), \end{cases} \tag{A.6}$$

where the nonlinearities $Q_{1,A}$ and $Q_{2,A}$ are given by

$$Q_{k,A}(u,v) := 3\lambda_{4k-3}|u|^2u + \lambda_{4k-2}(2|u|^2v + u^2\overline{v}) + \lambda_{4k-1}(2u|v|^2 + \overline{u}v^2) + 3\lambda_{4k}|v|^2v \tag{A.7}$$

($k = 1, 2$) with the parameters λ_j ($j = 1, \dots, 8$) given from A by the relation (1.9).

A.2.2 Representation of Systems in CNS

Given system $\sigma \in \mathrm{CNS}$, we introduce a matrix $C = C(\sigma)$ by

$$C = \begin{pmatrix} c_2 - c_3 & -c_1 + c_8 - c_9 & -c_7 \\ c_5 & -c_3 + c_{11} & -c_9 \\ c_6 & -c_4 + c_5 + c_{12} & -c_{10} + c_{11} \end{pmatrix} \in M_3(\mathbb{R}), \tag{A.8}$$

where $M_3(\mathbb{R})$ is a set of 3×3 matrices with real entries. Remark that the correspondence $\mathrm{CNS} \ni \sigma \mapsto C(\sigma) \in M_3(\mathbb{R})$ is not injective. To obtain a unique representation with $C(\sigma)$, let us take the kernel of the map $\mathfrak{G} : \sigma \mapsto C(\sigma)$ into account.

Lemma A.2 (Kernel of $\mathfrak{G}$) $\ker \mathfrak{G} = \mathrm{Span}(\mathbf{c}^{(1)}, \mathbf{c}^{(2)}, \mathbf{c}^{(3)})$, *where* $\mathbf{c}^{(k)} = (c_j^{(k)})_{1 \leqslant j \leqslant 12} \in \mathbb{R}^{12}$ ($k = 1, 2, 3$) *is given by*

$$c_j^{(1)} = \delta_{1,j} + \delta_{8,j}, \quad c_j^{(2)} = \delta_{2,j} + \delta_{3,j} + \delta_{10,j} + \delta_{11,j}, \quad c_j^{(3)} = \delta_{4,j} + \delta_{12,j},$$

respectively, with the Kronecker delta $\delta_{j,k}$.

Theorem A.3 (A matrix-kernel representation of CNS) *A system* $\sigma \in \mathrm{CNS}$ *is identified with* $(C, p, q, r) \in M_3(\mathbb{R}) \times \mathbb{R}^3$ *as follows: If* $\sigma = (c_j)_{1 \leqslant j \leqslant 12} \in \mathrm{CNS}$ *then* $C(\sigma)$ *is given by* (A.8) *and* $(p(\sigma), q(\sigma), r(\sigma))$ *is given by*

$$p(\sigma) = c_8 - 2c_9, \quad q(\sigma) = \tfrac{1}{2}(-c_2 + 2c_3 - c_{10} + 2c_{11}), \quad r(\sigma) = c_4 - 2c_5$$

respectively. On the other hand, if $((a_{ij})_{1 \leqslant i,j \leqslant 3}, p, q, r) \in M_3(\mathbb{R}) \times \mathbb{R}^3$ *is given then* $\sigma = (c_j)_{1 \leqslant j \leqslant 12} \in \mathrm{CNS}$ *is given by* $c_5 = a_{21}$, $c_6 = a_{31}$, $c_7 = -a_{13}$, $c_9 = -a_{23}$,

$$\begin{aligned} c_1 &= p - a_{12} - a_{23}, & c_8 &= p - 2a_{23}, \\ c_2 &= q + \tfrac{1}{2}(3a_{11} - a_{22} - a_{33}), & c_3 &= q + \tfrac{1}{2}(a_{11} - a_{22} - a_{33}), \\ c_{10} &= q + \tfrac{1}{2}(a_{11} + a_{22} - 3a_{33}), & c_{11} &= q + \tfrac{1}{2}(a_{11} + a_{22} - a_{33}), \\ c_4 &= r + 2a_{21}, & c_{12} &= r + a_{21} + a_{32}. \end{aligned}$$

Further, any system $\sigma \in \mathrm{CNS}$ *is uniquely written as*

$$\begin{cases} \mathcal{L}u_1 = Q_{1,C(\sigma)}(u_1, u_2) + V_{p(\sigma),q(\sigma),r(\sigma)}(u_1, u_2)u_1, \\ \mathcal{L}u_2 = Q_{2,C(\sigma)}(u_1, u_2) + V_{p(\sigma),q(\sigma),r(\sigma)}(u_1, u_2)u_2, \end{cases} \tag{A.9}$$

where

$$Q_{j,C}(u, v) := Q_{j,C-(\mathrm{tr}\, C)I_{22}}(u, v) + (-1)^j (\mathrm{tr}\, C) \operatorname{Re}(\overline{u}v)(\delta_{j,1}u + \delta_{j,2}v) \tag{A.10}$$

is an extension of $Q_{j,A}$ *define in* (A.7), $I_{22} = (\delta_{i,2}\delta_{j,2})_{1\leqslant i,j\leqslant 3} \in M_3(\mathbb{R})$, *and*

$$V_{p,q,r}(u, v) := p|u|^2 + 2q \operatorname{Re}(\overline{u}v) + r|v|^2 = (\overline{u}\ \overline{v}) \begin{pmatrix} p & q \\ q & r \end{pmatrix} \begin{pmatrix} u \\ v \end{pmatrix}. \tag{A.11}$$

The proof is straightforward. In the sequel, we freely use the identification σ with (C, p, q, r).

A.2.3 Characterization of CNS_A

To observe that the matrix-kernel expression for CNS given in Theorem A.3 is an extension of the matrix representation for CNS_A introduced in [22], we give a characterization for being a system in CNS_A in the matrix-kernel representation.

Proposition A.4 *A system* $\sigma \in$ CNS *belongs to* CNS_A *if and only if* $\mathrm{tr}\, C(\sigma) = 0$ *and* $p(\sigma) = q(\sigma) = r(\sigma) = 0$. *Further, if* $\sigma \in \mathrm{CNS}_A$ *then* $C(\sigma)$ *coincides with the matrix* A *given in* (1.9).

The proof is due to simple computation. By means of the form (A.9) and (A.10), it is also possible to identify $\sigma \in$ CNS with $(A, \tau, p, q, r) \in \mathcal{Z} \times \mathbb{R}^4$, where the parameter τ corresponds to $\mathrm{tr}\, C$.

A.3 Equivalence Relation in Terms of Matrices

We have introduced an equivalence relation between two systems in Sect. 3. By the matrix-kernel representation of systems in CNS given in Theorem A.3, an equivalence relation on $M_3(\mathbb{R}) \times \mathbb{R}^3$ is naturally induced. One benefit of the introduction of the matrix-kernel representation is that the equivalence relation can be explicitly described as matrix multiplications.

Theorem A.5 *Let* (C, p, q, r) *be a matrix-kernel representation of a system* $\sigma \in$ CNS *with unknowns* (u_1, u_2). *Define a new unknown* (v_1, v_2) *via*

$$\begin{pmatrix} v_1 \\ v_2 \end{pmatrix} = M \begin{pmatrix} u_1 \\ u_2 \end{pmatrix}, \quad M = \begin{pmatrix} a & b \\ c & d \end{pmatrix} \in GL_2(\mathbb{R}).$$

Let (C', p', q', r') *be the matrix-kernel representation of the system for* (v_1, v_2)*. Then,*

$$C' = \frac{1}{\det M} D(M) C D(M)^{-1} \tag{A.12}$$

and

$$\begin{pmatrix} p' \\ q' \\ r' \end{pmatrix} = \frac{1}{\det M} D(M) \begin{pmatrix} p \\ q \\ r \end{pmatrix} \tag{A.13}$$

hold, where

$$D(M) := \frac{1}{\det M} \begin{pmatrix} d^2 & -2cd & c^2 \\ -bd & ad + bc & -ac \\ b^2 & -2ab & a^2 \end{pmatrix} \in SL_3(\mathbb{R}),$$

$$D(M)^{-1} = \frac{1}{\det M} \begin{pmatrix} a^2 & 2ac & c^2 \\ ab & ad + bc & cd \\ b^2 & 2bd & d^2 \end{pmatrix} \in SL_3(\mathbb{R}).$$

In particular, $\operatorname{rank} C = \operatorname{rank} C'$. *Further,* $\operatorname{sign}((q')^2 - p'r') = \operatorname{sign}(q^2 - pr)$.

This can be shown by a direct computation. One sees from (A.12) that $\operatorname{tr} C' = (\det M)^{-1} \operatorname{tr} C$. Hence the property $\operatorname{tr} C = 0$ is conserved by any change of unknowns.

A.4 Matrix Representation and Conserved Quantities

We remark that there are several options on manners to represent a system with an explicit description of the equivalence relation. The simplest one would be the natural identification $\mathrm{CNS} = \mathbb{R}^{12}$, that is, identification with a column vector in $\mathbb{R}^{12}$. Then, there is a 12×12-matrix-valued function on $GL_2(\mathbb{R})$ such that the equivalence relation on CNS is understood as a multiplication of the matrix from the left. The good point of our choice is not only the capability of a clear description of the equivalence relation but also the fact that it reflects the characteristic property of systems. In particular, as for the concrete examples (1.6) and (A.2), our representation also clarifies the existence of conserved quantities.

A.4.1 Conserved Quantities for (A.2)

Let us consider (A.2). In this model, the kernel part of the nonlinearity is removed by a gauge transform: Let $I \subset \mathbb{R}$ be an interval such that $0 \in I$. Let $(u_1, u_2) \in C^1(I, \mathbb{C}^2)$ be a solution to

$$\begin{cases} i\partial_t u_1 = Q_{1,C}(u_1, u_2) + V_{p,q,r}(u_1, u_2)u_1, \\ i\partial_t u_2 = Q_{2,C}(u_1, u_2) + V_{p,q,r}(u_1, u_2)u_2 \end{cases} \tag{A.2$'$}$$

for some $(C, p, q, r) \in \mathrm{CNS}$. Then, $\tilde{u}_j(t) := u_j(t)e^{i\int_0^t V_{p,q,r}(u_1(s),u_2(s))ds}$ solves

$$\begin{cases} i\partial_t \tilde{u}_1 = Q_{1,A(\sigma)}(\tilde{u}_1, \tilde{u}_2), \\ i\partial_t \tilde{u}_2 = Q_{2,A(\sigma)}(\tilde{u}_1, \tilde{u}_2), \end{cases}$$

which is the system $(C, 0, 0, 0) \in \mathrm{CNS}$. This can be seen, for instance, by the fact that the nonlinearity is gauge-invariant and that $V_{p,q,r}$ is $\mathbb{R}$-valued. Since $V_{p,q,r}(u_1, u_2) = V_{p,q,r}(\tilde{u}_1, \tilde{u}_2)$, the inverse transform is given by $u_j(t) := \tilde{u}_j(t)e^{-i\int_0^t V_{p,q,r}(u_1(s),u_2(s))ds}$. Thus, as for this model it suffices to consider the case $p = q = r = 0$. Further, we have the following:

Proposition A.6 *Let $I \subset \mathbb{R}$ be an interval. Let $(u_1, u_2) \in C^1(I; \mathbb{C}^2)$ be a solution to* (A.2). *Then, for any* ${}^t(a, b, c) \in \mathbb{R}^3$,

$$\frac{d}{dt}\left(|u_1|^2\ 2\,\mathrm{Re}(\overline{u_1}u_2)\ |u_2|^2\right)\begin{pmatrix} a \\ b \\ c \end{pmatrix} = \mathrm{Im}(\overline{u_1}u_2)\left(|u_1|^2\ 2\,\mathrm{Re}(\overline{u_1}u_2)\ |u_2|^2\right) C \begin{pmatrix} a \\ b \\ c \end{pmatrix}.$$

In particular, if ${}^t(a, b, c) \in \ker C$ *then* $a|u_1|^2 + 2b\,\mathrm{Re}(\overline{u_1}u_2) + c|u_2|^2$ *is conserved. Further,* $3 - \operatorname{rank} C$ *is the number of linearly dependent conserved quantities of this form.*

Let us consider another type of conserved quantity. The following proposition makes the role of the matrix B defined in (1.10) clear.

Proposition A.7 *(1) Let $I \subset \mathbb{R}$ be an interval. Let $(u_1, u_2) \in C^1(I; \mathbb{C}^2)$ be a solution to* (A.2). *Then,*

$$\frac{d}{dt}\mathrm{Im}(\overline{u_1}u_2) = \frac{1}{4}B[|u_1|^2, 2\,\mathrm{Re}(\overline{u_1}u_2), |u_2|^2],$$

where $B[x] := {}^t x B x$ *is the quadratic form for* $x \in \mathbb{R}^3$ *given by the matrix*

$$B := \begin{pmatrix} -4c_7 & c_1 - c_8 - c_9 & 2(c_2 - c_3 - c_{10} + c_{11}) \\ c_1 - c_8 - c_9 & 2(c_3 - c_{11}) & c_4 + c_5 - c_{12} \\ 2(c_2 - c_3 - c_{10} + c_{11}) & c_4 + c_5 - c_{12} & 4c_6 \end{pmatrix}. \tag{A.14}$$

In particular, if $B = O$ then $2\,\mathrm{Im}(\overline{u_1}u_2)$ is conserved.

(2) If a system $\sigma \in \mathrm{CNS}$ satisfies $B(\sigma) = O$ and $p(\sigma) = q(\sigma) = r(\sigma) = 0$ then $\sigma \in \mathrm{CNS}_A$ and $C(\sigma)$ takes the form

$$\begin{pmatrix} \lambda_2 & -2\lambda_6 & 0 \\ \lambda_3 & 0 & -\lambda_6 \\ & 2\lambda_3 & -\lambda_2 \end{pmatrix} \in \mathcal{Z},$$

i.e., $C(\sigma)$ *is a matrix of the form* (1.9) *and satisfies* $\lambda_1 = \lambda_6$, $\lambda_4 = \lambda_5 = 0$, $\lambda_7 = \lambda_2$, *and* $\lambda_8 = \lambda_3$. *In particular,* $\operatorname{rank} C(\sigma) \in \{0, 2\}$.

The proof is due to a direct computation. Remark that the matrix B defined in (A.14) is an extension of the matrix given by (1.10). Namely if $\sigma \in \mathrm{CNS}_A$ then the two definitions give the same matrix.

A.4.2 Conserved Quantities for (1.6)

As for the nonlinear Schrödinger system (1.6), the mass-type conservation is well-described by the matrix C. Here we consider only one space dimension. However, it is just for the sake of simplicity. The conditions for the validity of the conservation laws for higher dimensions are the same.

Proposition A.8 *Let* $I \subset \mathbb{R}$ *be an interval. Let* $(u_1, u_2) \in (C(I; L^2(\mathbb{R})) \cap L^4(I; L^\infty(\mathbb{R})))^2$ *be a local* L^2*-solution to* (1.6). *Then, one has*[1]

$$\partial_t \left(\|u_1\|_{L^2}^2 \ 2\operatorname{Re}(u_1, u_2)_{L^2} \ \|u_2\|_{L^2}^2\right) \begin{pmatrix} a \\ b \\ c \end{pmatrix} = \int_{\mathbb{R}} 2\operatorname{Im}(\overline{u_1} u_2) \left(|u_1|^2 \ 2\operatorname{Re}(\overline{u_1} u_2) \ |u_2|^2\right) C \begin{pmatrix} a \\ b \\ c \end{pmatrix} dx.$$

In particular,

$$a \|u_1\|_{L^2}^2 + 2b \operatorname{Re}(u_1, u_2)_{L^2} + c \|u_2\|_{L^2}^2$$

is a conserved quantity if and only if ${}^t(a, b, c) \in \ker C$. *Moreover,* $3 - \operatorname{rank} C$ *is the number of linearly dependent conserved quantities of this form. Furthermore, if there exists* $(a_0, b_0, c_0) \in \ker C$ *such that* $b_0^2 < a_0 c_0$ *then any* L^2*-solution to* (1.6) *is global in time.*

Let us consider the energy conservation. We seek an energy of the form

$$a \|\partial_x u_1\|_{L^2}^2 + 2b \operatorname{Re}(\partial_x u_1, \partial_x u_2)_{L^2} + c \|\partial_x u_2\|_{L^2}^2 + (\text{quartic term}). \qquad \text{(A.15)}$$

It turns out that the kernel part must be taken into account.

Proposition A.9 *Let* $I \subset \mathbb{R}$ *be an interval. Let* $(u_1, u_2) \in (C(I; H^1(\mathbb{R})) \cap L^8(I; W^{1,4}(\mathbb{R})))^2$ *be a local* H^1*-solution to* (1.6). *There exists a conserved energy of the form* (A.15) *if and only if*

$$C \begin{pmatrix} a \\ b \\ c \end{pmatrix} = 0, \quad \begin{pmatrix} \operatorname{tr} C - 2q & 2p & 0 \\ -r & \operatorname{tr} C & p \\ 0 & -2r & \operatorname{tr} C + 2q \end{pmatrix} \begin{pmatrix} a \\ b \\ c \end{pmatrix} = 0.$$

Further, if the condition is satisfied then the quartic term of the energy is given by

[1] Rigorously speaking, it holds in the corresponding integral form.

$$\int_{\mathbb{R}}[(ac_1+bc_7)|u_1|^4+4(ac_3+bc_9)|u_1|^2\operatorname{Re}(\overline{u_1}u_2)+2(ac_4+bc_{10})|u_1|^2|u_2|^2$$
$$+2(ac_5+bc_{11})\operatorname{Re}(\overline{u_1}^2u_2^2)+4(bc_5+cc_{11})|u_2|^2\operatorname{Re}(\overline{u_1}u_2)+(bc_6+cc_{12})|u_2|^4]dx.$$

In particular, if the system is of the form (1.3) $\in \mathrm{CNS}_A$ *then the condition is simplified as* ${}^{\mathrm{t}}(a,b,c)\in\ker C$.

Remark A.10 We also remark that for given $a, b, c \in \mathbb{R}$ the conservation of momentum

$$\operatorname{Im}\int_{\mathbb{R}}(a\overline{u_1}\partial_x u_1+2b\overline{u_1}\partial_x u_2+c\overline{u_2}\partial_x u_2)dx=\int_{\mathbb{R}}\xi(a|\widehat{u_1}|^2+2b\operatorname{Re}(\overline{\widehat{u_1}}\widehat{u_2})+c|\widehat{u_2}|^2)d\xi$$

holds under the same condition as for the energy conservation.

A.5 Classification of Systems

We now turn to the main issue of the appendix, i.e., the classification of the systems. To this end, we introduce a notion. For a system $\sigma \in \mathrm{CNS}$, we let $[\sigma] := \{\sigma' \in \mathrm{CNS} \mid \sigma' \sim \sigma\} \in \mathrm{CNS}/\!\sim$ be an equivalent class contains σ, or an orbit of σ. For a subset $S \subset \mathrm{CNS}$, we define $[S] := \{[\sigma] \in \mathrm{CNS}/\!\sim \mid \sigma \in S\} \subset \mathrm{CNS}/\!\sim$.

Definition A.11 (*System of representatives (SR)*) Let $S \subset \mathrm{CNS}/\!\sim$ be a set of equivalent classes. We say a subset $T \subset \mathrm{CNS}$ is a *system of representatives (SR)* of S if T satisfies the following two properties:

(1) $[T] = S$;
(2) if $\sigma_1, \sigma_2 \in T$ satisfies $\sigma_1 \sim \sigma_2$ then $\sigma_1 = \sigma_2$.

In the present paper, we find an SR of $\mathrm{CNS}_0/\!\sim$ and $\mathrm{CNS}_1/\!\sim$, where

$$\mathrm{CNS}_j := \{\sigma \in \mathrm{CNS} \mid \operatorname{rank} C(\sigma) = j\}$$

for $j = 0, 1, 2, 3$.

One easily finds an SR of $\mathrm{CNS}_0/\!\sim$. Remark that $\operatorname{rank} C(\sigma) = 0$ implies $C(\sigma) = O$. Hence, the classification of CNS_0 boils down to the classification of a quadratic form $V_{p,q,r}$. By a standard invariant theory of quadratic forms, we have the following:

Theorem A.12 (System of representatives of $\mathrm{CNS}_0/\!\sim$) *The set*

$$\{\mathrm{O}\}\times\{(1,0,1),\ (-1,0,-1),\ (1,0,-1),\ (1,0,0),\ (-1,0,0),\ (0,0,0)\}$$

is a system of representatives of $\mathrm{CNS}_0/\!\sim$, *where* $\mathrm{O} \in M_3(\mathbb{R})$ *is the zero matrix.*

Let us turn to the SR of $\mathrm{CNS}_1/\!\sim$. We see from Theorem A.5 that the matrix part C and the kernel part (p, q, r) are transformed separately. According to this structure, we specify one SR in the following two steps. At the first step, we find an SR of the matrix part. More precisely, we define an equivalent relation on $M_3(\mathbb{R})$, which again denote by $\sim$, as follows: We say $M_3(\mathbb{R}) \ni C \sim C' \in M_3(\mathbb{R})$ holds if there exists $M \in GL_2(\mathbb{R})$ such that the relation (A.12) is valid. We will find a system T_Z of representatives of $\mathcal{M}/\sim$, where

$$\mathcal{M} := \{C \in M_3(\mathbb{R}) \mid \operatorname{rank} C = 1\}$$

and the notion of an SR is defined in a similar manner as in Definition A.11. This is the first step. Then, in the second step, we obtain a desired SR of $\mathrm{CNS}_1/\sim$ as follows. Let T_Z be as above. Given system $\sigma \in S$, there exists a unique representative $C' \in T_Z$ such that $C' \sim C(\sigma)$. For each $C' \in T_Z$, we let $G(C') \subset GL_2(\mathbb{R})$ a subgroup which leaves C' invariant. Then, one obtains a set $K(\sigma) \subset \mathbb{R}^3$ by dividing $\mathbb{R}^3$ by the action given by matrices in $G'(C')$. On sees that the set $K(\sigma)$ is the largest set of the kernel parts so that $C' \times K(C')$ does not contain two elements belong to the same equivalence class. Thus, we obtain an SR of the form

$$\{(C, p, q, r) \in M_3(\mathbb{R}) \times \mathbb{R}^3 \mid C \in T_Z,\ (p, q, r) \in K(C)\}.$$

The result is as follows.

Theorem A.13 (System of representatives of $\mathrm{CNS}_1/\sim$) *The set $T := \cup_{j=1}^{9}(Z_j \times K_j)$ is a system of representatives of* $\mathrm{CNS}_1/\sim$*, where*

$$Z_1 := \left\{ \begin{pmatrix} 0 & 1 & 0 \\ 0 & k & 0 \\ 0 & 1 & 0 \end{pmatrix} \middle|\, k > 0 \right\} \cup \left\{ \begin{pmatrix} 0 & 1 & 0 \\ 0 & 1 & 0 \\ 0 & 0 & 0 \end{pmatrix}, \begin{pmatrix} 0 & 0 & 0 \\ 0 & 1 & 0 \\ 0 & 1 & 0 \end{pmatrix}, \begin{pmatrix} 0 & 0 & 1 \\ 0 & 0 & 0 \\ 0 & 0 & 1 \end{pmatrix}, \begin{pmatrix} 0 & 0 & -1 \\ 0 & 0 & 0 \\ 0 & 0 & 1 \end{pmatrix} \right\}$$
$$\cup \left\{ \begin{pmatrix} \sin\theta & 0 & \sin\theta \\ \cos\theta & 0 & \cos\theta \\ \sin\theta & 0 & \sin\theta \end{pmatrix} \middle|\, \theta \in [0, \pi/2) \right\},$$

$$Z_2 := \left\{ \begin{pmatrix} 0 & \sigma & 0 \\ 0 & k & 0 \\ 0 & -\sigma & 0 \end{pmatrix} \middle|\, \sigma \in \{\pm 1\},\ k > 0 \right\},$$

$$Z_3 := \left\{ \begin{pmatrix} 0 & \sigma & 0 \\ 0 & 0 & 0 \\ 0 & -\sigma & 0 \end{pmatrix} \middle|\, \sigma \in \{\pm 1\} \right\},$$

$$Z_4 := \left\{ \begin{pmatrix} 0 & 0 & 0 \\ 0 & 0 & \sigma \\ 0 & 0 & 0 \end{pmatrix} \middle|\, \sigma \in \{\pm 1\} \right\} \cup \left\{ \begin{pmatrix} 0 & \sigma & 0 \\ 0 & 0 & 0 \\ 0 & 0 & 0 \end{pmatrix} \middle|\, \sigma \in \{\pm 1\} \right\},$$

$$Z_5 := \left\{ \begin{pmatrix} 0&1&0 \\ 0&0&0 \\ 0&1&0 \end{pmatrix} \right\}, \quad Z_6 := \left\{ \begin{pmatrix} 0&0&0 \\ 0&1&0 \\ 0&0&0 \end{pmatrix} \right\}, \quad Z_7 := \left\{ \begin{pmatrix} 0&0&1 \\ 0&0&0 \\ 0&0&0 \end{pmatrix} \right\},$$

$$Z_8 := \left\{ \begin{pmatrix} 0&0&0 \\ 0&0&0 \\ 0&0&1 \end{pmatrix} \right\}, \quad Z_9 := \left\{ \begin{pmatrix} 1&0&1 \\ 0&0&0 \\ 1&0&1 \end{pmatrix} \right\}$$

and $K_1 = \mathbb{R}^3$,

$$\begin{aligned}
K_2 &:= \left\{ {}^{t}(p,q,r) \in \mathbb{R}^3 \mid p > r \right\} \cup \left\{ {}^{t}(p,q,r) \in \mathbb{R}^3 \mid p = r,\ q \geqslant 0 \right\}, \\
K_3 &:= \left\{ {}^{t}(p,q,r) \in \mathbb{R}^3 \mid p \geqslant r,\ q \geqslant 0 \right\}, \\
K_4 &:= \left\{ {}^{t}(p,q,\sigma) \in \mathbb{R}^3 \mid q \geqslant 0,\ \sigma \in \{\pm 1\} \right\} \cup \left\{ {}^{t}(p,\tilde{\sigma},0) \in \mathbb{R}^3 \mid \tilde{\sigma} \in \{0,1\} \right\}, \\
K_5 &:= \left\{ {}^{t}(p,q,r) \in \mathbb{R}^3 \mid q \geqslant 0 \right\}, \\
K_6 &:= \left\{ {}^{t}(p,q,r) \in \mathbb{R}^3 \mid |p| = |r| \geqslant 0,\ q \geqslant 0 \right\} \cup \left\{ {}^{t}(p,q,0) \in \mathbb{R}^3 \mid |p| = 1 \right\}, \\
K_7 &:= \left\{ {}^{t}(p,q,r) \in \mathbb{R}^3 \mid |p| = |r| > 0 \right\} \cup \left\{ {}^{t}(p,q,0) \in \mathbb{R}^3 \mid |p| = 1 \right\} \\
&\qquad \cup \left\{ {}^{t}(0,q,r) \in \mathbb{R}^3 \mid |r| = 1 \right\} \cup \left\{ {}^{t}(0,1,0), {}^{t}(0,-1,0), {}^{t}(0,0,0) \right\}, \\
K_8 &:= \left\{ {}^{t}(p,q,r) \in \mathbb{R}^3 \mid |p| = |r| \geqslant 0 \right\} \cup \left\{ {}^{t}(p,q,0) \in \mathbb{R}^3 \mid |p| = 1 \right\} \\
&\qquad \cup \left\{ {}^{t}(0,q,r) \in \mathbb{R}^3 \mid |r| = 1 \right\}, \\
K_9 &:= \left\{ {}^{t}(p,q,r) \in \mathbb{R}^3 \mid p \geqslant r,\ q = 0 \right\}.
\end{aligned}$$

Here, a row vector in $\mathbb{R}^3$ *is denoted as* (a, b, c)*, with commas.*

Remark A.14 One obtains the asymptotic behavior of solutions to several representatives by applying known results. More precisely, let T be the SR of $\mathrm{CNS}_1/\sim$ given by the above theorem. If the matrix part of a representative $\sigma \in T$ satisfies

$$\begin{aligned}
Z_+ := &\left\{ \begin{pmatrix} 0&1&0 \\ 0&k&0 \\ 0&1&0 \end{pmatrix} \middle| \ k \geqslant 0 \right\} \cup \left\{ \begin{pmatrix} 0&\sigma&0 \\ 0&k&0 \\ 0&-\sigma&0 \end{pmatrix} \middle| \ \sigma \in \{\pm 1\},\ k \geqslant 0 \right\} \\
&\cup \left\{ \begin{pmatrix} 0&\sigma&0 \\ 0&0&0 \\ 0&0&0 \end{pmatrix} \middle| \ \sigma \in \{\pm 1\} \right\} \cup \left\{ \begin{pmatrix} 0&1&0 \\ 0&1&0 \\ 0&0&0 \end{pmatrix}, \begin{pmatrix} 0&0&0 \\ 0&1&0 \\ 0&1&0 \end{pmatrix}, \begin{pmatrix} 0&0&0 \\ 0&1&0 \\ 0&0&0 \end{pmatrix} \right\},
\end{aligned} \tag{A.16}$$

then [12, Theorem 2.1] is applicable, in view of Proposition A.6. Such an example is [12, Example 6.2], for instance.

Remark A.15 The case $\lambda_6 = 3\lambda_1 \in \{\pm 1\}$ of (1.1) belongs to $Z_4 \times K_4$, and the system (1.2) to $Z_7 \times K_7$. Notice that (1.1) belongs to CNS_2, and hence is out of the range of Theorem A.13, if $\lambda_6 \neq 3\lambda_1$.

A.6 Classification of Matrices of Rank One

In the rest of this section, we prove Theorem A.13. As mentioned above, what we do first is to specify $\mathcal{M}/\sim \subset M_3(\mathbb{R})/\sim$.

Theorem A.16 *The set $T_Z := Z_+ \cup Z_0 \cup Z_- \subset \mathcal{M}$ is a system of representatives of $\mathcal{M}/\sim$, where Z_+ is given by* (A.16),

$$Z_0 := \left\{ \begin{pmatrix} 0 & 0 & \sigma \\ 0 & 0 & 0 \\ 0 & 0 & 1 \end{pmatrix} \middle| \sigma \in \{0, \pm 1\} \right\} \cup \left\{ \begin{pmatrix} 0 & 0 & 0 \\ 0 & 0 & 1 \\ 0 & 0 & 0 \end{pmatrix}, \begin{pmatrix} 0 & 0 & 0 \\ 0 & 0 & -1 \\ 0 & 0 & 0 \end{pmatrix}, \begin{pmatrix} 0 & 0 & 1 \\ 0 & 0 & 0 \\ 0 & 0 & 0 \end{pmatrix} \right\},$$

and

$$Z_- := \left\{ \begin{pmatrix} \sin\theta & 0 & \sin\theta \\ \cos\theta & 0 & \cos\theta \\ \sin\theta & 0 & \sin\theta \end{pmatrix} \middle| \theta \in [0, \pi/2] \right\}.$$

Further $T_Z = \cup_{j=1}^{9} Z_j$ holds true.

Here, we give a constructive proof, i.e., we find a matrix $M \in GL_2(\mathbb{R})$ which transforms a given matrix into the corresponding representative. This construction is valuable when we apply the theory to concrete examples. This part is a generalization of a result in [22], where an SR of $(\mathcal{M} \cap \mathcal{Z})/\sim$ is given. We give a slightly different proof.

For $C \in \mathcal{M}$, let $\nu(C) \in \mathbb{R}^3$ be the unit vector such that C is reduced into ${}^{\mathrm{t}}\big(\nu(C)\ \mathbf{0}\ \mathbf{0}\big)$ by row operations, where $\mathbf{0} \in \mathbb{R}^3$ denotes the zero column vector. Remark that $\nu(C)$ is uniquely determined as an element of the projective space $\mathbb{RP}^2$.

Lemma A.17 *Let $C \in M_3(\mathbb{R})$. rank $C = 1$ if and only if C is written as $C = \mathbf{d}\,{}^{\mathrm{t}}\nu(C)$ with some nonzero vector $\mathbf{d} \neq 0$.*

Let $\mathcal{B}$ be the unit sphere of $\mathbb{R}^3$. Let $\Gamma_1, \Gamma_2 \subset \mathcal{B}$ be simple closed smooth curves given by

$$\begin{aligned} \Gamma_1 &:= \{{}^{\mathrm{t}}(a, b, c) \in \mathcal{B} \mid b^2 = ac,\ a \geqslant 0,\ c \geqslant 0\} \\ &= \left\{ {}^{\mathrm{t}}\left(\tfrac{\sqrt{2}(1+\sin\theta)}{\sqrt{7-\cos 2\theta}}, \tfrac{\sqrt{2}\cos\theta}{\sqrt{7-\cos 2\theta}}, \tfrac{\sqrt{2}(1-\sin\theta)}{\sqrt{7-\cos 2\theta}} \right) \in \mathcal{B} \,\middle|\, 0 \leqslant \theta \leqslant 2\pi \right\} \end{aligned} \tag{A.17}$$

and

$$\begin{aligned} \Gamma_2 &:= \{{}^{\mathrm{t}}(a, b, c) \in \mathcal{B} \mid b^2 = 4ac,\ a \geqslant 0,\ c \geqslant 0\} \\ &= \left\{ {}^{\mathrm{t}}\left(\tfrac{1-\sin\theta}{\sqrt{5+\cos 2\theta}}, -\tfrac{2\cos\theta}{\sqrt{5+\cos 2\theta}}, \tfrac{1+\sin\theta}{\sqrt{5+\cos 2\theta}} \right) \in \mathcal{B} \,\middle|\, 0 \leqslant \theta \leqslant 2\pi \right\}, \end{aligned} \tag{A.18}$$

respectively. Further, let $\mathscr{S}_1, \mathscr{S}_2 \subset \mathcal{B}$ be open connected surfaces given by

$$\mathscr{S}_1 := \{{}^{\mathrm{t}}(a,b,c) \in \mathcal{B} \,|\, b^2 > 4ac\} \quad \text{and} \quad \mathscr{S}_2 := \{{}^{\mathrm{t}}(a,b,c) \in \mathcal{B} \,|\, b^2 < 4ac,\ a > 0\}, \tag{A.19}$$

respectively. Remark that

$$\mathcal{B} = \mathscr{S}_2 \cup \Gamma_2 \cup \mathscr{S}_1 \cup (-\Gamma_2) \cup (-\mathscr{S}_2). \tag{A.20}$$

Further, $\partial\mathscr{S}_1 = \Gamma_2 \sqcup (-\Gamma_2)$ and $\partial\mathscr{S}_2 = \Gamma_2$. We define three subsets

$$\begin{aligned} \mathcal{M}_+ &:= \{C \in \mathcal{M} \mid \nu(C) \in \mathscr{S}_1\}, \\ \mathcal{M}_0 &:= \{C \in \mathcal{M} \mid \nu(C) \in \Gamma_2 \sqcup (-\Gamma_2)\}, \\ \mathcal{M}_- &:= \{C \in \mathcal{M} \mid \nu(C) \in \mathscr{S}_2 \sqcup (-\mathscr{S}_2)\}. \end{aligned}$$

By (A.20), we have $\mathcal{M} = \mathcal{M}_+ \cup \mathcal{M}_0 \cup \mathcal{M}_-$.

Let us now an intermediate sets.

$$\begin{aligned} Y_+ &:= \{C \in \mathcal{M}_+ \mid \nu(C) = {}^{\mathrm{t}}(0,1,0)\}, \\ Y_0 &:= \{C \in \mathcal{M}_0 \mid \nu(C) = {}^{\mathrm{t}}(0,0,1)\}, \\ Y_- &:= \{C \in \mathcal{M}_- \mid \nu(C) = 2^{-1/2\mathrm{t}}(1,0,1)\}. \end{aligned}$$

Our reduction is divided into two steps.

Proposition A.18 (First reduction) *Let $\varsigma \in \{0, \pm\}$. For any $C \in \mathcal{M}_\varsigma$, there exists $\tilde{C} \in Y_\varsigma$ such that $C \sim \tilde{C}$.*

Proof of Proposition A.18 First consider the case $\varsigma = +$. Pick $C_+ \in \mathcal{M}_+$. Since $\nu(C_+) \in \mathscr{S}_1$, the trichotomy in [22] implies that $\ker C_+$ has two intersection points with Γ_1. We name them as

$$\mathbf{p}_j = r(\theta_j) \begin{pmatrix} 1 + \sin\theta_j \\ \cos\theta_j \\ 1 - \sin\theta_j \end{pmatrix} \qquad (j = 1, 2)$$

with the normalizing factor $r(\theta)$ and $\theta_1, \theta_2 \in [0, 2\pi)$ such that $\theta_1 \neq \theta_2$. Note that (one of) $\nu(C_+) \in \mathcal{B}$ is written as

$$\nu(C_+) = cr(\theta_1)^{-1} r(\theta_2)^{-1} (\mathbf{p}_1 \times \mathbf{p}_2) = c \begin{pmatrix} \cos\theta_1(1 - \sin\theta_2) - \cos\theta_2(1 - \sin\theta_1) \\ -2(\sin\theta_1 - \sin\theta_2) \\ -\cos\theta_1(1 + \sin\theta_2) + \cos\theta_2(1 + \sin\theta_1) \end{pmatrix}$$

with a suitable normalizing factor $c = c(\theta_1, \theta_2) \neq 0$. Let

$$M_+ = \begin{pmatrix} \cos\eta_1 & \sin\eta_1 \\ \cos\eta_2 & \sin\eta_2 \end{pmatrix}, \qquad \eta_j = -\tfrac{\theta_j}{2} + \tfrac{\pi}{4}.$$

A computation shows that

$$D(M_+)^{-1} = \frac{1}{2\sin\frac{\theta_1-\theta_2}{2}} \begin{pmatrix} 1+\sin\theta_1 & 2(\sin\frac{\theta_1+\theta_2}{2}+\cos\frac{\theta_1-\theta_2}{2}) & 1+\sin\theta_2 \\ \cos\theta_1 & 2\cos\frac{\theta_1+\theta_2}{2} & \cos\theta_2 \\ 1-\sin\theta_1 & 2(-\sin\frac{\theta_1+\theta_2}{2}+\cos\frac{\theta_1-\theta_2}{2}) & 1-\sin\theta_2 \end{pmatrix}.$$

Note that

$${}^{t}D(M_+)^{-1}\nu(C_+) = -4c(1-\cos^2\tfrac{\theta_1-\theta_2}{2}) \begin{pmatrix} 0 \\ 1 \\ 0 \end{pmatrix}.$$

This shows that $\tilde{C}_+ := (\det M_+)^{-1} D(M_+) C_+ D(M_+)^{-1}$ satisfies $\nu(\tilde{C}_+) = {}^{t}(0, 1, 0)$, and hence $\tilde{C}_+ \in Y_+$.

Next consider the case $\varsigma = 0$. Pick $C_0 \in \mathcal{M}_0$. Since $\nu(C_0) \in \Gamma_2 \cup (-\Gamma_2)$, we deduce from the trichotomy that there exists $\theta \in \mathbb{R}/2\pi\mathbb{Z}$ such that

$$\nu(C_0) = c' \begin{pmatrix} 1-\sin\theta \\ -2\cos\theta \\ 1+\sin\theta \end{pmatrix}.$$

We define

$$M_0 = \begin{pmatrix} \cos\eta & \sin\eta \\ c & d \end{pmatrix} \in GL_2, \qquad \eta = -\tfrac{\theta}{2} + \tfrac{\pi}{4}.$$

One has

$$D(M_0)^{-1} = \frac{1}{2\det M_0} \begin{pmatrix} 1+\sin\theta & 4c\cos\eta & 2c^2 \\ \cos\theta & 2(d\cos\eta + c\sin\eta) & 2cd \\ 1-\sin\theta & 4d\sin\eta & 2d^2 \end{pmatrix}.$$

Hence, one has

$${}^{t}D(M_0)^{-1}\nu(C_0) = 2c' \det M_0 \begin{pmatrix} 0 \\ 0 \\ 1 \end{pmatrix}.$$

This shows that $\tilde{C}_0 := (\det M_0)^{-1} D(M_0) C_0 D(M_0)^{-1}$ satisfies $\nu(\tilde{C}_0) = {}^{t}(0, 0, 1)$, and hence $\tilde{C}_0 \in Y_0$.

Finally, let us consider the case $\varsigma = -$. Pick $C_- \in \mathcal{M}_-$. Since $\nu(C_-) = {}^{t}(u, v, w) \in \mathscr{S}_2 \cup (-\mathscr{S}_2)$, we have $v^2 < 4uw$. Take

$$M_- = \begin{pmatrix} 2w & -v \\ 0 & \sqrt{4uw - v^2} \end{pmatrix} \in GL_2.$$

Remark that

$$D(M_-)^{-1} = \frac{1}{\det M_-} \begin{pmatrix} 4w^2 & 0 & 0 \\ -2wv & 2w\sqrt{4uw - v^2} & 0 \\ v^2 & -2v\sqrt{4uw - v^2} & 4uw - v^2 \end{pmatrix}.$$

Hence,

$${}^{t}D(M_-)^{-1}\nu(C_-) = \frac{\sqrt{4wu - v^2}}{2} \begin{pmatrix} 1 \\ 0 \\ 1 \end{pmatrix}.$$

This shows that $\tilde{C}_- := (\det M_-)^{-1} D(M_-) C_- D(M_-)^{-1}$ satisfies $\nu(\tilde{C}_-) = 2^{-1/2}$ ${}^{t}(1, 0, 1)$, and hence $\tilde{C}_- \in Y_{-1}$. □

Proposition A.19 (Second reduction) *Let* $\varsigma \in \{0, \pm\}$. *For any* $C \in Y_\varsigma$, *there exists* $\tilde{C} \in Z_\varsigma$ *such that* $C \sim \tilde{C}$.

To show the proposition, we use the following:

Lemma A.20 *Let* $M \in GL_2(\mathbb{R})$. *Let* $\varsigma \in \{0, \pm\}$. *Let* $C, C' \in Y_\varsigma$. *Then,* $C \sim C'$ *is true if and only if* (A.12) *holds with the following matrices:*

- *If* $\varsigma = +$ *then*

$$M = \begin{pmatrix} 0 & 1 \\ 1 & 0 \end{pmatrix}^{\ell} \begin{pmatrix} p & 0 \\ 0 & s \end{pmatrix}$$

  for $p, s \in \mathbb{R} \setminus \{0\}$ *and* $\ell \in \{0, 1\}$;
- *If* $\varsigma = 0$ *then*

$$M = \begin{pmatrix} p & 0 \\ r & s \end{pmatrix}$$

  for $p, s \in \mathbb{R} \setminus \{0\}$ *and* $r \in \mathbb{R}$;
- *If* $\varsigma = -$ *then*

$$M = R \begin{pmatrix} 1 & 0 \\ 0 & -1 \end{pmatrix}^{\ell} \begin{pmatrix} \cos\eta & -\sin\eta \\ \sin\eta & \cos\eta \end{pmatrix}$$

  for some $R > 0$ *and* $\eta \in \mathbb{R}/2\pi\mathbb{Z}$ *and* $\ell \in \{0, 1\}$.

Proof of Lemma A.20 We prove the only if part. Suppose that $C \sim C'$. Then, (A.12) is valid for some

$$M = \begin{pmatrix} p & q \\ r & s \end{pmatrix} \in GL_2(\mathbb{R}).$$

Let us first consider the case $\varsigma = +$. Since $\nu(C) = \nu(C') = {}^{t}(0, 1, 0)$, we see from (A.12) that ${}^{t}(0, 1, 0)$ must be an eigenvector of

$${}^{t}(D(M)^{-1}) = \frac{1}{\det M} \begin{pmatrix} p^2 & pq & q^2 \\ 2pr & ps + qr & 2qs \\ r^2 & rs & s^2 \end{pmatrix}$$

associated with a nonzero eigenvalue. To this end, we need $pq = rs = 0$ and $ps + qr \neq 0$. If $p = 0$ then we have $qr \neq 0$ and hence $s = 0$. Similarly, if $q = 0$ then we have $ps \neq 0$ and hence $r = 0$.

Let us next consider the case $\varsigma = 0$. By the same argument as in the case $\sigma = 1$, one sees from $\nu(C) = \nu(C') = {}^t(0, 0, 1)$ that ${}^t(0, 0, 1)$ is an eigenvector of ${}^t(D(M)^{-1})$. From this, we obtain $q = 0$ and $s \neq 0$.

Finally, we consider the case $\varsigma = -$. In this case, ${}^t(1, 0, 1)$ is an eigenvector of ${}^t(D(M)^{-1})$. To this end, we need $p^2 + q^2 = r^2 + s^2 > 0$ and $pr + qs = 0$. One sees from the first equation that there exists $R > 0$, η, and ζ such that

$$p = R\cos\eta, \quad q = -R\sin\eta, \quad r = R\cos\zeta, \quad s = -R\sin\zeta.$$

Plugging these formulas to the second identity, we obtain $\cos(\eta - \zeta) = 0$, which shows $\zeta = \eta + \frac{\pi}{2} + \ell\pi$ ($\ell \in \{0, 1\}$). Thus, we obtain the result.

The if part is obvious from the argument above. □

Proof of Proposition A.19 We first consider the case $\varsigma = +$. Pick $C \in Y_+$. Then, C is written as

$$C = \begin{pmatrix} 0 & k_1 & 0 \\ 0 & k_2 & 0 \\ 0 & k_3 & 0 \end{pmatrix} = \begin{pmatrix} k_1 \\ k_2 \\ k_3 \end{pmatrix} (0\ 1\ 0)$$

with a nonzero vector ${}^t(k_1, k_2, k_3) \in \mathbb{R}^3$. By Lemma A.20, $C' \in Y_+$ satisfies $C' \sim C$ if and only if there exist nonzero p, s and $\ell \in \{0, 1\}$ such that

$$C' = \begin{pmatrix} p^{-2}k_1 \\ (ps)^{-1}k_2 \\ s^{-2}k_3 \end{pmatrix} (0\ 1\ 0)$$

if $\ell = 0$ and

$$C' = -\begin{pmatrix} s^{-2}k_3 \\ (ps)^{-1}k_2 \\ p^{-2}k_1 \end{pmatrix} (0\ 1\ 0)$$

if $\ell = 1$. Let us show that $C' \in Z_+$ holds for a suitable choice of p, r and ℓ.

When $k_1 > 0$ and $k_3 > 0$ then we choose $\ell = 0$, $p = \sqrt{k_1}$, and $s = \sqrt{k_3}$ if $k_2 \geqslant 0$ and $s = -\sqrt{k_3}$ if $k_2 < 0$. Then, one sees that

$$C' = \begin{pmatrix} 1 \\ |k_2|/\sqrt{k_1 k_3} \\ 1 \end{pmatrix} (0\ 1\ 0) \in Z_+.$$

The case $k_1 < 0$ and $k_3 < 0$ is handled similarly, we take $\ell = 1$, $p = \sqrt{|k_1|}$, and $s = -\sqrt{|k_3|}$ if $k_2 \geqslant 0$ and $s = \sqrt{|k_3|}$ if $k_2 < 0$. The choice gives us the same reduced matrix. If $k_1 > 0 > k_3$ then one can choose parameters so that

$$C' = \begin{pmatrix} 1 \\ |k_2|/\sqrt{k_1|k_3|} \\ -1 \end{pmatrix} (0\ 1\ 0) \in Z_+.$$

If $k_3 > 0 > k_1$ then we have

$$C' = \begin{pmatrix} -1 \\ |k_2|/\sqrt{|k_1|k_3} \\ 1 \end{pmatrix} (0\ 1\ 0) \in Z_+.$$

for a suitable choice of the parameters. We left the precise choice for the readers.

If $k_1 > 0 = k_3$ and $k_2 \neq 0$ then we take $\ell = 0$, $p = \sqrt{k_1}$ and $s = p^{-1}k_2$ to get

$$C' = \begin{pmatrix} 1 \\ 1 \\ 0 \end{pmatrix} (0\ 1\ 0) \in Z_+.$$

If $k_1 < 0 = k_3$ and $k_2 \neq 0$ then we let $\ell = 1$, $p = \sqrt{|k_1|}$ and $s = -p^{-1}k_2$ to get

$$C' = \begin{pmatrix} 0 \\ 1 \\ 1 \end{pmatrix} (0\ 1\ 0) \in Z_+.$$

If $k_1 \neq 0 = k_2 = k_3$ then we take $\ell = 0$, $p = \sqrt{k_1}$ and $s = 1$ to get

$$C' = \begin{pmatrix} \operatorname{sign} k_1 \\ 0 \\ 0 \end{pmatrix} (0\ 1\ 0) \in Z_+.$$

The case $k_1 = 0 \neq k_3$ are handled similarly. Finally, in the case $k_1 = k_3 = 0 \neq k_2$, we choose $\ell = 0$, $p = 1$, and $s = k_2$ to get

$$C' = \begin{pmatrix} 0 \\ 1 \\ 0 \end{pmatrix} (0\ 1\ 0) \in Z_+.$$

Hence, we have the results for $\varsigma = +$.

Let us turn to the case $\varsigma = 0$. Pick $C \in Y_0$. Then, C is written as

$$C = \begin{pmatrix} 0 & 0 & k_1 \\ 0 & 0 & k_2 \\ 0 & 0 & k_3 \end{pmatrix} = \begin{pmatrix} k_1 \\ k_2 \\ k_3 \end{pmatrix} (0\ 0\ 1)$$

with a nonzero vector ${}^{t}(k_1, k_2, k_3) \in \mathbb{R}^3$.

By Lemma A.20, if $C' \in Y_0$ satisfies $C' \sim C$ then C' is written as

$$C' = \frac{1}{(ps)^3} \begin{pmatrix} s^2k_1 - 2rsk_2 + r^2k_3 \\ p(sk_2 - rk_3) \\ p^2k_3 \end{pmatrix} (0\ 0\ s^2)$$

for $ps \neq 0$ and $r \in \mathbb{R}$. When $k_3 \neq 0$, we choose $p = k_3/s$ and $r = sk_2/k_3$. Then,

$$C' = \begin{pmatrix} \frac{s^4}{k_3^3}(k_1 - \frac{k_2^2}{k_3}) \\ 0 \\ 1 \end{pmatrix} (0\ 0\ 1).$$

One can choose $s > 0$ so that $\frac{s^4}{k_3^3}(k_1 - \frac{k_2^2}{k_3}) \in \{0, \pm 1\}$. For this choice, we have $C' \in Z_0$. When $k_3 = 0$ and $k_2 \neq 0$, we choose $p = |k_2|^{1/2}$ and $r = sk_1/2k_2$. Then

$$C' = \begin{pmatrix} 0 \\ \operatorname{sign} k_2 \\ 0 \end{pmatrix} (0\ 0\ 1) \in Z_0.$$

When $k_3 = k_2 = 0$ and $k_1 \neq 0$, we choose $p = 1$ and $s = 1/k_1$ to obtain

$$C' = \begin{pmatrix} 1 \\ 0 \\ 0 \end{pmatrix} (0\ 0\ 1) \in Z_0.$$

For all cases, we reach to the desired conclusion.

Finally consider the case $\varsigma = -$. Pick $C \in Y_-$. Then, C is written as

$$C = \begin{pmatrix} k_1 & 0 & k_1 \\ k_2 & 0 & k_2 \\ k_3 & 0 & k_3 \end{pmatrix} = \begin{pmatrix} k_1 \\ k_2 \\ k_3 \end{pmatrix} (1\ 0\ 1)$$

with a nonzero vector ${}^t(k_1, k_2, k_3) \in \mathbb{R}^3$.

By Lemma A.20, $C' \in Y_-$ satisfies $C' \sim C$ if and only if there exists $R > 0$, $\eta \in \mathbb{R}/2\pi\mathbb{Z}$, and $\ell \in \{0, 1\}$ such that

$$C' = \frac{1}{2R^2} \begin{pmatrix} (k_1 + k_3) + [(k_1 - k_3)\cos 2\eta - 2k_2 \sin 2\eta] \\ (k_1 - k_3)\sin 2\eta + 2k_2 \cos 2\eta \\ (k_1 + k_3) - [(k_1 - k_3)\cos 2\eta - 2k_2 \sin 2\eta] \end{pmatrix} (1\ 0\ 1)$$

if $\ell = 0$ and

$$C' = \frac{1}{2R^2} \begin{pmatrix} -(k_1 + k_3) + [-(k_1 - k_3)\cos 2\eta + 2k_2 \sin 2\eta] \\ -(k_1 - k_3)\sin 2\eta + 2k_2 \cos 2\eta \\ -(k_1 + k_3) - [-(k_1 - k_3)\cos 2\eta + 2k_2 \sin 2\eta] \end{pmatrix} (1\ 0\ 1)$$

if $\ell = 1$.

We let $\ell = 0$ if $k_1 + k_3 \geqslant 0$ and $\ell = 1$ if $k_1 + k_3 < 0$. Choose η so that $(k_1 - k_3)\cos 2\eta - 2(-1)^\ell k_2 \sin 2\eta = 0$ and

$$(-1)^\ell (k_1 - k_3)\sin 2\eta + 2k_2 \cos 2\eta = \sqrt{(k_1 - k_3)^2 + 4k_2^2}$$

hold. Then, we have

$$C' = \frac{1}{2R^2}\begin{pmatrix} |k_1 + k_3| \\ \sqrt{(k_1 - k_3)^2 + 4k_2^2} \\ |k_1 + k_3| \end{pmatrix} (1\ 0\ 1).$$

Now we choose $R > 0$ so that $(2R^2)^2 = (k_1 + k_3)^2 + (k_1 - k_3)^2 + 4k_2^2 > 0$ to obtain

$$C' = \begin{pmatrix} y_1 \\ y_2 \\ y_1 \end{pmatrix} (1\ 0\ 1)$$

with $y_1, y_2 \geqslant 0$ and $y_1^2 + y_2^2 = 1$. This shows $C' \in Z_-$. Thus, we reach to the desired conclusion in all cases. □

Now we are in a position to complete the proof of Theorem A.16.

Proof of Theorem A.16 We have shown that $\mathcal{M}/\sim = [Z_+] \cup [Z_0] \cup [Z_-]$. Hence, it suffices to show that if $C, C' \in Z_+ \cup Z_0 \cup Z_-$ satisfies $C \sim C'$ then $C = C'$. Suppose that $C, C' \in Z_+ \cup Z_0 \cup Z_-$ satisfies $C \sim C'$. First note that if $C \in Z_\varsigma$ for $\varsigma \in \{0, \pm\}$ then $C' \in Z_\varsigma$ with the same ς.

Suppose that $\varsigma = +$. By Lemma A.20, C' is given by the formula (A.12) with

$$M = \begin{pmatrix} 0 & 1 \\ 1 & 0 \end{pmatrix}^\ell \begin{pmatrix} p & 0 \\ 0 & s \end{pmatrix}$$

for $p, s \in \mathbb{R} \setminus \{0\}$ and $\ell \in \{0, 1\}$. If

$$C = \begin{pmatrix} 1 \\ k \\ 1 \end{pmatrix} (0\ 0\ 1)$$

for some $k \geqslant 0$ then one has

$$C' = \begin{pmatrix} p^{-2} \\ (ps)^{-1}k \\ s^{-2} \end{pmatrix} (0\ 1\ 0)$$

if $\ell = 0$ and

$$C' = -\begin{pmatrix} s^{-2} \\ (ps)^{-1}k \\ p^{-2} \end{pmatrix} (0\ 1\ 0)$$

if $\ell = 1$. One then sees from $C' \in Z_1$ that $\ell = 0$ and $p^2 = s^2 = 1$. Further, since $k \geqslant 0$, we have $ps = 1$. This shows $C' = C$. If

$$C = \begin{pmatrix} \tilde{\sigma} \\ k \\ -\tilde{\sigma} \end{pmatrix} (0\ 0\ 1)$$

for some $\tilde{\sigma} \in \{\pm 1\}$ and $k \geqslant 0$ then one sees that $p^2 = s^2 = 1$ from the first and the third row of C'. Then, looking at the second row, we have $ps = (-1)^\ell$. This shows $C = C'$. The other cases are handled similarly.

Let us proceed to the case $\varsigma = 0$. Suppose that

$$C = \begin{pmatrix} \tilde{\sigma} \\ 0 \\ 1 \end{pmatrix} (0\ 0\ 1)$$

with $\tilde{\sigma} \in \{0, \pm 1\}$. Then, any $C' \in [C]$ is written as

$$C' = \begin{pmatrix} \frac{s}{p^3}\tilde{\sigma} + \frac{r^2}{p^3 s} \\ -\frac{r}{p^2 s} \\ \frac{1}{ps} \end{pmatrix} (0\ 0\ 1).$$

Suppose that $C' \in Z_0$. By noticing that the third row of C' is nonzero, we see that

$$C = \begin{pmatrix} \tilde{\sigma}' \\ 0 \\ 1 \end{pmatrix} (0\ 0\ 1)$$

with $\tilde{\sigma}' \in \{0, \pm 1\}$. Let us show $\tilde{\sigma}' = \tilde{\sigma}$. Comparing the two representations, one sees that $ps = 1$ and $r = 0$ are necessary. With the choice, the former representation reads as

$$C' = \begin{pmatrix} \frac{s^2}{p^2}\tilde{\sigma} \\ 0 \\ 1 \end{pmatrix} (0\ 0\ 1).$$

Since $\frac{s^2}{p^2}\tilde{\sigma} = \tilde{\sigma}'$, we conclude that $\tilde{\sigma} = \tilde{\sigma}'$. The other cases are handled similarly.

The case $\varsigma = -$. Suppose that

$$C = \begin{pmatrix} \sin\theta \\ \cos\theta \\ \sin\theta \end{pmatrix} (1\ 0\ 1)$$

with $\theta \in [0, \pi/2]$. Then, any $C' \in [C]$ is written as

$$C' = \frac{1}{R^2} \begin{pmatrix} \sin\theta - \cos\theta \sin 2\eta \\ \cos\theta \cos 2\eta \\ \sin\theta + \cos\theta \sin 2\eta \end{pmatrix} (1\ 0\ 1)$$

or

$$C' = \frac{1}{R^2} \begin{pmatrix} -\sin\theta - \cos\theta \sin 2\eta \\ \cos\theta \cos 2\eta \\ -\sin\theta + \cos\theta \sin 2\eta \end{pmatrix} (1\ 0\ 1)$$

It is easy to see that $C' \in Z_-$ implies that C' is of the first form and $\eta = 0$ and $R = 1$. This shows $C' = C$. □

A.7 Reduction of the Kernel Part

We next specify the possible kernel part for each element of $C \in T_Z$. To this end, for a given $C \in T_Z$, we find a subgroup of $GL_2(\mathbb{R})$ which leaves C invariant.

Proposition A.21 *Let $C \in Z_j$ for some $1 \leqslant j \leqslant 9$. Let $M \in GL_2(\mathbb{R})$ and let C' be the matrix defined by* (A.12). *If $C' = C$ then M belongs to G_j, where $G_1 := \{\pm E_2\}$,*

$$G_2 := \left\{ \pm E_2, \begin{pmatrix} 0 & 1 \\ -1 & 0 \end{pmatrix}, \begin{pmatrix} 0 & -1 \\ 1 & 0 \end{pmatrix} \right\},$$

$$G_3 := \left\{ \begin{pmatrix} \sigma_1 & 0 \\ 0 & \sigma_2 \end{pmatrix} \middle| \sigma_1, \sigma_2 \in \{\pm 1\} \right\} \cup \left\{ \begin{pmatrix} 0 & \sigma_1 \\ \sigma_2 & 0 \end{pmatrix} \middle| \sigma_1, \sigma_2 \in \{\pm 1\} \right\},$$

$$G_4 := \left\{ \begin{pmatrix} \sigma & 0 \\ 0 & s \end{pmatrix} \middle| s \neq 0,\ \sigma \in \{\pm 1\} \right\},$$

$$G_5 := \left\{ \begin{pmatrix} \sigma_1 & 0 \\ 0 & \sigma_2 \end{pmatrix} \middle| \sigma_1, \sigma_2 \in \{\pm 1\} \right\},$$

$$G_6 := \left\{ \begin{pmatrix} p & 0 \\ 0 & 1/p \end{pmatrix} \middle| p \neq 0 \right\} \cup \left\{ \begin{pmatrix} 0 & p \\ -1/p & 0 \end{pmatrix} \middle| p \neq 0 \right\},$$

$$G_7 := \left\{ \begin{pmatrix} p & 0 \\ 0 & p^3 \end{pmatrix} \middle| p \neq 0 \right\},$$

$$G_8 := \left\{ \begin{pmatrix} p & 0 \\ 0 & 1/p \end{pmatrix} \middle| p \neq 0 \right\},$$

$$G_9 := \left\{ \begin{pmatrix} \cos\eta & -\sin\eta \\ \sin\eta & \cos\eta \end{pmatrix} \middle| \eta \in \mathbb{R}/2\pi\mathbb{Z} \right\}.$$

Proof of Proposition A.21 Let $C \in T_Z$. Let $M \in GL_2(\mathbb{R})$ and let C' be the matrix defined by (A.12). Suppose that $C' = C$.

Let us begin with the case $C \in Z_+$. Then, C is of the form

$$C = \begin{pmatrix} 0 & k_1 & 0 \\ 0 & k_2 & 0 \\ 0 & k_3 & 0 \end{pmatrix} = \begin{pmatrix} k_1 \\ k_2 \\ k_3 \end{pmatrix} (0\ 1\ 0)$$

with a nonzero vector ${}^t(k_1, k_2, k_3) \in \mathbb{R}^3$. By Lemma A.20, M is of the form

$$M = \begin{pmatrix} 0 & 1 \\ 1 & 0 \end{pmatrix}^{\ell} \begin{pmatrix} p & 0 \\ 0 & s \end{pmatrix}$$

for $p, s \in \mathbb{R} \setminus \{0\}$ and $\ell \in \{0, 1\}$. Remark that

$$C' = \begin{pmatrix} p^{-2}k_1 \\ (ps)^{-1}k_2 \\ s^{-2}k_3 \end{pmatrix} (0\ 1\ 0)$$

if $\ell = 0$ and

$$C' = -\begin{pmatrix} s^{-2}k_3 \\ (ps)^{-1}k_2 \\ p^{-2}k_1 \end{pmatrix} (0\ 1\ 0)$$

if $\ell = 1$.

If $k_1 = k_3 = 1$ and $k_2 \geqslant 0$, i.e., if $C \in (Z_+ \cap Z_1) \cup Z_5$ then we see from the first row and the third row of C' that $\ell = 0$, $p^2 = s^2 = 1$. When $k_2 > 0$ ($C \in Z_+ \cap Z_1$), we also have $ps = 1$ by the second row. Hence, $(p, s) = \pm(1, 1)$, which implies that $M \in G_1$. When $k_2 = 0$ ($C \in Z_5$), there is no additional condition. Hence, $M \in G_5$.

If $(k_1, k_2, k_3) = (1, 1, 0)$ then we see that $\ell = 0$, $p^2 = 1$, and $ps = 1$. Hence, $M \in G_1$. The case $(k_1, k_2, k_3) = (0, 1, 1)$ is handled similarly. This completes the proof of the case of $C \in Z_+ \cap Z_1$.

If $k_1 = -k_3 \in \{\pm 1\}$ and $k_2 \geqslant 0$, i.e., if $C \in Z_2 \cup Z_3$ then we see from the first row and the third row of C' that $p^2 = s^2 = 1$. When $k_2 > 0$ ($C \in Z_2$), we also have $ps = (-1)^{\ell}$ by the second row. Hence, $M \in G_2$. When $k_2 = 0$ ($C \in Z_3$), there is no additional condition. We obtain $M \in G_3$.

If $(k_1, k_2, k_3) = (\pm 1, 0, 0)$, i.e., if $C \in Z_+ \cap Z_4$ then we see that $\ell = 0$ and $p^2 = 1$. Recall that $s \neq 0$ follows from $M \in GL_2(\mathbb{R})$. Hence, $M \in G_4$.

If $(k_1, k_2, k_3) = (0, 1, 0)$, i.e., if $C \in Z_6$ then we see that $ps = (-1)^{\ell}$. Hence, $M \in G_6$.

We turn to the case $C \in Z_0$. In this case, C is of the form

$$C = \begin{pmatrix} 0 & 0 & k_1 \\ 0 & 0 & k_2 \\ 0 & 0 & k_3 \end{pmatrix} = \begin{pmatrix} k_1 \\ k_2 \\ k_3 \end{pmatrix} (0\ 0\ 1)$$

with a nonzero vector ${}^{\mathrm{t}}(k_1, k_2, k_3) \in \mathbb{R}^3$. By Lemma A.20, M is of the form

$$M = \begin{pmatrix} p & 0 \\ r & s \end{pmatrix}$$

for $p, s, r \in \mathbb{R}$ with $ps \neq 0$. Remark that

$$C' = \frac{1}{p^3 s} \begin{pmatrix} s^2 k_1 - 2rsk_2 + r^2 k_3 \\ p(sk_2 - rk_3) \\ p^2 k_3 \end{pmatrix} (0\ 0\ 1)\,.$$

If $k_1 \in \{0, \pm 1\}$, $k_2 = 0$, and $k_3 = 1$, i.e., if $C \in (Z_0 \cap Z_1) \cup Z_8$ then we see from the second row and the third row of C' that $ps = 1$ and $r = 0$. When $k_1 \neq 0$ ($C \in Z_0 \cap Z_1$), we further need $s/p^3 = 1$. This implies $(p, s) = \pm(1, 1)$. Hence, $M \in G_1$. When $k_1 = 0$ ($C \in Z_8$), we need no additional condition. Hence $M \in G_8$.

If $k_1 = k_3 = 0$ and $k_2 \in \{\pm 1\}$, i.e., if $C \in Z_0 \cap Z_4$ then we see from the first row and the second row of C' that $r = 0$ and $p^2 = 1$. Hence, $M \in G_4$.

If $k_1 = 1$ and $k_2 = k_3 = 0$, i.e., if $C \in Z_7$ then we see that $s/p^3 = 1$, showing $M \in G_7$.

We finally consider the case $C \in Z_-$. Then, C is of the form

$$C = \begin{pmatrix} \sin\theta \\ \cos\theta \\ \sin\theta \end{pmatrix} (1\ 0\ 1)$$

with $\theta \in [0, \pi/2]$. By Lemma A.20, M is of the form

$$M = R \begin{pmatrix} 1 & 0 \\ 0 & -1 \end{pmatrix}^{\ell} \begin{pmatrix} \cos\eta & -\sin\eta \\ \sin\eta & \cos\eta \end{pmatrix}$$

for some $R > 0$ and $\eta \in \mathbb{R}/2\pi\mathbb{Z}$ and $\ell \in \{0, 1\}$. Remark that

$$C' = \frac{1}{R^2} \begin{pmatrix} \sin\theta - \cos\theta \sin 2\eta \\ \cos\theta \cos 2\eta \\ \sin\theta + \cos\theta \sin 2\eta \end{pmatrix} (1\ 0\ 1)$$

if $\ell = 0$ and

$$C' = \frac{1}{R^2} \begin{pmatrix} -\sin\theta + \cos\theta \sin 2\eta \\ \cos\theta \cos 2\eta \\ -\sin\theta - \cos\theta \sin 2\eta \end{pmatrix} (1\ 0\ 1)$$

if $\ell = 1$.

If $\theta < \pi/2$, i.e., if $C \in Z_- \cap Z_1$ then we see that $\ell = 0$, $R = 1$, $\sin 2\eta = 0$, and $\cos 2\eta = 1$. Hence, $M \in G_1$.

If $\theta = \pi/2$, i.e., if $C \in Z_9$ then we see that $\ell = 0$ and R=1. Hence, $M \in G_9$. □

We are now in a position to complete the proof of Theorem A.13.

Proposition A.22 *For $1 \leqslant j \leqslant 9$, K_j is a system of representatives of*

$$\mathbb{R}^3/G_j := \{\{(\det M)^{-1} D(M)x \mid M \in G_j\} \subset \mathbb{R}^3 \mid x \in \mathbb{R}^3\}.$$

Proof of Proposition A.22 We divide the proof into three steps.

Step 1. We treat the cases $j = 1, 4, 5, 7, 8$. Note that if $M \in GL_2(\mathbb{R})$ is of the form

$$M = \begin{pmatrix} p & 0 \\ 0 & s \end{pmatrix}$$

with $ps \neq 0$ then

$$\frac{1}{\det M} D(M) = \begin{pmatrix} 1/p^2 & 0 & 0 \\ 0 & 1/ps & 0 \\ 0 & 0 & 1/s^2 \end{pmatrix}. \tag{A.21}$$

Let us consider the case $j = 1$. For any $M \in G_1$, $\frac{1}{\det M} D(M)$ is the identity matrix. Hence, by definition, we have $\mathbb{R}^3/G_1 := \{\{x\} \subset \mathbb{R}^3 \mid x \in \mathbb{R}^3\}$. It then follows that, for each $x \in \mathbb{R}^3$, the representative of $\{x\}$ is x. Namely, the system of representatives K_1 is $\mathbb{R}^3$.

Let us consider the case $j = 4$. We show that the representative of $\mathbb{R}^3/G_4$ is a subset of K_4. To this end, let us prove that for any $x = {}^t(p, q, r) \in \mathbb{R}^3$ there exists $M \in G_4$ such that $\frac{1}{\det M} D(M)x \in K_4$. One sees from (A.21) if $M \in G_4$ then

$$\frac{1}{\det M} D(M) = \begin{pmatrix} 1 & 0 & 0 \\ 0 & 1/\sigma s & 0 \\ 0 & 0 & 1/s^2 \end{pmatrix}, \tag{A.22}$$

where $s \neq 0$ and $\sigma \in \{\pm 1\}$. When $r \neq 0$, we choose $s = \sqrt{|r|}$ and $\sigma = \operatorname{sign} q$ with the convention $\operatorname{sign} 0 = 1$. Then,

$$\frac{1}{\det M} D(M)x = \begin{pmatrix} p \\ |q|/\sqrt{|r|} \\ \operatorname{sign} r \end{pmatrix} \in K_4.$$

When $r = 0$ and $q \neq 0$, we choose $s = q$ and $\sigma = 1$. Then,

$$\frac{1}{\det M} D(M)x = \begin{pmatrix} p \\ 1 \\ 0 \end{pmatrix} \in K_4.$$

When $q = r = 0$, we choose $s = 1$ and $\sigma = 1$. Then,

$$\frac{1}{\det M}D(M)x = \begin{pmatrix} p \\ 0 \\ 0 \end{pmatrix} \in K_4.$$

Hence, the representative of $\mathbb{R}^3/G_4$ is a subset of K_4. To prove the other inclusion, let us prove the uniqueness. Pick $x \in K_4$ and $M \in G_4$. It suffices to show if $y := \frac{1}{\det M}D(M)x \in K_4$ then $y = x$. Consider the case

$$x = \begin{pmatrix} p \\ q \\ \tilde{\sigma} \end{pmatrix}, \quad q \geqslant 0., \ \tilde{\sigma} = \pm 1.$$

Since $\frac{1}{\det M}D(M)$ is of the form (A.22), we have

$$y = \frac{1}{\det M}D(M)x = \begin{pmatrix} p \\ q/\sigma s \\ \tilde{\sigma}/s^2 \end{pmatrix}.$$

Suppose that $y \in K_4$. Then, the third exponent of y is either 0, 1, or -1. Hence, $s^2 = 1$. Further, since $q/\sigma s \geqslant 0$, either $\sigma s = 1$ or $q = 0$. In the both cases, we have $q/\sigma s = q$. Thus,

$$y = \frac{1}{\det M}D(M)x = \begin{pmatrix} p \\ q/\sigma s \\ \tilde{\sigma}/s^2 \end{pmatrix} = \begin{pmatrix} p \\ q \\ \tilde{\sigma} \end{pmatrix} = x.$$

The uniqueness holds. Consider the case

$$x = \begin{pmatrix} p \\ \tilde{\sigma} \\ 0 \end{pmatrix}, \quad \tilde{\sigma} = 0, 1.$$

Since $\frac{1}{\det M}D(M)$ is of the form

$$\frac{1}{\det M}D(M) = \begin{pmatrix} 1 & 0 & 0 \\ 0 & 1/\sigma s & 0 \\ 0 & 0 & 1/s^2 \end{pmatrix},$$

we have

$$y = \frac{1}{\det M}D(M)x = \begin{pmatrix} p \\ \tilde{\sigma}/\sigma s \\ 0 \end{pmatrix}.$$

Suppose that $y \in K_4$. Then, since the third exponent is zero, the second exponent is either 0 or 1. Since $\sigma s \neq 0$, we see that $\tilde{\sigma}/\sigma s$ must be equal to $\tilde{\sigma}$. Thus, $y = x$. The uniqueness holds.

Let us consider the case $j = 5$. As in the previous case, we shall first show that the representative of $\mathbb{R}^3/G_5$ is a subset of K_5. To this end, let us prove that for any $x = {}^t(p, q, r) \in \mathbb{R}^3$ there exists $M \in G_5$ such that $\frac{1}{\det M} D(M)x \in K_5$. Thanks to (A.21), if $M \in G_5$ then one has

$$\frac{1}{\det M} D(M) = \begin{pmatrix} 1 & 0 & 0 \\ 0 & 1/\sigma_1\sigma_2 & 0 \\ 0 & 0 & 1 \end{pmatrix}, \tag{A.23}$$

where $\sigma_1, \sigma_2 \in \{\pm 1\}$. We set $\sigma_1\sigma_2 = \operatorname{sign} q$ with the convention $\operatorname{sign} 0 = 1$. Then,

$$\frac{1}{\det M} D(M)x = \begin{pmatrix} p \\ |q| \\ r \end{pmatrix} \in K_5.$$

This shows the representative of $\mathbb{R}^3/G_5$ is a subset of K_5. To prove the other inclusion, let us prove the uniqueness. Pick $x \in K_5$ and $M \in G_5$. It suffices to show if $y := \frac{1}{\det M} D(M)x \in K_5$ then $y = x$. Since $\frac{1}{\det M} D(M)$ is of the form (A.23), we have

$$y = \frac{1}{\det M} D(M)x = \begin{pmatrix} p \\ q/\sigma_1\sigma_2 \\ r \end{pmatrix}.$$

Suppose that $y \in K_5$. Then, the second exponent of y must be nonnegative. Hence, we have $q = 0$ or $\sigma_1\sigma_2 = 1$. In the both cases, we have $y = x$. The uniqueness holds.

Let us consider the case $j = 7$. We begin with showing that the representative of $\mathbb{R}^3/G_7$ is a subset of K_7. To this end, let us prove that for any $x = {}^t(p, q, r) \in \mathbb{R}^3$ there exists $M \in G_7$ such that $\frac{1}{\det M} D(M)x \in K_7$. Note that (A.21) yields

$$\frac{1}{\det M} D(M) = \begin{pmatrix} 1/a^2 & 0 & 0 \\ 0 & 1/a^4 & 0 \\ 0 & 0 & 1/a^6 \end{pmatrix} \tag{A.24}$$

for any $M \in G_7$, where $a \neq 0$. We remark that the sign of a does not matter. When $pr \neq 0$, we choose $a > 0$ so that $|p|/a^2 = |r|/a^6$, i.e., $a^4 = |r/p|$. Then,

$$\frac{1}{\det M} D(M)x = \begin{pmatrix} p/a^2 \\ q/a^4 \\ (\operatorname{sign} r)|p|/a^2 \end{pmatrix} \in K_7.$$

When $p \neq 0$ and $r = 0$, we choose $a > 0$ so that $|p|/a^2 = 1$. Then,

$$\frac{1}{\det M} D(M)x = \begin{pmatrix} \operatorname{sign} p \\ q/a^4 \\ 0 \end{pmatrix} \in K_7.$$

When $p = 0$ and $r \neq 0$, we choose $a > 0$ so that $|r|/a^6 = 1$. Then,

$$\frac{1}{\det M} D(M)x = \begin{pmatrix} 0 \\ q/a^4 \\ \operatorname{sign} r \end{pmatrix} \in K_7.$$

When $p = r = 0$ and $q \neq 0$, we choose $a > 0$ so that $|q|/a^4 = 1$. Then,

$$\frac{1}{\det M} D(M)x = \begin{pmatrix} 0 \\ \operatorname{sign} q \\ 0 \end{pmatrix} \in K_7.$$

Finally, when $x = 0$ then $\frac{1}{\det M} D(M)x = 0 \in K_7$. This shows the representative of $\mathbb{R}^3/G_7$ is a subset of K_7. Let us proceed to the uniqueness part. Pick $x \in K_7$ and $M \in G_7$. It suffices to show if $y := \frac{1}{\det M} D(M)x \in K_7$ then $y = x$. Consider the case

$$x = \begin{pmatrix} p \\ q \\ r \end{pmatrix}, \quad |p| = |r| > 0.$$

Since $\frac{1}{\det M} D(M)$ is of the form (A.24), one sees that

$$y = \frac{1}{\det M} D(M)x = \begin{pmatrix} p/a^2 \\ q/a^4 \\ r/a^6 \end{pmatrix}.$$

Suppose that $y \in K_7$. Then, since the first and the third components of y are nonzero, we have $|p/a^2| = |r/a^6|$. This implies that $a^2 = 1$ and hence $y = x$. The uniqueness holds. Consider the case

$$x = \begin{pmatrix} p \\ q \\ 0 \end{pmatrix}, \quad |p| = 1.$$

Since $\frac{1}{\det M} D(M)$ is of the form (A.24), we have

$$y = \frac{1}{\det M} D(M)x = \begin{pmatrix} p/a^2 \\ q/a^4 \\ 0 \end{pmatrix}.$$

$$y = \frac{1}{\det M} D(M)x = \begin{pmatrix} 0 \\ q \\ ra^2 \end{pmatrix}.$$

Suppose that $y \in K_8$. Then, since the first component is zero and the third component is nonzero, we have $|ra^2| = 1$. Thus $a^2 = 1$ and so $y = x$. The uniqueness holds. Let us next consider the case

$$x = \begin{pmatrix} 0 \\ q \\ 0 \end{pmatrix}.$$

The formula (A.25) implies that

$$y = \frac{1}{\det M} D(M)x = \begin{pmatrix} 0 \\ q \\ 0 \end{pmatrix} = x.$$

The uniqueness holds. This completes the proof of the case $j = 8$.

Step 2. In the second step, we treat the case $j = 2, 3, 6$. To this end, we note that if $M \in GL_2(\mathbb{R})$ is of the form

$$M = \begin{pmatrix} 0 & q \\ r & 0 \end{pmatrix}$$

with $qr \neq 0$ then

$$\frac{1}{\det M} D(M) = \begin{pmatrix} 0 & 0 & 1/q^2 \\ 0 & 1/qr & 0 \\ 1/r^2 & 0 & 0 \end{pmatrix}. \tag{A.26}$$

We use this formula and (A.21) to handle these three cases.

Let us begin with the proof of the case $j = 2$. We prove that the representative of $\mathbb{R}^3/G_2$ is a subset of K_2. It suffices to prove that for any $x = {}^t(p, q, r) \in \mathbb{R}^3$ there exists $M \in G_2$ such that $\frac{1}{\det M} D(M)x \in K_2$. By means of (A.21) and (A.26), if $M \in G_2$ then

$$\frac{1}{\det M} D(M) = \begin{pmatrix} 1 & 0 & 0 \\ 0 & 1 & 0 \\ 0 & 0 & 1 \end{pmatrix} \tag{A.27a}$$

or

$$\frac{1}{\det M} D(M) = \begin{pmatrix} 0 & 0 & 1 \\ 0 & -1 & 0 \\ 1 & 0 & 0 \end{pmatrix}. \tag{A.27b}$$

If $p > r$ or if $p = r$ and $q \geqslant 0$ then we choose the former form to obtain $\frac{1}{\det M} D(M)x = x \in K_2$. If $p < r$ or if $p = r$ and $q < 0$ then we choose the other one. Then,

$$\frac{1}{\det M}D(M)x = \begin{pmatrix} r \\ -q \\ p \end{pmatrix} \in K_2.$$

Hence, the representative of $\mathbb{R}^3/G_2$ is a subset of K_2. To prove the other inclusion, let us prove the uniqueness. Pick $x \in K_2$ and $M \in G_2$. We show if $y := \frac{1}{\det M}D(M)x \in K_2$ then $y = x$. Let us consider the case

$$x = \begin{pmatrix} p \\ q \\ r \end{pmatrix}, \quad p > r.$$

Recalling that $\frac{1}{\det M}D(M)$ is of the form (A.27a) or (A.27b), we have

$$y = \frac{1}{\det M}D(M)x = \begin{pmatrix} p \\ q \\ r \end{pmatrix} \quad \text{or} \quad \begin{pmatrix} r \\ -q \\ p \end{pmatrix}.$$

Suppose that $y \in K_2$. Then, the latter one is excluded. Hence, $y = x$. The uniqueness holds. Next consider the case

$$x = \begin{pmatrix} p \\ q \\ p \end{pmatrix}, \quad q \geqslant 0.$$

We see from (A.27a) and (A.27b) that

$$y = \frac{1}{\det M}D(M)x = \begin{pmatrix} p \\ q \\ p \end{pmatrix} \quad \text{or} \quad \begin{pmatrix} p \\ -q \\ p \end{pmatrix}.$$

Suppose that $y \in K_2$. Then, the latter is excluded if $q > 0$. The both give the same vector if $q = 0$. In any case, we have $y = x$. The uniqueness holds.

Let us move on to the case $j = 3$. As above, in order to see that the representative of $\mathbb{R}^3/G_3$ is a subset of K_3, we prove that for any $x = {}^t(p, q, r) \in \mathbb{R}^3$ there exists $M \in G_3$ such that $\frac{1}{\det M}D(M)x \in K_3$. It follows from (A.21) and (A.26) that if $M \in G_3$ then

$$\frac{1}{\det M}D(M) = \begin{pmatrix} 1 & 0 & 0 \\ 0 & 1/\sigma_1\sigma_2 & 0 \\ 0 & 0 & 1 \end{pmatrix}. \tag{A.28a}$$

or

$$\frac{1}{\det M}D(M) = \begin{pmatrix} 0 & 0 & 1 \\ 0 & 1/\sigma_1\sigma_2 & 0 \\ 1 & 0 & 0 \end{pmatrix}, \tag{A.28b}$$

where $\sigma_1, \sigma_2 \in \{\pm 1\}$. If $p \geqslant r$ then we choose (A.28a) with $\sigma_1\sigma_2 = \operatorname{sign} q$ with the convention $\operatorname{sign} 0 = 1$. Then, we obtain

$$\frac{1}{\det M} D(M)x = \begin{pmatrix} p \\ |q| \\ r \end{pmatrix} \in K_3.$$

If $p < r$ then we choose (A.28b) with $\sigma_1\sigma_2 = \operatorname{sign} q$ to obtain

$$\frac{1}{\det M} D(M)x = \begin{pmatrix} r \\ |q| \\ p \end{pmatrix} \in K_3.$$

Thus, the representative of $\mathbb{R}^3/G_3$ is a subset of K_3. We show the uniqueness property. Pick $x \in K_3$ and $M \in G_3$. We shall establish if $y := \frac{1}{\det M} D(M)x \in K_3$ then $y = x$. It holds that

$$x = \begin{pmatrix} p \\ q \\ r \end{pmatrix}, \quad p \geqslant r,\ q \geqslant 0.$$

Since $\frac{1}{\det M} D(M)$ is of the form (A.28a) or (A.28b), one sees that

$$y = \frac{1}{\det M} D(M)x = \begin{pmatrix} p \\ q/\sigma_1\sigma_2 \\ r \end{pmatrix} \text{ or } \begin{pmatrix} r \\ q/\sigma_1\sigma_2 \\ p \end{pmatrix}.$$

Suppose that $y \in K_3$. If $p > r$ then the latter is excluded. If $p = r$ then the both choices give the same vector. Further, if $q > 0$ then $\sigma_1\sigma_2 = 1$ holds and if $q = 0$ then the choice of σ_1 and of σ_2 do not matter. In conclusion, we have $y = x$. The uniqueness holds.

Let us next handle the case $j = 6$. We shall prove that for any $x = {}^t(p, q, r) \in \mathbb{R}^3$ there exists $M \in G_6$ such that $\frac{1}{\det M} D(M)x \in K_6$. The formulas (A.21) and (A.26) give us that if $M \in G_6$ then

$$\frac{1}{\det M} D(M) = \begin{pmatrix} 1/a^2 & 0 & 0 \\ 0 & 1 & 0 \\ 0 & 0 & a^2 \end{pmatrix}. \tag{A.29a}$$

or

$$\frac{1}{\det M} D(M) = \begin{pmatrix} 0 & 0 & a^2 \\ 0 & -1 & 0 \\ 1/a^2 & 0 & 0 \end{pmatrix}, \tag{A.29b}$$

for some $a \neq 0$, where the choice of a differs in the two forms, i.e., a is replaced by a^{-1} in the latter form. Suppose that $pr \neq 0$. We choose the former form (A.29a) if

$q \geqslant 0$ and the latter form (A.29b) if $q < 0$. Further, we choose a so that $|p/a^2| = |ra^2|$, i.e., $a^4 = |p/r|$. Then,

$$\frac{1}{\det M} D(M)x = \begin{pmatrix} |p/a^2| \operatorname{sign} p \\ q \\ |p/a^2| \operatorname{sign} r \end{pmatrix} \in K_6$$

for $q \geqslant 0$ and

$$\frac{1}{\det M} D(M)x = \begin{pmatrix} |p/a^2| \operatorname{sign} r \\ -q \\ |p/a^2| \operatorname{sign} p \end{pmatrix} \in K_6$$

for $q < 0$. If $p \neq 0$ and $r = 0$ then we choose (A.29a) with $a = |p|^{1/2}$. Then,

$$\frac{1}{\det M} D(M)x = \begin{pmatrix} \operatorname{sign} p \\ q \\ 0 \end{pmatrix} \in K_6.$$

If $p = 0$ and $r \neq 0$ then we choose (A.29b) with $a = |r|^{-1/2}$. Then,

$$\frac{1}{\det M} D(M)x = \begin{pmatrix} \operatorname{sign} r \\ -q \\ 0 \end{pmatrix} \in K_6.$$

If $p = 0$ and $r = 0$ then we choose the former form if $q \geqslant 0$ and the latter form if $q < 0$. Then,

$$\frac{1}{\det M} D(M)x = \begin{pmatrix} 0 \\ |q| \\ 0 \end{pmatrix} \in K_6.$$

This shows the representative of $\mathbb{R}^3/G_6$ is a subset of K_6. To prove the other inclusion, let us prove uniqueness. Pick $x \in K_6$ and $M \in G_6$. It suffices to show if $y := \frac{1}{\det M} D(M)x \in K_6$ then $y = x$. First consider the case

$$x = \begin{pmatrix} p \\ q \\ r \end{pmatrix}, \quad |p| = |r| \geqslant 0,\ q \geqslant 0.$$

Since $\frac{1}{\det M} D(M)$ is of the form (A.29a) or (A.29b), one obtains

$$y = \frac{1}{\det M} D(M)x = \begin{pmatrix} p/a^2 \\ q \\ ra^2 \end{pmatrix} \text{ or } \begin{pmatrix} ra^2 \\ -q \\ p/a^2 \end{pmatrix}.$$

Suppose that $y \in K_6$. If $pr \neq 0$ then $|p/a^2| = |ra^2|$ gives us $a^2 = 1$. (If $p = r = 0$ then the choice of a makes no difference.) Further, if $q > 0$ then the latter form is excluded and if $q = 0$ then the both forms give the same vector. In all cases, we conclude that $y = x$. The uniqueness holds. Next, consider the case

$$x = \begin{pmatrix} p \\ q \\ 0 \end{pmatrix}, \quad |p| = 1.$$

In this case, we have

$$y = \frac{1}{\det M} D(M)x = \begin{pmatrix} p/a^2 \\ q \\ 0 \end{pmatrix} \text{ or } \begin{pmatrix} 0 \\ -q \\ p/a^2 \end{pmatrix}.$$

Suppose that $y \in K_6$. Then the latter choice is excluded. Further, we see from $|p/a^2| = 1$ that $a^2 = 1$. Hence, we have $y = x$. The uniqueness holds.

Step 3. Finally, we consider the case $j = 9$. Let us show that the representative of $\mathbb{R}^3/G_9$ is a subset of K_9. The goal is to prove that for any $x = {}^t(p, q, r) \in \mathbb{R}^3$ there exists $M \in G_9$ such that $\frac{1}{\det M} D(M)x \in K_9$. Since $M \in G_9$ is of the form

$$M = \begin{pmatrix} \cos\eta & -\sin\eta \\ \sin\eta & \cos\eta \end{pmatrix}$$

for some $\eta \in \mathbb{R}/2\pi\mathbb{Z}$, we have

$$\frac{1}{\det M} D(M) = \begin{pmatrix} \cos^2\eta & -2\sin\eta\cos\eta & \sin^2\eta \\ \sin\eta\cos\eta & \cos^2\eta - \sin^2\eta & -\sin\eta\cos\eta \\ \sin^2\eta & 2\sin\eta\cos\eta & \cos^2\eta. \end{pmatrix}. \tag{A.30}$$

Hence,

$$\frac{1}{\det M} D(M)x = \begin{pmatrix} \frac{p+r}{2} + (\frac{p-r}{2}\cos 2\eta - q\sin 2\eta) \\ \frac{p-r}{2}\sin 2\eta + q\cos 2\eta \\ \frac{p+r}{2} - (\frac{p-r}{2}\cos 2\eta - q\sin 2\eta) \end{pmatrix}.$$

If $p - r \neq 0$ or $q \neq 0$ there exists η_0 such that

$$\tfrac{p-r}{2}\sin 2\eta_0 + q\cos 2\eta_0 = 0.$$

Notice that there are multiple choices of η_0 in $\mathbb{R}/2\pi\mathbb{Z}$. This is because if η_0 satisfies the above the $\eta_0 + \pi/2$ also works. If $\frac{p-r}{2}\cos 2\eta_0 - q\sin 2\eta_0 \geqslant 0$ then we choose $\eta = \eta_0$ to obtain

$$\frac{1}{\det M} D(M)x = \begin{pmatrix} \frac{p+r}{2} + (\frac{p-r}{2}\cos 2\eta_0 - q\sin 2\eta_0) \\ 0 \\ \frac{p+r}{2} - (\frac{p-r}{2}\cos 2\eta_0 - q\sin 2\eta_0) \end{pmatrix} \in K_9.$$

If $\frac{p-r}{2}\cos 2\eta_0 - q\sin 2\eta_0 < 0$ then we choose $\eta = \eta_0 + \pi/2$ to obtain

$$\frac{1}{\det M} D(M)x = \begin{pmatrix} \frac{p+r}{2} - (\frac{p-r}{2}\cos 2\eta_0 - q\sin 2\eta_0) \\ 0 \\ \frac{p+r}{2} + (\frac{p-r}{2}\cos 2\eta_0 - q\sin 2\eta_0) \end{pmatrix} \in K_9.$$

If $p - r = 0$ and $q = 0$ then we have

$$\frac{1}{\det M} D(M)x = \begin{pmatrix} \frac{p+r}{2} \\ 0 \\ \frac{p+r}{2} \end{pmatrix} \in K_9.$$

This shows the representative of $\mathbb{R}^3/G_9$ is a subset of K_9. To prove the other inclusion, let us prove uniqueness. Pick $x \in K_9$ and $M \in G_9$. It suffices to show if $y := \frac{1}{\det M} D(M)x \in K_9$ then $y = x$. Since x is of the form ${}^t(p, 0, r)$ with $p \geqslant r$, we see from (A.30) that

$$\frac{1}{\det M} D(M)x = \begin{pmatrix} \frac{p+r}{2} + \frac{p-r}{2}\cos 2\eta \\ \frac{p-r}{2}\sin 2\eta \\ \frac{p+r}{2} - \frac{p-r}{2}\cos 2\eta \end{pmatrix}.$$

If $p = r$ then any choice of η gives $y = x$. Let us consider $p > r$. Suppose that $y \in K_9$. Then, we have $\sin 2\eta = 0$. In this case, $\cos 2\eta \in \{\pm 1\}$. $\cos 2\eta = -1$ is excluded by comparing the first and the third components of y. Thus, we have $\cos 2\eta = 1$ and then $y = x$. The uniqueness holds. □

Appendix B Examples

In this appendix we exhibit two examples of system of the form (1.6). We see that the nonlinear amplification/dissipation phenomenon takes place for the systems although the coefficients of the nonlinearities are real. Further, the system admits several types of behaviors.

B.1 Typical Types of Behavior for the Single NLS Equation

To begin with, we quickly recall several results on the single cubic NLS equation in one dimension.

B.1.1 Modified Scattering

Let us consider

$$i\partial_t u + \frac{1}{2}\partial_x^2 u = \lambda |u|^2 u, \tag{B.1}$$

where $\lambda \in \mathbb{R}$. Due to Hayashi–Naumkin [8], if $u(0)$ is sufficiently small in $H^1 \cap H^{0,1}$ and $0 < \gamma < 1/100$, then there exists a global in time solution to (B.1) satisfying

$$u(t) = t^{-\frac{1}{2}} W\left(\frac{x}{t}\right) e^{\frac{ix^2}{2t} - \lambda i |W(\frac{x}{t})|^2 \log t - i\frac{\pi}{4}} + O(t^{-\frac{3}{4}+\gamma})$$

in L^∞ as $t \to \infty$ for some $W \in L^\infty$. The asymptotic profile is that of the free solution with a phase correction by nonlinear effect.

B.1.2 Nonlinear Dissipation

Let us consider

$$i\partial_t u + \frac{1}{2}\partial_x^2 u = -i|u|^2 u. \tag{B.2}$$

By Shimomura [28], if $u(0)$ is sufficiently small in $H^1 \cap H^{0,1}$ and $0 < \gamma < 1/100$, then there exists a global in time solution to (B.2) satisfying

$$u(t) = \frac{t^{-\frac{1}{2}}}{\sqrt{1 + \left|W\left(\frac{x}{t}\right)\right|^2 \log t}} W\left(\frac{x}{t}\right) e^{\frac{ix^2}{2t} - i\frac{\pi}{4}} + O(t^{-\frac{3}{4}+\gamma}) \tag{B.3}$$

in L^∞ as $t \to \infty$ for some $W \in L^\infty$. Note that the rate of time decay of u is faster than that of free solution due to nonlinear dissipation.

B.1.3 Nonlinear Amplification

Let us consider

$$i\partial_t u + \frac{1}{2}\partial_x^2 u = i|u|^2 u. \tag{B.4}$$

By Kita [15], for any $\varepsilon > 0$ there exists a solution $u_\varepsilon \in C([0, T^*), H^1 \cap H^{0,1})$ to (B.4) which satisfies $\|u_\varepsilon(0)\|_{L^2} < \varepsilon$ and blows up in finite time, i.e.,

$$\lim_{t \nearrow T^*} \|u_\varepsilon(t)\|_{L^2} = \infty$$

for some $T^* < \infty$. This is due to nonlinear amplification. Note that for any solution to (B.4), $\overline{u(-t, x)}$ is a solution to (B.2). Hence, solutions to (B.4) have a dissipative structure for the negative time direction. Here we remark that the above forward-blowing-up solution u_ε is global backward in time and satisfies $\|u_\varepsilon(t)\|_{L^2} = O((\log |t|)^{-1/3})$ as $t \to -\infty$. It is not known whether the asymptotic profile of the solution is like the one given (B.3). Nevertheless, one can find a dissipative effect in the decay of the mass.

B.2 Example 1

The first example is the system which admits solution of the following three types; modified scattering, nonlinear dissipation, nonlinear amplification. Let us consider

$$\begin{cases} i\partial_t u_1 + \frac{1}{2}\partial_x^2 u_1 = -2(|u_1|^2 - |u_2|^2)u_1 + \overline{u_1}u_2^2, & t \in \mathbb{R},\ x \in \mathbb{R}, \\ i\partial_t u_2 + \frac{1}{2}\partial_x^2 u_2 = -2(|u_1|^2 - |u_2|^2)u_2 - u_1^2\overline{u_2}, & t \in \mathbb{R},\ x \in \mathbb{R}. \end{cases} \tag{B.5}$$

In the terminology introduced in Appendix A, the system belongs to CNS_A, that is, the corresponding matrix-kernel representation (C, p, q, r) given in Theorem A.3 satisfies $\operatorname{tr} C = 0$ and $p = q = r = 0$ and

$$C = A = \begin{pmatrix} 0 & 1 & 0 \\ 1 & 0 & 1 \\ 0 & 1 & 0 \end{pmatrix}.$$

One easily sees that a pair $(u_1(t), 0)$ is a solution to the system if and only if $u_1(t)$ solves

$$i\partial_t u_1 + \frac{1}{2}\partial_x^2 u_1 = -2|u_1|^2 u_1.$$

Hence, the asymptotic behavior of a small solution of this form is a type of the modified scattering for both time directions. Similarly, a pair $(0, u_2(t))$ is also a solution with the same kind of behavior. However, the sign of the nonlinearity is opposite, which corresponds to the time direction is reversed.

One also sees that a pair $(u(t), e^{-i\pi/4}u(t))$ becomes a solution if and only if $u(t)$ solves

$$i\partial_t u + \frac{1}{2}\partial_x^2 u = -i|u|^2 u.$$

Hence, the solution to (B.5) of this form enjoys the dissipative decay forward in time. If it is sufficiently small (in $H^1 \cap H^{0,1}$) at some time then the asymptotic behavior is as in (B.3) as $t \to \infty$. Further, there exists an arbitrarily small (in L^2) initial data of the form $(u(0), e^{-i\pi/4}u(0))$ such that the corresponding solution blows up for the negative time direction in a finite time.

Finally, one verifies that a pair $(u(t), e^{i\pi/4}u(t))$ becomes a solution if and only if $u(t)$ solves

$$i\partial_t u + \frac{1}{2}\partial_x^2 u = i|u|^2 u.$$

Hence, the solution to (B.5) of this form enjoys the same property as above with the time directions switched.

B.3 Example 2

The second example is

$$\begin{cases} i\partial_t u_1 + \frac{1}{2}\partial_x^2 u_1 = 2|u_1|^2 u_2 - u_1^2\overline{u_2} + |u_2|^2 u_2, \\ i\partial_t u_2 + \frac{1}{2}\partial_x^2 u_2 = -|u_1|^2 u_1 - 2u_1|u_2|^2 + \overline{u_1}u_2^2. \end{cases} \tag{B.6}$$

The matrix-kernel representation of the system is

$$\left(\begin{pmatrix} 3 & 0 & 1 \\ 0 & 2 & 0 \\ 1 & 0 & 3 \end{pmatrix}, 0, 0, 0 \right) \in M_3(\mathbb{R}) \times \mathbb{R}^3.$$

We apply the following linear transform of unknowns with *complex coefficients*:

$$\begin{pmatrix} v_1 \\ v_2 \end{pmatrix} = \begin{pmatrix} 1 & -i \\ -1 & -i \end{pmatrix} \begin{pmatrix} u_1 \\ u_2 \end{pmatrix}.$$

Then, one sees that the system for (v_1, v_2) becomes

$$\begin{cases} i\partial_t v_1 + \frac{1}{2}\partial_x^2 v_1 = i|v_1|^2 v_1, \\ i\partial_t v_2 + \frac{1}{2}\partial_x^2 v_2 = -i|v_2|^2 v_2. \end{cases}$$

This is a decoupled system. It is easy to see that there exist a solution of the form $(v_1(t), 0)$ which blows up forward in time and decays dissipatively backward in time and a solution of the form $(0, v_2(t))$ which blows up backward in time and decays dissipatively forward in time. Remark that a sum of these two solutions are also a solution to the system. The solution blows up for the both time directions.

Remark B.23 The second example suggests that the property $M \in GL_2(\mathbb{R})$ in the linear transformation (A.5) is restrictive for revealing the dissipation/amplification phenomenon of the system (1.6), or that, since the matrix-kernel representation (Theorem A.3) is based on the connection to the conserved quantities of the systems, it is not well-suited to for the purpose. It is not hard to construct variants of the matrix-kernel representation for which the change of (possibly complex-valued) coefficients caused by a linear transformation of unknowns with $M \in GL_2(\mathbb{C})$ is formulated as a similar matrix manipulation. However, as far as the authors know, a more useful one is not found.

References

1. J.E. Barab, Nonexistence of asymptotically free solutions for a nonlinear Schrödinger equation. J. Math. Phys. **25**(11), 3270–3273 (1984). MR 761850
2. T. Cazenave, Semilinear Schrödinger equations, in *Courant Lecture Notes in Mathematics*, vol. 10 (New York University, Courant Institute of Mathematical Sciences, New York; American Mathematical Society, Providence, RI, 2003). MR 2002047
3. J. Ginibre, T. Ozawa, Long range scattering for nonlinear Schrödinger and Hartree equations in space dimension $n \geqslant 2$. Comm. Math. Phys. **151**(3), 619–645 (1993). MR 1207269
4. N. Hayashi, C. Li, P.I. Naumkin, On a system of nonlinear Schrödinger equations in 2D. Differ. Integr. Equ. **24**(5–6), 417–434 (2011). MR 2809614
5. N. Hayashi, C. Li, P.I. Naumkin, Modified wave operator for a system of nonlinear Schrödinger equations in 2d. Comm. Partial Differ. Equ. **37**(6), 947–968 (2012). MR 2924463
6. N. Hayashi, C. Li, P.I. Naumkin, Nonlinear Schrödinger systems in 2d with nondecaying final data. J. Differ. Equ. **260**(2), 1472–1495 (2016). MR 3419736
7. N. Hayashi, C. Li, P.I. Naumkin, Time decay for nonlinear dissipative Schrödinger equations in optical fields. Adv. Math. Phys. (2016), Art. ID 3702738, 7. MR 3465033
8. N. Hayashi, P.I. Naumkin, Asymptotics for large time of solutions to the nonlinear Schrödinger and Hartree equations. Amer. J. Math. **120**(2), 369–389 (1998). MR 1613646
9. G. Hoshino, Dissipative nonlinear Schrödinger equations for large data in one space dimension. Commun. Pure Appl. Anal. **19**(2), 967–981 (2020). MR 4043774
10. J. Ieda, T. Miyakawa, M. Wadati, Matter-wave solitons in an f=1 spinor bose-einstein condensate. J. Phys. Soc. Japan **73**(11), 2996–3007 (2004)
11. S. Katayama, C. Li, H. Sunagawa, A remark on decay rates of solutions for a system of quadratic nonlinear Schrödinger equations in 2D. Differ. Integr. Equ. **27**(3–4), 301–312 (2014). MR 3161606
12. S. Katayama, D. Sakoda, Asymptotic behavior for a class of derivative nonlinear Schrödinger systems. Partial Differ. Equ. Appl. **1**(3), Paper No. 12 (2020). MR 4336288
13. Y. Kawahara, H. Sunagawa, Remarks on global behavior of solutions to nonlinear Schrödinger equations. Proc. Japan Acad. Ser. A Math. Sci. **82**(8), 117–122 (2006). MR 2279276
14. D. Kim, A note on decay rates of solutions to a system of cubic nonlinear Schrödinger equations in one space dimension. Asymptot. Anal. **98**(1–2), 79–90 (2016). MR 3502373
15. N. Kita, Existence of blowing-up solutions to some nonlinear Schrödinger equations including nonlinear amplification with small initial data. Preprint. OCAMI Preprint Series 2020, 20-2
16. N. Kita, Optimal decay rate of solutions to 1d Schrödinger equation with cubic dissipative nonlinearity. J. Appl. Sci. Eng. A **1**(1), 15–18

17. N. Kita, Y. Nakamura, Large time behavior of small solutions to multi-component nonlinear Schrödinger equations related with spinor Bose-Einstein condensate. Linear Nonlinear Anal. **5**(1), 73–85 (2019). MR 4054437
18. C. Li, Y. Nishii, Y. Sagawa, H. Sunagawa, Large time asymptotics for a cubic nonlinear Schrödinger system in one space dimension. Funkcialaj Ekvacioj **64**(3), 361–377 (2021). MR 4360614
19. C. Li, Y. Nishii, Y. Sagawa, H. Sunagawa, Large time asymptotics for a cubic nonlinear Schrödinger system in one space dimension, II. Tokyo J. Math. **44**(2), 411–416 (2021). MR 4379734
20. C. Li, H. Sunagawa, On Schrödinger systems with cubic dissipative nonlinearities of derivative type. Nonlinearity **29**(5), 1537–1563 (2016). MR 3481342
21. S.V. Manakov, On the theory of two-dimensional stationary self-focusing of electromagnetic waves. JETP Sov. Phys. **38**(2) (1974)
22. S. Masaki, J.-I. Segata, K. Uriya, On asymptotic behavior of solutions to cubic nonlinear Klein-Gordon systems in one space dimension. Trans. Amer. Math. Soc. Ser. B **9**, 517–563 (2022). MR 4439505
23. Y. Nakamura, A. Shimomura, S. Tonegawa, Global existence and asymptotic behavior of solutions to some nonlinear systems of Schrödinger equations. J. Math. Sci. Univ. Tokyo **22**(3), 771–792 (2015). MR 3408074
24. T. Ogawa, T. Sato, L^2-decay rate for the critical nonlinear Schrödinger equation with a small smooth data. NoDEA Nonlinear Differ. Equ. Appl. **27**(2), Paper No. 18, 20 (2020). MR 4069852
25. T. Ozawa, Long range scattering for nonlinear Schrödinger equations in one space dimension. Comm. Math. Phys. **139**(3), 479–493 (1991). MR 1121130
26. D. Sakoda, H. Sunagawa, Small data global existence for a class of quadratic derivative nonlinear Schrödinger systems in two space dimensions. J. Differ. Equ. **268**(4), 1722–1749 (2020). MR 4042357
27. T. Sato, L^2-decay estimate for the dissipative nonlinear Schrödinger equation in the Gevrey class. Arch. Math. (Basel) **115**(5), 575–588 (2020). MR 4154572
28. A. Shimomura, Asymptotic behavior of solutions for Schrödinger equations with dissipative nonlinearities. Comm. Partial Differ. Equ. **31**(7–9), 1407–1423 (2006). MR 2254620
29. W.A. Strauss, Nonlinear scattering theory, in *Scattering Theory in Mathematical Physics*. ed. by J.A. Lavita, J.-P. Marchand (Springer, Netherlands, Dordrecht, 1974), pp.53–78
30. H. Sunagawa, Large time asymptotics of solutions to nonlinear Klein-Gordon systems. Osaka J. Math. **42**(1), 65–83 (2005). MR 2130963
31. Y. Tsutsumi, The null gauge condition and the one-dimensional nonlinear Schrödinger equation with cubic nonlinearity. Indiana Univ. Math. J. **43**(1), 241–254 (1994). MR 1275461
32. Y. Tsutsumi, K. Yajima, The asymptotic behavior of nonlinear Schrödinger equations. Bull. Amer. Math. Soc. (N.S.) **11**(1), 186–188 (1984). MR 741737
33. K. Uriya, Final state problem for systems of cubic nonlinear Schrödinger equations in one dimension. Ann. Henri Poincaré **18**(7), 2523–2542 (2017). MR 3665222

Remarks on Blow up of Solutions of Nonlinear Wave Equations in Friedmann-Lemaître-Robertson-Walker Spacetime

Kimitoshi Tsutaya and Yuta Wakasugi

Dedicated to Professor Tohru Ozawa on the occasion of his sixtieth birthday.

Abstract Consider nonlinear wave equations in the spatially flat Friedmann-Lemaître-Robertson-Walker (FLRW) spacetimes. We improve some upper bounds of the lifespan of blow-up solutions for the power nonlinearity. We also show upper bounds of the lifespan in the critical cases for the time derivative nonlinearity.

Keywords Wave equation · Blow-up · Lifespan · FLRW spacetime

AMS Subject Classifications 35Q85 · 35L05 · 35L70

1 Introduction

The purpose of this paper is to supplement and improve our earlier results for nonlinear wave equations in the Friedmann-Lemaître-Robertson-Walker (FLRW) spacetimes. The spatially flat FLRW metric is given by

$$g: \ ds^2 = -dt^2 + a(t)^2 d\sigma^2,$$

The research was supported by JSPS KAKENHI Grant Number JP18K03351 and JP19H01795

K. Tsutaya (✉)
Graduate School of Science and Technology, Hirosaki University, Hirosaki 036-8561, Japan
e-mail: tsutaya@hirosaki-u.ac.jp

Y. Wakasugi
Graduate School of Advanced Science and Engineering, Hiroshima University, Higashi-Hiroshima 739-8527, Japan

S. Machihara (ed.), *Mathematical Physics and Its Interactions*, Springer Proceedings in Mathematics & Statistics 451, https://doi.org/10.1007/978-981-97-0364-7_6

where the speed of light is equal to 1, $d\sigma^2$ is the line element of an n-dimensional Euclidean space and $a(t)$ is the scale factor, which describes expansion or contraction of the spatial metric. As in our earlier work [13–16], we have treated the scale factor as

$$a(t) = ct^{\frac{2}{n(1+w)}} \tag{1.1}$$

where c is a positive constant, and w is the proportionality constant in the range $-1 < w \le 1$. The constant w appears in the equation of state relating the pressure to the density for the perfect fluid. See [13].

For the flat FLRW metric with (1.1), the semilinear wave equation $\Box_g u = |g|^{-1/2}\partial_\alpha(|g|^{1/2}g^{\alpha\beta}\partial_\beta)u = -G(u, \partial u)$ becomes

$$u_{tt} - \frac{1}{t^{4/(n(1+w))}}\Delta u + \frac{2}{(1+w)t}u_t = G(u, \partial u), \quad x \in \mathbf{R}^n, \tag{1.2}$$

where $\Delta = \partial_1^2 + \cdots + \partial_n^2$, $\partial_j = \partial/\partial x^j$, $j = 1, \cdots, n$, $G(u, \partial u) = |u|^p$, $|\nabla_x u|^p$ or $|u_t|^p$ with $p > 1$, $\nabla_x u = (\partial u/\partial x^1, \cdots, \partial u/\partial x^n)$, $(x^1, \cdots, x^n) \in \mathbf{R}^n$.

Generalizing Eq. (1.2), we first consider the Cauchy problem

$$\begin{cases} u_{tt} - \dfrac{1}{t^{2\alpha}}\Delta u + \dfrac{\mu}{t}u_t = |u|^p, & t > 1,\ x \in \mathbf{R}^n, \\ u(1,x) = \varepsilon u_0(x),\ u_t(1,x) = \varepsilon u_1(x), & x \in \mathbf{R}^n, \end{cases} \tag{1.3}$$

where α and μ are non-negative constants, and $\varepsilon > 0$ is a small parameter.

Let T_ε be the lifespan of solutions of (1.3), that is, T_ε is the supremum of T such that (1.3) have a solution for $x \in \mathbf{R}^n$ and $1 \le t < T$.

The authors have recently proved blow-up in finite time and upper estimates of the lifespan for the case $0 \le \alpha < 1$. Let

$$\gamma(n, p, \alpha, \mu) = -p^2\left(n - 1 + \frac{\mu - \alpha}{1 - \alpha}\right) + p\left(n + 1 + \frac{\mu + 3\alpha}{1 - \alpha}\right) + 2 \tag{1.4}$$

and $p_c(n, \alpha, \mu)$ be the positive root of the equation $\gamma(n, p, \alpha, \mu) = 0$. We note that if $\alpha = \mu = 0$, then p_c coincides with the Strauss exponent. Let $p_F(n)$ denote the Fujita exponent $p_F(n) = 1 + 2/n$.

For the case $n \ge 2$ and $0 \le \alpha < 1$, we have shown in [13, 15] that blow-up in finite time occurs if $1 < p \le p_c(n, \alpha, \mu)$ or if $1 < p \le p_F(n(1-\alpha)) = 1 + 2/\{n(1-\alpha)\}$. Moreover, the following upper bounds of the lifespan are proved:

$$T_\varepsilon \le C\varepsilon^{\frac{-2p(p-1)}{(1-\alpha)\gamma(n,p,\alpha,\mu)}} \quad \text{if} \quad 1 < p < p_c(n, \alpha, \mu), \tag{1.5}$$

$$T_\varepsilon \le \exp(C\varepsilon^{-p(p-1)}) \quad \text{if} \quad p = p_c(n, \alpha, \mu) > p_F(n(1-\alpha)) = 1 + 2/(n(1-\alpha)), \tag{1.6}$$

$$T_\varepsilon \le C\varepsilon^{-\frac{p-1}{2-n(1-\alpha)(p-1)}} \quad \text{if} \quad 1 < p < p_F(n(1-\alpha)), \tag{1.7}$$

$$T_\varepsilon \le \exp\left(C\varepsilon^{-(p-1)}\right) \qquad \text{if} \quad p = p_F(n(1-\alpha)). \tag{1.8}$$

Recently, Palmieri [11] has independently obtained similar results to above by using different methods. See also [1] for special cases. For the special case $\alpha = 0$, we refer to [7] and references therein.

We can say from (1.5)–(1.8) that if the power p is dominated by $p_c(n, \alpha, \mu)$ related to the Strauss exponent or by the Fujita exponent $p_F(n(1-\alpha))$, the lifespan of solutions is *wavelike* or *heatlike*, respectively, as mentioned in [10]. On the other hand, for $\alpha \ge 1$, blow-up in finite time can happen to occur for all $p > 1$. This is in contrast to the case $0 \le \alpha < 1$. In fact, we have proved in [16] that the lifespan satisfies

$$T_\varepsilon^2 (\ln T_\varepsilon)^{-n(p-1)} \le C\varepsilon^{-(p-1)} \qquad \text{if } p > 1 \text{ and } \alpha = 1, \tag{1.9}$$

$$T_\varepsilon \le C\varepsilon^{-(p-1)/2} \qquad \text{if } p > 1 \text{ and } \alpha > 1. \tag{1.10}$$

For the case of the space derivative nonlinearity $|\nabla_x u|^p$, the same or similar estimates hold. In fact, it has been proved in [14] that (1.10) and

$$T_\varepsilon^2 (\ln T_\varepsilon)^{-(p+n(p-1))} \le C\varepsilon^{-(p-1)} \quad \text{if } p > 1 \text{ and } \alpha = 1 \tag{1.11}$$

for $|\nabla_x u|^p$. The results (1.9)–(1.11) hold true also for $n = 1$. One first aim of this paper is to show that the upper bounds of the lifespan (1.9)–(1.11) can be improved if $\mu \le 1$.

We next consider the Cauchy problem

$$\begin{cases} u_{tt} - \dfrac{1}{t^{2\alpha}}\Delta u + \dfrac{\mu}{t} u_t = |u_t|^p, & t > 1, \; x \in \mathbf{R}^n, \\ u(1, x) = \varepsilon u_0(x), \; u_t(1, x) = \varepsilon u_1(x), & x \in \mathbf{R}^n. \end{cases} \tag{1.12}$$

Let

$$p_G(n, \alpha, \mu) = 1 + \frac{2}{(1-\alpha)(n-1) + \mu + \alpha},$$

$$p_0(n, \alpha, \mu) = 1 + \frac{1}{n(1-\alpha) + \mu},$$

and $p_G(n) = p_G(n, 0, 0) = 1 + 2/(n-1)$. If $\alpha = \mu = 0$, the so-called Glassey's conjecture [2] asserts that if $p > p_G(n)$, then there exist global solutions in time for small initial data, on the other hand, if $1 < p \le p_G(n)$ with $n \ge 2$ or if $p > 1$ for $n = 1$, then blow-up in finite time occurs. This conjecture is proved to be almost true. Actually, after several blow-up results had been obtained, Zhou [18] finally gave a simple proof of the blow-up result for $1 < p \le p_G(n)$ and $n \ge 2$ as well as for $p > 1$ and $n = 1$. Global existence of solutions in low dimensions ($n = 2, 3$) has been proved in, e.g., [5, 12, 17]. For high dimensions ($n \ge 3$), we refer to [6] which treats the radial case for $p > p_G(n)$.

For the case $\alpha = 0$ and $\mu \geq 0$, it is recently shown by Hamouda and Hamza [3] that blow-up in finite time occurs and the lifespan of the blow-up solutions satisfies

$$T_\varepsilon \leq C\varepsilon^{\frac{-(p-1)}{1-(n+\mu-1)(p-1)/2}} \quad \text{if} \quad 1 < p < p_G(n, 0, \mu),\ n \geq 1, \tag{1.13}$$

$$T_\varepsilon \leq \exp(C\varepsilon^{-(p-1)}) \quad \text{if} \quad p = p_G(n, 0, \mu),\ n \geq 1. \tag{1.14}$$

These results improve the ones in [9].

The authors have recently shown blow-up in finite time and upper estimates of the lifespan of solutions of (1.12) in the case $n \geq 2,\ 0 \leq \alpha < 1$ and $\mu \geq 0$:

$$T_\varepsilon \leq C\varepsilon^{\frac{-(p-1)}{1-\{(1-\alpha)(n-1)+\mu+\alpha\}(p-1)/2}} \quad \text{if } p < p_G(n, \alpha, \mu), \tag{1.15}$$

$$T_\varepsilon \leq \exp(C\varepsilon^{-(p-1)}) \quad \text{if } p = p_G(n, \alpha, \mu), \tag{1.16}$$

$$T_\varepsilon \leq C\varepsilon^{\frac{-(p-1)}{1-(p-1)\{n(1-\alpha)+\mu\}}} \quad \text{if } p < p_0(n, \alpha, \mu). \tag{1.17}$$

If $\alpha = 0$, our upper bounds of the lifespan coincide with the results above by [3]. Similar results are independently shown by [4] where energy solutions are treated. In our results, however, another exponent appears as a blow-up condition in some case. This is different from the results by [4]. We emphasize that the generalized exponent of $p_G(n, \alpha, \mu)$ cannot always be the critical exponent for the global existence of solutions. We also have shown in [14] that if $\alpha \geq 1$ and $1 < p < 1 + 1/\mu$, then blow-up in finite time occurs and the lifespan of the blow-up solutions satisfies

$$T_\varepsilon^{1-\mu(p-1)}(\ln T_\varepsilon)^{-n(p-1)} \leq C\varepsilon^{-(p-1)} \quad \text{if } \alpha = 1, \tag{1.18}$$

$$T_\varepsilon \leq C\varepsilon^{\frac{-(p-1)}{1-\mu(p-1)}} \quad \text{if } \alpha > 1. \tag{1.19}$$

The results (1.18) and (1.19) hold true also for $n = 1$. The present paper treats the critical cases $p = p_0(n, \alpha, \mu)$ for $0 \leq \alpha < 1$ and $p = 1 + 1/\mu$ for $\alpha \geq 1$. Our second aim of this paper is to show upper bounds of the lifespan for these critical cases. Our result shows that if $p = p_0(n, \alpha, \mu)$ and $0 \leq \alpha < 1$, then upper bounds of the lifespan is the same as (1.16).

The paper is organized as follows. In Sect. 2 we state our first main result for (1.3) in the case $\alpha \geq 1$ and $\mu \leq 1$. Theorem 2.1 presents that it is possible in some cases to improve the estimates of the lifespan (1.9) and (1.10). To prove the theorem, we establish a new type of Kato's lemma. This is proved by John's iteration argument [8]. Similar results to Theorem 2.1 hold for the equation with the space derivative nonlinear term $|\nabla_x u|^p$. In Sect. 3 we treat (1.12) and show blow-up in finite time and upper bounds of the lifespan for the remaining critical cases. We establish new generalized Kato's lemma for a first-order differential inequality, which is applied to the wave equation with the scale-invariant damping. Finally in Sect. 4, we apply the theorems in Sects. 2 and 3 to the original Eq. (1.2). We discuss the upper bounds for the lifespan of blow-up solutions and compare those with the case of the Minkowski spacetime.

2 Power Nonlinearity Case

Consider problem (1.3). We state our first results which improve the estimates of the lifespan (1.9) and (1.10).

Theorem 2.1 *Let $n \geq 1,\ \alpha \geq 1,\ 0 \leq \mu \leq 1$ and $p > 1$. Assume that $u_0 \in C^2(\mathbf{R}^n)$ and $u_1 \in C^1(\mathbf{R}^n)$ are nonnegative, nontrivial and supp u_0, supp $u_1 \subset \{|x| \leq R\}$ with $R > 0$. Suppose that the problem (1.3) has a classical solution $u \in C^2([1, T) \times \mathbf{R}^n)$. Then $T < \infty$, and there exists a constant $\varepsilon_0 > 0$ depending on $p, \alpha, \mu, R, u_0, u_1$ such that T_ε has to satisfy*

$$T_\varepsilon^{2+(p-1)(1-\mu)}(\ln T_\varepsilon)^{-n(p-1)} \leq C\varepsilon^{-(p-1)} \quad \text{if } \alpha = 1 \text{ and } \mu < 1, \tag{2.1}$$

$$T_\varepsilon^2(\ln T_\varepsilon)^{-(n-1)(p-1)} \leq C\varepsilon^{-(p-1)} \quad \text{if } \alpha = 1, \mu = 1 \text{ and } p \geq 1 + 1/(n-1), \tag{2.2}$$

$$T_\varepsilon^{2+(p-1)(1-\mu)} \leq C\varepsilon^{-(p-1)} \quad \text{if } \alpha > 1 \text{ and } \mu < 1, \tag{2.3}$$

$$T_\varepsilon^2(\ln T_\varepsilon)^{p-1} \leq C\varepsilon^{-(p-1)} \quad \text{if } \alpha > 1 \text{ and } \mu = 1 \tag{2.4}$$

for $0 < \varepsilon \leq \varepsilon_0$, where $C > 0$ is a constant independent of ε.

Remark It still remains open if the estimate (1.9) for the case $\alpha = 1$, $\mu = 1$ and $p < 1 + 1/(n-1)$ can be improved to (2.2).

To prove the theorem above, we use the following generalized Kato's lemma.

Lemma 2.2 *Let $p > 1,\ a \geq 0,\ b \geq 0,\ c > 0,\ q \geq 0,\ \mu \geq 0$ and*

$$M \equiv (p-1)(c-a) + 2 > 0.$$

Let $T \geq T_1 > T_0 \geq 1$. Assume that $F \in C^2([T_0, T))$ satisfies the following three conditions:

$$\begin{aligned}
&(i) \quad F(t) \geq A_0 t^{-a}(\ln t)^{-b}(t - T_1)^c \quad \text{for } t \geq T_1,\\
&(ii) \quad F''(t) + \frac{\mu F'(t)}{t} \geq A_1(\ln t)^{-q}|F(t)|^p \quad \text{for } t \geq T_0,\\
&(iii) \quad F(T_0) \geq 0, \quad F'(T_0) > 0,
\end{aligned}$$

where A_0 and A_1 are positive constants. Then T has to satisfy

$$T^{M/(p-1)}(\ln T)^{-b-q/(p-1)} < CA_0^{-1},$$

where C is a constant depending on A_1, μ, p and c.

Proof See [16].

Corollary *Let* $p > 1,\ a \geq 0,\ b \geq 0,\ c > 0,\ \mu \geq 0$ *and*

$$M \equiv (p-1)(c-a) + 2 > 0.$$

Let $T \geq \tilde{T} > T_0 \geq 1$. *Assume that* $F \in C^2([T_0, T))$ *satisfies the following three conditions:*

$$\begin{aligned}
&(i)\quad F(t) \geq A_0(\ln \tilde{T})^b t^{-a}(t-\tilde{T})^c \quad \text{for } t \geq \tilde{T},\\
&(ii)\quad F''(t) + \frac{\mu F'(t)}{t} \geq A_1|F(t)|^p \quad \text{for } t \geq T_0,\\
&(iii)\quad F(T_0) \geq 0, \quad F'(T_0) > 0,
\end{aligned}$$

where A_0 *and* A_1 *are positive constants. Let* $\tilde{T}$ *satisfy*

$$A_0 \tilde{T}^{M/(p-1)} (\ln \tilde{T})^b = 1. \tag{2.5}$$

Then T *has to satisfy*

$$T < C\tilde{T}$$

for a positive constant C *depending only on* A_1, p, a *and* c.

Moreover, if A_0 *is sufficiently small, then there holds*

$$T^{M/(p-1)} (\ln T)^b < C A_0^{-1}, \tag{2.6}$$

where C *is a constant depending on* A_1, μ, p, a, b *and* c.

Proof From the proof of Lemma 2.2 (see [16, (2.11)]), we see that

$$\begin{aligned}
F(t) \geq\ & t^{\mu/(p-1)} (t-\tilde{T})^{-(\mu+2)/(p-1)}\\
&\cdot \exp\left[\left\{E + \left(c + \frac{\mu+2}{p-1}\right)\ln(t-\tilde{T}) - \left(a + \frac{\mu}{p-1}\right)\ln t + b\ln(\ln \tilde{T})\right\} p^j\right]
\end{aligned} \tag{2.7}$$

for $t \geq \tilde{T}$, where

$$E = \frac{1}{p-1}\min(0, \ln B) - 2\sum_{k=0}^{\infty} \frac{k}{p^k} \ln p + \ln A_0.$$

Let $t \geq C\tilde{T}$. Then

$$
\begin{aligned}
&E + \left(c + \frac{\mu+2}{p-1}\right)\ln(t - \tilde{T}) - \left(a + \frac{\mu}{p-1}\right)\ln t + b\ln(\ln\tilde{T}) \\
&\geq E + \left(c + \frac{\mu+2}{p-1}\right)\ln\left(1 - \frac{1}{C}\right) + \left(c - a + \frac{2}{p-1}\right)\ln t + b\ln(\ln\tilde{T}) \\
&\geq E + \left(c + \frac{\mu+2}{p-1}\right)\ln(C-1) - \left(a + \frac{\mu}{p-1}\right)\ln C + \left(c - a + \frac{2}{p-1}\right)\ln\tilde{T} \\
&\qquad + b\ln(\ln\tilde{T}) \\
&= \frac{1}{p-1}\min(0, \ln B) - 2\sum_{k=0}^{\infty}\frac{k}{p^k}\ln p + \left(c + \frac{\mu+2}{p-1}\right)\ln(C-1) - \left(a + \frac{\mu}{p-1}\right)\ln C \\
&\quad + \ln A_0 + \left(c - a + \frac{2}{p-1}\right)\ln\tilde{T} + b\ln(\ln\tilde{T}) \\
&> 0
\end{aligned}
$$

if we take $C > 0$ so large depending on A_1, μ, p, a and c, where $B = \{c + (\mu + 2)/(p-1)\}^{-2}A_1$ since (2.5) and

$$
\left(c + \frac{\mu+2}{p-1}\right) - \left(a + \frac{\mu}{p-1}\right) = c - a + \frac{2}{p-1} > 0
$$

by assumption. It then follows from (2.7) that $F(t) \longrightarrow \infty$ as $j \to \infty$ for $t \geq C\tilde{T}$ with sufficiently large $C > 0$. Thus, we obtain $T < C\tilde{T}$. Moreover, we see from (2.5) that (2.6) holds for sufficiently small A_0. In fact, we observe that $f(T) = T^{M/(p-1)}(\ln T)^b$ is an increasing function for $T \geq 1$. This completes the proof of the corollary. □

Proof of Theorem 2.1. We now prove Theorem 2.1 by applying Lemma 2.2 and Corollary. Set

$$
F(t) = \int u(t, x)dx. \tag{2.8}
$$

We remark that the C^2-solution u of (1.12) has the property of finite speed of propagation, and satisfies

$$
\operatorname{supp} u(t, \cdot) \subset \{|x| \leq A(t) + R\}, \tag{2.9}
$$

where

$$
A(t) = \begin{cases}
\displaystyle\int_1^t s^{-\alpha}ds = \frac{t^{1-\alpha} - 1}{1-\alpha} & \text{if } 0 \leq \alpha < 1, \\
\displaystyle\int_1^t s^{-1}ds = \ln t & \text{if } \alpha = 1, \\
\displaystyle\int_1^t s^{-\alpha}ds = \frac{1}{\alpha - 1}(1 - t^{1-\alpha}) & \text{if } \alpha > 1,
\end{cases} \tag{2.10}
$$

provided that supp u_0, supp $u_1 \subset \{|x| \le R\}$. See [13, 16] for its proof.
Case $\alpha = 1$. Integrating the equation in (1.3) and using Hölder's inequality, we have by (2.9) and (2.10),

$$F''(t) + \frac{\mu}{t} F'(t) \ge \frac{1}{(A(t)+R)^{n(p-1)}} |F(t)|^p \ge \frac{C}{(\ln t)^{n(p-1)}} |F(t)|^p. \tag{2.11}$$

Multiplying the inequality above by t^μ and integrating imply

$$t^\mu F'(t) - F'(1) \gtrsim \int_1^t \frac{s^\mu}{(\ln s)^{n(p-1)}} |F(s)|^p ds \ge 0. \tag{2.12}$$

Note that

$$F'(t) > 0 \quad \text{for } t \ge 1$$

since $F'(1) > 0$ by assumption. Moreover, from (2.12),

$$F'(t) \gtrsim t^{-\mu} F'(1).$$

Integrating again over $[1, t]$, we have from $F(1) > 0$ by assumption

$$F(t) \gtrsim F'(1) t^{1-\mu} \qquad \text{for } t \ge T' \qquad \text{if } \mu < 1, \tag{2.13}$$

$$F(t) \gtrsim F'(1) \ln t \qquad \text{for } t \ge 1 \qquad \text{if } \mu = 1 \tag{2.14}$$

with some $T' > 1$. Going back to (2.12), since $F'(1) > 0$, we have

$$F'(t) \ge C t^{-\mu} \int_1^t \frac{s^\mu}{(\ln s)^{n(p-1)}} |F(s)|^p ds.$$

Integrating again, we have from $F(1) > 0$ by assumption,

$$F(t) \ge C \int_1^t \tau^{-\mu} \int_1^\tau \frac{s^\mu}{(\ln s)^{n(p-1)}} |F(s)|^p ds d\tau. \tag{2.15}$$

Let $\mu < 1$. From (2.13) and (2.15),

$$\begin{aligned} F(t) &\ge C\varepsilon^p \int_{T'}^t (\ln \tau)^{-n(p-1)} \tau^{-\mu} (\tau - T')^{\mu+1+p(1-\mu)} d\tau \\ &\ge C\varepsilon^p (\ln t)^{-n(p-1)} t^{-\mu} \int_{T'}^t (\tau - T')^{\mu+1+p(1-\mu)} d\tau \\ &\ge C\varepsilon^p (\ln t)^{-n(p-1)} t^{-\mu} (t - T')^{\mu+2+p(1-\mu)} \qquad \text{for } t \ge T'. \end{aligned} \tag{2.16}$$

By (2.11) and (2.16), applying Lemma 2.2 with $a = \mu$, $b = q = n(p-1)$, $c = \mu + 2 + p(1-\mu)$ and $A_0 = C\varepsilon^p$, we obtain the desired result (2.1) since

$$M = (p-1)(\mu+2+p(1-\mu)-\mu)+2 = p(2+(p-1)(1-\mu)) > 0.$$

Let $\mu = 1$ and $p \geq 1 + 1/(n-1)$. Note that $p - n(p-1) \leq 0$. We have from (2.14) and (2.15),

$$\begin{aligned} F(t) &\geq C\varepsilon^p \int_1^t \tau^{-1} \int_1^\tau s(\ln s)^{p-n(p-1)} ds d\tau \\ &\geq C\varepsilon^p (\ln t)^{p-n(p-1)} t^{-1} \int_1^t (\tau-1)^2 d\tau \\ &\geq C\varepsilon^p (\ln t)^{p-n(p-1)} t^{-1} (t-1)^3 \qquad \text{for } t \geq 1. \end{aligned} \tag{2.17}$$

By (2.11) and (2.17), applying Lemma 2.2 with $a = 1, b = -p + n(p-1), c = 3$, $q = n(p-1)$ and $A_0 = C\varepsilon^p$, we obtain the desired result (2.2) since $M = (p-1)(3-1)+2 = 2p > 0$.

Case $\alpha > 1$. Proceeding as before, we obtain

$$F''(t) + \frac{\mu}{t} F'(t) \geq \frac{1}{(A(t)+R)^{n(p-1)}} |F(t)|^p \geq C|F(t)|^p, \tag{2.18}$$

(2.13), (2.14) and

$$F(t) \geq C \int_1^t \tau^{-\mu} \int_1^\tau s^\mu |F(s)|^p ds d\tau. \tag{2.19}$$

If $\mu < 1$, then proceeding as in (2.16), we have from (2.13),

$$F(t) \geq C\varepsilon^p t^{-\mu} (t-T')^{\mu+2+p(1-\mu)} \qquad \text{for } t \geq T'. \tag{2.20}$$

By (2.18) and (2.20), applying Lemma 2.2 with $a = \mu$, $b = q = 0$, $c = \mu + 2 + p(1-\mu)$, and $A_0 = C\varepsilon^p$, we obtain the desired result (2.3) since

$$M = p(2+(p-1)(1-\mu)) > 0.$$

If $\mu = 1$, taking $T > \tilde{T} > 1$, we have from (2.14) and (2.19),

$$\begin{aligned} F(t) &\geq C\varepsilon^p (\ln \tilde{T})^p \int_{\tilde{T}}^t \tau^{-1} (\tau - \tilde{T})^2 d\tau \\ &\geq C\varepsilon^p (\ln \tilde{T})^p t^{-1} \int_{\tilde{T}}^t (\tau - \tilde{T})^2 d\tau \\ &\geq C\varepsilon^p (\ln \tilde{T})^p t^{-1} (t-\tilde{T})^3 \qquad \text{for } t \geq \tilde{T}. \end{aligned} \tag{2.21}$$

By (2.18) and (2.21), applying Corollary with $a = 1$, $b = p$, $c = 3$, $q = 0$ and $A_0 = C\varepsilon^p$, we obtain the desired result (2.4) since $M = 2p$. This completes the proof of Theorem 2.1. □

We next consider the Cauchy problem

$$\begin{cases} u_{tt} - \dfrac{1}{t^{2\alpha}}\Delta u + \dfrac{\mu}{t}u_t = |\nabla_x u|^p, & t > 1,\ x \in \mathbf{R}^n, \\ u(1,x) = \varepsilon u_0(x),\ u_t(1,x) = \varepsilon u_1(x), & x \in \mathbf{R}^n. \end{cases} \tag{2.22}$$

Similar results to Theorem 2.1 hold for (2.22).

Theorem 2.3 *Let $n \ge 1,\ \alpha \ge 1,\ 0 \le \mu \le 1$ and $p > 1$. Assume that $u_0 \in C^2(\mathbf{R}^n)$ and $u_1 \in C^1(\mathbf{R}^n)$ are nonnegative, nontrivial and supp u_0, supp $u_1 \subset \{|x| \le R\}$ with $R > 0$. Suppose that the problem (2.22) has a classical solution $u \in C^2([1,T) \times \mathbf{R}^n)$. Then $T < \infty$, and there exists a constant $\varepsilon_0 > 0$ depending on $p, \alpha, \mu, R, u_0, u_1$ such that T_ε has to satisfy*

$$T_\varepsilon^{2+(p-1)(1-\mu)}(\ln T_\varepsilon)^{-p-n(p-1)} \le C\varepsilon^{-(p-1)} \quad \text{if } \alpha = 1 \text{ and } \mu < 1, \tag{2.23}$$

$$T_\varepsilon^{2}(\ln T_\varepsilon)^{-1-n(p-1)} \le C\varepsilon^{-(p-1)} \quad \text{if } \alpha = 1 \text{ and } \mu = 1, \tag{2.24}$$

$$T_\varepsilon^{2+(p-1)(1-\mu)} \le C\varepsilon^{-(p-1)} \quad \text{if } \alpha > 1 \text{ and } \mu < 1, \tag{2.25}$$

$$T_\varepsilon^{2}(\ln T_\varepsilon)^{p-1} \le C\varepsilon^{-(p-1)} \quad \text{if } \alpha > 1 \text{ and } \mu = 1 \tag{2.26}$$

for $0 < \varepsilon \le \varepsilon_0$, where $C > 0$ is a constant independent of ε.

Proof Define $F(t)$ by (2.8). As in the proof of Theorem 3.9 in [14], there holds that

$$F''(t) + \frac{\mu}{t}F'(t) \ge \frac{C}{(\ln t)^{p+n(p-1)}}|F(t)|^p \quad \text{if } \alpha = 1,$$

$$F''(t) + \frac{\mu}{t}F'(t) \ge C|F(t)|^p \quad \text{if } \alpha > 1.$$

Let $\alpha = 1$. From (2.13) and (2.14), we have

$$F(t) \ge C\varepsilon^p(\ln t)^{-p-n(p-1)}t^{-\mu}(t-T')^{\mu+2+p(1-\mu)} \quad \text{for } t \ge T' \quad \text{if } \mu < 1,$$

$$F(t) \ge C\varepsilon^p(\ln t)^{-n(p-1)}t^{-1}(t-1)^3 \quad \text{for } t \ge 1 \quad \text{if } \mu = 1.$$

Using Lemma 2.2, we obtain the desired results.

The proof for the case $\alpha > 1$ is the same as that of Theorem 2.1. □

3 Time Derivative Nonlinearity Case

We consider problem (1.12).

Theorem 3.1 *Let $n \ge 1,\ \alpha \ge 0,\ \mu \ge 0$ and*

$$p = p_0(n, \alpha, \mu) = 1 + \frac{1}{n(1-\alpha)+\mu} \qquad \text{if } 0 \le \alpha < 1,$$

$$p = 1 + \frac{1}{\mu} < 1 + \frac{1}{n} \qquad \text{if } \alpha = 1,$$

$$p = 1 + \frac{1}{\mu} \qquad \text{if } \alpha > 1.$$

Assume that $u_0 \in C^2(\mathbf{R}^n)$ *and* $u_1 \in C^1(\mathbf{R}^n)$ *are nontrivial and satisfy* $u_1(x) \ge 0$, supp u_0, supp $u_1 \subset \{|x| \le R\}$ *with* $R > 0$. *Suppose that problem* (1.12) *has a classical solution* $u \in C^2([1, T) \times \mathbf{R}^n)$. *Then* $T < \infty$, *and there exists a constant* $\varepsilon_0 > 0$ *depending on* $p, \alpha, \mu, R, u_0, u_1$ *such that* T_ε *has to satisfy*

$$T_\varepsilon \le \exp(C\varepsilon^{-(p-1)}) \qquad \text{if } \alpha \ne 1, \tag{3.1}$$

$$T_\varepsilon \le \exp(C\varepsilon^{-(p-1)/\{1-n(p-1)\}}) \qquad \text{if } \alpha = 1 \tag{3.2}$$

for $0 < \varepsilon \le \varepsilon_0$, *where* $C > 0$ *is a constant independent of* ε.

To prove the theorem above, we use another generalized Kato's lemma.

Lemma 3.2 *Let* $p > 1,\ a \ge 0,\ b \ge 0,\ q \ge 0,\ r \ge 0,\ \mu \ge 0,$

$$\mu(p-1) + q = 1, \quad \text{and} \quad (a-b)(p-1) + 1 - r > 0.$$

Let $T \ge T_1 \ge T_0 \ge 1$. *Assume that* $F \in C^1([T_0, T))$ *satisfies the following three conditions:*

$$(i)\quad t^\mu F(t) \ge A_0 \left(\ln \frac{t}{T_1}\right)^a (\ln t)^{-b} \quad \text{for } t \ge T_1,$$

$$(ii)\quad F'(t) + \frac{\mu}{t}F(t) \ge A_1 (t+R)^{-q}(\ln t)^{-r}|F(t)|^p \quad \text{for } t \ge T_0,$$

$$(iii)\quad F(T_0) > 0,$$

where A_0, A_1 *and* R *are positive constants. Then* T *has to satisfy*

$$T < \exp\left(C A_0^{-(p-1)/\{(a-b)(p-1)+1-r\}}\right),$$

where C *is a constant depending on* R, A_1, μ, p, q, r, a *and* b.

Proof Multiplying assumption (ii) by t^μ, we have

$$t^\mu F' + \mu t^{\mu-1} F \ge A_1 t^\mu (t+R)^{-q} (\ln t)^{-r} |F|^p.$$

Integrating the above inequality over $[T_0, t]$ yields

$$t^{\mu}F(t) - F(T_0) \geq A_1 C_{R,q} \int_{T_0}^{t} s^{\mu-q}(\ln s)^{-r}|F(s)|^p ds \geq 0, \quad C_{R,q} = (1+R)^{-q}. \tag{3.3}$$

By assumption (iii), we see that $F(t) > 0$ for $t \geq T_0$. Hence, we have from assumption (i) and $\mu(p-1) + q = 1$

$$\begin{aligned} t^{\mu}F(t) &\geq A_0^p A_1 C_{R,q} \int_{T_1}^{t} s^{-q-\mu(p-1)}(\ln s)^{-bp-r}\left(\ln \frac{s}{T_1}\right)^{ap} ds \\ &\geq \frac{A_0^p A_1 C_{R,q}}{ap+1}(\ln t)^{-bp-r}\left(\ln \frac{t}{T_1}\right)^{ap+1} \qquad \text{for } t \geq T_1. \end{aligned}$$

Based on the fact above, we define the sequences $a_j,\ b_j,\ C_j$ for $j = 0, 1, 2, \cdots$ by

$$a_{j+1} = pa_j + 1, \qquad b_{j+1} = pb_j + r, \qquad C_{j+1} = \frac{A_1 C_{R,q} C_j^p}{pa_j + 1}, \tag{3.4}$$

$$a_0 = a, \qquad b_0 = b, \qquad C_0 = A_0. \tag{3.5}$$

Solving (3.4) and (3.5), we obtain

$$a_j = p^j\left(a + \frac{1}{p-1}\right) - \frac{1}{p-1}, \qquad b_j = p^j\left(b + \frac{r}{p-1}\right) - \frac{r}{p-1},$$

and thus

$$C_{j+1} = \frac{A_1 C_{R,q} C_j^p}{a_{j+1}} \geq \left(a + \frac{1}{p-1}\right)^{-1} \frac{A_1 C_{R,q} C_j^p}{p^{j+1}}.$$

Then, as in the proof of Lemma 2.2 in [14],

$$C_j \geq \frac{B^{1+p+p^2+\cdots+p^{j-1}}}{p^{j+p(j-1)+p^2(j-2)+\cdots+p^{j-1}}} C_0^{p^j},$$

and

$$\ln C_j \geq \frac{\ln B}{p-1}(p^j - 1) - p^j \sum_{k=0}^{j} \frac{k}{p^k} \ln p + p^j \ln C_0,$$

where $B = \{a + 1/(p-1)\}^{-1} A_1 C_{R,q}$. For sufficiently large j, we have

$$C_j \geq \exp(Ep^j),$$

where

$$E = \frac{1}{p-1} \min(0, \ln B) - \sum_{k=0}^{\infty} \frac{k}{p^k} \ln p + \ln A_0. \tag{3.6}$$

Thus, since $t^\mu F(t) \geq C_j(\ln t/T_1)^{a_j}(\ln t)^{-b_j}$ holds for $t \geq T_1$, we obtain

$$F(t) \geq t^{-\mu}\left(\ln \frac{t}{T_1}\right)^{-1/(p-1)}(\ln t)^{r/(p-1)} \cdot \exp\left[\left\{E + \left(a + \frac{1}{p-1}\right)\ln\left(\ln \frac{t}{T_1}\right) - \left(b + \frac{r}{p-1}\right)\ln(\ln t)\right\}p^j\right] \tag{3.7}$$

for $t \geq T_1$. Choosing t large enough, we can find a positive δ such that

$$E + \left(a + \frac{1}{p-1}\right)\ln\left(\ln \frac{t}{T_1}\right) - \left(b + \frac{r}{p-1}\right)\ln(\ln t) \geq \delta > 0$$

since $a - b + (1-r)/(p-1) > 0$ by assumption. It then follows from (2.7) that $F(t) \longrightarrow \infty$ as $j \to \infty$ for sufficiently large t. We therefore see that the lifespan T of $F(t)$ has to satisfy

$$T < \exp\left(CA_0^{-(p-1)/\{(a-b)(p-1)+1-r\}}\right),$$

where C is a constant depending on R, A_1, μ, p, q, r, a and b. This completes the proof of the lemma. □

We now prove Theorem 3.1 by applying Lemma 3.2. Set

$$F(t) = \int u_t(t,x)dx. \tag{3.8}$$

Integrating the equation in (1.12), and using Hölder's inequality and (2.9), we have

$$\begin{aligned} F'(t) + \frac{\mu}{t}F(t) &= \int |u_t|^p dx \\ &\geq \frac{1}{(A(t)+R)^{n(p-1)}}|F(t)|^p. \end{aligned} \tag{3.9}$$

Multiplying (3.9) by t^μ and integrating imply

$$t^\mu F(t) - F(1) \gtrsim \int_1^t s^\mu (A(s)+R)^{-n(p-1)}|F(s)|^p ds \geq 0. \tag{3.10}$$

We hence see that

$$F(t) \geq t^{-\mu}F(1) = C\varepsilon t^{-\mu} > 0 \quad \text{for } t \geq 1 \tag{3.11}$$

by assumption.

Let $0 \le \alpha < 1$. By (2.10) and (3.9), we have

$$F'(t) + \frac{\mu}{t} F(t) \ge C(t+R)^{-n(1-\alpha)(p-1)} |F(t)|^p. \tag{3.12}$$

It follows from (2.10), (3.10) and (3.11) that

$$\begin{aligned} t^\mu F(t) &\ge C\varepsilon^p \int_1^t s^{\mu - n(1-\alpha)(p-1) - \mu p} ds \\ &= C\varepsilon^p \int_1^t s^{-1} ds \end{aligned}$$

by assumption. Therefore, we obtain

$$t^\mu F(t) \ge C\varepsilon^p \ln t \qquad \text{for } t \ge 1. \tag{3.13}$$

By (3.12) and (3.13), applying Lemma 3.2 with $q = n(1-\alpha)(p-1),\ r = 0,\ a = 1,\ b = 0$ and $A_0 = C\varepsilon^p$, we obtain the desired result for the case $0 \le \alpha < 1$.

Let $\alpha \ge 1$. By (2.10) and (3.9), we have

$$F'(t) + \frac{\mu}{t} F(t) \ge C(\ln t)^{-n(p-1)\delta_{1\alpha}} |F(t)|^p, \tag{3.14}$$

where $\delta_{1\alpha}$ is the Kronecker delta. It follows from (2.10), (3.10) and (3.11) that

$$\begin{aligned} t^\mu F(t) &\ge C\varepsilon^p \int_1^t s^{\mu - \mu p} (\ln s)^{-n(p-1)\delta_{1\alpha}} ds \\ &\ge C\varepsilon^p (\ln t)^{-n(p-1)\delta_{1\alpha}} \int_1^t s^{-1} ds \end{aligned}$$

by assumption. Therefore, we obtain

$$t^\mu F(t) \ge C\varepsilon^p (\ln t)^{1 - n(p-1)\delta_{1\alpha}} \qquad \text{for } t \ge 1. \tag{3.15}$$

We note that $1 - n(p-1)\delta_{1\alpha} > 0$ by assumption. Finally, by (3.14) and (3.15), applying Lemma 3.2 with $q = 0,\ a = 1,\ b = r = n(p-1)\delta_{1\alpha}$, and $A_0 = C\varepsilon^p$, we obtain the desired result for the case $\alpha \ge 1$. This completes the proof of Theorem 3.1. □

4 Wave Equations in FLRW

We now consider the original Eq. (1.2), which is equivalent to the equations in (1.3) and (1.12) with $\alpha = 2/(n(1+w))$ and $\mu = 2/(1+w)$. We treat the case $-1 < w \le 1$ and $n \ge 2$ so that $\alpha \ge 1/n$ and $\mu \ge 1$ in (1.3) and (1.12). Observe that the cases $-1 < w \le 2/n - 1$ and $2/n - 1 < w \le 1$ correspond to accelerating and decelerating expanding universes, respectively.

We first see that Theorem 2.1 is not applied to (1.2) with $G(u, \partial u) = |u|^p$. In fact, if we try to apply Theorem 2.1, only $\mu = 1$ is a possible case. However, if $\mu = 1$, then $w = 1$, hence $\alpha = 1/n < 1$, which does not satisfy the conditions of the theorem.

For (1.2) with $G(u, \partial u) = |\nabla_x u|^p$, Theorem 2.3 is also not applicable except for the case $n = w = 1$. Since $\alpha = \mu = 1$ in this case, we can apply (2.24) to (1.2) with $G(u, \partial u) = |\nabla_x u|^p$ and obtain

$$T_\varepsilon^2(\ln T_\varepsilon)^{-p} \le C\varepsilon^{-(p-1)} \qquad \text{if } n = w = 1.$$

We next consider (1.2) with the time derivative nonlinear term $|u_t|^p$. For $-1 < w \le 1$ and $w \ne 2/n - 1$, we obtain the following upper bound of the lifespan from Theorem 3.1:

$$T_\varepsilon \le \exp(C\varepsilon^{-(p-1)}) \quad \text{if} \begin{cases} p = 1 + \frac{1}{n},\ 2/n - 1 < w \le 1, \text{ and } n \ge 2, \\ \text{or} \\ p = \frac{3+w}{2},\ -1 < w < 2/n - 1, \text{ and } n \ge 1. \end{cases} \tag{4.1}$$

We note that if $w = 2/n - 1$, then $\alpha = 1$ and $p = 1 + 1/\mu = 1 + 1/n$, which does not satisfy the conditions of Theorem 3.1.

Denote $p_G(n, 2/(n(1+w)), 2/(1+w))$ by $p_G(n, w)$. We remark that (1.15) with $\alpha = 2/(n(1+w))$ and $\mu = 2/(1+w)$ is better than (4.1) if $p = 1 + 1/n$ and $2/n - 1 < w \le 1$ since $p = 1 + 1/n < p_G(n, w)$ in this case.

We summarize the upper bounds of the lifespan for (1.2) with $|u_t|^p$ including the ones obtained in [14]:

$$T_\varepsilon \le C\varepsilon^{\frac{-(p-1)}{1 - \{n - 1 + 4/(n(1+w))\}(p-1)/2}} \qquad \text{if } 1 < p < p_G(n, w) \text{ and } 2/n - 1 < w \le 1,$$

$$T_\varepsilon \le \exp(C\varepsilon^{-(p-1)}) \qquad \text{if } p = p_G(n, w) \text{ and } 2/n - 1 < w \le 1,$$

$$T_\varepsilon^{1-2(p-1)/(1+w)}(\ln T_\varepsilon)^{-n(p-1)} \le C\varepsilon^{-(p-1)} \qquad \text{if } 1 < p < 1 + 1/n \text{ and } w = 2/n - 1,$$

$$T_\varepsilon \le C\varepsilon^{-(p-1)/\{1-2(p-1)/(1+w)\}} \qquad \text{if } 1 < p < (3+w)/2 \text{ and } -1 < w < 2/n - 1,$$

$$T_\varepsilon \le \exp(C\varepsilon^{-(p-1)}) \qquad \text{if } p = (3+w)/2 \text{ and } -1 < w < 2/n - 1.$$

We see that the critical cases except for $w = 2/n - 1$ have the same upper bound $\exp(C\varepsilon^{-(p-1)})$. Figure 1 below shows the range of blow-up conditions in terms of w and p in the case $n = 3$. From these results, we see that the blow-up range of p in the flat FLRW spacetime is smaller than that in the Minkowski spacetime because $p_G(n, w) < p_G(n) = 1 + 2/(n-1)$ for $2/n - 1 < w \le 1$ and $(3+w)/2 < p_G(n)$ for $-1 < w < 2/n - 1$. Moreover, in the subcritical cases $p < p_G(n, w)$ and $p < (3+w)/2$, the lifespan of the blow-up solutions in the FLRW spacetime is longer than that in the Minkowski spacetime since $\varepsilon^{-(p-1)/\{1-(n-1)(p-1)/2\}} <$

Fig. 1 Range of blow-up conditions in case $n = 3$

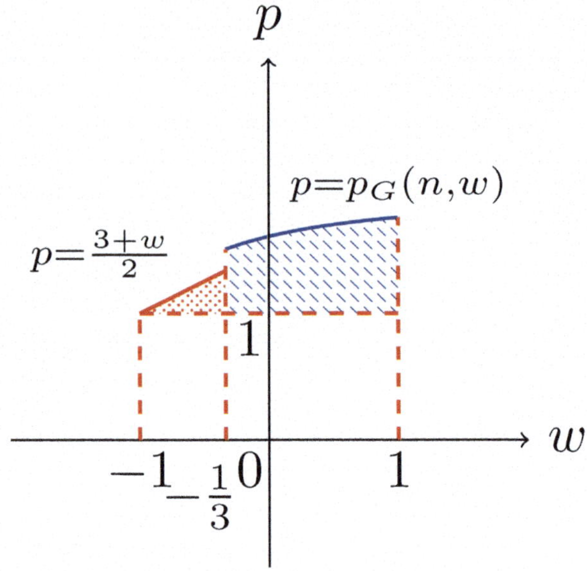

$\varepsilon^{-(p-1)/\{1-(n-1+4/(n(1+w)))(p-1)/2\}}$ and $\varepsilon^{-(p-1)/\{1-(n-1)(p-1)/2\}} < \varepsilon^{-(p-1)/\{1-2(p-1)/(1+w)\}}$ for sufficiently small ε. Although the critical value has not been established, we can say at the present time that global solutions in the FLRW spacetime exist more easily than in the Minkowski spacetime.

References

1. A. Galstian, K. Yagdjian, Finite lifespan of solutions of the semilinear wave equation in the Einstein-de Sitter spacetime. Rev. Math. Phys. **32**(2050018), 31 (2020)
2. R.T. Glassey, Mathreview to “Global behavior of solutions to nonlinear wave equations in three space dimensions” of Sideris. Commun. Partial. Differ. Equ (1983). 1985. AMS, MR711440 (85e:35081)
3. M. Hamouda, M.A. Hamza, Improvement on the blow-up of the wave equation with the scale-invariant damping and combined nonlinearities. Nonlinear Anal. Real World Appl. **59**, 103275 (2021)
4. M. Hamouda, M.A. Hamza, A. Palmieri, Blow-up and lifespan estimates for a damped wave equation in the Einstein-de Sitter spacetime with nonlinearity of derivative type. Nonlinear Differ. Equ. Appl. **29**, (19) (2022)
5. K. Hidano, K. Tsutaya, Global existence and asymptotic behavior of solutions for nonlinear wave equations. Indiana Univ. Math. J. **44**, 1273–1305 (1995)
6. K. Hidano, C. Wang, K. Yokoyama, The Glassey conjecture with radially symmetric data. J. Math. Pures Appl. **98**, 518–541 (2012)
7. M. Ikeda, M. Sobajima, Life-span of solutions to semilinear wave equation with time-dependent critical damping for specially localized initial data. Math. Ann. **372**, 1017–1040 (2018)
8. F. John, Blow-up of solutions of nonlinear wave equations in three space dimensions. Manuscr. Math. **28**, 235–268 (1979)

9. N.-A. Lai, H. Takamura, Nonexistence of global solutions of nonlinear wave equations with weak time-dependent damping related to Glassey's conjecture. Differ. Integr. Equ. **32**, 37–48 (2019)
10. N.-A. Lai, H. Takamura, K. Wakasa, Blow-up for semilinear wave equations with the scale invariant damping and super-Fujita exponent. J. Differ. Equ. **263**, 5377–5394 (2017)
11. A. Palmieri, Blow-up results for semilinear damped wave equations in Einstein-de Sitter spacetime. Z. Angew. Math. Phys. **72**(64), (2021)
12. T.C. Sideris, Global behavior of solutions to nonlinear wave equations in three space dimensions. Commun. Partial. Differ. Equ. **8**, 1291–1323 (1983)
13. K. Tsutaya, Y. Wakasugi, Blow up of solutions of semilinear wave equations in Friedmann-Lemaître-Robertson-Walker spacetime. J. Math. Phys. **61**, 091503 (2020). https://doi.org/10.1063/1.5139301
14. K. Tsutaya, Y. Wakasugi, On Glassey's conjecture for semilinear wave equations in Friedmann-Lemaître-Robertson-Walker spacetime. Bound. Value Probl. **2021**, 94 (2021). https://doi.org/10.1186/s13661-021-01571-0
15. K. Tsutaya, Y. Wakasugi, On heatlike lifespan of solutions of semilinear wave equations in Friedmann-Lemaître-Robertson-Walker spacetime. J. Math. Anal. Appl. **500**, 125133 (2021). https://doi.org/10.1016/j.jmaa.2021.125133
16. K. Tsutaya, Y. Wakasugi, Blow up of solutions of semilinear wave equations in accelerated expanding Friedmann-Lemaître-Robertson-Walker spacetime. Rev. Math. Phys. **33**, 2250003 (2022). https://doi.org/10.1142/S0129055X22500039
17. N. Tzvetkov, Existence of global solutions to nonlinear massless Dirac system and wave equation with small data. Tsukuba J. Math. **22**, 193–211 (1998)
18. Y. Zhou, Blow-up of solutions to the Cauchy problem for nonlinear wave equations. Chin. Ann. Math. **22B**, 275–280 (2001)

The Frisch–Parisi Formalism for Fluctuations of the Schrödinger Equation

Sandeep Kumar, Felipe Ponce-Vanegas, Luz Roncal, and Luis Vega

Dedicated to Professor Tohru Ozawa on the occasion of his 60th birthday.

Abstract We consider the solution of the Schrödinger equation u in $\mathbb{R}$ when the initial datum tends to the Dirac comb. Let $h_{\mathrm{p},\delta}(t)$ be the fluctuations in time of $\int |x|^{2\delta} |u(x,t)|^2 \, dx$, for $0 < \delta < 1$, after removing a smooth background. We prove that the Frisch–Parisi formalism holds for $H_\delta(t) = \int_{[0,t]} h_{\mathrm{p},\delta}(2s)\, ds$, which is morally a simplification of Riemann's non-differentiable curve R. Our motivation is to understand the evolution of the vortex filament equation of polygonal filaments, which are related to R.

Keywords Schrödinger equation · Vortex filament equation · Talbot effect · Frisch–Parisi formalism · Multifractals

2010 Mathematics Subject Classification Primary 35Q41 · Secondary 26A27 · 76B47

S. Kumar
Department of Quantitative Methods, CUNEF Universidad, 28040 Madrid, Spain
e-mail: sandeep.kumar@cunef.edu

F. Ponce-Vanegas · L. Roncal
BCAM–Basque Center for Applied Mathematics, 48009 Bilbao, Spain
e-mail: fponce@bcamath.org

L. Roncal
e-mail: lroncal@bcamath.org

L. Roncal
Ikerbasque, Basque Foundation for Science, 48011 Bilbao, Spain

L. Vega (✉)
Department of Mathematics UPV/EHU, Apto. 644, 48080 Bilbao, Spain
e-mail: luis.vega@ehu.es

S. Machihara (ed.), *Mathematical Physics and Its Interactions*, Springer Proceedings in Mathematics & Statistics 451, https://doi.org/10.1007/978-981-97-0364-7_7

1 Introduction

The binormal curvature flow, also known as the vortex filament equation,

$$\chi_t = \chi_x \wedge \chi_{xx}, \tag{1}$$

is a model for the dynamics of vortex filaments in Euler equations. Here $\wedge$ denotes the cross product. The function $\chi(t, x)$ describes a family of curves in 3d that move with time t and are parametrized by arclength x. Using the Frenet equations one easily concludes that the right-hand side of (1) is a vector whose modulus equals the curvature and whose direction is the binormal vector. By differentiating both sides by x, we get the one dimensional Schrödinger map

$$T_t = T \wedge T_{xx}, \qquad \text{where } T := \chi_x \in \mathbb{S}^2. \tag{2}$$

Our interest in this paper is in curves that can develop corners in finite time. For that purpose, it is better to use the so-called parallel frame (T, e_1, e_2) instead of the usual Frenet frame, where the former is defined by

$$\begin{pmatrix} T_x \\ (e_1)_x \\ (e_2)_x \end{pmatrix} = \begin{pmatrix} 0 & \alpha & \beta \\ -\alpha & 0 & 0 \\ -\beta & 0 & 0 \end{pmatrix} \cdot \begin{pmatrix} T \\ e_1 \\ e_2 \end{pmatrix}.$$

Hasimoto proved in [11] that for T to be a solution of (2), $u := \alpha + i\beta$ has to solve the 1d cubic non-linear Schrödinger equation

$$iu_t + u_{xx} + \frac{1}{2}((|u|^2 - A(t))u = 0,$$

for some real function $A(t)$; Hasimoto used the Frenet frame, but the proof admits more general frames.

In [6], de la Hoz and the fourth author studied the evolution of regular planar polygons χ_M, with M denoting the number of sides. In particular, they were interested in the trajectories described by any of the corners. That is to say, if we assume that at time $t = 0$ there is a corner at $x = 0$, then they studied the curve in 3d

$$R_M(t) := \chi_M(t, 0). \tag{3}$$

These curves show a characteristic fractal behavior which is reminiscent of the so-called Riemann's non-differentiable function. In fact, they found compelling numerical evidence that $\lim_{M\to\infty} R_M(t) = R(t)$ with

$$R(t) := \int_0^t u_D(0, r)\, dr = t + i \sum_{n\in\mathbb{Z}\setminus\{0\}} \frac{e^{-i\pi n^2 t} - 1}{\pi n^2},$$

where u_D is the solution of the linear Schödinger equation $u_t = \frac{1}{4\pi} i u_{xx}$ with initial datum $F_D = \sum_{n\in\mathbb{Z}} \delta_n$, that is, F_D is the Dirac's comb. It turns out that R is a small modification of the complex version of Riemann's function

$$\phi(t) := \sum_{n=1}^{\infty} \frac{e^{i\pi n^2 t}}{i\pi n^2} = -\frac{1}{2} R(-t) - \frac{t}{2} - \frac{i\pi}{6}.$$

We notice that u_D exhibits the Talbot effect, that is, the appearence of rescaled and weighted Dirac combs at rational times, which easily justifies the fractal appearence of R. There is a rich literature about the Talbot effect; see for example [3, 8, 18–20].

Recently, Banica and the fourth author [2] tightened the connection between R and the binormal curvature flow. They proved that the evolution of a corner of a suitably chosen sequence of polygonal vortex filaments approaches $R(t)$ in the limit when the number of sides is infinite. Additionally, inspired by the work of Jaffard [12], they showed that the limiting behavior of the corners falls within the multifractal formalism of Frisch and Parisi (see [9, Appendix] for the multifractal formalism), which is conjectured to govern turbulent fluids. By analogy with turbulence, we would expect that R could be understood as the outcome of some stochastic process; such an interpretation still seems to be missing.

Yet another interesting physical phenomenon, which is closely related to multifractality, is the intermittency. Roughly speaking, the idea is that the velocity of a fluid in fully developed turbulence may erratically change in very small distances, suggesting a very irregular structure. This phenomenon, called *intermittency in small scales* is related to the Frisch–Parisi multifractal formalism, but it does not seem to be well-defined in the literature. In [4], by adapting the physical concept of intermittency to the setting of functions and giving a precise definition, the authors gave quantitative estimates of the intermittency of Riemann's non-differentiable function.

Within this circle of ideas, in [17], during an investigation of the dispersive properties of the free Schrödinger equation, an interesting behavior was discovered for the functional

$$h_\delta[f](t) := \int |x|^{2\delta} \, |u(x,t)|^2 \, dx, \qquad x \in \mathbb{R}^d \qquad \text{for } 0 < \delta < 1,$$

where u is the solution of the linear Schrödinger equation with initial datum f. First, the authors extended the definition of h_δ to smooth periodic initial data F by approaching it with $f_{\varepsilon_1}(t) = N_{\varepsilon_1}^{-1} \psi(\varepsilon_1 t) F(t)$, where ψ is a smooth function with $\psi(0) = 1$ and N_{ε_1} a normalization constant, and then renormalized $h_\delta[f_{\varepsilon_1}]$ (that is, remove an infinite and rescale) leaving behind the fluctuations of $h_\delta[f_{\varepsilon_1}]$, which in the limit $\varepsilon_1 \to 0^+$ tend to a function we call $h_{\mathrm{p},\delta}[F]$ (p for periodic); for details, see [17, Theorem 18].

In the case of the Dirac comb F_D, if we approach it through smooth periodic functions F_{ε_2}, then the limit (in the sense of distributions) of $h_{\mathrm{p},\delta}[F_{\varepsilon_2}]$ exists; see Fig. 1. The function $h_{\mathrm{p},\delta}[F_D]$ is supported in the rational numbers and admits an explicit expression [17, Theorem 3]:

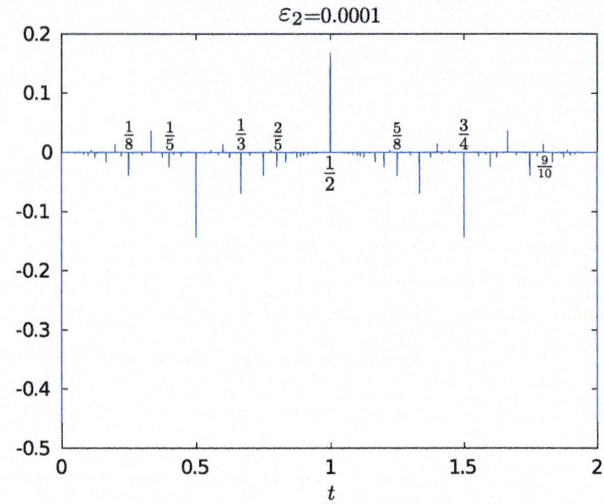

Fig. 1 Evolution of $h_{\mathrm{p},\delta}[F_{\varepsilon_2}]$, where F_{ε_2} is a smooth periodic function that approaches the Dirac comb in the sense of distributions as $\varepsilon_2 \to 0^+$. At $t = 0$ and 2 there are two large negative peaks

$$h_{\mathrm{p},\delta}[F_D](2t) = \sum_{\substack{p/q \\ (p,q)=1, q>0}} \frac{a_{q,\delta}}{q^{2(1+\delta)}} \delta_{p/q}(t), \tag{4}$$

$$a_{q,\delta} = -\frac{2b_{1,\delta}}{\|\psi\|_2^2} \zeta(2(1+\delta)) \begin{cases} 1 & \text{if } q \text{ is odd}, \\ -2(2^{1+2\delta} - 1) & \text{if } q \equiv 2 \pmod 4, \\ 2^{2(1+\delta)} & \text{if } q \equiv 0 \pmod 4; \end{cases} \tag{5}$$

here, ζ is the Riemann zeta function and

$$b_{1,\delta} := \frac{1}{(2\pi)^{2\delta}} \frac{\Gamma(2\delta)}{|\Gamma(-\delta)|\,\Gamma(\delta)}.$$

Let us compare the functions $h_{\mathrm{p},\delta}[F_D](2t)$ and $u_D(x, 2t)$. For $t = 2p/q$, with $(p, q) = 1$, we have

$$u_D(x, 2p/q) = \frac{1}{q} \sum_{m \in \mathbb{Z}} G(-p, m, q) \delta\left(x - \frac{m}{q}\right),$$

where G refers to the Gauss quadratic sums

$$G(a, b, c) := \sum_{r=0}^{c-1} e^{2\pi i (ar^2 + br)/c}, \qquad \text{for } a, b, c \in \mathbb{Z} \text{ and } c > 0;$$

see [7, Appendices]. At rational times u_D concentrates in a discrete set, and $h_{\mathrm{p},\delta}[F_D]$, being a "weighted average" of $|u_D|^2$, records this phenomenon as fluctuations at rational times. On the other hand, at irrational times u_D behaves as a noisy function; however, after taking expected values, that is, computing $h_\delta[f_\varepsilon]$, the noisy part becomes a smooth function in time which ends up being removed during the renormalization. The solution u_D is hard to understand at irrational times, whereas $h_{\mathrm{p},\delta}[F_D]$ vanishes right there. This simplification offers the possibility of understanding hard questions about u_D by first considering $h_{\mathrm{p},\delta}[F_D]$.

In [17] the authors exposed evidences showing that

$$H_\delta(t) = \int_{[0,t]} h_{\mathrm{p},\delta}[F_D](2r)\,dr, \qquad \text{for } 0 < \delta < 1, \tag{6}$$

can be seen as the outcome of a $(2/s)$-Lévy process with $s := 2(1+\delta)$. Unbeknownst to the authors, similarities between Riemann's non-differentiable function (which is behind u_D) and Lévy processes had already been pointed out by Jaffard in [13, Sects. 2.3 and 4.4].

The velocity of turbulent flows differs widely from point to point, so in this context the spectrum of singularities of a function was introduced, which measures the size of the sets with different Hölder exponents.

Definition 1.1 (*Hölder exponent*) Let f be a function, $l \geq 0$ and $t \in \mathbb{R}$. We write $f \in C^l(t)$ if there is a polynomial P_t of degree less than l such that in a neighborhood of t

$$|f(r) - P_t(r)| \lesssim |r-t|^l\,.$$

The Hölder exponent of f at t is

$$h_f(t) := \sup\{l \mid f \in C^l(t)\}.$$

To measure the size of a set, we use the concept of Hausdorff dimension.

Definition 1.2 (*Hausdorff dimension*) Let $A \subset \mathbb{R}^n$ and R_ε be the set of all coverings of A by sets A_i of diameter at most ε. Let

$$\mathcal{H}^d_\varepsilon(A) := \inf_{r\in R_\varepsilon} \sum_{A_i \in r} (\operatorname{diam} A_i)^d.$$

Then,

$$\mathcal{H}^d(A) := \lim_{\varepsilon\to 0} \mathcal{H}^d_\varepsilon(A)$$

is the d-dimensional Hausdorff measure of A. The *Hausdorff dimension* of A is

$$\dim_{\mathcal{H}} A := \inf\{d : \mathcal{H}^d(A) = 0\} = \sup\{d : \mathcal{H}^d(A) = +\infty\}.$$

Now we can define the spectrum of singularities of a function.

Definition 1.3 (*Spectrum of singularities*) Let f be a function and define the set

$$\Gamma_\alpha := \{t \in \mathbb{R} \mid h_f(t) = \alpha\}.$$

The spectrum of singularities is the function

$$D_f(\alpha) := \dim_{\mathcal{H}} \Gamma_\alpha.$$

If $\Gamma_\alpha = \emptyset$, then $D_f(\alpha) = -\infty$.

Jaffard proved in [12] that the spectrum of singularities of Riemann's non-differentiable curve is

$$D_R(\alpha) = 4\alpha - 2, \qquad \text{for all } \alpha \in \left[\frac{1}{2}, \frac{3}{4}\right].$$

Later, it was proved in [2, Proposition 4.1] that the spectrum of singularities of R (and modifications of it) is equal to that of Riemann's non-differentiable curve.

Concerning H_δ, it was proved in [17, Theorem 4] that the spectrum of singularities of H_δ is

$$D_{H_\delta}(\alpha) = \begin{cases} 2\alpha/s, & \text{if } \alpha \in [0, s/2], \\ -\infty, & \text{if } \alpha > s/2, \end{cases} \tag{7}$$

where $s = 2(1 + \delta)$, for $0 < \delta < 1$. Surprisingly, for $(2/s)$-Lévy processes Jaffard proved in [15] that the spectrum of singularities is almost surely equal to (7). Before stating our main result, we describe briefly the Frisch–Parisi formalism.

1.1 Frisch–Parisi Formalism

The so-called multifactral formalism for functions relates some functional norms of a function to its spectrum of singularities. This formalism was introduced by Frisch and Parisi in order to numerically determine the spectrum of fully turbulent fluids [9]. Even though the Frisch–Parisi formalism has several versions, we decided to use the *wavelet–transform integral method* as described in the introduction of [14]. First we must define the wavelet transform of a function.

Definition 1.4 The *wavelet transform* of a function f is defined as $\psi_N * f$, where $\psi_N(t) := N\psi(Nt)$. The wavelet ψ is a function whose smoothness and decay are adjusted depending on the problem, and such that

$$\int t^k \psi(t)\, dt = 0, \qquad \text{for } k = 0, \ldots, L \text{ and some suitable } L.$$

The Frisch–Parisi formalism suggests that the spectrum of singularities can be computed through the *scaling exponent* η_f, which is defined by

$$\eta_f(p) := -\liminf_{N\to\infty} \frac{\log \|\psi_N * f\|_{L^p}^p}{\log N}; \tag{8}$$

informally, this means that $\|\psi_N * f\|_{L^p}^p$ behaves like $N^{-\eta_f(p)}$ as $N \to \infty$. The Legendre transform provides a link between D_f and η_f through the conjectured relationship:

$$D_f(\alpha) = \inf_{p>0}(p\alpha - \eta_f(p) + 1).$$

1.2 Main Result

Since $H_\delta[F_D]$ might be seen as a simplification of Riemann's non-differentiable function, for which the Frisch–Parisi formalism has been proved (see [12] or [2]), then we should be able to prove the Frisch–Parisi formalism for $H_\delta[F_D]$ in the range $[0, s/2]$. We confirm this in our main theorem below.

Theorem 1.5 (Frisch–Parisi formalism) *Let $0 < \delta < 1$ and $s := 2(1+\delta)$. Let ψ be an integrable function such that:*

(i) $\int_{\mathbb{R}} \psi = 0$.
(ii) $|\psi(t)| \lesssim |t|^{-L}$, *for* $L > 1 + s$.

Let η_{H_δ} be the scaling exponent defined in (8) *for the function H_δ. Then,*

$$\eta_{H_\delta}(p) = \begin{cases} sp/2, & \text{if } 0 < p \le 2/s, \\ 1, & \text{if } p \ge 2/s. \end{cases}$$

In particular, the Frisch–Parisi formalism holds in the range of Hölder exponents $[0, s/2]$; see (7).

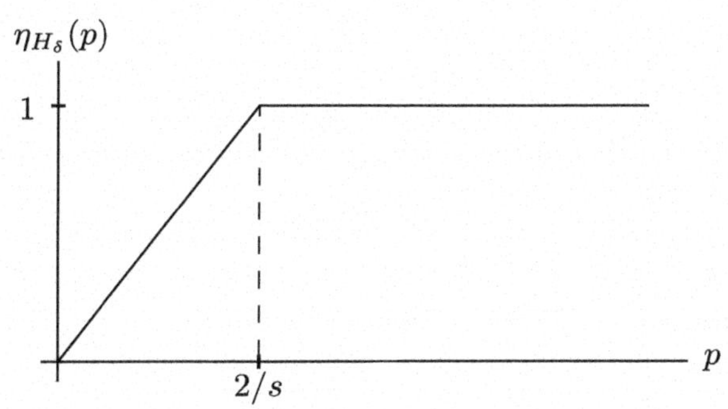

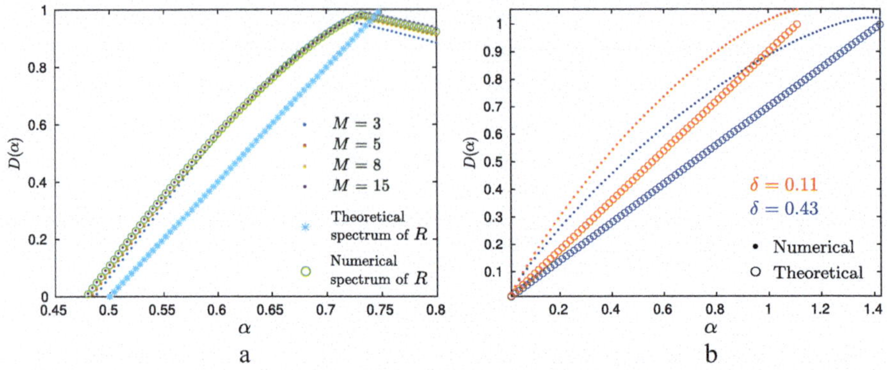

Fig. 2 The spectrum of singularities $D(\alpha)$ estimated using the *wavelet–transform modulus maxima* method for: **a** M-sided polygons R_M with different M values and Riemann's function R, and **b** H_δ with two values of δ. Clearly, up to the numerical errors, it captures well the behaviour in the part corresponding to the support of $D(\alpha)$ in both cases. See Appendix A for more details about the numerical methods

As already said, we see $H_\delta[F_D]$ as a simplification of Riemann's non-differentiable function R. Therefore, a natural question is to determine whether the multifractal formalism holds true for the non-linear trajectories R_M given in (3). This seems to be a very challenging question at the theoretical level, so, to gain insight into the subject, we computed numerically the spectrum of singularities of R_M for several values of M; see Fig. 2a. As a matter of comparision, we also computed numerically the spectrum of R (Fig. 2a) and H_δ (Fig. 2b), for which the theoretical values are known. In Appendix A we describe the methods used to compute the spectrum of singularities. Although more careful experiments are needed, they suggest that the spectrum of singularities of R_M should be equal to that of R.

We wonder whether it is possible to define H_δ replacing u_D by the solution of the non-linear Schrödinger equation, and in that case, whether the resulting H_δ and its spectrum of singularities is more amenable to theoretical studies.

Notation

We write $A \lesssim B$ if $A \leq CB$ for some constant $C > 0$; the relations $\gtrsim$ and $\simeq$ are similar. We also write $\|f\|_p = \|f\|_{L^p([0,1])}$.

The Hölder exponent is given in Definition 1.1; the Hausdorff dimension is Definition 1.2; the spectrum of singularities is Definition 1.3; and the scaling exponent is (8).

Funding

This work was supported by the Basque Government (BERC 2022–2025 program) and by the Spanish State Research Agency (Severo Ochoa SEV-2017-0718). The second author was funded by the project PID2021-122156NB-I00/AEI/10.13039/501100011033 funded by Agencia Estatal de Investigación and acronym "HAMIP" and by a Juan de la Cierva–Formation grant FJC2019-039804-I. The third author acknowledges the project PID2020-113156GB-I00, the RyC project RYC2018-025477-I, and Ikerbasque. The fourth author is supported by ERCEA Advanced Grant 2014 669689—HADE and the project PID2021-126813NB-I00 funded by MCIN/AEI/10.13039/50110001103.

2 Proof of Theorem 1.5

To prove Theorem 1.5 it is more convenient to work with $h_{\mathrm{p},\delta}$ rather than directly with H_δ. Since ψ has vanishing mean, we prefer instead $\phi(t) := \int_{-\infty}^{t} \psi(y)\,dy$. Thus, $\psi_N * H_\delta$ can be expressed as

$$\int \psi_N(t-y)H_\delta(y)\,dy = -\int \frac{d}{dy}[\phi(N(t-y))]\,H_\delta(y)\,dy = \int \phi(N(t-y))h_{\mathrm{p},\delta}(2y)\,dy. \tag{9}$$

Due to the lack of regularity of the functions involved, the integration by parts in (9) requires further justification. We can think of $h_{p,\delta}\,dr$ as a measure, and by the periodicity of it we can see that $\int_{[0,t]} |h_{p,\delta}(2r)|\,dr \le C(|t|+1)$. We want to apply Fubini to the left hand side of (9), so we have to verify that

$$\int |\psi_N(t-y)| \int_{[0,y]} |h_{p,\delta}(2r)|\,dr dy \lesssim \int |\psi_N(t-y)|(1+|y|)\,dy < \infty,$$

which holds true because of the decay properties of ψ. Now, by Fubini and $\int \psi = 0$ we can see that

$$\int \psi_N(t-y) \int_{[0,y]} h_{p,\delta}(2r)\,dr dy = \int \phi_N(t-y)h_{p,\delta}(2y)\,dy,$$

as desired.

The function ϕ has the following properties: for some $c_1, c_2 > 0$,

- $|\phi(t)| \lesssim 1$, for $|t| \le c_1$, because ψ is integrable;
- decay of the tails: for some $\beta > s$,

$$|t|^\beta\,|\phi(t)| \le c_2, \qquad \text{for } |t| \ge c_1, \tag{10}$$

  because of the decay of ψ;

- the L^p-norm is concentrated

$$\int_{|t|\le c_1} |\phi|^p \, dt \ge \frac{1}{2}\int |\phi|^p \, dt.$$

Let us denote the last integral in (9) by $P_N h_{p,\delta}[F_D]$, or simply $P_N h_{p,\delta}$, so in view of (4)

$$P_N h_{p,\delta}(2t) = \sum_{p/q} \frac{a_{q,\delta}}{q^s}\phi(N(t - p/q)), \qquad \text{for } s = 2(1+\delta),$$

where $\sum_{p/q}$ stands for the sum over all pairs of integers p, q such that $(p, q) = 1$; we will adhere to this notation from here on. There should not be confusion between the appearance of p as an integer and in the L^p norms. The only property of $a_{q,\delta}$ we will make use of is $|a_{q,\delta}| \simeq_\delta 1$. We aim to prove the next theorem, from which our main theorems follow.

Theorem 2.1 *Let* $0 < p \le \infty$, $0 < \delta < 1$ *and* $s = 2(1+\delta)$. *Then, for* $N \gg 1$,

$$\|P_N h_{p,\delta}\|_{L^p([0,1])} \simeq_\delta \begin{cases} N^{-1/p}, & \text{if } 2/s \le p \le \infty, \\ N^{-s/2}(\log N)^{s/2}, & \text{if } p = 2/s, \\ N^{-s/2}, & \text{if } 0 < p \le 2/s. \end{cases}$$

To estimate the L^p norm of $P_N h_{p,\delta}$, the idea is to split this function as

$$\begin{aligned} P_N h_{p,\delta}(2t) = & \sum_{\substack{p/q \\ q \le c_0\sqrt{N}}} \frac{a_{q,\delta}}{q^s}\phi(N(t - p/q)) + \quad (\text{main term } \boldsymbol{M}(t)) \\ & \sum_{\substack{p/q \\ q > c_0\sqrt{N}}} \frac{a_{q,\delta}}{q^s}\phi(N(t - p/q)). \quad (\text{error term } \boldsymbol{E}(t)) \end{aligned} \tag{11}$$

Here, $0 < c_0 \ll 1$ is a constant to be fixed later. In light of the inequality

$$\big| \|P_N h_{p,\delta}\|_p - \|\boldsymbol{M}\|_p \big| \le \|\boldsymbol{E}\|_p,$$

the goal is to get an estimate for $\|\boldsymbol{M}\|_p$ and a suitable upper bound of $\|\boldsymbol{E}\|_p$.

2.1 The Role of M

This subsection is philosophical in nature. We want to discuss what is the relationship between $\boldsymbol{M}$ and the spectrum of singularities of H_δ.

In [17, Theorem 4], during the proof of (7), it is actually shown that the Hölder exponent of H_δ is s/μ, for $\mu \ge 2$, exactly in the set of numbers T_μ with irrationality

μ; in the notation of Definition 1.3 we have $\Gamma_{s/\mu} = T_\mu$. Let us recall the definition of irrationality.

Definition 2.2 (*Irrationality Measure*) A number t has irrationality μ if for every $\eta < \mu$ there are infinitely many rationals p/q such that

$$0 < \left|t - \frac{p}{q}\right| < \frac{1}{q^\eta},$$

but for $\eta > \mu$ there are at most finitely many.

The set T_μ of numbers with irrationality μ has Hausdorff dimension $2/\mu$, which is consequence of Jarník's [16, Theorem 1], that is,

$$\begin{aligned} \dim W = 2/\mu \qquad &\text{and} \qquad \mathcal{H}^{2/\mu}(W) = +\infty, \\ \text{where} \quad W = \Big\{t \mid \left|t - \frac{p}{q}\right| &< \frac{1}{q^\mu} \text{ for infinitely many rationals } p/q\Big\}. \end{aligned}$$

Now if we consider the function $\boldsymbol{M}$, it is essentially supported around fractions p/q with $q \le \sqrt{N}$, for N very large; let us forget about the parameter c_0, which is introduced for technical reasons. We can decompose $\boldsymbol{M}$ in a dyadic parameter λ as

$$\boldsymbol{M}(t) = \sum_{\lambda \le \sqrt{N}} \sum_{p/q: q \simeq \lambda} \frac{a_{q,\delta}}{q^s} \phi(N(t - p/q)) =: \sum_{\lambda \le \sqrt{N}} \boldsymbol{M}_\lambda(t).$$

Hence, ignoring the tails of ϕ and assuming that $\phi(N(t - p/q))$ is supported in $p/q + I_N$, for $I_N := (-1/N, 1/N)$, we see that $\boldsymbol{M}_\lambda$ is supported in

$$V_\lambda = \bigcup_{p/q: q \simeq \lambda} (p/q + I_N),$$

which is a union of $\sum_{p/q: \lambda/2 < q \le \lambda} 1 = \sum_{\lambda/2 < n \le \lambda} \varphi(n) \simeq \lambda^2$ pairwise disjoint intervals[1]; moreover, the sets $\{V_\lambda\}_{\lambda \le \sqrt{N}}$ are morally pairwise disjoint, so $|\boldsymbol{M}_\lambda| \simeq \lambda^{-s}$ in V_λ. For each $t \in V_\lambda$ we can find a fraction p/q such that

$$\left|t - \frac{p}{q}\right| < \frac{1}{N},$$

where $q \simeq \lambda = N^{1/\mu}$, that is, $\mu := \log N / \log \lambda \ge 2$. Since $1/N \simeq 1/q^\mu$, we might see V_λ as a blurring of $T_\mu = \Gamma_{s/\mu}$ at scale $1/N$, so let us rename V_λ as "$\Gamma_{s/\mu}$" and notice that $|\text{“}\Gamma_{s/\mu}\text{”}| \simeq \lambda^2/N = N^{2/\mu - 1}$, which is what we would expect of a blurring at scale $1/N$ of a set of dimension $2/\mu$.

[1] Here $\varphi(n)$ is the Euler's totient function, namely the number of integers k, $1 \le k \le n$, such that $(n, k) = 1$.

We can compute heuristically the Hölder exponent of H_δ in "$\Gamma_{s/\mu}$" blurred at scale $1/N$, where H_δ was defined in (6). Recall that $P_N h_{\mathrm{p},\delta}[F_D]$ is by definition (9). Hence, for $t \in$ "$\Gamma_{s/\mu}$":$= V_\lambda$, we have

$$|H_\delta(t) - H_\delta(t + 1/N)| \sim |\psi_N * H_\delta(t)| = \left|P_N h_{\mathrm{p},\delta}[F_D]\right| \sim |\boldsymbol{M}(t)|\,;$$

here $\sim$ means "is like", without rigorous meaning attached to it. Since $t \in V_\lambda =$"$\Gamma_{s/\mu}$" and these sets are morally pairwise disjoint (so t belongs to a unique V_λ, or $\sum_{\lambda \le \sqrt{N}} \boldsymbol{M}_\lambda(t) \sim \sum_{\lambda \le \sqrt{N}} \boldsymbol{M}_\lambda(t) \mathbb{1}_{V_\lambda}(t)$ with V_λ pairwise disjoint), we have that

$$|H_\delta(t) - H_\delta(t + 1/N)| \sim |\boldsymbol{M}_\lambda(t)| \simeq \frac{1}{\lambda^s} = N^{-s/\mu}.$$

Hence, in "$\Gamma_{s/\mu}$" the Hölder exponent would be s/μ, which gives a heuristic proof of [17, Theorem 4] and shows the predominant role of $\boldsymbol{M}$.

We conclude that *the Frisch–Parisi formalism is reflected in the L^p norm of* $\boldsymbol{M}$; see for instance (13). The strength of the relationship between $\boldsymbol{M}$ and D_{H_δ} fades away as $\mu \to 2$, and at $\mu = 2$ the "error term" $\boldsymbol{E}$ takes on the main role.

2.2 *Proof of Theorem 2.1: The Upper Bound*

First, we bound $P_N h_{\mathrm{p},\delta}$ pointwise with a simpler function. Since ϕ decays strongly, then to control it pointwise we can tile $\mathbb{R}$ with intervals J with some suitable length $|J| = c_1$ so that

$$\phi(t) \le \sum_J b_J \mathbb{1}_J(t),$$

where b_J are coefficients decaying very fast; the tiling is so that one of the intervals, call it J_0, is centered at the origin. Hence, we can write

$$\|P_N h_{\mathrm{p},\delta}(\cdot)\|_p \lesssim \sum_J b_J \left\| \sum_{p/q} \frac{1}{q^s} \mathbb{1}_J(N(\cdot - p/q)) \right\|_p;$$

notice that by (5) we have $\left|a_{q,\delta}\right| \lesssim 1$ uniformly in q. By translation symmetry, it suffices to consider $J = J_0$.

Let $\{K\}$ be a tiling of $\mathbb{R}$ with intervals of length $4c_1/N$, one of them centered at zero, and let $\{K'\}$ be another tiling equal to $\{K\}$ but shifted by $2c_1/N$. Since every interval $I_{p/q}$ of length $2c_1/N$ and centered at p/q is contained in one interval either from $\{K\}$ or from $\{K'\}$, then

$$\sum_{p/q} \frac{1}{q^s} \mathbb{1}_{I_{p/q}}(t) \le \sum_K \sum_{p/q \in K} \frac{1}{q^s} \mathbb{1}_K(t) + \sum_{K'} \sum_{p/q \in K'} \frac{1}{q^s} \mathbb{1}_{K'}(t).$$

Let us assume that $c_0 < (2c_1)^{-1/2}$. Hence, it suffices to control the L^p norm of

$$\sum_K \sum_{p/q \in K} \frac{1}{q^s} \mathbb{1}_K = \sum_K \sum_{\substack{p/q \in K \\ q \le c_0\sqrt{N}}} \frac{1}{q^s} \mathbb{1}_K + \sum_K \sum_{\substack{p/q \in K \\ q > c_0\sqrt{N}}} \frac{1}{q^s} \mathbb{1}_K =: \boldsymbol{M} + \boldsymbol{E}.$$

Here, with abuse of notation, we choose the symbols $\boldsymbol{M}$ and $\boldsymbol{E}$ for the terms in the descomposition, like in (11), because they are morally the same.

Now we prove an upper bound for $\boldsymbol{E}$.

Lemma 2.3 *Let* $0 < p \le \infty$. *Let* $0 < \delta < 1$ *and* $s = 2(1+\delta)$. *Then, for* $N \gg 1$,

$$\|\boldsymbol{E}\|_p \lesssim_\delta \begin{cases} N^{-\frac{s}{2}+\frac{1}{2p'}}, & \text{if } 1 \le p \le \infty \\ N^{-\frac{s}{2}}, & \text{if } 0 < p < 1. \end{cases}$$

Proof We begin with the range $1 \le p \le \infty$. We first estimate the L^1 norm. On the one hand,

$$\begin{aligned} \|\boldsymbol{E}\|_1 &\lesssim_\delta \sum_K \sum_{\substack{p/q \in K \\ q > c_0\sqrt{N}}} \frac{1}{q^s} |K| \\ &\lesssim_\delta \frac{1}{N} \sum_{q > c_0\sqrt{N}} \frac{\varphi(q)}{q^s} \\ &\lesssim_\delta \frac{1}{N} \sum_{q > c_0\sqrt{N}} \frac{1}{q^{s-1}} \qquad (\text{by } \varphi(q) \le q) \\ &\lesssim_\delta \frac{1}{N} (\sqrt{N})^{-s+2} = N^{-s/2}. \end{aligned}$$

On the other hand,

$$\|\boldsymbol{E}\|_\infty \lesssim_\delta \sup_{|K| = 2c_1/N} \sum_{\substack{p/q \in K \\ q > c_0\sqrt{N}}} \frac{1}{q^s}.$$

Fix an interval K. If $kN/(2c_1) \le q < (k+1)N/(2c_1)$, then there are at most $k+1$ rationals $p/q \in K$, so

$$\begin{aligned} \sum_{\substack{p/q \in K \\ q > c_0\sqrt{N}}} \frac{1}{q^s} &\le \sum_{c_0\sqrt{N} < q \le N/(2c_1)} \frac{1}{q^s} + \sum_{k \ge 1} \sum_{k \le 2c_1 q/N < k+1} \frac{k+1}{q^s} \\ &\lesssim N^{(1-s)/2} + N^{1-s} \\ &\lesssim N^{(1-s)/2}. \end{aligned}$$

Now, interpolate the two above estimates: since $p > 1$,

$$\begin{aligned}\|\boldsymbol{E}\|_p &\lesssim \|\boldsymbol{E}\|_1^{1/p}\|\boldsymbol{E}\|_\infty^{1-1/p}\\ &\lesssim (N^{-s/2})^{1/p}(N^{\frac{1-s}{2}})^{1-1/p}\\ &= N^{-\frac{s}{2}+\frac{1-1/p}{2}} = N^{-\frac{s}{2}+\frac{1}{2p'}},\end{aligned}$$

as desired.

For the range $0 < p < 1$ we use Hölder with $r = 1/p$ so that $\|\boldsymbol{E}\|_p = \|\boldsymbol{E}\|_{L^p([0,1])} \le \|\boldsymbol{E}\|_1 \lesssim N^{-s/2}$. □

It remains to bound the main term $\boldsymbol{M}$. Notice that each interval K with $|K| = \frac{2c_1}{N}$ contains at most one rational p/q with $q \le c_0\sqrt{N}$ whenever

$$c_0 < (2c_1)^{-1/2}. \tag{12}$$

Indeed, this follows by contradiction, since $\left|\frac{p_1}{q_1} - \frac{p_2}{q_2}\right| \ge \frac{1}{q_1q_2} \ge \frac{1}{c_0^2N}$, assuming $q_1, q_2 \le c_0\sqrt{N}$. Hence, we can control the L^p norm as

$$\|\boldsymbol{M}\|_p^p = \sum_K \sum_{\substack{p/q\in K\\ q\le c_0\sqrt{N}}} \frac{1}{q^{ps}}|K| \lesssim \frac{1}{N}\sum_{\substack{p/q\\ q\le c_0\sqrt{N}}} \frac{1}{q^{ps}} \lesssim \frac{1}{N}\sum_{q\le c_0\sqrt{N}} \frac{1}{q^{ps-1}},$$

where we used $\varphi(q) \le q$. We compute the last sum to get

$$\|\boldsymbol{M}\|_p \lesssim \begin{cases} N^{-1/p}, & \text{if } p > 2/s,\\ N^{-s/2}(\log N)^{s/2}, & \text{if } p = 2/s,\\ N^{-s/2}, & \text{if } p < 2/s.\end{cases} \tag{13}$$

Inequality (13) and Lemma 2.3 imply the upper bound in Theorem 2.1.

2.3 *Proof of Theorem 2.1: The Lower Bound*

Before getting to the proof of the lower bound, we need a quite technical lemma that says that the tail of ϕ can be safely ignored.

Recall that for $q \le c_0\sqrt{N}$ with $c_0 \ll 1$, we denote by $I_{p/q}$ the interval of length $2c_1/N$ centered at p/q; we assume that $c_0 < (2c_1)^{-1/2}$ to ensure that the intervals are disjoint. We pick a rational p_0/q_0 and define the error function

$$e(t) = \sum_{p/q\notin 2I_{p_0/q_0}} \frac{a_{q,\delta}}{q^s}\phi(N(t-p/q)), \qquad t \in I_{p_0/q_0}. \tag{14}$$

We will show that e, which is the sum of the tails in I_{p_0/q_0}, is small. The reader can skip the next lemma under the assumption $\operatorname{supp}\phi \subset [-c_1, c_1]$.

Lemma 2.4 (Tails are negligible) *Let $s > 2$ and $q_0 \leq c_0\sqrt{N}$ for $c_0 \ll 1$. Then,*

$$|e(t)| \lesssim \frac{q_0^{s-2}}{N^{s-1}}, \qquad \textit{for } t \in I_{p_0/q_0}.$$

Proof Recall from (10) that $|\phi(t)| \leq c_2 |t|^{-\beta}$ for $|t| \geq c_1$. Since $|t - p/q| \geq c_1/N$ then

$$|e(t)| \leq C\frac{c_2}{N^\beta} \sum_{\substack{p/q \\ p/q \notin 2I_{p_0/q_0}}} \frac{1}{q^s} \frac{1}{|t - p/q|^\beta}. \tag{15}$$

Moreover, we can write $t = p_0/q_0 + \epsilon$, with $|\epsilon| \leq c_1/N$, then

$$\left|t - \frac{p}{q}\right| \geq \left|\frac{p_0}{q_0} - \frac{p}{q}\right| - \frac{c_1}{N} \geq \frac{1}{2}\left|\frac{p_0}{q_0} - \frac{p}{q}\right|.$$

We replace the above in (15) so that

$$|e(t)| \leq C\frac{q_0^\beta}{N^\beta} \sum_{\substack{p/q \\ p/q \notin 2I_{p_0/q_0}}} \frac{q^{\beta-s}}{|qp_0 - q_0 p|^\beta}.$$

Since $p/q \notin 2I_{p_0/q_0}$ is the same as

$$|qp_0 - q_0 p| \geq 2c_1\frac{q_0 q}{N},$$

then it is sensible to break the sum above as

$$|e(t)| \leq C\frac{q_0^\beta}{N^\beta} \sum_{k \geq 0} \sum_{\{q\,:\,k < 2c_1 q_0 q/N \leq k+1\}} q^{\beta-s} \sum_{\{p\,:\,|qp_0 - q_0 p| > k\}} \frac{1}{|qp_0 - q_0 p|^\beta}. \tag{16}$$

To estimate the very last sum in (16), let $\overline{a}$ denote a residue of a mod q_0 such that $|\overline{a}| \leq q_0/2$, so we can rewrite this sum as

$$\sum_{\{p\,:\,|qp_0 - q_0 p| > k\}} \frac{1}{|qp_0 - pq_0|^\beta} = \sum_{\{l\,:\,|\overline{qp_0} + lq_0| > k\}} \frac{1}{\left|\overline{qp_0} + lq_0\right|^\beta}.$$

Notice that $\overline{qp_0} + lq_0 \in lq_0 + [-q_0/2, q_0/2]$, so if $l \neq 0$ then $\left|\overline{qp_0} + lq_0\right| \geq q_0\,||l| - 1/2| \geq q_0\,|l|\,/2$. Now, we break the analysis of the last sum into three cases:

- If $\left|\overline{qp_0}\right| > k$ then $\{l : \left|\overline{qp_0} + lq_0\right| > k\} = \mathbb{Z}$. Hence, assuming that $\beta > 1$, we have

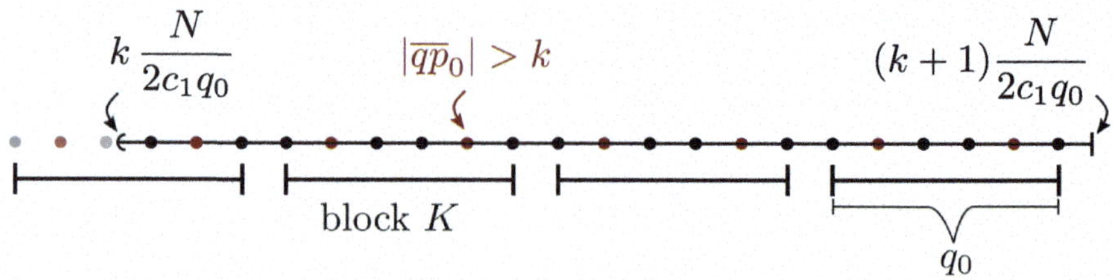

Fig. 3 Blocks K of length q_0

$$\sum_{\{l:|\overline{q}p_0+lq_0|>k\}} \frac{1}{|\overline{q}p_0+lq_0|^\beta} = \frac{1}{|\overline{q}p_0|^\beta} + \frac{2}{q_0^\beta}\sum_{l\neq 0}\frac{1}{|l|^\beta} \lesssim \frac{1}{|\overline{q}p_0|^\beta};$$

we used that $q_0 \geq 2\,|\overline{q}p_0|$.

- If $|\overline{q}p_0| \leq k < q_0/2$ then $\{l \,:\, |\overline{q}p_0 + lq_0| > k\} = \mathbb{Z}\setminus\{0\}$ and we have

$$\sum_{\{l:|\overline{q}p_0+lq_0|>k\}} \frac{1}{|\overline{q}p_0+lq_0|^\beta} \leq \frac{2}{q_0^\beta}\sum_{l\neq 0}\frac{1}{|l|^\beta} \lesssim \frac{1}{q_0^\beta}.$$

- If $|\overline{q}p_0| \leq q_0/2 \leq k$ then $\{l \,:\, |\overline{q}p_0 + lq_0| > k\} \subseteq \{l \,:\, |l| > k/q_0 - 1/2 \geq 0\}$ and

$$\sum_{\{l:|\overline{q}p_0+lq_0|>k\}} \frac{1}{|\overline{q}p_0+lq_0|^\beta} \leq \frac{2}{q_0^\beta}\sum_{|l|>k/q_0-1/2}\frac{1}{|l|^\beta} \lesssim \frac{1}{q_0 k^{\beta-1}}.$$

Now we have to compute the contribution of each case to the sum in (16).

The contribution of the case $|\overline{q}p_0| > k$ is less than

$$A_1 := \frac{q_0^\beta}{N^\beta}\sum_{0\leq k<q_0/2}\ \sum_{\{q:k<2c_1q_0q/N\leq k+1\}} \mathbb{1}_{\{|\overline{q}p_0|>k\}}(q)\frac{q^{\beta-s}}{|\overline{q}p_0|^\beta}. \tag{17}$$

The last sum in q runs over an interval of length $N/(2c_1q_0) \geq q_0$ if $c_0 \ll 1$ (recall that $q_0 \leq c_0\sqrt{N}$), so we can break it into a number $\simeq N/(2c_1q_0^2)$ of blocks K of length q_0 so that

$$\sum_{\{q:k<2c_1q_0q/N\leq k+1\}} \mathbb{1}_{\{|\overline{q}p_0|>k\}}(q)\frac{q^{\beta-s}}{|\overline{q}p_0|^\beta} \leq \sum_K\sum_{q\in K}\mathbb{1}_{\{|\overline{q}p_0|>k\}}(q)\frac{q^{\beta-s}}{|\overline{q}p_0|^\beta};$$

see Fig. 3.

Since $(p_0, q_0) = 1$, then $\overline{q}p_0$ runs over all residues r mod q_0, so we may bound this case as

$$\sum_{\{q:k<2c_1q_0q/N\le k+1\}} \mathbb{1}_{\{|\overline{qp_0}|>k\}}(q)\frac{q^{\beta-s}}{|\overline{qp_0}|^{\beta}} \lesssim \left((k+1)\frac{N}{c_1q_0}\right)^{\beta-s}\sum_K\sum_{\substack{r\in K\\|r|>k}}\frac{1}{|r|^{\beta}}$$
$$\lesssim \frac{1}{q_0(k+1)^{s-1}}\left(\frac{N}{c_1q_0}\right)^{\beta+1-s}.$$

We replace it in (17) so that

$$A_1 \lesssim \frac{q_0^{\beta}}{N^{\beta}}\sum_{0\le k<q_0/2}\frac{1}{q_0(k+1)^{s-1}}\left(\frac{N}{c_1q_0}\right)^{\beta+1-s} \lesssim \frac{q_0^{s-2}}{N^{s-1}}. \tag{18}$$

The contribution of the case $|\overline{qp_0}| \le k < q_0/2$ is less than

$$A_2 := \frac{q_0^{\beta}}{N^{\beta}}\sum_{0\le k<q_0/2}\sum_{\{q:k<2c_1q_0q/N\le k+1\}}\mathbb{1}_{\{|\overline{qp_0}|\le k\}}(q)\frac{q^{\beta-s}}{q_0^{\beta}}$$
$$\lesssim \frac{1}{N^{\beta}}\sum_{0\le k<q_0/2}\left[(k+1)\frac{N}{q_0}\right]^{\beta-s}\sum_{\{q:k<2c_1q_0q/N\le k+1\}}\mathbb{1}_{\{|\overline{qp_0}|\le k\}}(q).$$

We estimate the last sum in q as before, breaking the sum into blocks of length q_0, so that

$$A_2 \lesssim \frac{1}{N^s q_0^{\beta-s}}\sum_{0\le k<q_0/2}(k+1)^{\beta-s}\sum_K\sum_{\substack{r\in K\\|r|\le k}}\mathbb{1}_{\{|r|\le k\}}(r)$$
$$\lesssim \frac{1}{N^s q_0^{\beta-s}}\sum_{0\le k<q_0/2}(k+1)^{\beta-s}(k+1)\frac{N}{2c_1q_0^2}$$
$$\lesssim \frac{1}{N^{s-1}}. \tag{19}$$

The contribution of the case $k \ge q_0/2$ is less than

$$A_3 := \frac{q_0^{\beta}}{N^{\beta}}\sum_{k\ge q_0/2}\sum_{\{q:k<2c_1q_0q/N\le k+1\}}\frac{q^{\beta-s}}{q_0k^{\beta-1}}$$
$$\lesssim \frac{q_0^{\beta}}{N^{\beta}}\cdot\frac{N^{\beta+1-s}}{q_0^{\beta+2-s}}\sum_{k\ge q_0/2}\frac{1}{k^{s-1}}$$
$$\lesssim \frac{1}{N^{s-1}}. \tag{20}$$

We sum up all the contributions (18), (19) and (20) to the error term (16) to find out

$$|e(t)| \lesssim A_1 + A_2 + A_3 \lesssim \frac{q_0^{s-2}}{N^{s-1}}, \qquad \text{for } t \in I_{p_0/q_0},$$

which is what we wanted. □

After this lemma, for each p_0/q_0 with $q_0 \leq c_0\sqrt{N}$, we can split $P_N h_{\mathrm{p},\delta}$ as

$$\begin{aligned} P_N h_{\mathrm{p},\delta}(2t) = & \sum_{\substack{p/q \in 2I_{p_0/q_0} \\ q \leq c_0\sqrt{N}}} \frac{a_{q,\delta}}{q^s}\phi(N(t-p/q)) \\ & + \sum_{\substack{p/q \in 2I_{p_0/q_0} \\ q > c_0\sqrt{N}}} \frac{a_{q,\delta}}{q^s}\phi(N(t-p/q)) + e(t), \qquad \text{if } t \in I_{p_0/q_0}. \end{aligned}$$

Since the only fraction $p/q \in 2I_{p_0/q_0}$ with $q \leq c_0\sqrt{N}$, for $c_0 \ll 1$, is p_0/q_0 itself, then we can write this decomposition as

$$P_N h_{\mathrm{p},\delta} = \boldsymbol{M} + \boldsymbol{E} + e \tag{21}$$

where

$$\begin{aligned} \boldsymbol{M}(t) &= \frac{a_{q_0,\delta}}{q_0^s}\phi(N(t-p_0/q_0)), \qquad t \in I_{p_0/q_0}, \\ \boldsymbol{E}(t) &= \sum_{\substack{p/q \in 2I_{p_0/q_0} \\ q > c_0\sqrt{N}}} \frac{a_{q,\delta}}{q^s}\phi(N(t-p/q)), \qquad t \in I_{p_0/q_0}, \end{aligned}$$

and e is error due to the tails in (14).

2.3.1 The Range $p > 2/s$

The peak of $P_N h_{\mathrm{p},\delta}$ at zero is the largest one (see Fig. 1), and to estimate the L^p norm when $p > 2/s$ it suffices to integrate over the interval I_0, that is, around the fraction $p_0/q_0 = 0$.

Let us start with $\boldsymbol{M}$. According to (21) $\boldsymbol{M}(t) = a_{1,\delta}\phi(Nt)$ for $t \in I_0$, so

$$\|\boldsymbol{M}\|_{L^p(I_0)} \gtrsim N^{-1/p}.$$

The L^p norm of $\boldsymbol{E}$ is

$$\|\boldsymbol{E}\|_{L^p(I_0)} \lesssim \frac{1}{N^{1/p}} \sum_{\substack{p/q \in 2I_0 \\ q > c_0\sqrt{N}}} \frac{1}{q^s}.$$

Since $|p/q| \leq 2c_1/N$ then necessarily $q \geq N/(2c_1) > c_0\sqrt{N}$, and the number of fractions with denominator q in $2I_0$ is $\leq 2c_1 q/N$, so

$$\|\boldsymbol{E}\|_{L^p(I_0)} \lesssim \frac{1}{N^{1/p}} \sum_{q \geq N/(2c_1)} \frac{1}{Nq^{s-1}} \lesssim N^{-1/p-s+1}.$$

By Lemma 2.4 we have

$$\|e\|_{L^p(I_0)} \lesssim N^{-1/p-s+1}.$$

This leads us to the conclusion

$$\|P_N h_{\mathrm{p},\delta}\|_p \geq \|P_N h_{\mathrm{p},\delta}\|_{L^p(I_0)} \gtrsim N^{-1/p},$$

which proves the lower bound in Theorem 2.1 in the range $p > 2/s$.

2.3.2 The Range $0 < p \leq 2/s$

The lower bound will be estimated by integrating $P_N h_{\mathrm{p},\delta}$ over

$$U := \bigcup_{\substack{p/q \\ q \leq c_0\sqrt{N}}} I_{p/q}. \tag{22}$$

Since the intervals are pairwise disjoint (see (12))

$$|U| = \sum_{\substack{p/q \\ q \leq c_0\sqrt{N}}} |I_{p/q}| \leq \frac{2c_1}{N} \sum_{\substack{p/q \\ q \leq c_0\sqrt{N}}} 1 \leq \frac{2c_1}{N} \sum_{q \leq c_0\sqrt{N}} \varphi(q) \leq \frac{2c_1}{N} \sum_{q \leq c_0\sqrt{N}} q \leq 2c_0^2 c_1. \tag{23}$$

By Lemma 2.4 the L^p norm of e is small

$$\begin{aligned} \|e\|^p_{L^p(U)} &\lesssim \sum_{\substack{p/q \\ q \leq c_0\sqrt{N}}} \frac{q^{p(s-2)}}{N^{p(s-1)}} \left|I_{p/q}\right| \\ &\lesssim \frac{1}{N^{p(s-1)+1}} \sum_{q \leq c_0\sqrt{N}} q^{p(s-2)+1} \lesssim c_0^{p(s-2)+2} N^{-ps/2}. \end{aligned} \tag{24}$$

In the decomposition (21), the L^p norm of the main term is

$$\begin{aligned} \|\boldsymbol{M}\|^p_{L^p(U)} &\simeq \sum_{\substack{p/q \\ q \leq c_0\sqrt{N}}} \int_{t \in I_{p/q}} \frac{1}{q^{ps}} \left|\phi(N(t - p/q))\right|^p \, dt \\ &\simeq \frac{1}{N} \sum_{q \leq c_0\sqrt{N}} \frac{\varphi(q)}{q^{ps}}, \end{aligned} \tag{25}$$

by using the properties of ϕ. We estimate the last sum in the next lemma.

Lemma 2.5 *Let* $0 < \beta \le 2$ *and* $M \gg 1$. *Then,*

$$\sum_{1\le q\le M} \frac{\varphi(q)}{q^\beta} \gtrsim \begin{cases} \log M & \text{if } \beta = 2, \\ M^{2-\beta} & \text{if } 0 < \beta < 2. \end{cases} \tag{26}$$

Proof Recall the identity

$$\varphi(q) = q \sum_{d|q} \frac{\mu(d)}{d}, \tag{27}$$

where μ is the Möbius function; see [10, Sect. 16.3] for the definition. The only property of μ we need is that $\mu(n) \in \{-1, 0, 1\}$.

Let $M_0 \gg 1$ and replace (27) into the left-hand side of (26) so that

$$\begin{aligned} \sum_{1\le q\le M} \frac{\varphi(q)}{q^\beta} &\ge \sum_{M_0\le q\le M} \frac{\varphi(q)}{q^\beta} \\ &\simeq \sum_{M_0\le m\le M} \frac{1}{m^{\beta-1}} \sum_{m/2\le q\le m} \sum_{d|q} \frac{\mu(d)}{d}, \end{aligned} \tag{28}$$

where $m \in 2^{\mathbb{N}}$. For each dyadic block we have

$$\begin{aligned} \sum_{m/2\le q\le m} \sum_{d|q} \frac{\mu(d)}{d} &= \sum_{d\ge 1} \frac{\mu(d)}{d} \sum_{m/2\le q\le m} \mathbb{1}_{d|q}(q) \\ &= \sum_{1\le d\le m} \frac{\mu(d)}{d} \sum_{m/(2d)\le k\le m/d} 1 \\ &= \sum_{1\le d\le m} \frac{\mu(d)}{d} \Big(\frac{m}{2d}\Big) + \sum_{1\le d\le m} \frac{\mu(d)}{d} \Big(-\frac{m}{2d} + \sum_{m/(2d)\le k\le m/d} 1\Big). \end{aligned}$$

Since $|m/(2d) - |\{k \mid m/(2d) \le k \le m/d\}|| \lesssim 1$ and $|\mu(d)| \le 1$, then

$$\begin{aligned} \sum_{m/2\le q\le m} \sum_{d|q} \frac{\mu(d)}{d} &= \frac{m}{2} \sum_{1\le d\le m} \frac{\mu(d)}{d^2} + \mathcal{O}(\log m) \\ &= \frac{m}{2\zeta(2)} - \frac{m}{2} \sum_{d>m} \frac{\mu(d)}{d^2} + \mathcal{O}(\log m) \\ &= \frac{m}{2\zeta(2)} + \mathcal{O}(\log m). \end{aligned}$$

Here, we used the identity $\sum_{d\ge 1} \mu(d)/d^2 = 1/\zeta(2)$, where ζ is the Riemann zeta function, see [10, Theorem 287].

Going back to (28), for $M_0 \gg 1$ we get

$$\sum_{1\le q\le M} \frac{\varphi(q)}{q^\beta} \gtrsim \sum_{M_0\le m\le M} \frac{1}{m^{\beta-2}},$$

which yields (26). □

We apply Lemma 2.5 to (25) with $c_0 \ll 1$ to find out that for $N \gg 1$ we have

$$\|\boldsymbol{M}\|_p \gtrsim \begin{cases} N^{-s/2}(\log c_0 N)^{1/p}, & \text{if } p = 2/s \\ c_0^{2/p-s} N^{-s/2}, & \text{if } p < 2/s. \end{cases} \tag{29}$$

It remains to bound the error term

$$\boldsymbol{E}(t) = \sum_{\substack{p/q\in 2I_{p_0/q_0} \\ q>c_0\sqrt{N}}} \frac{a_{q,\delta}}{q^s}\phi(N(t-p/q)), \qquad \text{if } t \in I_{p_0/q_0}. \tag{30}$$

To compute the L^p norm we use Hölder with exponent $r = 1/p$ so that, in view of (23),

$$\|\boldsymbol{E}\|_{L^p(U)} \le |U|^{1/(pr')}\|\boldsymbol{E}\|_{L^1(U)} \lesssim c_0^{2/p-2}\|\boldsymbol{E}\|_{L^1(U)}.$$

The L^1 norm is

$$\|\boldsymbol{E}\|_{L^1(U)} \lesssim \sum_{\substack{p'/q' \\ q'\le c_0\sqrt{N}}} \sum_{\substack{p/q\in 2I_{p'/q'} \\ q>c_0\sqrt{N}}} \frac{1}{q^s}|I_{p'/q'}| \tag{31}$$

$$\lesssim \frac{1}{N}\sum_{\substack{p/q \\ q>c_0\sqrt{N}}} \frac{1}{q^s} \sum_{\substack{p'/q' \\ q'\le c_0\sqrt{N}}} \mathbb{1}_{\{|p/q-p'/q'|\le 2c_1/N\}}(p'/q').$$

The condition $q' \le c_0\sqrt{N}$ forces the last sum over p'/q' to be at most one. To see this, suppose that there are two distinct fractions p_1'/q_1' and p_2'/q_2' such that $|p/q - p_i'/q_i'| \le 2c_1/N$, then $1/(q_1'q_2') \le |p_1'/q_2' - p_1'/q_2'| \le 4c_1/N$. On the other hand, the condition $q' \le c_0\sqrt{N}$ implies $1/(q_1'q_2') \ge 1/(c_0^2 N)$, so $c_0 \ge 1/(2c_1^{1/2})$ which contradicts (12).

In (31), if the last sum over p'/q' equals one, then notice that $p'/q' \ne p/q$ because $q' \le c_0\sqrt{N}$ but $q > c_0\sqrt{N}$. Then, $1/(qq') \le |p/q - p'/q'| \le 2c_1/N$ and the condition $q' \le c_0\sqrt{N}$ allows us to conclude that $q \ge N/(2c_1q') \ge \sqrt{N}/(2c_0c_1)$. Thus, we bound $\|\boldsymbol{E}\|_{L^1(U)}$ as

$$\|\boldsymbol{E}\|_{L^1(U)} \lesssim \frac{1}{N}\sum_{\substack{p/q \\ q\ge\sqrt{N}/(2c_0c_1)}} \frac{1}{q^s} \lesssim \frac{1}{N}\sum_{q\ge\sqrt{N}/(2c_0c_1)} \frac{\varphi(q)}{q^s} \lesssim c_0^{s-2}N^{-s/2}. \tag{32}$$

In Appendix B we prove a better upper bound, but for our present purposes this is enough. We get thus

$$\|\boldsymbol{E}\|_{L^p(U)} \lesssim c_0^{2/p+s-4} N^{-s/2}. \tag{33}$$

We can now conclude the lower bound. We have

$$\|P_N h_{\mathrm{p},\delta}\|_{L^p(U)} \geq \|\boldsymbol{M}\|_{L^p(U)} - \|\boldsymbol{E}\|_{L^p(U)} - \|e\|_{L^p(U)},$$

so, when $p < 2/s$, from (29), (33) and (24) we get

$$\begin{aligned}\|P_N h_{\mathrm{p},\delta}\|_p &\geq \|P_N h_{\mathrm{p},\delta}\|_{L^p(U)} \\ &\gtrsim c_0^{2/p}(c_0^{-s} - Cc_0^s - Cc_0^{s-4})N^{-s/2} \gtrsim N^{-s/2} \quad (\text{by } s > 2).\end{aligned}$$

For the critical exponent $p = 2/s$ we get (observe that the only logarithmic term below comes from $\boldsymbol{M}$)

$$\|P_N h_{\mathrm{p},\delta}\|_p \geq \|P_N h_{\mathrm{p},\delta}\|_{L^p(U)} \gtrsim N^{-s/2}(\log c_0 N)^{s/2},$$

which concludes the proof of Theorem 2.1 when $0 < p \leq 2/s$.

Acknowledgements We thank the referee for pointing out several inaccuracies and helping us to greatly improve the exposition of our work.

Appendix A. Numerical Simulations

For a given signal/function, we calculate its spectrum of singularities $D(\alpha)$ numerically using the *wavelet transform modulus maxima (WTMM) method* implemented in MATLAB using the Wavelab 850 toolbox [5]. With a suitable choice of a wavelet, through the wavelet coefficients, we compute the partition function, scaling exponent $\eta(p)$ and thus, $D(\alpha)$ is estimated using the Legendre transform (see [21] for their precise definition as they are different from the ones mentioned earlier). The input parameters consist of the signal X with length $N = 2^J$, the number of scales, range of parameters p and α. Thus, for $X = H_\delta$ in Fig. 2b, we choose $J = 13$, $\delta = 0.11, 0.43$, $p \in [-5, 5]$, $\alpha \in [\alpha_{\min}, \alpha_{\max}]$, where $\alpha_{\min} = 0$, $\alpha_{\max} = 1/(1+\delta)$, that is, the support of H_δ, and the wavelet used is the first derivative of a gaussian. To further compare them quantitatively, we calculate the error as defined in [21, (32)] and obtain the values 0.0956 and 0.1166 for $\delta = 0.43$ and $\delta = 0.11$, respectively. We notice that these results are indeed comparable with the ones obtained in [21] and can be reduced further with a larger N.

Next, we estimate $D(\alpha)$ for R and R_M in the context of vortex filament equation. More precisely, we consider the input signal X as the trajectory of the third component (without the vertical height) of the M-sided filament curve [6]. With

$p \in [-5, 5]$, $\alpha_{\min} = 0.4$ and $\alpha_{\max} = 0.8$, we plot the Riemann's function R and R_M, for $M = 3, 5, 8, 15$, in Fig. 2a. The plots show the estimated values of $D(\alpha)$ where its maximum value varies with M and converges to that of R (circled points). Indeed, for $M = 15$ the agreement is remarkable and deviations from the theoretical values (starred points), including the values for $\alpha > 0.75$, are a result of a numerical error which is the minimum when the wavelet chosen is the second derivative of a gaussian. In each case, the left part of $D(\alpha)$ lies in an interval with a length very close to 0.25, which is the same as the measure of the support in the theoretical case while the right part is erroneous which is related to one of the limitations of the WTMM method (see [21, Sects. 5.2, 8]).

Appendix B. Counting Rationals

In the next proposition we improve the upper bound $\|\boldsymbol{E}\|_{L^1(U)} \lesssim c_0^{s-2} N^{-s/2}$ we proved in (32).

Proposition 2.6 *Let U be the set* (22) *and $\boldsymbol{E}$ the function* (30). *Then,*

$$\|\boldsymbol{E}\|_{L^1(U)} \lesssim c_0^s N^{-s/2}.$$

Proof The L^1 norm is

$$\|\boldsymbol{E}\|_{L^1(U)} \lesssim \frac{1}{N} \sum_{\substack{p/q \\ q > c_0\sqrt{N}}} \sum_{\substack{p'/q' \\ q' \le c_0\sqrt{N}}} \frac{1}{q^s} \mathbb{1}_{\{|p/q - p'/q'| \le 2c_1/N\}}(p/q, p'/q').$$

Now we break the ranges $q > c_0\sqrt{N}$ and $q' \le c_0\sqrt{N}$ dyadically into parameters $\lambda, \mu \in 2^{\mathbb{N}}$, respectively, so that

$$\|\boldsymbol{E}\|_{L^1(U)} \lesssim \frac{1}{N} \sum_{\substack{\lambda < c_0\sqrt{N} \\ \mu \ge c_0\sqrt{N}}} \mu^{-s} \sum_{\substack{p/q,\ p'/q' \\ q \simeq \lambda,\ q' \simeq \mu}} \mathbb{1}_{\{|p/q - p'/q'| \le 2c_1/N\}}(p/q, p'/q').$$

Equivalently, we have to count at most how many pairs of rationals $(p/q, p'/q')$ satisfy

$$0 < \left|q'p - qp'\right| \le 2c_1 \frac{\lambda\mu}{N}; \tag{34}$$

for that, we use the arguments in [1, Proposition 4.2].

For $m \in \mathbb{Z} \setminus \{0\}$, the goal is to count how many representations has m as $m = q'p - qp'$ with $0 \le p < q$ and $0 \le p' < q'$, so let us fix q and q'. If $(q, q') = d$ then necessarily $m = d\tilde{m}$, so let us clear out d from the representation of m and write $\tilde{m} = \tilde{q}'p - \tilde{q}p'$, where $q' = d\tilde{q}'$, $q = d\tilde{q}$ and $(\tilde{q}, \tilde{q}') = 1$. Now assume that

$\tilde{q}'p - \tilde{q}p' = \tilde{q}'r - \tilde{q}r'$, or after reordering, $\tilde{q}'(p-r) = \tilde{q}(p'-r')$. This implies that $\tilde{q} \mid (p-r)$ so, for some $l \in \mathbb{Z}$, we have $0 \le r = p + l\tilde{q} < q$ (recall that $0 \le p < q$ and $q = d\tilde{q}$) and then the number of different r's is at most d. In conclusion, for every m divisible by d there are d representations $m = q'p - qp'$ with $0 \le p < q$ and $0 \le p' < q'$.

The above paragraph shows that, for fixed q and q', the number of choice of pairs (p, p') satisfying (34) is $\le 2c_1\lambda\mu/N$, so the total number of fractions satisfying (34) is $\lesssim 2c_1\lambda^2\mu^2/N$. Notice that the collection of representations is empty unless $\lambda\mu \ge N/(2c_1)$, so

$$\begin{aligned}
\|\boldsymbol{E}\|_{L^1(U)} &\lesssim \frac{1}{N^2} \sum_{\substack{\lambda < c_0\sqrt{N} \\ \mu \ge c_0\sqrt{N}}} \mu^{2-s}\lambda^2 \mathbb{1}_{\{\lambda\mu \ge N/(2c_1)\}}(\lambda, \mu) \\
&= \frac{1}{N^2} \sum_{\lambda < c_0\sqrt{N}} \lambda^2 \sum_{\mu \ge N/(2c_1\lambda)} \mu^{2-s} \qquad (\text{by } N/(2c_1\lambda) \ge c_0\sqrt{N} \text{ for } c_0 \ll 1) \\
&\lesssim \frac{1}{N^s} \sum_{\lambda < c_0\sqrt{N}} \lambda^s \lesssim c_0^s N^{-s/2},
\end{aligned}$$

where we used that μ and λ are dyadic. The proof is completed. □

References

1. C. An, R. Chu, L.B. Pierce, Counterexamples for high-degree generalizations of the Schrödinger maximal operator. Int. Math. Res. Not. (2022). rnac088
2. V. Banica, L. Vega, Riemann's non-differentiable function and the binormal curvature flow. Arch. Rat. Mech. Anal. **244**(2), 501–540 (2022)
3. M.V. Berry, S. Klein, Integer, fractional and fractal Talbot effects. J. Mod. Opt. **43**(10), 2139–2164 (1996)
4. A. Boritchev, D. Eceizabarrena, V.V. Da Rocha, Intermittency of Riemann's non-differentiable function through the fourth-order flatness. J. Math. Phys. **62**(9), 14 (2021). Paper No. 093101
5. J. Buckheit, S. Chen, D. Donoho, I. Johnstone, J. Scargle, About wavelab. Handbook of WaveLab Version **850**, 1–37 (1995)
6. F. de la Hoz, L. Vega, Vortex filament equation for a regular polygon. Nonlinearity **27**(12), 3031–3057 (2014)
7. D. Eceizabarrena, The Talbot effect as the fundamental solution to the free Schrödinger equation. Port. Math. **78**(2), 233–253 (2021)
8. M.B. Erdoğan, N. Tzirakis, *Dispersive Partial Differential Equations. Wellposedness and Applications*, vol. 86 (Cambridge University Press, Cambridge, 2016)
9. U. Frisch, G. Parisi, *Fully developed turbulence and intermittency, in Proc* (Enrico Fermi International Summer School in Physics, North-Holland, Amsterdam, 1985)
10. G.H. Hardy, E.M. Wright, *An Introduction to the Theory of Numbers*, 6th edn. Edited and revised by D.R. Heath-Brown, J.H. Silverman. With a foreword by Andrew Wiles (Oxford University Press, Oxford, 2008)
11. H. Hasimoto, A soliton on a vortex filament. J. Fluid Mech. **51**(3), 477–485 (1972)
12. S. Jaffard, The spectrum of singularities of Riemann's function. Rev. Mat. Iberoam. **12**(2), 441–460 (1996)

13. S. Jaffard, Some open problems about multifractal functions, in*Fractals in Engineering: from Theory to Industrial Applications* (Springer, 1997)
14. S. Jaffard, Multifractal formalism for functions. I. Results valid for all functions. SIAM J. Math. Anal. **28**(4), 944–970 (1997)
15. S. Jaffard, The multifractal nature of Lévy processes. Probab. Theory Relat. Fields **114**(2), 207–227 (1999)
16. V. Jarník, Über die simultanen diophantischen Approximationen. Math. Z. **33**(1), 505–543 (1931)
17. S. Kumar, F. Ponce Vanegas, L. Vega, Static and dynamical, fractional uncertainty principles. Trans. Am. Math. Soc. **375**(8), 5691–5725 (2022)
18. K.I. Oskolkov, M.A. Chakhkiev, On the "nondifferentiable" Riemann function and the Schrödinger equation. Tr. Mat. Inst. Steklova **269**, Teoriya Funktsiĭ i Differentsial' nye Uravneniya, 193–203 (2010)
19. I. Rodnianski, Fractal solutions of the Schrödinger equation, in *Nonlinear PDE's, Dynamics and Continuum Physics* (South Hadley, MA, vol. 255 of Contemp. Math. Amer. Math. Soc. Providence, RI) **2000**, 181–187 (1998)
20. M. Taylor, The Schrödinger equation on spheres. Pacific J. Math. **209**(1), 145–155 (2003)
21. A. Turiel, C.J. Pérez-Vicente, J. Grazzini, Numerical methods for the estimation of multifractal singularity spectra on sampled data: a comparative study. J. Comput. Phys. **216**(1), 362–390 (2006)

Recent Developments in Spectral Theory for Non-self-adjoint Hamiltonians

Lucrezia Cossetti, Luca Fanelli, and Nico M. Schiavone

Abstract The objective of this survey is to collect and elaborate on different tools, both well-established and more recent ones, which have been developed in the last decades to investigate spectral properties of non-self-adjoint operators of the form $H = H_0 + V$. More specifically, we will show how Hardy-type and Sobolev inequalities, together with Virial theorems and Birman-Schwinger principles enter into play in the analysis of the spectrum of these Hamiltonians.

Keywords Spectrum · Self-adjoint operators · Non self-adjoint operators · Hardy-type inequalities · Mourre theory · Birman-Schwinger principle

2020 Mathematical Subject Classification 47A55 · 47A25 · 81Q15

Dedicated to Prof. Tohru Ozawa for his birthday, with deep gratitude for all he has done and sincere friendship.

L. Cossetti
Fakultät Für Mathematik, Institut Für Analysis, Karlsruher Institut Für Technologie (KIT), Englerstraße 2, 76131 Karlsruhe, Germany
e-mail: lucrezia.cossetti@kit.edu

L. Fanelli (✉)
Ikerbasque & Departamento de Matematicas, Universidad del País Vasco/Euskal Herriko Unibertsitatea (UPV/EHU), Aptdo. 644, 48080 Bilbao, Spain
e-mail: luca.fanelli@ehu.es

N. M. Schiavone
Department of Mathematics, Faculty of Nuclear Sciences and Physical Engineering, Czech Technical University in Prague, Trojanova 13, 12000 Prague 2, Czechia
e-mail: schianic@fjfi.cvut.cz

S. Machihara (ed.), *Mathematical Physics and Its Interactions*, Springer Proceedings in Mathematics & Statistics 451, https://doi.org/10.1007/978-981-97-0364-7_8

1 Introduction

In this survey we consider the following Hamiltonian

$$H := H_0 + V.$$

This can be seen as a perturbation through the operator V of a reference operator H_0. Hereafter, unless explicitly mentioned, H_0 will represent a *general* self-adjoint operator in $L^2(\mathbb{R}^d)$, whereas V will denote a potential, i.e. a multiplication operator by a generating function $V : \mathbb{R}^d \to \mathbb{C}$. Notice that as V is possibly-complex valued, H does not need to be self-adjoint.[1]

If one is interested in the spectrum of the perturbed Hamiltonian H, then the following natural questions arise:

1. *Under which perturbations V the spectral properties of the free operator H_0 are preserved?*

Or, from a different point of view:

2. *How and to what extend spectral properties of H deviate from the ones of H_0?*

In this paper, without attempting to be complete, we collect several techniques that have been developed in the last decades to give a satisfactory answer to the questions posed above, both in a self-adjoint and non-self-adjoint context. As we will see in details below, some of the tools available in the self-adjoint framework do not carry over to the non-self-adjoint setting. This has required developing entirely new methods suitably adapted to the non-self-adjoint context.

To fix the notation we start recalling the following basic definitions. The spectrum $\sigma(H)$ of a closed linear operator H with dense domain in a Hilbert space $\mathfrak{H}$ is the set of complex numbers z for which the operator $H - z\colon \operatorname{dom}(H) \to \mathfrak{H}$ is not bijective. The resolvent set is the complement of the spectrum, namely $\rho(H) := \mathbb{C} \setminus \sigma(H)$. The set of all eigenvalues z of H of finite multiplicities which are isolated points of the spectrum and such that the range of $H - z$ is closed is the *discrete* spectrum $\sigma_d(H)$. The *essential* spectrum $\sigma_{\text{ess}}(H)$ of H is the set of complex numbers z such that $H - z$ is not Fredholm.[2] If H is a self-adjoint operator, $\sigma_{\text{ess}}(H)$ coincides with the complement set of $\sigma_d(H)$, i.e. $\sigma_{\text{ess}}(H) = \sigma(H) \setminus \sigma_d(H)$. In general, namely for non-self-adjoint operators, this property does not need to be true. Nevertheless, as we shall see, for the non-self-adjoint operators H considered in this paper, this will still be the case. The set of *all* eigenvalues is the *point* spectrum $\sigma_p(H)$.

---

[1] In classical quantum mechanics the notion of Hamiltonian is customarily connected to a purely self-adjoint setting (it is the observable which represents the total energy of the system). Nevertheless, it turned out that adopting a less fundamental approach one can provide a relaxation of the notion which is meaningful also for non-self-adjoint operators (see the introduction in [48] and references therein.).

[2] Actually, this is just *one* of the many possible definition for the essential spectrum, which are all equivalent only in the self-adjoint setting. In particular, the one we are using here is σ_{e_3} in the monograph [31] by Edmunds and Evans.

Throughout the paper we make the standing assumption that the spectrum of the free Hamiltonian H_0 is purely essential and that there are no eigenvalues embedded in the essential spectrum, namely

$$\sigma(H_0) = \sigma_{\text{ess}}(H_0), \qquad \sigma_p(H_0) = \varnothing.$$

This assumption does not represent any loss of generality, as all the physically relevant examples we will give below enjoy this properties.

When available, the *disjoint* decomposition of the spectrum into discrete and essential part makes it possible to get information on the spectrum as *a whole*: locating the essential spectrum as a set excludes regions where the discrete spectrum can be located and vice-versa. This, as we will see in the first part of our analysis (Sect. 2), allows to find perturbations V such that the spectrum of the free Hamiltonian does not change: if V is *local* as dictated by the Weyl theorem then the essential spectrum remains stable; if, in addition, V is *small* as suggested by Hardy-type inequalities (when available) then no eigenvalues are created outside the essential spectrum and thus one has full spectral stability, i.e. $\sigma(H_0) = \sigma(H_0 + V)$.

This approach already provides useful information on the spectrum, but, as said, only on the spectrum in its entirety. However, not investigating the intrinsic nature of the single parts of the spectrum undermines the possibility of getting a satisfactory understanding of the system described by our Hamiltonian. For instance, it is a very crucial point in scattering theory to identify which part of the spectrum corresponds to the so-called *bound states* (states which remain localised near a scattering center) and which one corresponds to *scattering states* (states which move freely after the interaction with the potential). In atomic physics this distinction is identified with the separation between discrete and essential spectrum, respectively. However, according to the definition of the essential spectrum, this is only true if one can exclude a priori the presence of embedded eigenvalues. This is notoriously a very difficult problem especially because they have been constructed many examples of Hamiltonians which do admit embedded eigenvalues even though the physical intuitions behind the quantum mechanical tunnelling would like to exclude this possibility. Thus, in the second part of our analysis (Sect. 3), we will show how different, but also intertwined, techniques as Mourre theory, Birman-Schwinger principle and the method of multipliers can be used for the purpose of excluding embedded eigenvalues in the essential spectrum. In the case when the perturbation of the free Hamiltonian does create eigenvalues, it is an interesting problem understanding where these eigenvalues are located in the complex plane. A very powerful tool to address such a problem is represented by the Birman-Schwinger principle that we discuss in the third part of this survey.

We emphasise that throughout the paper we decided to keep almost always the reasoning on a formal level and to avoid all technical details. This to make the reader familiar with the ideas behind the techniques we will present and that could easily get lost within a too technical presentation. Nevertheless all the rigorous proofs of the results that we mention can be found in the related recalled references.

2 Qualitative Spectral Stability

As far as stability of the essential spectrum is concerned, it is well known from Weyl's theorem (see [58, Theorem XIII.14]) that perturbations V (possibly-complex valued) which decay sufficiently fast at infinity (or, more in general, which are relatively compact with respect to the free Hamiltonian H_0) do not change the essential spectrum, i.e. $\sigma_{\mathrm{ess}}(H_0 + V) = \sigma_{\mathrm{ess}}(H_0)$. On the other hand, Weyl's theorem alone does not exclude the possibility of these perturbations creating discrete eigenvalues outside the essential spectrum.

This can be explained easily for bounded from below H_0 and considering self-adjoint perturbations, namely for real-valued potentials $V : \mathbb{R}^d \to \mathbb{R}$. In this case, the min/max principle says that, in principle, if V is an attractive perturbation, i.e. a non-trivial and non-positive function, then

$$\inf \sigma(H_0 + V) \leq \inf \sigma(H_0). \tag{2.1}$$

If the inequality above is strict, then the attractive nature of the potential *does* affect the spectrum creating discrete eigenvalues below the essential spectrum. In passing, observe that if the potential is repulsive, namely V is non-trivial and non-negative, then $H_0 + V \geq 0$, thus $\inf \sigma(H_0 + V) = \inf \sigma(H_0)$ and no discrete eigenvalues can be created.

Nevertheless, it is a well known fact that the existence of Hardy-type inequalities for the free Hamiltonian H_0 represents a powerful source of sufficient conditions (intrinsically smallness) on the attractive perturbation to ensure that this strict bound does not occur. We elaborate more on this consequence of Hardy-inequalities in the next section.

2.1 *Hardy-Type Inequalities*

We start this subsection recalling the following definition (see [47]): we say that a self-adjoint and bounded from below operator H_0 on $L^2(\mathbb{R}^d)$ satisfies an Hardy-type inequality if there exists a non-trivial, measurable and almost everywhere positive function ρ such that

$$H_0 - \inf \sigma(H_0) \geq \rho. \tag{2.2}$$

We take $V := -\varepsilon|W|$, $\varepsilon > 0$ and assume, for simplicity, $W \in C_0^\infty(\mathbb{R}^d)$. From (2.2) one has

$$H_0 + V = H_0 - \varepsilon|W| = H_0 - \rho + \rho - \varepsilon|W| \geq \inf \sigma(H_0) + \rho - \varepsilon|W|.$$

If ε is sufficiently small, then $\rho - \varepsilon|W|$ is non-negative. Thus, from the variational characterisation of the spectrum and from (2.1), $\inf \sigma(H_0 + V) = \inf \sigma(H_0)$ and therefore no discrete eigenvalues are created. This simple argument shows the relevance of proving Hardy-type inequalities and, more importantly, of proving optimality of such inequalities: if inequality (2.2) is available, the "larger" (i.e. the sharper) is the right-hand-side, the "stronger" can be the attractive perturbation that still can be proven not affecting the free Hamiltonian. For reader's convenience we give below two concrete examples.

Example 2.1 (*Schrödinger and Hardy inequality*) Let $H = -\Delta + V$ be the Schrödinger operator with $V\colon \mathbb{R}^d \to (-\infty, 0]$. It is well know that $\sigma(-\Delta) = \sigma_{\mathrm{ess}}(-\Delta) = [0, \infty)$. In particular, $\inf \sigma(-\Delta) = 0$.

If $d \geq 3$, the operator $-\Delta$ satisfies the Hardy inequality

$$\int_{\mathbb{R}^d} |\nabla\psi(x)|^2\, dx \geq \frac{(d-2)^2}{4} \int_{\mathbb{R}^d} \frac{|\psi(x)|^2}{|x|^2}\, dx$$

for any $\psi \in H^1(\mathbb{R}^d)$, which can be alternatively written as

$$-\Delta \geq \frac{(d-2)^2}{4} \frac{1}{|x|^2}, \tag{2.3}$$

(cf. (2.3) with (2.2) above), where (2.3) has to be interpreted in a form sense. From (2.3) and the general reasoning above one sees that if our attractive potential decays at least as fast as $|x|^{-2}$ at infinity and it is sufficiently small, then $-\Delta + V \geq 0$ which gives $\sigma(-\Delta + V) \geq 0$, thus no negative eigenvalues can occur.

Example 2.2 (*Biharmonic operator and Rellich inequality*) Let $H = \Delta^2 + V$ be the perturbed biharmonic operator with $V\colon \mathbb{R}^d \to (-\infty, 0]$. It is well known that $\sigma(\Delta^2) = \sigma_{\mathrm{ess}}(\Delta^2) = [0, \infty)$. In particular, $\inf \sigma(\Delta^2) = 0$. Moreover, if $d \geq 5$, Δ^2 satisfies the Rellich inequality

$$\int_{\mathbb{R}^d} |\Delta\psi(x)|^2\, dx \geq \frac{d^2(d-4)^2}{16} \int_{\mathbb{R}^d} \frac{|\psi(x)|^2}{|x|^4}\, dx,$$

for any $\psi \in H^2(\mathbb{R}^d)$, which can be alternatively written as

$$\Delta^2 \geq \frac{d^2(d-4)^2}{16} \frac{1}{|x|^4},$$

(cf. again (2.2)). Reasoning as above one can see that for suitably small attractive potentials decaying at infinity at least as $|x|^{-4}$, the discrete spectrum of the biharmonic operator remains empty.

If one looks at the general argument presented above, it is clear that, due to the consistent use of the variational principle, this cannot be generalised to cover complex-valued potentials. Nevertheless, in specific situations, for instance in the case of

Schrödinger operators, it has been shown that a suitable usage of the method of multipliers allows to still exploit Hardy-type inequalities to detect smallness/repulsivity conditions which ensure that the discrete spectrum remains empty (actually with this approach one can even prove that the entire point spectrum is empty) also in a non-self-adjoint setting. We briefly explain below the method taking the Schrödinger operator as toy model. Afterwards we will give references to similar results for other relevant models.

2.2 *The Method of Multipliers: Absence of Discrete Eigenvalues*

As anticipated, to explain the method of multipliers we take as toy model the non-self-adjoint Schrödinger operator

$$H = -\Delta + V, \qquad V : \mathbb{R}^d \to \mathbb{C}, \qquad d \geq 3.$$

Here, for simplicity, we assume $V \in C_0^\infty(\mathbb{R}^d)$, nevertheless the method we shall illustrate below works under weaker assumption on the potential V (we refer to [17, 36, 37] where the method of multipliers for this model is developed in full details). We assume that there exists $\psi \in H^1(\mathbb{R}^d)$ which satisfies (weakly) the eigenvalue equation associated to H, namely

$$-\Delta\psi + V\psi = z\psi,$$

namely, for any $v \in H^1(\mathbb{R}^d)$, ψ satisfies

$$(v, -\Delta\psi) + (v, V\psi) = z(v, \psi). \tag{2.4}$$

In a nutshell, the method is based on producing several integral identities by choosing various test functions in (2.4) and later combining them in a way that one can ultimately detect conditions on V which imply $\psi = 0$. This would contradict ψ being an eigenfunction corresponding to z and thus $\sigma_p(H) = \varnothing$. Thus, in principle, the method can be used to totally exclude eigenvalues (not only discrete).

As a warm-up we start considering the self-adjoint case, namely the case of V being real-valued. Since we are interested in excluding discrete eigenvalues only, we assume that $z \in \mathbb{R}$ is *negative* (recall that since $V \in C_0^\infty(\mathbb{R}^d)$ then $\sigma_{\text{ess}}(-\Delta + V) = \sigma_{\text{ess}}(-\Delta) = [0, \infty)$). In this easy situation we consider as test function $v \in H^1(\mathbb{R}^d)$ in (2.4) the solution itself, namely $v := \psi$. Thus one easily obtains

$$\int_{\mathbb{R}^d} |\nabla\psi|^2 \, dx + \int_{\mathbb{R}^d} V|\psi|^2 \, dx = z \int_{\mathbb{R}^d} |\psi|^2 \, dx \leq 0,$$

where the last inequality follows from z being negative. Now one notices that in the term involving the potential it is the sign of the potential which gives the sign to the integral. This fact allows to easily detect repulsivity conditions on V which exclude discrete eigenvalues. Indeed, if $V_- = 0$ or the Hardy-type smallness condition

$$\int_{\mathbb{R}^d} V_-|\psi|^2\,dx \leq a^2 \int_{\mathbb{R}^d} |\nabla\psi|^2\,dx, \qquad a^2 < 1 \tag{2.5}$$

holds for any $\psi \in H^1(\mathbb{R}^d)$, then in both case one gets $\int_{\mathbb{R}^d} |\nabla\psi|^2\,dx \leq 0$, which entails $\psi = 0$. Thus, no discrete (negative) eigenvalues can occur for repulsive ($V_- = 0$) potentials or for attractive but small (in the sense of (2.5)) potentials.

Now, we see how to handle the non-self-adjoint situation. Our purpose is to exclude discrete eigenvalues, i.e. we want to show that if z is an eigenvalue of H, then $z \notin \mathbb{C} \setminus [0, \infty)$. For simplicity, instead of considering this general case, we shall see how to exclude eigenvalues *only* outside a fixed "half-cone" containing $[0, \infty)$, namely in the set $\mathbb{C} \setminus \mathcal{C}$, where

$$\mathcal{C} := \{z \in \mathbb{C} \colon |\Im z| \leq \Re z, \Im z \neq 0\}. \tag{2.6}$$

The general situation will be considered in the next Subsect. 3.2. By contradiction we assume $z \in \mathbb{C} \setminus \mathcal{C}$. Without loss of generality we assume also $\Im z \geq 0$, the case $\Im z < 0$ can be done similarly. Using as test function $v := \psi$ and taking the real and the imaginary part of the resulting identity one has

$$\int_{\mathbb{R}^d} |\nabla\psi|^2\,dx + \int_{\mathbb{R}^d} \Re V|\psi|^2\,dx = \Re z \int_{\mathbb{R}^d} |\psi|^2\,dx, \tag{2.7}$$

and

$$\int_{\mathbb{R}^d} \Im V|\psi|^2\,dx = \Im z \int_{\mathbb{R}^d} |\psi|^2\,dx, \tag{2.8}$$

respectively. Subtracting (2.8) from (2.7) one gets

$$\int_{\mathbb{R}^d} |\nabla\psi|^2\,dx + \int_{\mathbb{R}^d} (\Re V - \Im V)|\psi|^2\,dx = (\Re z - \Im z) \int_{\mathbb{R}^d} |\psi|^2\,dx \leq 0,$$

where the last inequality follows from $\Im z > \Re z$ (cf. (2.6)). Now, if $\Re V$ is sufficiently repulsive (i.e. its negative part $[\Re V]_-$ is sufficiently small) and if $\Im V$ is sufficiently small, then a similar argument as in the self-adjoint case applies which entails $\psi = 0$. More precisely, the conditions that one assumes are

$$\int_{\mathbb{R}^d} [\Re V]_-|\psi|^2\,dx \leq b_1^2 \int_{\mathbb{R}^d} |\nabla\psi|^2\,dx, \qquad \int_{\mathbb{R}^d} |\Im V||\psi|^2\,dx \leq b_2^2 \int_{\mathbb{R}^d} |\nabla\psi|^2\,dx, \tag{2.9}$$

for $b_1^2 + b_2^2 < 1$.

3 Absence of Embedded Eigenvalues

In the previous section we have seen that the presence of Hardy-type inequalities for the free Hamiltonian H_0 provides sources of sufficient conditions on the potentials V which guarantee absence of *discrete* eigenvalues of H. The question of whether H can have eigenvalues *embedded* in the essential spectrum is in general more difficult and, as a matter of fact, the results that do exist require rather subtle tools. In the self-adjoint case, one successful theory is the so-called Mourre theory or positive commutator theory of Quantum Mechanics whose classical formulation we shortly recall below.

3.1 The Mourre Theory and the Virial Theorem

The starting point of the Mourre theory is the Virial theorem for self-adjoint operators. Here, for simplicity, we give just a formal statement of the result (we disregard domain issues, etc.). We refer to [3, Proposition 6.6], [25, Sect. 4]) and also [58, Theorem XIII.59 and notes on Sect. XIII.13] for the rigorous statements. See also [57].

Theorem 3.1 (Virial Theorem) *Let H and A be self-adjoint operators on a Hilbert space $\mathcal{H}$. If ψ is an eigenfunction of H then*

$$(\psi, i[H, A]\psi) = 0. \tag{3.1}$$

Formally, the statement is obtained just expanding the commutator and using self-adjointness of the Hamiltonian. However, for unbounded operators a rigorous proof requires more care.

The following result is an immediate consequence of the Virial Theorem 3.1.

Theorem 3.2 (Mourre Theorem) *Let H and A be self-adjoint operators on on a Hilbert space $\mathcal{H}$. Assume that the commutator $i[H, A]$ between H and A is positive, i.e. there exists a positive number a such that*

$$i[H, A] \geq aI. \tag{3.2}$$

Then $\sigma_{\mathrm{p}}(H) = \varnothing$.

Proof Even though the proof is trivial, we sketch it here for the sake of completeness. If ψ were an eigenfunction, then from Theorem 3.1 in the Virial theorem and from assumption (3.2) one would get the contradiction

$$0 = (\psi, i[H, A]\psi) \geq a > 0.$$

Then the point spectrum of H is empty. □

Remark 3.1 In 1980 Mourre [57] showed that the *global* positivity hypothesis (3.2) in Thm 3.2 can be relaxed. More precisely, he proved that *spectrally localized* positive commutator estimates (known as Mourre's inequalities) are sufficient to get the finiteness of embedded eigenvalues and the absence of singular continuous spectrum. In particular, Theorem 3.2 follows from his result.

Remark 3.2 Notice that Theorem 3.2 provides a condition to prove *total* absence of eigenvalues (not only embedded).

Applied to particular situations, the abstract positive-commutator condition (3.2) in Theorem 3.2 translates into natural repulsivity conditions on the potential V for Hamiltonians of the form $H = H_0 + V$. As above, we take as reference model the Schrödinger and perturbed bharmonic operators on $L^2(\mathbb{R}^d)$, namely we take $H_0 = -\Delta$ and $H_0 = \Delta^2$, respectively.

Example 3.1 (*Schrödinger and perturbed biharmonic operators*) Let A be the generator of dilation, namely the self-adjoint operator

$$A := -\frac{i}{2}(x \cdot \nabla + \nabla \cdot x), \tag{3.3}$$

then the following commutation relations hold:

$$i[-\Delta, A] = -2\Delta, \qquad i[\Delta^2, A] = 4\Delta^2, \qquad i[V, A] = -x \cdot \nabla V.$$

Assuming the pointwise repulsivity condition $x \cdot \nabla V \leq 0$ gives

$$i[-\Delta + V, A] \geq -2\Delta, \qquad i[\Delta^2 + V, A] \geq 4\Delta^2. \tag{3.4}$$

Since $-\Delta$ and Δ^2 are just a non-negative operator, the inequalities above do not imply the validity of the positive commutator condition as stated in (3.2) and thus the Mourre theorem 3.2 does not directly apply. Nevertheless, since the right-hand-sides in (3.4) do have a sign, analogous reasoning as in Theorem 3.2 allow to conclude: Indeed, let $\psi \in H^1(\mathbb{R}^d)$ be an eigenfunction of $H = -\Delta + V$ and let $\phi \in H^2(\mathbb{R}^d)$ be an eigenfunction of $H = \Delta^2 + V$, then from Theorem 3.1 and from inequalities (3.4) one has

$$0 = (\psi, i[-\Delta + V, A]\psi) \geq \|\nabla\psi\|_2^2, \qquad 0 = (\phi, i[\Delta^2 + V, A]\phi) \geq 4\|\Delta\phi\|_2^2.$$

From these and from the decay assumptions on the eigenfunctions, it immediately follows that $\psi = \phi = 0$.

Remark 3.3 Notice that the pointwise repulsivity condition $x \cdot \nabla V \leq 0$ on the potential can be easily weakened asking the validity of a variational smallness condition (in the spirit of (2.5) and (2.9)) for the positive part of the radial derivative of the potential, namely

$$(\psi, [x \cdot \nabla V]_+ \psi) \le c \|H_0^{1/2} \psi\|_2^2, \qquad \text{for all } \psi \in \operatorname{dom}(H_0^{1/2}). \tag{3.5}$$

If c in (3.5) is sufficiently small, then the absence of point spectrum still follows under the weaker assumption (3.5).

Notice that, as already mentioned, the possibility to exclude bound states in the previous two situations worked out only because both $-\Delta$ and Δ^2 are bounded below (actually non-negative) operators. For operators that are not bounded below, Mourre-type results still apply but they allow to exclude embedded eigenvalues only in part of the essential spectrum. As an explanatory example we consider below the perturbed Dirac operator.

Example 3.2 (*Dirac operator*) Let $H = \mathcal{D}_m + V I_{\mathbb{C}^{N \times N}}$, where $\mathcal{D}_m$ is the free Dirac operator

$$\mathcal{D}_m := -i\boldsymbol{\alpha} \cdot \nabla + m\alpha_{d+1} = -i \sum_{k=1}^{d} \alpha_k \partial_k + m\alpha_{d+1}. \tag{3.6}$$

Here $m \ge 0$ is the mass, $N := 2^{\lceil d/2 \rceil}$ and $V : \mathbb{R}^d \to \mathbb{R}$. The Dirac matrices $\alpha_k \in \mathbb{C}^{N \times N}$, for $k \in \{1, \ldots, d+1\}$ satisfy the anti-commutation relations

$$\alpha_j \alpha_k + \alpha_k \alpha_j = 2\delta_j^k$$

where δ_j^k is the Kronecker symbol.

If $m > 0$ the spectrum of the free Dirac operator $\mathcal{D}_m$ is the union of two disjoint semi-axis, namely

$$\sigma(\mathcal{D}_m) = \sigma_{\text{ess}}(\mathcal{D}_m) = (-\infty, -m] \cup [m, +\infty),$$

otherwise if $m = 0$ then its spectrum is the whole real line

$$\sigma(\mathcal{D}_0) = \sigma_{\text{ess}}(\mathcal{D}_0) = \mathbb{R}.$$

Computing the commutator with the dilation operator A defined in (3.3) one has

$$i[\mathcal{D}, A] = -i\boldsymbol{\alpha} \cdot \nabla, \qquad i[V, A] = -x \cdot \nabla V.$$

Differently from the previous case, since the commutator $i[\mathcal{D}, A] = -i\alpha \cdot \nabla$ does not have a sign, the full commutator $i[H, A] = -i\boldsymbol{\alpha} \cdot \nabla - x \cdot \nabla V I_{\mathbb{C}^{N \times N}}$ is also unsigned (even under the repulsivity condition $x \cdot \nabla V \le 0$) and therefore the Mourre theorem does not apply as such. Nevertheless, a slight modification of the argument above gives the following analogous (but weaker) implications: if $V + x \cdot \nabla V \le 0$, then $\sigma_p(H) \cap (m, \infty) = \varnothing$ (if $V + x \cdot \nabla V < 0$, then $\sigma_p(H) \cap [m, \infty) = \varnothing$). Analogously, if $V + x \cdot \nabla V \ge 0$, then $\sigma_p(H) \cap (-\infty, -m) = \varnothing$ (if $V + x \cdot \nabla V > 0$, then $\sigma_p(H) \cap (-\infty, m] = \varnothing$). If $V + x \cdot \nabla V = 0$ (as in the case of Coulomb potential or, more elementary, in the free case $V = 0$), then $\sigma_p(H) \cap [-m, m]^c =$

$\varnothing$. Finally, if $M_1 \leq V + x \cdot \nabla V \leq M_2$, then $\sigma_p(H) \cap [-m - M_1, m + M_2]^c = \varnothing$. These are consequences of the following modified Virial theorem for the Dirac operator.

Theorem 3.3 (Virial for Dirac) *Let $H = \mathcal{D}_m + V I_{\mathbb{C}^{N\times N}}$ be the perturbed Dirac operator, with $\mathcal{D}_m$ as in* (3.6) *and $V : \mathbb{R}^d \to \mathbb{R}$ and let A be the dilation operator defined in* (3.3). *If $\psi \in H^1(\mathbb{R}^d)^N$ is an eigenfunction of H corresponding to the eigenvalue $E \in \mathbb{R}$, then*

$$E\|\psi\|_2^2 = (\psi, (V + x \cdot \nabla V)\psi) + m(\psi, \alpha_{d+1}\psi). \tag{3.7}$$

Proof Formally, the statement is obtained just observing that from Theorem 3.1 one has

$$0 = (\psi, i[H, A]\psi) = (\psi, -i\alpha \cdot \nabla\psi) - (\psi, x \cdot \nabla V\psi).$$

Then (3.7) is obtained just using the eigenvalue equation satisfied by ψ. □

Remark 3.4 Observe that the statement of Theorem 3.3 is formal (no hypotheses on the regularity/integrability of V are given). The rigorous result can be found, for instance, in [2, Theorems 4.1.1, 4.1.2].

Now if one chooses for α_{d+1} the following representation

$$\alpha_{d+1} = \begin{pmatrix} I_{\mathbb{C}^{N/2\times N/2}} & 0 \\ 0 & -I_{\mathbb{C}^{N/2\times N/2}} \end{pmatrix},$$

it is easy to see that

$$-m \leq \langle \alpha_{d+1} m\psi, \psi \rangle \leq m,$$

and thus the implications above follow easily from (3.7). See [2] and [62] for further details).

To sum up, we have seen that in the case of the Dirac operator the lack of positivity of the commutator does not entail total absence of eigenvalues, nevertheless the underlying idea of the virial theorem can be still used to exclude eigenvalues embedded in (part of) the essential spectrum.

The beauty as well as the power of the Mourre theory relies on the fact that it provides a completely abstract condition (see (3.2)), i.e. not depending on the specific Hamiltonian of the system, to obtain information on eigenvalues embedded (and not) in the essential spectrum, which moreover, as shown in the examples above, is also easy to verify in the particular cases (at least on a formal level).

Nevertheless, if one considers non-self-adjoint Hamiltonians of the form $H = H_0 + V$, with $V : \mathbb{R}^d \to \mathbb{C}$ (complex-valued) several difficulties arise. As a matter of fact, already the Virial Theorem 3.1 ceases to be valid: indeed if ψ is an eigenfunction of the non-self-adjoint Hamiltonian $H = H_0 + V$, i.e. ψ satisfies $H\psi = z\psi$, with $z \in \mathbb{C}$ (the spectrum is no more necessarily real), an easy computation gives

$$(\psi, i[H, A]\psi) = 2i\Im z(\psi, A\psi) - 2(\Im V\psi, A\psi). \tag{3.8}$$

Thus the mean value of the commutator over the state ψ is no longer zero and, as a matter of fact, it depends on the imaginary part of the eigenvalue and of the complex potential. Since the right-hand-side is no longer zero and it is not even defined in sign, finding conditions such that $i[H, A] > 0$ would not lead to any contradiction. In other words, this shows that the standard Mourre theory does not easily carry over the non-self-adjoint framework.

Leaving aside for now the Mourre theory, it turned out in the last decade that in the non-self-adjoint case a successful technique to get absence of bound states (both discrete and embedded) is the method of multipliers which, so far we have seen at work only to show absence of discrete eigenvalues. In the next section we show how to modify the strategy developed in Sect. 2.2 to obtain absence of embedded eigenvalues as well.

3.2 *The Method of Multipliers: Absence of Embedded Eigenvalues*

We recall that the core of the method of multipliers already introduced in Sect. 2.2 relies on finding suitable test functions in the weak formulation of the corresponding eigenvalue equation such that a consequent manipulation of the resulting identities leads to a contradiction. We consider again as toy model the Schrödinger operator. In Sect. 2.2 we have seen how to exclude eigenvalues outside the conical sector $\mathcal{C}$ defined in (2.6), i.e. far away from the essential spectrum of $-\Delta$. Now we consider the case of $z \in \mathcal{C}$ (this in particular covers embedded eigenvalues). Here, for simplicity, we consider $V \in C_0^\infty(\mathbb{R}^d)$. Nevertheless the result presented below holds under less restrictive hypotheses about V (see [17]). Considering as test function the skew-symmetric multiplier $v := [-\Delta, |x|^2]\psi$ and taking the real part of the resulting identity, one gets

$$\begin{aligned}\Re(H\psi, [-\Delta, |x|^2]\psi) &= \Re z(\psi, [-\Delta, |x|^2]\psi)\\ &= \Re z\Re(\psi, [-\Delta, |x|^2]\psi) - \Im z\Im(\psi, [-\Delta, |x|^2]\psi)\\ &= -\Im z\Im(\psi, [-\Delta, |x|^2]\psi),\end{aligned} \tag{3.9}$$

where the last identity follows from $(\psi, [-\Delta, |x|^2]\psi)$ being purely imaginary. Computing explicitly the L^2-products above gives

$$2\int_{\mathbb{R}^d} |\nabla\psi|^2\,dx + d\int_{\mathbb{R}^d} \Re V|\psi|^2\,dx + 2\Re\int_{\mathbb{R}^d} x\cdot V\psi\overline{\nabla\psi}\,dx = -2\Im z\Im\int_{\mathbb{R}^d} x\cdot\psi\overline{\nabla\psi}\,dx. \tag{3.10}$$

As a warm-up let suppose that V is real-valued, then $\Im z = 0$ and an integration by parts gives

$$2\int_{\mathbb{R}^d} |\nabla\psi|^2\,dx - \int_{\mathbb{R}^d} x\cdot\nabla V|\psi|^2\,dx = 0.$$

Notice that in the term involving V it is the sign of the radial derivative $x\cdot\nabla V$ that gives the sign to the integral. Thus, one can either assume the repulsivity condition $[x\cdot\nabla V]_+ = 0$ or the Hardy-type smallness condition

$$\int_{\mathbb{R}^d} [x\cdot\nabla V]_+|\psi|^2\,dx \le a^2\int_{\mathbb{R}^d} |\nabla\psi|^2\,dx, \qquad a^2 < 2,$$

valid for all $\psi \in H^1(\mathbb{R}^d)$. From both these conditions one has $\int_{\mathbb{R}^d} |\nabla\psi|^2\,dx \le 0$, which entails $\psi = 0$.

If V is complex-valued one immediately realizes that the sole identity (3.10) is not enough to obtain the contradiction as in the self-adjoint case. Indeed, the term on the right-hand-side which depends on the eigenvalue (or, better, on its imaginary part) does not have a sign and has to be controlled exploiting some suitable positive quantities. From this one understands that, in the non-self-adjoint case, more identities are needed to get positivity. More precisely, given $g\colon \mathbb{R}^d \to \mathbb{R}$ to be chosen later, one takes as test function $v = g\psi$ and then takes the real and the imaginary parts of the resulting identity. A suitable algebra of these three identities and suitable choices of g (see [17, 37] for details) give the following identity

$$\begin{aligned}\int_{\mathbb{R}^d} |\nabla\psi^-|^2\,dx + \frac{|\Im z|}{\sqrt{\Re z}}\int_{\mathbb{R}^d} |x||\nabla\psi^-|^2\,dx - \frac{d-1}{2}\frac{|\Im z|}{\sqrt{\Re z}}\int_{\mathbb{R}^d} \frac{|\psi|^2}{|x|}\,dx\\ - \int_{\mathbb{R}^d} \partial_r(r\Re V)|\psi|^2\,dx - 2\Im\int_{\mathbb{R}^d} x\Im V\psi\overline{\nabla\psi^-}\,dx + \frac{|\Im z|}{\sqrt{\Re z}}\int_{\mathbb{R}^d} r\Re V|\psi|^2\,dx = 0,\end{aligned} \tag{3.11}$$

where $\psi^-(x) := e^{-i\sqrt{\Re z}\,\mathrm{sgn}(\Im z)|x|}\psi(x)$. The auxiliary function ψ^- is related to the Sommerfeld radiation conditions

$$\lim_{r\to\infty}\int_{|x|=r} \left|\nabla\psi - i\lambda^{1/2}\frac{x}{|x|}\psi\right|^2 d\sigma(r) = 0$$

for unique solvability of the Helmholtz equation $-\Delta\psi + V\psi = \lambda\psi$, $\lambda \ge 0$, once one observes that $|\nabla\psi^-|^2 = \left|\nabla\psi - i\lambda^{1/2}\frac{x}{|x|}\psi\right|^2$ (see [34] where the auxiliary function ψ^- was introduced for the first time). At this point, from identity (3.11), it is reasonable to believe that if $[\Re V]_-$, $[\partial_r(r\Re V)]_+$ and $\Im V$ are sufficiently small, then a similar argument as in the self-adjoint case applies which entails $\psi = 0$. Again, the smallness conditions are of Hardy-type, namely one asks

$$\int_{\mathbb{R}^d} r^2[\Re V]_-^2|\psi|^2\,dx \le b_1^2\int_{\mathbb{R}^d} |\nabla\psi|^2\,dx, \qquad \int_{\mathbb{R}^d} [\partial_r(r\Re V)]_+|\psi|^2\,dx \le b_2^2\int_{\mathbb{R}^d} |\nabla\psi|^2\,dx,$$

and

$$\int_{\mathbb{R}^d} |\Im V|^2 |\psi|^2 \, dx \le b_3^2 \int_{\mathbb{R}^d} |\nabla \psi|^2 \, dx,$$

for suitably small b_1, b_2 and b_3. As for the negative term in the first line of (3.11), this can be estimated using the weighted Hardy inequality

$$\int_{\mathbb{R}^d} \frac{|\psi|^2}{|x|} \, dx \le \frac{4}{(d-1)^2} \int_{\mathbb{R}^d} |x| |\nabla \psi|^2 \, dx, \qquad \psi \in H^1(\mathbb{R}^d), \quad d \ge 2,$$

and controlled in terms of the second one.

Remark 3.5 (*Comparison between Mourre theory and method of multipliers*) In the two sections above we have seen that the abstract Mourre theory does not seem to be immediately applicable to non-self-adjoint Hamiltonian, instead, the method of multipliers, relying just on a direct manipulation of the eigenvalue equations does provide results in this more challenging setting. Nevertheless, one can easily show that when we restrict to the self-adjoint setting, the two approaches reduce one another. We clarify this fact again taking as toy model the Schrödinger operator $H = -\Delta + V$, with real potential $V : \mathbb{R}^d \to \mathbb{R}$.
As an outcome of the method of multipliers we had the following result.

Theorem 3.4 *Let $H = H_0 + V$, with $H_0 = -\Delta$ and $V : \mathbb{R}^d \to \mathbb{R}$. If $\psi \in H^1(\mathbb{R}^d)$ is an eigenfunction of H then*

$$(\psi, [H, [H_0, |x|^2]]\psi) = 0. \tag{3.12}$$

Proof This follows from (3.9) with $\Im z = 0$ and using the simple fact

$$\begin{aligned} 2\Re (H\psi, [H_0, |x|^2]\psi) &= (H\psi, [H_0, |x|^2]\psi) + \overline{(H\psi, [H_0, |x|^2]\psi)} \\ &= (\psi, [H, [H_0, |x|^2]]\psi), \end{aligned}$$

where in the last identity we have used that H is self-adjoint and that $[H_0, |x|^2]$ is skew-symmetric. □

Then, the next step in the method was computing explicitly the double commutator and finding conditions on the potential such that the double commutator $(\psi, [H, [H_0, |x|^2]]\psi)$ had a sign. Looking at the Mourre theory, one notices that there it was crucial instead to give a sign to the commutator $(\psi, [H, A]\psi)$ of the Hamiltonian with the dilation operator A. On the other hand, an easy computation shows that

$$[H_0, |x|^2] = -4iA.$$

Thus the two methods are in fact the same method.

The ideas presented above related to the method of multipliers are developed in full details in a series of papers [15, 17, 18, 36, 37]. See also [8, 9] for some related results on Dirac operators. Refer to [15] for an adaptation of the method to the Lamé operator of elasticity. Instead for some recent result which relies on the positive commutator theory we refer to [4].

4 Spectral Enclosures

So far we have considered only tools for ensuring spectral stability, in particular we have investigated only situations where the potentials perturbing the free Hamiltonian do not create eigenvalues (the point spectrum remains empty). Nevertheless, in the case when the perturbation does create eigenvalues, it is an interesting problem understanding where these eigenvalues are located in the complex plane. A very powerful tool to address such a problem is represented by the Birman-Schwinger principle that we discuss in the next section.

4.1 The Birman-Schwinger Principle

The principle, originated in the work [5] by Birman and [61] by Schwinger, has become nowadays a fundamental and widely employed tool in spectral analysis. The easiest statement of this principle can be given as follows:

Theorem 4.1 *Let H be a perturbation of H_0 by a factorizable bounded potential $V = B^*A$, i.e. $H = H_0 + B^*A$ and let $z \in \rho(H_0)$. Then $z \in \sigma_p(H)$ if and only if $-1 \in \sigma_p(K_z)$, where $K_z := A(H_0 - z)^{-1}B^*$ denotes the so-called Birman-Schwinger operator.*

Proof Let $z \in \sigma_p(H) \cap \rho(H_0)$. Then there exists a non-trivial $\psi \in \operatorname{dom}(H_0)$ such that $H\psi = z\psi$. Thus

$$H_0\psi + V\psi = z\psi \iff (H_0 - z)\psi = -V\psi \iff -\psi = (H_0 - z)^{-1}B^*A\psi$$

and so, denoting $\phi = A\psi$ and applying to both side the operator A we get

$$A(H_0 - z)^{-1}B^*\phi = -\phi,$$

so $-1 \in \sigma_p(K_z)$. The "if part" is achieved simply performing the previous few steps in reverse order. □

If -1 is an eigenvalue of K_z, this means that the norm of the Birman-Schwinger operator is at least 1, and consequently the eigenvalues of the perturbed operator H are confined in the complex region defined by $1 \leq \|K_z\|$; if in particular $\|K_z\| < 1$ uniformly with respect to z, then the point spectrum of H is empty. Thus the Birman-Schwinger principle allows either to locate the point spectrum in the complex-plane or to show its emptyness.

The advantage of studying the norm of the Birman-Schwinger operator is evident already from its definition: if we are able to bound the resolvent $(H_0 - z)^{-1}$, it is usually an easy matter setting A and B in a suitable normed space, and then obtain an estimate for K_z. This approach reduces to establishing suitable resolvent estimates for the unperturbed operator H_0. Each of them corresponds, via the Birman-Schwinger principle, to a localization estimate for the eigenvalues of the perturbed operator.

Before moving on to see the Birman-Schwinger principle at work in specific situations, we emphasise that Theorem 4.1 was stated under the convenient assumption of $V = B^*A$ being bounded (then H is well defined as operator sum of H_0 and V on $\mathrm{dom}(H_0)$) and considering only the case $z \in \rho(H_0)$ (so that $(H_0 - z)^{-1}$ is well defined). This simplified setting led to the trivial proof of the result given above. Nevertheless a more general and deeper result than Theorem 4.1 can be proved and it will be considered in the next sections. We emphasise that no proofs will be given there, as this would go beyond the scope of this survey, nevertheless all the detailed references will be provided to the reader.

4.1.1 Birman-Schwinger Principle: Kato's Approach

As we have observed before, requiring $V = B^*A$ to be a bounded operator ensured that $H = H_0 + B^*A$ is well-defined as a sum operator on $\mathrm{dom}(H) = \mathrm{dom}(H_0)$. For unbounded potentials, the Hamiltonian H is still well-defined on $\mathrm{dom}(H_0) \cap \mathrm{dom}(B^*A)$, but the latter intersection domain may consist of $\{0\}$ only. In this section, following the approach that dates back to Kato [43] and Konno and Kuroda [46], we recall how to rigorously construct an extension of this operator. Then, with this rigorous definition at hand, we give the general statement of the Birman-Schwinger principle.

Let $\mathfrak{H}, \mathfrak{H}'$ be Hilbert spaces and consider the densely defined, closed linear operators

$$H_0\colon \mathrm{dom}(H_0) \subseteq \mathfrak{H} \to \mathfrak{H}, \quad A\colon \mathrm{dom}(A) \subseteq \mathfrak{H} \to \mathfrak{H}', \quad B\colon \mathrm{dom}(B) \subseteq \mathfrak{H} \to \mathfrak{H}',$$

such that $\rho(H_0) \neq \varnothing$ and

$$\mathrm{dom}(H_0) \subseteq \mathrm{dom}(A), \quad \mathrm{dom}(H_0^*) \subseteq \mathrm{dom}(B).$$

Assume also, for simplicity, that $\sigma(H_0) \subset \mathbb{R}$. By $R_{H_0}(z) := (H_0 - z)^{-1}$, we denote the resolvent operator of H_0 for any $z \in \rho(H_0)$.

We assume the following set of assumptions.

Assumption A For some, and hence for all, $z \in \rho(H_0)$, the operator $AR_{H_0}(z)B^*$, densely defined on $\mathrm{dom}(B^*)$, has a closed extension K_z in $\mathfrak{H}'$,

$$K_z = \overline{AR_{H_0}(z)B^*},$$

which we call the *Birman-Schwinger operator*, with norm bounded by

$$\|K_z\|_{\mathfrak{H}' \to \mathfrak{H}'} \leq \Lambda(z) \tag{4.1}$$

for some function $\Lambda\colon \rho(H_0) \to \mathbb{R}_+$.

Assumption B There exists $z_0 \in \rho(H_0)$ such that $-1 \in \rho(K_{z_0})$.

The last assumption is a bit tricky to check in the application, but luckily for us is implied by the following, easier, one:

Assumption B′ There exists $z_0 \in \rho(H_0)$ such that $\Lambda(z_0) < 1$.
Indeed, assuming Assumptions A and B', we get that $\|K_{z_0}\|_{\mathfrak{H}' \to \mathfrak{H}'} < 1$. Thus, expanding in a Neumann series, we see that $(1 + K_{z_0})^{-1}$ exists and hence $-1 \in \rho(K_{z_0})$.

The extension of the perturbed operator $H_0 + B^*A$ is then constructed as follow.

Lemma 4.1 (Extension of operators with factorizable potential) *Suppose Assumptions A and B. Let $z_0 \in \rho(H_0)$ such that $-1 \in \rho(K_{z_0})$. Then the operator*

$$R_H(z_0) = R_{H_0}(z_0) - \overline{R_{H_0}(z_0)B^*}(1 + K_{z_0})^{-1} A R_{H_0}(z_0) \tag{4.2}$$

defines a densely defined, closed, linear operator H in $\mathfrak{H}$ which has $R_H(z_0)$ as resolvent and which is an extension of $(H_0 + B^*A)|_{\mathrm{dom}(H_0)\cap \mathrm{dom}(B^*A)}$.

We can finally formulate the abstract Birman-Schwinger principle.

Lemma 4.2 (Birman-Schwinger principle) *Suppose Assumptions A and B. Let $z_0 \in \rho(H_0)$ such that $-1 \in \rho(K_{z_0})$ and H be the extension of $H_0 + B^*A$ given by Lemma 4.1. Fix $z \in \sigma_p(H)$ with eigenfunction $0 \neq \psi \in \mathrm{dom}(H)$, i.e. $H\psi = z\psi$, and set $\phi := A\psi$.*

Then $\phi \neq 0$, and in addition

(i) if $z \in \rho(H_0)$ then

$$K_z\phi = -\phi$$

and in particular

$$1 \leq \|K_z\|_{\mathfrak{H}' \to \mathfrak{H}'} \leq \Lambda(z);$$

(ii) if $z \in \sigma(H_0) \setminus \sigma_p(H_0)$ and if H_0 is self-adjoint, then

$$\lim_{\varepsilon \to 0^\pm} (\varphi, K_{z+i\varepsilon}\phi)_{\mathfrak{H}'} = -(\varphi, \phi)_{\mathfrak{H}'} \tag{4.3}$$

for every $\varphi \in \mathfrak{H}'$. In particular

$$1 \leq \liminf_{\varepsilon \to 0^\pm} \|K_{z+i\varepsilon}\|_{\mathfrak{H}' \to \mathfrak{H}'} \leq \liminf_{\varepsilon \to 0^\pm} \Lambda(z + i\varepsilon). \tag{4.4}$$

Remark 4.1 Notice that differently from Theorem 4.1, Lemma 4.2 covers also the case of $z \notin \rho(H_0)$. (The proof of Lemma 4.2 in this case follows adapting the limiting argument introduced in [40, Theorem 8].)

4.1.2 Birman-Schwinger Principle: Abstract Approach

In this subsection, we recall again the technicalities for the Birman-Schwinger principle and for properly define an operator perturbed by a factorizable potential, but with a different approach. This time we completely rely on the abstract analysis carried out by Hansmann and Krejčiřík in [40], to which we refer for more results and background. There, in addition to the point spectrum, appropriate versions of the principle are stated even for the residual, essential and continuous spectra.

Here we collect the set of hypotheses we need.

Assumption I Let $\mathfrak{H}$ and $\mathfrak{H}'$ be complex separable Hilbert spaces, H_0 be a self-adjoint operator in $\mathfrak{H}$ and $|H_0| := (H_0^2)^{1/2}$ its absolute value. Also, let $A\colon \mathfrak{H} \supseteq \operatorname{dom}(A) \to \mathfrak{H}'$ and $B\colon \mathfrak{H} \supseteq \operatorname{dom}(B) \to \mathfrak{H}'$ be linear operators such that $\operatorname{dom}(|H_0|^{1/2}) \subseteq \operatorname{dom}(A) \cap \operatorname{dom}(B)$. We assume that for some (and hence for all) $b > 0$ the operators $A(|H_0| + b)^{-1/2}$ and $B(|H_0| + b)^{-1/2}$ are bounded and linear from $\mathfrak{H}$ to $\mathfrak{H}$.

At this point, defining $G_0 := |H_0| + 1$, we can consider, for any $z \in \rho(H_0)$, the *Birman-Schwinger operator*

$$K_z := [AG_0^{-1/2}][G_0(H_0 - z)^{-1}][BG_0^{-1/2}]^*, \tag{4.5}$$

which is linear and bounded from $\mathfrak{H}'$ to $\mathfrak{H}'$.

The second assumption we need is stated below.

Assumption II There exists $z_0 \in \rho(H_0)$ such that $-1 \notin \sigma(K_{z_0})$.

Again, recurring to Neumnn series, we can replace it with the following hypothesis, stronger but more manageable.

Assumption II′ There exists $z_0 \in \rho(H_0)$ such that $\|K_{z_0}\|_{\mathfrak{H}' \to \mathfrak{H}'} < 1$.

Before recalling the Birman-Schwinger principle, we properly define the formal perturbed operator $H_0 + V$ with $V = B^*A$.

Theorem 4.2 *Under Assumptions I and II, there exists a unique closed extension H of $H_0 + V$ such that* $\operatorname{dom}(H) \subseteq \operatorname{dom}(|H_0|^{1/2})$ *and the following representation formula holds true:*

$$(\phi, H\psi)_{\mathfrak{H}\to\mathfrak{H}} = (G_0^{1/2}\phi, (H_0G_0^{-1} + [BG_0^{-1/2}]^*AG_0^{-1/2})G_0^{1/2}\psi)_{\mathfrak{H}\to\mathfrak{H}}$$

for $\phi \in \operatorname{dom}(|H_0|^{1/2})$, $\psi \in \operatorname{dom}(H)$.

This result correspond to Theorem 5 in [40], where the operator H is obtained via the pseudo-Friedrichs extension. Note that following the alternative approach by Kato [43], the extension of $H_0 + B^*A$ is not only closed, but also quasi-selfadjoint ([50]). We refer to the paper of Hansmann and Krejčiřík [40] for a cost-benefit comparison of the two methods, and for a list of cases when the two extensions coincide.

Finally, we can exhibit the abstract Birman-Schwinger principle, for the proof of which we refer to Theorem 6, 7, 8 and Corollary 4 of [40].

Theorem 4.3 *Under Assumption I and II, we have:*

(i) if $z \in \rho(H_0)$, *then* $z \in \sigma_p(H)$ *if and only if* $-1 \in \sigma_p(K_z)$;
(ii) if $z \in \sigma_c(H_0) \cap \sigma_p(H)$ *and* $H\psi = z\psi$ *for* $0 \neq \psi \in \operatorname{dom}(H)$, *then* $A\psi \neq 0$ *and*

$$\lim_{\varepsilon \to 0^{\pm}} (K_{z+i\varepsilon} A\psi, \phi)_{\mathfrak{H}' \to \mathfrak{H}'} = -(A\psi, \phi)_{\mathfrak{H}' \to \mathfrak{H}'}$$

for all $\phi \in \mathfrak{H}'$.

In particular

(i) if $z \in \sigma_p(H) \cap \rho(H_0)$, *then* $\|K_z\|_{\mathfrak{H}' \to \mathfrak{H}'} \geq 1$;
(ii) if $z \in \sigma_p(H) \cap \sigma_c(H_0)$, *then* $\liminf_{\varepsilon \to 0^{\pm}} \|K_{z+i\varepsilon}\|_{\mathfrak{H} \to \mathfrak{H}'} \geq 1$.

While from the "in particular" part of the previous theorem one could infer a localization for the eigenvalues of H, the principle can be employed in a "negative" way to prove their absence when the norm of the Birman-Schwinger operator is strictly less than 1 uniformly respect to $z \in \rho(H_0)$. This is precisely stated in the next concluding result, corresponding to Theorem 3 in [40], which is even richer: it not only gives information on the absence of the eigenvalues, but also on the invariance of the spectrum of the perturbed operator.

Theorem 4.4 *Suppose Assumption I and that* $\sup_{z \in \rho(H_0)} \|K_z\|_{\mathfrak{H}' \to \mathfrak{H}'} < 1$. *Then we have:*

(i) $\sigma(H_0) = \sigma(H)$;
(ii) $\sigma_p(H) \cup \sigma_r(H) \subseteq \sigma_p(H_0)$ *and* $\sigma_c(H_0) \subseteq \sigma_c(H)$.

In particular, if $\sigma(H_0) = \sigma_c(H_0)$, *then* $\sigma(H) = \sigma_c(H) = \sigma_c(H_0)$.

4.2 Case Studies

We now specialize the situation from the abstract to more concrete settings.

Suppose that $\mathfrak{H} = \mathfrak{H}' = L^2(\mathbb{R}^d; \mathbb{C}^{N \times N})$, $N \in \mathbb{N}$, and V is the multiplication operator generated in $\mathfrak{H}$ by a matrix-valued (scalar-valued if $N = 1$) function $V : \mathbb{R}^d \to \mathbb{C}^{N \times N}$, with initial domain $\operatorname{dom}(V) = C_0^\infty(\mathbb{R}^d; \mathbb{C}^N)$. As customary, we consider the factorization of V given by the polar decomposition $V = UW$, where $W = \sqrt{V^*V}$ and the unitary matrix U is a partial isometry. Therefore we may set $A = \sqrt{W}$,

$B = \sqrt{W}U^*$ and consider the corresponding multiplication operators generated by A and B^* in $\mathfrak{H}$ with initial domain $C_0^\infty(\mathbb{R}^d; \mathbb{C}^N)$, denoted by the same symbols. In the end, we can factorize the potential V in two closed operators A and B^*. Via the Closed Graph Theorem, Assumption I is verified.

Furthermore, in general the operator K_z defined in (4.5) is a bounded extension of the classical Birman-Schwinger operator $A(H_0 - z)^{-1}B^*$ defined on $\mathrm{dom}(B^*)$. Since in our case the initial domain of B^* is $C_0^\infty(\mathbb{R}^d; \mathbb{C}^N)$, hence dense in $\mathfrak{H}$, we get that K_z is exactly the closure of $A(H_0 - z)^{-1}B^*$.

In conclusion, everything reduces to the study of $\|A(H_0 - z)^{-1}B^*\|_{\mathfrak{H}\to\mathfrak{H}}$: if there exists $z_0 \in \rho(H_0)$ such that this norm is strictly less than 1, then Theorem 4.3 holds; if this is true uniformly respect to $z \in \rho(H_0)$, then also Theorem 4.4 holds true.

Example 4.1 (*Non-self-adjoint Schrödinger operator*) Consider the operator $H = -\Delta + V$ with $V : \mathbb{R}^d \to \mathbb{C}$. For V we use the decomposition

$$V(x) = \operatorname{sgn} V(x)|V(x)| = \operatorname{sgn} V(x)|V(x)|^{1/2}|V(x)|^{1/2}$$

for almost every $x \in \mathbb{R}^d$. Here $\operatorname{sgn} z := z/|z|$ if $z \neq 0$ and $\operatorname{sgn} z := 0$ if $z = 0$. Thus we choose

$$A(x) := |V(x)|^{1/2} \quad \text{and} \quad B(x) := \operatorname{sgn} \overline{V}(x)|V(x)|^{1/2}. \tag{4.6}$$

One has the following result on the bound of the norm of the Birman-Schwinger operator.

Lemma 4.3 *Let $H := -\Delta + V$, with $V : \mathbb{R}^d \to \mathbb{C}$. For V we use the decomposition $V = B^*A$ with A and B as in* (4.6)*. Assume that $V \in L^{\gamma+d/2}(\mathbb{R}^d)$. Then there exists a positive constant $D_{\gamma,d}$ such that*

$$\|A(-\Delta - z)^{-1}B^*\|_{L^2\to L^2}^{\gamma+d/2} \leq D_{\gamma,d}|z|^{-\gamma}\|V\|_{L^{\gamma+d/2}}^{\gamma+d/2}, \quad \text{where} \begin{cases} \gamma = 1/2, & \text{if } d = 1, \\ 0 < \gamma \leq 1/2, & \text{if } d = 2, \\ 0 \leq \gamma \leq 1/2 & \text{if } d \geq 3. \end{cases} \tag{4.7}$$

From the bound (4.7) and using the Birman-Schwinger principle (as stated in Theorem 4.3 or Theorem 4.4) one immediately has the following result on the localisation of the point spectrum of H.

Theorem 4.5 *Let $H = -\Delta + V$, with $V : \mathbb{R}^d \to \mathbb{C}$. Assume that $V \in L^{\gamma+d/2}(\mathbb{R}^d)$. Then, there exists a positive constant $D_{\gamma,d}$ such that*

$$|z|^\gamma \leq D_{\gamma,d}\|V\|_{L^{\gamma+d/2}}^{\gamma+d/2} \quad \text{where} \begin{cases} \gamma = 1/2, & \text{if } d = 1, \\ 0 < \gamma \leq 1/2, & \text{if } d \geq 2. \end{cases} \tag{4.8}$$

Moreover, when $d \geq 3$ there exists a positive constant $D_{0,d}$ such that if

$$\|V\|_{L^{d/2}} < D_{0,d}$$

then $\sigma_p(H) = \emptyset$.

The first contribution to the proof of Theorem 4.5 goes back to the work by Abramov, Aslanyan and Davies [1] covering the case $d = 1$, $\gamma = 1/2$, where they found the optimal constant $D_{1/2,1} = 1/2$. The higher dimensional framework was first addressed by Frank in his seminal paper [38]. Later, together with Simon [39], they extended Theorem 4.5 to cover long-range potentials, namely considering $1/2 < \gamma < d/2$, but under a radial symmetry assumption. Very recently Bögli and Cuenin [7] (see also [6]) proved that the result in [38] was optimal, thus ultimately disproving the celebrated conjecture by Laptev and Safronov [51] claiming that the bound (4.8) had to be true in $d \geq 2$ also for $1/2 < \gamma < d/2$.

In order to extend to higher dimensions the result in [1], based on a pointwise bound for the Green function of $-\frac{d^2}{dx^2} - z$, and clearly highly sensitive of the one dimensional framework, Frank in [38] introduces a new method which has permeated all the subsequent works on the topic and which relies on the combination of the Birman-Schwinger principle together with suitable resolvent estimates. More specifically, he used the Kenig-Ruiz-Sogge estimates on the conjugate line for the resolvent of the Laplacian from [45], namely

$$\|(-\Delta - z)^{-1}\|_{L^p \to L^{p'}} \leq C|z|^{-d/2+d/p-1}, \quad \frac{2}{d+1} \leq \frac{1}{p} - \frac{1}{p'} \leq \frac{2}{d}, \tag{4.9}$$

where $1/p + 1/p' = 1$ and $2/(d+1) \leq 1/p - 1/p' \leq 2/d$ if $d \geq 3$ and $2/3 \leq 1/p - 1/p' < 1$ if $d = 2$. With estimate (4.9) at hands, the proof of Lemma 4.3 is immediate.

Proof of Lemma 4.3 Let $d \geq 3$. From Hölder inequality and using (4.9) we have

$$\begin{aligned}\|A(H_0 - z)^{-1}B^*\|_{L^2 \to L^2} &\leq \|A\|_{L^{2p/(2-p)}}\|(H_0 - z)^{-1}\|_{L^p \to L^{p'}}\|B^*\|_{L^{2p/(2-p)}} \\ &\leq C|z|^{-d/2+d/p-1}\|V\|_{L^{p/(2-p)}};\end{aligned}$$

setting $p/(2-p) = \gamma + d/2$ and observing that

$$\frac{2}{d+1} \leq \frac{1}{p} - \frac{1}{p'} \leq \frac{2}{d} \iff \frac{2d}{d+2} \leq p \leq \frac{2(d+1)}{d+3} \iff 0 \leq \gamma \leq \frac{1}{2},$$

then bound (4.7) follows. The lower dimensional setting $d = 2$ can be shown similarly. □

Example 4.2 (*Self-adjoint Schrödinger operator*) Now we consider the easier case of self-adjoint Schrödinger operators. Namely we consider $H = -\Delta + V$ with V real-valued. In this setting a much stronger result than Theorem 4.5 holds: this is represented by the so-called Lieb-Thirring inequalities, a fundamental tool developed to prove stability of matter (see, for instance, [54]).

Theorem 4.6 (Lieb-Thirring inequalities) *Let $H = -\Delta + V$ be the Schrödinger operator where $V : \mathbb{R}^d \to \mathbb{R}$ is a real-valued potential. Assume that the negative part of the potential $V_- := -\min\{0, V\}$ lies in $L^{\gamma+d/2}(\mathbb{R}^d)$ for some non-negative γ to be specified below. Let $E_0 \leq E_1 \leq \cdots \leq 0$ the non-positive eigenvalues of H. Then, there exists a positive constant $L_{\gamma,d}$ such that*

$$\sum_{j\geq 0} |E_j|^\gamma \leq L_{\gamma,d} \|V_-\|_{L^{\gamma+d/2}}^{\gamma+d/2}, \qquad \text{where} \begin{cases} \gamma \geq 1/2 & \text{if } d = 1, \\ \gamma > 0 & \text{if } d = 2, \\ \gamma \geq 0 & \text{if } d \geq 3. \end{cases} \tag{4.10}$$

The above estimates were proved in the non-endpoint case, namely for $\gamma > 1/2$ if $d = 1$ and for $\gamma > 0$ if $d \geq 2$, by Lieb and Thirring in the 1975 work [55]. The extension $\gamma = 1/2$ in the 1-dimensional case was obtained by Weidl [63], whereas the case $\gamma = 0$ for $d \geq 3$ was independently reached by Cwikel [24], Lieb [52] and Rozenbljum [59], and it is known as the CLR bound, which estimates from above the number of eigenvalues.

Remark 4.2 Some comments on this celebrated result follow. First of all the range of γ for which the inequalities (4.10) hold is optimal. In particular, one sees that $\gamma = 0$ is not allowed in lower dimensions $d = 1, 2$. This is in agreement with the criticality of $H_0 = -\Delta$ in lower dimensions or, what is the same, with the fact that any arbitrarily small non-trivial negative potential V always creates eigenvalues for the operator $H = -\Delta + V$. In light of this comment, inequalities (4.10) cannot hold for $\gamma = 0$ and $d = 1, 2$ as this would provide instead a smallness condition on the potential for excluding eigenvalues.

The main open question about the Lieb-Thirring inequalities regards the optimal constant $L_{\gamma,d}$. In particular it is still unknown its value in one of the most interesting physical case, namely when $d = 3$ and $\gamma = 1$. As a matter of fact, inequalities (4.10) for $\gamma = 1$ gives a bound for the sum of negative eigenvalues which was crucial for the proof of the stability of matter.

To conclude we just mention that, as in the non-self-adjoint setting, properties and structure of the Birman-Schwinger operator play a crucial role also in the proof of these inequalities. In particular, here the self-adjoint setting permits to employ the following stronger version of the principle: under suitable integrability conditions on the potential, the number N_z of negative eigenvalues of $-\Delta + V_-$ smaller than $-|z|$ equals the number B_z of eigenvalues greater than 1 of the Birman-Schwinger operator $K_z = \sqrt{V_-}(-\Delta + |z|)^{-1}\sqrt{V_-}$. In passing, note that in the self-adjoint framework only the negative part V_- of the potential counts indeed, via the variational min-max principle, a bound for the sum of eigenvalues powers of the full Hamiltonian $H = -\Delta + V$ is obtained from the analogous bound for the Hamiltonian $H = -\Delta + V_-$.

From the validity of Theorem 4.6 on has immediately the following bound for the single eigenvalue.

Theorem 4.7 *Let $H = -\Delta + V$ be the Schrödinger operator $V : \mathbb{R}^d \to \mathbb{R}$ a real-valued potential. Assume that the negative part of the potential $V_- := -\min\{0, V\}$ lies in $L^{\gamma+d/2}(\mathbb{R}^d)$ for some non-negative γ to be specified below. Then, there exists a positive constant $L_{\gamma,d}$ such that any non-positive eigenvalue E satisfies*

$$|E|^\gamma \leq L_{\gamma,d} \|V_-\|_{L^{\gamma+d/2}}^{\gamma+d/2}, \qquad \textit{where} \begin{cases} \gamma \geq 1/2 & \textit{if } d = 1, \\ \gamma > 0 & \textit{if } d \geq 2. \end{cases} \tag{4.11}$$

Moreover, when $d \geq 3$ there exists a positive constant $L_{0,d}$ such that if

$$\|V_-\|_{L^{d/2}} < L_{0,d}$$

there are no non-positive eigenvalue.

Estimates (4.11) for the single eigenvalue are known as Keller-type estimates as the first appearance of such bounds goes back to Keller [44] (see also [10, 55] for more recent results).

Even though inequalities (4.11) are immediate consequence of the Lieb-Thirring inequalities (4.10), one could give an alternative proof of (4.11) which is simply based on Sobolev inequalities. For the sake of completeness, we give a sketch of this proof below.

Proof of Theorem 4.7 We will see the proof only for $d \geq 3$, the lower dimensional case can be treated similarly. Since we are in the self-adjoint framework, the variational characterisation of the spectrum states that for any $u \in H^1(\mathbb{R}^d)$

$$\inf \sigma(-\Delta + V_-) = \inf_{\substack{\|u\|_{L^2}=1 \\ u \in H^1(\mathbb{R}^d)}} (u, (-\Delta - V_-)u).$$

Therefore, in order to get (4.11), it is sufficient to prove the following lower bound

$$(u, (-\Delta - V_-)u) \geq -L_{\gamma,d}^{1/\gamma} \|V_-\|_{\gamma+d/2}^{1+d/2\gamma} \tag{4.12}$$

for any $u \in H^1(\mathbb{R}^d)$ with $\|u\|_2 = 1$. Using Hölder and the interpolation inequality we have

$$\begin{aligned} (u, V_- u) &\leq \|V_-\|_{\gamma+d/2} \|u\|_{2(2\gamma+d)/(2\gamma+d-2)}^2 \\ &\leq \|V_-\|_{\gamma+d/2} \|u\|_2^{4\gamma/(2\gamma+d)} \|u\|_{2d/(d-2)}^{2d/(2\gamma+d)} \\ &\leq \varepsilon^{-1-d/2\gamma} \frac{2\gamma}{2\gamma+d} \|V_-\|_{\gamma+d/2}^{1+d/2\gamma} + \varepsilon^{1+2\gamma/d} \frac{d}{2\gamma+d} \|u\|_{2d/(d-2)}^2 \end{aligned}$$

for some ε to be chosen later. In the last bound we have used the Young inequality $ab \leq a^p/p + b^q/q$ valid for all positive a, b and for $1/p + 1/q = 1$.

As for the kinetic term, using the Sobolev inequality one has

$$(u, -\Delta u) \geq S_d \|u\|_{2d/(d-2)}^2,$$

where S_d denotes the optimal Sobolev constant (see [53]). Plugging the two estimates together one has

$$(u, (-\Delta - V_-)u) \geq -\varepsilon^{-1-d/2\gamma} \frac{2\gamma}{2\gamma+d} \|V_-\|_{\gamma+d/2}^{1+d/2\gamma} + \left(S_d - \varepsilon^{1+2\gamma/d} \frac{d}{2\gamma+d}\right) \|u\|_{2d/(d-2)}^2,$$

which gives bound (4.12) after choosing a suitable small ε. □

Example 4.3 (*Non-self-adjoint Dirac operator*) Let us consider now the non-self-adjoint Dirac operator $\mathcal{D}_{m,V} = \mathcal{D}_m + V$, where $\mathcal{D}_m$ is the free Dirac operator defined in (3.6) and V is a matrix-valued potential $V : \mathbb{R}^d \to \mathbb{C}^{N \times N}$, with $N = 2^{\lceil d/2 \rceil}$. The spectral studies for $\mathcal{D}_{m,V}$ were started by Cuenin, Laptev and Tretter in their celebrated work [22], for the 1-dimensional case. There they proved that if $V \in \mathbb{C}^{2\times 2}$ is a potential such that

$$\|V\|_{L^1(\mathbb{R})} = \int_{\mathbb{R}} |V(x)| dx < 1,$$

where $|V(\cdot)|$ is the operator norm of $V(\cdot)$ in $\mathbb{C}^2$ with the Euclidean norm, then every non-embedded eigenvalue $z \in \mathbb{C} \setminus \{(-\infty, -m] \cup [m, +\infty)\}$ of $\mathcal{D}_{m,V}$ lies in the union

$$z \in \overline{B}_{R_0}(x_0^-) \cup \overline{B}_{R_0}(x_0^+) \tag{4.13}$$

of two disjoint closed disks in the complex plane, with centers and radius respectively

$$x_0^{\pm} = \pm m \sqrt{\frac{\|V\|_1^4 - 2\|V\|_1^2 + 2}{4(1 - \|V\|_1^2)} + \frac{1}{2}}, \quad R_0 = m \sqrt{\frac{\|V\|_1^4 - 2\|V\|_1^2 + 2}{4(1 - \|V\|_1^2)} - \frac{1}{2}}.$$

In particular, in the massless case the spectrum is $\sigma(\mathcal{D}_{0,V}) = \mathbb{R}$. Moreover, this inclusion is shown to be optimal. Again, the proof relies on the combination of the Birman-Schwinger principle with a resolvent estimate for the free Dirac operator, namely

$$\|(\mathcal{D}_m - z)^{-1}\|_{L^1(\mathbb{R}) \to L^\infty(\mathbb{R})} \leq \sqrt{\frac{1}{2} + \frac{1}{4}\left|\frac{z+m}{z-m}\right| + \frac{1}{4}\left|\frac{z-m}{z+m}\right|}.$$

In some sense, this is the counterpart for the Dirac operator of the above-cited Abramov-Aslanyan-Davies inequality for the Schrödinger operator in 1-dimension.

One could ask if, in the same fashion of Frank's argument in [38], one can combine the Birman-Schwinger principle with $L^p - L^{p'}$ resolvent estimates for the free Dirac operator, to derive Keller-type inequalities for the perturbed Dirac operator. Unfortunately, these reasoning can not be straightforwardly applied, since such Kenig-Ruiz-

Sogge-type estimates does not exists in the case of Dirac for dimension $d \geq 2$, as observed by Cuenin in [19]. Indeed, due to the Stein-Thomas restriction theorem and standard estimates for Bessel potentials, the resolvent $(\mathcal{D}_m - z)^{-1} : L^p(\mathbb{R}^d) \to L^{p'}(\mathbb{R}^d)$ is bounded uniformly for $|z| > 1$ if and only if

$$\frac{2}{d+1} \leq \frac{1}{p} + \frac{1}{p'} \leq \frac{1}{d},$$

hence the only possible choice is $(d, p, p') = (1, 1, \infty)$. For the Schrödinger operator the situation is much better since the right-hand side of the above range is replaced by $2/d$, as per the Kenig-Ruiz-Sogge estimates.

For the higher dimensional case $d \geq 2$, we may refer among others to the works [20, 23, 28, 35] where the eigenvalues are localized in terms of L^p-norm of the potential, but, differently to the 1-dimensional case, the confinement region is unbounded around the spectrum of the free operator, namely $\sigma(\mathcal{D}_m) = (-\infty, -m] \cup [m, +\infty)$.

Recently, two of the authors of this survey in [30], together with D'Ancona, were able to get in higher dimension a bounded enclosure region asking smallness of the potential in a suitable mixed Lebesgue space. More precisely, introducing the functional space

$$Y \equiv Y(\mathbb{R}^d) := \bigcap_{j=1}^{d} L^1_{x_j} L^\infty_{\hat{x}_j}(\mathbb{R}^d), \qquad \hat{x}_j := (x_1, \dots, x_{j-1}, x_{j+1}, \dots, x_d) \in \mathbb{R}^{d-1},$$

they showed that for suitably Y-small V, namely $\|V\|_Y < C_0$ for suitable positive constant C_0, then any eigenvalue $z \in \sigma_p(\mathcal{D}_{m,V})$ of $\mathcal{D}_{m,V}$ lies again in the union of two closed disks in $\mathbb{C}$ (cf. (4.13)) with centers and radius given by

$$x_0^\pm := \pm m \frac{\nu^2 + 1}{\nu^2 - 1}, \qquad R_0 := m \frac{2\nu}{\nu^2 - 1}, \qquad \nu \equiv \nu(V) := \left[\frac{(d+1)C_0}{\|V\|_Y} - d \right]^2 > 1.$$

In the massless case they proved spectral stability under the same smallness condition about V, namely $\sigma(\mathcal{D}_{0,V}) = \mathbb{R}$. As above, the proof comes again as a consequence of the machinery combining the Birman-Schwinger principle with uniform resolvent estimates. In particular, the following Agmon-Hörmander-type estimates were crucial for proving the result:

$$\|(\mathcal{D}_m - z)^{-1}\|_{X \to X^*} \leq \left[d + \left| \frac{z+m}{z-m} \right|^{\operatorname{sgn}(\Re z)/2} \right] \tag{4.14}$$

where the functional space X and its dual X^* are

$$X \equiv X(\mathbb{R}^d) := \bigcap_{j=1}^{d} L^1_{x_j} L^2_{\hat{x}_j}(\mathbb{R}^d), \qquad X^* \equiv X(\mathbb{R}^d) := \sum_{j=1}^{d} L^\infty_{x_j} L^2_{\hat{x}_j}(\mathbb{R}^d).$$

For previous resolvent estimates in the spirit of (4.14) we refer to [26, 27, 33]. In the direction of spectral stability result we should also mention the recent work [29] where the authors was able to prove spectral stability of perturbed Dirac and Klein-Gordon operators under suitable pointwise smallness/decay at infinity conditions for the potential. Last but not least in [56] the matricial structure of the Dirac operator was fully exploited to obtain Keller-type estimates dropping the smallness assumption on the potential. More precisely, the authors required the potentials to be of the form $V = vW$, where $v\colon \mathbb{R}^d \to \mathbb{C}$ is a scalar function with the desired integrability, whereas W is a constant matrix satisfying suitable rigidity conditions. Other remarkable results can be found in [14, 20, 21, 60].

This technique presented above is very robust and has produced interesting results also for other models than Schrödinger and Dirac operators. Among second order operators, we mention the Lamé operator of elasticity and we refer to [11, 12, 16] for the corresponding results in the spirit of the ones available for Schrodinger. For higher order operators, we refer to [41] for the application of the Birman-Schwinger principle to the bilaplacian in lower dimensions, see also [32]. We also mention that the same technique has been applied successfully to a discrete setting. In this respect we refer to [13, 42] and the recent preprint [49].

Acknowledgements The research of the first author (L.C.) is Funded by the Deutsche Forschungsgemeinschaft (DFG, German Research Foundation) - Project-ID 258734477 - SFB 1173. The third author (N.M.S.) is supported by the EXPRO grant No. 20-17749X of the Czech Science Foundation. He is also member of the Gruppo Nazionale per L'Analisi Matematica, la Probabilità e le loro Applicazioni (GNAMPA) of the Istituto Nazionale di Alta Matematica (INdAM).

References

1. A.A. Abramov, A. Aslanyan, E.B. Davies, Bounds on complex eigenvalues and resonances. J. Phys. A **34**(1), 57–72 (2001)
2. S. Albeverio, On bound states in the continuum of N-body systems and the virial theorem. Ann. Phys. **71**, 167–276 (1972)
3. W.O. Amrein, *Hilbert Space Methods in Quantum Mechanics* (EPFL Press, 2009)
4. S. Avramska-Lukarska, D. Hundertmark, H. Kovarik. Absence of positive eigenvalues of magnetic Schrödinger operators. Calc. Var. **62**(63) (2023)
5. M.V. Birman, On the spectrum of singular boundary-value problems. Mat. Sb. (N.S.) **55**(97), 125–174 (1961)
6. S. Bögli, Schrödinger operator with non-zero accumulation points of complex eigenvalues. Comm. Math. Phys. **352**(2), 629–639 (2017)
7. S. Bögli, J.-C. Cuenin, Counterexample to the Laptev-Safronov conjecture. Comm. Math. Phys. **398**, 1349–1370 (2023)
8. N. Boussaid, P. D'Ancona, L. Fanelli, Virial identity and weak dispersion for the magnetic Dirac equation. J. Math. Pures Appl. (9), **95**(2), 137–150 (2011)
9. F. Cacciafesta. Virial identity and dispersive estimates for the n-dimensional Dirac equation. J. Math. Sci. Univ. Tokyo **18**(4), 441–463 (2011, 2012)
10. E.A. Carlen, R.L. Frank, E.H. Lieb, Stability estimates for the lowest eigenvalue of a Schrödinger operator. Geom. Funct. Anal. **24**(1), 63–84 (2014)

11. B. Cassano, L. Cossetti, L. Fanelli, Eigenvalue bounds and spectral stability of Lamé operators with complex potentials. J. Differ. Equ. **298**, 528–559 (2021)
12. B. Cassano, L. Cossetti, L. Fanelli, Spectral enclosures for the damped elastic wave equation. Math. Eng. **4**(6), 1–10 (2022)
13. B. Cassano, O.O. Ibrogimov, D. Krejčiřík, F. Štampach, Location of eigenvalues of non-self-adjoint discrete Dirac operators. Ann. Henri Poincaré **21**(7), 2193–2217 (2020)
14. B. Cassano, F. Pizzichillo, L. Vega, A Hardy-type inequality and some spectral characterizations for the Dirac-Coulomb operator. Rev. Mat. Complut. **33**(1), 1–18 (2020)
15. L. Cossetti, Uniform resolvent estimates and absence of eigenvalues for Lamé operators with complex potentials. J. Math. Anal. Appl. **455**(1), 336–360 (2017)
16. L. Cossetti. Bounds on eigenvalues of perturbed Lamé operators with complex potentials. Math. Eng. **4**(5), Paper No. 037, 29 (2022)
17. L. Cossetti, L. Fanelli, D. Krejčiřík, Absence of eigenvalues of Dirac and Pauli Hamiltonians via the method of multipliers. Comm. Math. Phys. **379**(2), 633–691 (2020)
18. L. Cossetti, D. Krejčiřík, Absence of eigenvalues of non-self-adjoint Robin Laplacians on the half-space. Proc. Lond. Math. Soc. **121**(3), 584–616 (2020)
19. J.-C. Cuenin, Estimates on complex eigenvalues for Dirac operators on the half-line. Integral Equ. Oper. Theory **79**(3), 377–388 (2014)
20. J.-C. Cuenin, Eigenvalue bounds for Dirac and fractional Schrödinger operators with complex potentials. J. Funct. Anal. **272**(7), 2987–3018 (2017)
21. J.-C. Cuenin, Eigenvalue estimates for bilayer graphene. Ann. Henri Poincaré **20**(5), 1501–1516 (2019)
22. J.-C. Cuenin, A. Laptev, C. Tretter, Eigenvalue estimates for non-selfadjoint Dirac operators on the real line. Ann. Henri Poincaré **15**(4), 707–736 (2014)
23. J.-C. Cuenin, C. Tretter, Non-symmetric perturbations of self-adjoint operators. J. Math. Anal. Appl. **441**(1), 235–258 (2016)
24. M. Cwikel, Weak type estimates for singular values and the number of bound states of Schrödinger operators. Ann. Math. (2), **106**(1), 93–100 (1977)
25. H.L. Cycon, R.G. Froese, W. Kirsch, B. Simon, *Schrödinger Operators with Application to Quantum Mechanics and Global Geometry, study* (Texts and Monographs in Physics (Springer, Berlin, 1987)
26. P. D'Ancona, L. Fanelli, Decay estimates for the wave and Dirac equations with a magnetic potential. Comm. Pure Appl. Math. **60**(3), 357–392 (2007)
27. P. D'Ancona, L. Fanelli, Strichartz and smoothing estimates of dispersive equations with magnetic potentials. Comm. Partial Differ. Equ. **33**(6), 1082–1112 (2008)
28. C. Dubuisson, On quantitative bounds on eigenvalues of a complex perturbation of a Dirac operator. Integral Equ. Oper. Theory **78**(2), 249–269 (2014)
29. P. D'Ancona, L. Fanelli, D. Krejčiřík, N.M. Schiavone, Localization of eigenvalues for non-self-adjoint Dirac and Klein-Gordon operators. Nonlinear Anal. **214**, 112565 (2022)
30. P. D'Ancona, L. Fanelli, N.M. Schiavone, Eigenvalue bounds for non-selfadjoint Dirac operators. Math. Ann. **383**, 621–644 (2022)
31. D.E. Edmunds, W.D. Evans, *Spectral Theory and Differential Operators Oxford Mathematical Monographs* (Oxford University Press, Oxford, 2018)
32. A. Enblom, Estimates for eigenvalues of Schrödinger operators with complex-valued potentials. Lett. Math. Phys. **106**(2), 197–220 (2016)
33. M.B. Erdoğan, M. Goldberg, W.R. Green, Limiting absorption principle and Strichartz estimates for Dirac operators in two and higher dimensions. Comm. Math. Phys. **367**(1), 241–263 (2019)
34. D.M. Èĭdus. On the principle of limiting absorption. Mat. Sb. (N.S.) **57**(99), 13–44 (1962)
35. L. Fanelli, D. Krejčiřík, Location of eigenvalues of three-dimensional non-self-adjoint Dirac operators. Lett. Math. Phys. **109**(7), 1473–1485 (2019)
36. L. Fanelli, D. Krejčiřík, L. Vega, Absence of eigenvalues of two-dimensional magnetic Schrödinger operators. J. Funct. Anal. **275**(9), 2453–2472 (2018)

37. L. Fanelli, D. Krejčiřík, L. Vega, Spectral stability of Schrödinger operators with subordinated complex potentials. J. Spectr. Theory **8**(2), 575–604 (2018)
38. R.L. Frank, Eigenvalue bounds for Schrödinger operators with complex potentials. Bull. Lond. Math. Soc. **43**(4), 745–750 (2011)
39. R.L. Frank, B. Simon, Eigenvalue bounds for Schrödinger operators with complex potentials. II. J. Spectr. Theory **7**(3), 633–658 (2017)
40. M. Hansmann, D. Krejčiřík, The abstract Birman-Schwinger principle and spectral stability. JAMA **148**, 361–398 (2022)
41. O.O. Ibrogimov, D. Krejčiřík, A. Laptev, Sharp bounds for eigenvalues of biharmonic operators with complex potentials in low dimensions. Math. Nachr. **294**(7), 1333–1349 (2021)
42. O.O. Ibrogimov, F. Štampach, Spectral enclosures for non-self-adjoint discrete Schrödinger operators. Integr. Eqn. Oper. Theory **91**(6), 1–15 (2019)
43. T. Kato, Wave operators and similarity for some non-selfadjoint operators. Math. Ann. **162**, 258–279 (1965/1966)
44. J.B. Keller, Lower bounds and isoperimetric inequalities for eigenvalues of the Schrödinger equation. J. Math. Phys. **2**, 262–266 (1961)
45. C.E. Kenig, A. Ruiz, C.D. Sogge, Uniform Sobolev inequalities and unique continuation for second order constant coefficient differential operators. Duke Math. J. **55**(2), 329–347 (1987)
46. R. Konno, S.T. Kuroda, On the finiteness of perturbed eigenvalues. J. Fac. Sci. Univ. Tokyo Sect. **I**(13), 55–63 (1966)
47. D. Krejčiřík, Geometrical aspects of spectral theory, http://nsa.fjfi.cvut.cz/david/
48. D. Krejčiřík, Mathematical aspects of quantum mechanics with non-self-adjoint operators, http://nsa.fjfi.cvut.cz/david/
49. D. Krejčiřík, A. Laptev, F. Stampach, Spectral enclosures and stability for non-self-adjoint discrete Schrödinger operators on the half-line. Bull. London Math. Soc. **54**, 2379–2403 (2022)
50. D. Krejčiřík, P. Siegl, Elements of spectral theory without the spectral theorem, in *Non-selfadjoint Operators in Quantum Physics: Mathematical Aspects*, ed. by F. Bagarello, J.-P. Gazeau, F.H. Szafraniec, M. Znojil (Wiley-Interscience, 2015), p. 432
51. A. Laptev, O. Safronov, Eigenvalue estimates for Schrödinger operators with complex potentials. Comm. Math. Phys. **292**(1), 29–54 (2009)
52. E.H. Lieb, The number of bound states of one-body Schroedinger operators and the Weyl problem, in *Geometry of the Laplace Operator (Proceedings of Symposia in Pure Mathematics, University. Hawaii, Honolulu, Hawaii, 1979)*, Proceedings of Symposia in Pure Mathematics, vol. XXXVI (American Mathematical Society, Providence, 1980), pp. 241–252
53. E.H. Lieb, M. Loss, *Analysis*, Graduate Studies in Mathematics, vol. 14, 2nd end. (American Mathematical Society, Providence, 2001)
54. E.H. Lieb, R. Seiringer, *The Stability of Matter in Quantum Mechanics* (Cambridge University Press, Cambridge, 2010)
55. E.H. Lieb, W.E. Thirring, Inequalities for the moments of the eigenvalues of the Schrödinger Hamiltonian and their relation to Sobolev inequalities, in *The Stability of Matter: From Atoms to Stars* (Springer, 1997), pp. 203–237
56. H. Mizutani, N.M. Schiavone, Spectral enclosures for Dirac operators perturbed by rigid potentials. Rev. Math. Phys. **34**(08), 2250023 (2022)
57. E. Mourre, Absence of singular continuous spectrum for certain selfadjoint operators. Comm. Math. Phys. **78**(3), 391–408 (1980/81)
58. M. Reed, B. Simon, *Methods of Modern Mathematical Physics IV*. Analysis of Operators (Academic [Harcourt Brace Jovanovich, Publishers], New York, 1978)
59. G.V. Rozenbljum, Distribution of the discrete spectrum of singular differential operators. Izv. Vysš. Učebn. Zaved. Matematika **1**(164), 75–86 (1976)
60. O. Safronov, A. Laptev, F. Ferrulli, Eigenvalues of the bilayer graphene operator with a complex valued potential. Anal. Math. Phys. **9**(3), 1535–1546 (2019)
61. J. Schwinger, On the bound states of a given potential. Proc. Nat. Acad. Sci. U.S.A. **47**, 122–129 (1961)

62. B. Thaller, *The Dirac Equation, Texts and Monographs in Physics* (Springer, Berlin, 1992)
63. T. Weidl, On the Lieb-Thirring constants $L_{\gamma,1}$ for $\gamma \geq 1/2$. Comm. Math. Phys. **178**(1), 135–146 (1996)

Boussinesq, Schrödinger and Euler-Korteweg

Jean-Claude Saut and Li Xu

Abstract The aim of this paper of survey nature is to emphasize the links between two specific strongly dispersive "abcd" Boussinesq systems and nonlinear Schrödinger equations, one of those systems being a particular case of the Euler-Korteweg (EK) systems. We will in particular translate results for the later the known results for the EK system. We also comment on a one-dimensional integrable version of those systems, known as the Kaup-Broer-Kupperschmidt system. Finally we discuss another Boussinesq system that can be viewed as a (weakly) dispersive perturbation of the Saint-Venant (shallow-water) system.

1 Introduction: Boussinesq

We are interested in this paper in mathematical issues for particular cases of the so-called abcd Boussinesq systems for surface water[1] waves, see [15–17]

$$\begin{cases} \eta_t + \nabla \cdot \mathbf{v} + \epsilon \nabla \cdot (\eta \mathbf{v}) + \mu[a\nabla \cdot \Delta \mathbf{v} - b\Delta \eta_t] = 0 \\ \mathbf{v}_t + \nabla \eta + \epsilon \frac{1}{2} \nabla |\mathbf{v}|^2 + \mu[c\nabla \Delta \eta - d\Delta \mathbf{v}_t] = 0. \end{cases} \tag{1.1}$$

[1]Boussinesq [18] was the first to derive a particular Boussinesq system, not in the class of those studied here though. We refer to [27–29] for details and for an excellent history of Hydrodynamic in the nineteenth century.

To Professor Tohru Ozawa with admiration and friendship.

J.-C. Saut (✉)
Laboratoire de MathÉmatiques, UMR 8628, Université Paris-Saclay et CNRS, 91405 Orsay, France
e-mail: jean-claude.saut@universite-paris-saclay.fr

L. Xu
School of Mathematical Sciences, Beihang University, Beijing 100191, China
e-mail: xuliice@buaa.edu.cn

S. Machihara (ed.), *Mathematical Physics and Its Interactions*, Springer Proceedings in Mathematics & Statistics 451, https://doi.org/10.1007/978-981-97-0364-7_9

Here η is the elevation of the wave, $\mathbf{v}$ is a measure of the horizontal velocity, μ and ϵ are the small parameters (shallowness and nonlinearity parameters respectively) defined as

$$\mu = \frac{h^2}{\lambda^2}, \quad \epsilon = \frac{\alpha}{h}$$

where α is a typical amplitude of the wave, h a typical depth and λ a typical horizontal wavelength.

In the Boussinesq regime, ϵ and μ are supposed to be of same order, $\epsilon \sim \mu \ll 1$, and we will take for simplicity $\epsilon = \mu$, writing (1.1) as[2]

$$\begin{cases} \eta_t + \nabla \cdot \mathbf{v} + \epsilon[\nabla \cdot (\eta \mathbf{v}) + a\nabla \cdot \Delta \mathbf{v} - b\Delta \eta_t] = 0 \\ \mathbf{v}_t + \nabla \eta + \epsilon[\frac{1}{2}\nabla |\mathbf{v}|^2 + c\nabla \Delta \eta - d\Delta \mathbf{v}_t] = 0, \end{cases} \tag{1.2}$$

Remark 1 When $\mathbf{v}$ is curl free, i.e., curl $\mathbf{v} = 0$, a natural assumption in the context of Boussinesq systems, one has $\mathbf{v} \cdot \nabla \mathbf{v} = \frac{1}{2}\nabla |\mathbf{v}|^2$.

The coefficients (a, b, c, d) are restricted by the condition

$$a + b + c + d = \frac{1}{3} - \tau,$$

where $\tau \geq 0$ is the surface tension coefficient.

When restricted to one-dimensional, unidirectional motions, (1.2) leads to the same KdV equation, see [44]

$$u_t + u_x + \epsilon \left(\frac{1}{3} - \tau \right) u_{xxx} + \epsilon u u_x = 0.$$

The class of systems (1.1), (1.2) models water waves on a flat bottom propagating in both directions in the aforementioned regime (see [15–17]). From a modeling viewpoint those systems have the advantage with respect to the Korteweg-de Vries equations to allow bi-directional waves and to be possibly two-dimensional. They also allow, for instance via Padé approximants, to fit better their dispersion relation (by choosing appropriate parameters a,b,c,d) with the dispersion of the original water waves system.

On the mathematical side, they display a fascinating richness because, although their dispersion relations are essentially the same for long waves (small wave numbers), leading to the same error estimates with the full water wave system on time scales of order $O(1/\epsilon)$, see [15], they differ drastically in a short wave regime, with dispersion of order between 3 and -1, leading to very different dispersive properties and though long time dynamics. Actually the mathematical study of such systems is far from being complete. We refer for instance to [42, 43, 49] for recent results.

---

[2] particular cases are formally derived in [19, 30].

Eliminating n from the second equation yields

$$-\epsilon\phi'' + (1-c^2)\phi + \frac{3}{2}c\epsilon\phi^2 - \frac{\epsilon^2}{2}\phi^3 = 0. \tag{2.10}$$

From Theorem 5 in [14] non trivial solitary waves exist if and only if $|c| < 1$. Actually (2.10) can be integrated by standard methods, leading, for $|c| < 1$, a unique even solution (up to translations) which reads when $\epsilon = 1$, see [4]:

$$v_{c,1}(\xi) = \frac{2(1-c^2)}{\cosh(\sqrt{1-c^2}\xi) - c}, \tag{2.11}$$

from which we deduce the expression of $u_{c,\epsilon}$:

$$v_{c,\epsilon}(\xi) = \frac{2(1-c^2)}{\epsilon[\cosh(\sqrt{1-c^2}\epsilon^{-1/2}\xi) - c]} \tag{2.12}$$

We now turn to stability issues for the above solitary wave, following Angulo [4] who solved the problem when $\epsilon = 1$ and actually proved using the method of [34] that the solitary wave is orbitally stable in $H^1(\mathbb{R}) \times L^2(\mathbb{R})$ for $|c| < 1$. This uses the fact that $\mathbf{U}_c$ can be viewed as a critical point of the functional $F = \Phi_1 - c\Phi_0$, that is it satisfies

$$(\Phi_1 - c\Phi_0)'(\mathbf{U}_c) = 0$$

where

$$\Phi_1(v,\eta) = \frac{1}{2}\int_{\mathbb{R}} [v^2 + \eta^2 + v^2\eta + (v_x)^2]dx,$$

and

$$\Phi_0(v,\eta) = \int_{\mathbb{R}} v\eta dx.$$

Using the non trivial conservation law, Angulo proved moreover that the Caux-chy problem is globally well-posed for initial data close to the solitary wave.

2.1.2 Remarks on the Stationary Solutions

We comment here on the stationary solutions of (1.4). Assuming that such a solution vanishes at infinity. In one spatial dimension (η, v) should satisfy

$$-v'' + v - \frac{\epsilon^2}{2}|v|^2 v = 0, \tag{2.13}$$

$$\eta = -\frac{\epsilon}{2}|v|^2.$$

and v can be expressed in terms of the profile of the soliton of the focusing cubic nonlinear Schrödinger equation, leading to the explicit form

$$v(x) = \frac{2}{\epsilon \cosh(\epsilon^{-1/2}x)}.$$

In the two-dimensional case, one has

$$\nabla \cdot [-\Delta \mathbf{v} + \mathbf{v} - \frac{\epsilon^2}{2}|\mathbf{v}|^2 \mathbf{v}] = 0, \tag{2.14}$$

$$\eta = -\frac{\epsilon}{2}|\mathbf{v}|^2.$$

Any solution $\mathbf{v}$ of

$$-\Delta \mathbf{v} + \mathbf{v} - \frac{\epsilon^2}{2}|\mathbf{v}|^2 \mathbf{v} = 0, \tag{2.15}$$

solves (2.14).

Equation (2.15) is a particular case of the equation of a bound state solution of a vector nonlinear Schrödinger equation. It is proven in [24] that (2.15) has a solution whose components are constant multiples of the ground state of the corresponding *scalar* nonlinear Schrödinger equation.

In both cases it would be interesting to investigate the stability of those solutions with respect to (1.4).

2.1.3 More on the Kaup-Broer-Kuperschmidt System

We give more details to a system which is the one-dimensional version of (1.2) when $a = 1, b = c = d = 0$, that is

$$\begin{cases} \eta_t + \nabla \cdot \mathbf{v} + \epsilon \nabla \cdot (\eta \mathbf{v}) + \alpha \epsilon \Delta \nabla \cdot \mathbf{v} = 0, \;\; \alpha = \pm 1 \\ \mathbf{v}_t + \nabla \eta + \frac{\epsilon}{2} \nabla |\mathbf{v}|^2 = 0, \end{cases} \tag{2.16}$$

The integrable system is the one-dimensional version of (2.16)[8] namely, when $\epsilon = 1$

$$\begin{cases} \eta_t + v_x + (\eta v)_x + \alpha v_{xxx} = 0 \\ v_t + \eta_x + v v_x = 0, \end{cases} \tag{2.17}$$

[8] The Kaup system is usually written with $a = 1/3, b = c = d = 0$ but one reduces to this case by a trivial change of scale in x and t.

This system known as the Kaup-Broer-Kuperschmidt system, see [19, 37, 41] and below.

With the plus sign, $\alpha = 1$, (2.17) has a Hadamard ill-posed linear part, making it useless as an asymptotic model.

More precisely it is immediate to check that its linearization at zero is Hadamard ill-posed since the dispersion relation is $\omega^2 = k^2 - k^4$, showing the short wave ill-posedness. It was established in [2] that nonlinearity does not erase the problem, making the Cauchy problem ill-posed in all Sobolev spaces in the sense that arbitrary small smooth solutions can blow-up in arbitrary short time in Sobolev spaces norms.[9]

Remark 7 Despite the fact that the Kaup-Broer-Kupperschmidt system with the plus sign is linearly ill-posed, it possesses solitary wave solutions $(\eta(x-kt), v(x-kt))$ when $k > 1$.[10] Actually following [23], such a solitary wave satisfies the equations

$$(v')^2 = R_1(u) \equiv \frac{3}{4}v^2(v-(2k-2))(v-(2k+2)), \quad \eta = v\left(k - \frac{v}{2}\right),$$

where the derivative is taken with respect to $\xi = x - kt$.

By studying the function R_1, one shows that there exists a unique solution $v(\xi)$ which is even and monotone for $\xi > 0$ for any $k > 1$. Actually, the solution $v(\xi)$ reads

$$v(\xi) = \frac{2(k^2-1)}{\cosh\left(\sqrt{3(k^2-1)}\xi\right) + k}.$$

And we also have

$$\eta(\xi) = \frac{2(k^2-1)\left(k\cosh\left(\sqrt{3(k^2-1)}\xi\right)+1\right)}{\left(\cosh\left(\sqrt{3(k^2-1)}\xi\right)+k\right)^2}.$$

When $1 < k \le 2$, the corresponding $\eta(\xi)$ is also monotonically decreasing for $\xi > 0$ while this is no longer true when $k > 2$.

One notices that for $1 < k \le 2$, $\|v\|_{L^\infty} = \|\eta\|_{L^\infty} = 2(k-1)$.

Note however that contrary to (1.4), (2.17) has no H^1 stationary solutions.

Remark 8 Again following [41], the change of variable $v = v, \eta = h - v_x$ transforms the Kaup-Broer-Kupperschmidt system with $\alpha = 1$ into the (ill-posed) system (compare to (2.8)):

$$\begin{cases} h_t + v_x + h_{xx} + (vh)_x = 0, \\ v_t + h_x - v_{xx} + vv_x = 0 \end{cases} \tag{2.18}$$

with dispersion $\omega(k) = \pm ik\sqrt{1-k^2}$.

---

[9] Similar ill-posedness results are proven in [2] for other, linearly ill-posed, members of the abcd class of Boussinesq systems.

[10] In the classical version $a = 1/3$.

On the other hand, as it was already noticed in a previous section, in the well-posed case $\alpha = -1$, the change of variable $v = v, \eta = h + iv_x$ leads to the (linearly well-posed) system

$$\begin{cases} h_t + v_x - ih_{xx} + (vh)_x = 0, \\ v_t + h_x + iv_{xx} + vv_x = 0, \end{cases} \tag{2.19}$$

with dispersion $\omega(k) = \pm ik(1+k^2)^{1/2}$.

Remark 9 Observe however that (2.16) can be turned in a well-posed system by using the BBM trick in the first equation, yielding the system:

$$\begin{cases} \eta_t + \nabla \cdot \mathbf{v} + \epsilon \nabla \cdot (\eta \mathbf{v}) - \epsilon \Delta \eta_t = 0 \\ \mathbf{v}_t + \nabla \eta + \epsilon \frac{1}{2} \nabla |\mathbf{v}|^2 = 0, \end{cases} \tag{2.20}$$

New considerations on the Kaup-Broer-Kupperschmidt system are given in the recent paper [52] where it is recalled that Broer [19] and Kaup [37] derived the system as an asymptotic model of surface waves in dimensional form

$$\begin{cases} \eta_t + h_0 \phi_{xx} + (\eta \phi_x)_x + \left(\frac{h_0^3}{3} - \frac{h_0 \tau}{\rho g} \right) \phi_{xxx} = 0 \\ \phi_t + \frac{1}{2}(\phi_x)^2 + g\eta = 0, \end{cases} \tag{2.21}$$

where η is the elevation of the wave, ϕ the velocity potential evaluated at the free surface, the constant h_0 is the quiescent water depth, $g > 0$ is the gravitational acceleration, $\tau \geq 0$ is the surface tension coefficient and ρ the density of the fluid.

For pure gravity waves ($\tau = 0$) or gravity-capillarity waves with small surface tension ($\tau/(g\rho) < h_0^2/3$, corresponding to $b = d = c = 0, a > 0$ in (1.2)) Kaup has shown that (2.21) can be formally solved by the Inverse Scattering Method, Kuperschmidt proving the complete integrability. Nevertheless, considering that this Boussinesq system is ill-posed, this property seems irrelevant to solve the Cauchy problem, but it allows the construction of one and N-solitons, periodic solitons, undular bores see for instance [33, 37, 52]. On the other hand the case with strong surface tension, that is $\tau/(g\rho) > h_0^2/3$ corresponds to $b = d = c = 0, a < 0$ in (1.2) and thus to (1.4). It also have an integrable structure, for instance Kaup [37] fonds the Lax pair (see also [51]), Matveev and Yavor [47] computed finite gap solutions (also in the case of small surface tension).

Two-dimensional integrable generalizations of the Kaup-Broer-Kuperschmidt system are given in [52]. None of them seems physically relevant however.

On the other hand for capillary waves such that $\tau/(g\rho) > h_0^2/3$, (2.21) is analogous to (1.4).

Remark 10 The Kaup-Broer-Kuperschmidt system was studied from the Inverse Scattering point of view in a recent paper [63] under the form

$$\begin{cases} v_t = \frac{1}{2}(v^2 + 2w - v_x)_x \\ w_t = (vw + \frac{1}{2}w_x)_x, \end{cases} \tag{2.22}$$

which is an alternative form of (2.18). Using this version of the system, (2.22) is the compatibility condition of the following Lax pair

$$\psi_x = U\psi, \quad \psi_t = V\psi, \tag{2.23}$$

with

$$\begin{cases} U = -ik\sigma_3 + Q, \quad Q = \frac{v}{2}\sigma_3 + \sigma_+ - w\sigma_-, \\ V = k^2\sigma_3 + \widehat{Q}, \quad \widehat{Q} = -\frac{1}{4}(v_x - v^2)\sigma_3 + (ik + \frac{v}{2})\sigma_+ - \frac{1}{2}(w_x + wv)\sigma_- \end{cases} \tag{2.24}$$

where k is a spectral parameter, σ_j, $j = 1, 2, 3$ are the classical Pauli matrices and $\sigma_+ = \frac{1}{2}(\sigma_1 + i\sigma_2)$, $\sigma_- = \frac{1}{2}(\sigma_1 - i\sigma_2)$.

The Inverse Scattering mechanism described in [63] allows the construction of special solutions such as kinks, see [63] for details.

Remark 11 There is so far no rigorous result on the complete resolution of the Cauchy problem for the well-posed Kaup-Broer-Kuperschmidt system ($\alpha = -1$) (under an appropriate functional setting for the initial data and the solution) by using inverse scattering techniques, and giving possibly insight on the qualitative behavior of the solutions.This would include a rigorous theory for the direct and inverse scattering problem. We refer to [51] for progress in this direction.

On the other hand, the Kaup-Broer-Kuperschmidt system in the space periodic case is studied in the very recent paper of Arlen, Fromcke and Nabelek [1]. A matrix Lax pair is introduced and used to determine an infinite number of conservation laws. The matrix Lax pair is also used to compute all finite genus periodic solutions.

2.2 A Second Schrödinger System

We now turn to (1.3) which, as will be developed in the next Section is a particular case of an Euler-Korteweg system.

We note that in the irrotational case (that will be considered from now on), one has $\mathbf{v} = \nabla\phi$ so that the second equation in (1.3) can be rewritten as

$$\phi_t + \eta + \epsilon[\frac{1}{2}|\nabla\phi|^2 - \Delta\eta] = 0. \tag{2.25}$$

System (1.3) has an Hamiltonian structure, namely

$$\partial_t \begin{pmatrix} \eta \\ \mathbf{v} \end{pmatrix} + J \text{grad } H_\epsilon(\eta, \mathbf{v}) = 0$$

where

$$J = \begin{pmatrix} 0 & \partial_x & \partial_y \\ \partial_x & 0 & 0 \\ \partial_y & 0 & 0 \end{pmatrix}.$$

and

$$H(\eta, \mathbf{v}) = \frac{1}{2}\int_{\mathbb{R}^2} (\epsilon|\nabla\eta|^2 + \eta^2 + |\mathbf{v}|^2(1+\epsilon\eta))dxdy.$$

The hamiltonian is inherited from that of the Euler-Korteweg system (see (3.2) in the next section).

Unfortunately, by lack of control on the derivatives of $\mathbf{v}$, one cannot use the Hamiltonian $H(\eta, \mathbf{v})$ to get a global bound on $(\eta, \mathbf{v})$, even for small solutions.

In the one-dimensional case, (1.3) has another conserved quantities, namely

$$M(\eta, v)(t) = \int_{\mathbb{R}} \eta v(x,t)dx = const.$$

Again, to get a further insight of the system we look for an equivalent "diagonal" system, starting with the one-dimensional case, as was done for (1.4), setting now

$$\widehat{A}(\xi) = i\xi \begin{pmatrix} 0 & 1 \\ 1+\epsilon|\xi|^2 & 0 \end{pmatrix}.$$

with the same eigenvalues $\pm i\xi(1+\epsilon|\xi|^2)^{1/2}$.

In the notation of the previous section, one has still

$$\mathcal{D} = \begin{pmatrix} J_\epsilon & 0 \\ 0 & -J_\epsilon \end{pmatrix}.$$

and now

$$P = \begin{pmatrix} J_\epsilon^{-1} & -J_\epsilon^{-1} \\ 1 & 1 \end{pmatrix},$$

$$P^{-1} = \frac{1}{2}\begin{pmatrix} J_\epsilon & 1 \\ -J_\epsilon & 1 \end{pmatrix},$$

Setting again

$$U = \begin{pmatrix} \eta \\ v \end{pmatrix}$$

and

$$W = \begin{pmatrix} \zeta \\ w \end{pmatrix} = P^{-1}U,$$

one can therefore reduce (1.3) to the equivalent form

$$\begin{cases} \zeta_t + J_\epsilon\zeta_x + \frac{\epsilon}{2}N_1(\zeta, w) = 0, \\ w_t - J_\epsilon w_x + \frac{\epsilon}{2}N_2(\zeta, w) = 0, \end{cases} \tag{2.26}$$

where

$$N_1(\zeta, w) = \partial_x J_\epsilon[(\zeta + w)J_\epsilon^{-1}(\zeta - w)] + (\zeta + w)(\zeta + w)_x$$

and

$$N_2(\zeta, w) = -\partial_x J_\epsilon[(\zeta + w)J_\epsilon^{-1}(\zeta - w)] + (\zeta + w)(\zeta + w)_x.$$

We can also write (2.26) as

$$\begin{cases} \zeta_t + \epsilon^{1/2}\mathcal{H}\zeta_{xx} + R_\epsilon\zeta + \frac{\epsilon}{2}N_1(\zeta, w) = 0, \\ w_t - \epsilon^{1/2}\mathcal{H}w_{xx} - R_\epsilon w + \frac{\epsilon}{2}N_2(\zeta, w) = 0, \end{cases} \tag{2.27}$$

where again R_ϵ is the order zero skew-adjoint operator with symbol $\frac{i\xi}{(1+\epsilon\xi^2)^{1/2}+\epsilon^{1/2}|\xi|}$.

Note that the nonlinearity is worse than in the case $a = -1$ and even the local theory does not seem to be straightforward using this formulation.

As in the one-dimensional case, one can express (1.3) on the equivalent form

$$V_t + \mathcal{D}V + \tilde{N}(V) = 0, \tag{2.28}$$

where again $U = (\eta, \mathbf{v})^T$, $V = (\zeta, \mathbf{w})^T$,

$$\mathcal{D} = \begin{pmatrix} 0 & 0 & 0 \\ 0 & i(-\Delta)^{1/2}J_\epsilon & 0 \\ 0 & 0 & -i(-\Delta)^{1/2}J_\epsilon \end{pmatrix}$$

and $\tilde{N}(V)$ is expressed as

$$\frac{1}{2}\begin{pmatrix} 0 \\ -iJ_\epsilon[\partial_x(v_1\eta) + \partial_y(v_2\eta)] - \frac{1}{2}[R_1\partial_x|\mathbf{v}|^2 + R_2\partial_y|\mathbf{v}|^2] \\ iJ_\epsilon[\partial_x(v_1\eta) + \partial_y(v_2\eta)] - \frac{1}{2}[R_1\partial_x|\mathbf{v}|^2 + R_2\partial_y|\mathbf{v}|^2] \end{pmatrix}$$

with

$$\eta = iJ_\epsilon^{-1}(w_1 - w_2), \quad v_1 = -R_2\zeta + R_1(w_1 + w_2), \quad v_2 = R_1\zeta + R_2(w_1 + w_2).$$

To state precisely the large time result for (1.3) obtained in [55] which leads serious difficulties, not present in the other Boussinesq systems, we need introduce the following function space

Definition 1 For any $s \in \mathbb{R}$, $k \in \mathbb{N}$, $\epsilon \in (0, 1)$, the Banach space $X^s_{\epsilon^k}(\mathbb{R}^n)$ is defined as $H^{s+k}(\mathbb{R}^n)$ equipped with the norm:

$$|u|^2_{X^s_{\epsilon^k}} = |u|^2_{H^s} + \epsilon^k|u|^2_{H^{s+k}}. \tag{2.29}$$

Then

Theorem 3 ([55]) *Assume that* $\eta_0 \in X^{1+n}_{\epsilon^{2+n}}(\mathbb{R}^n)$, $\mathbf{v}_0 \in X^{1+n}_{\epsilon^{1+n}}(\mathbb{R}^n)$ *with* $n = 1, 2$ *and curl* $\mathbf{v}_0 = 0$ *when* $n = 2$, *satisfy the (non-cavitation) condition*

$$1 + \epsilon\eta_0 \geq H > 0, \quad H \in (0, 1), \tag{2.30}$$

Then there exists a constant $\tilde{c}_0 > 0$ *such that for any* $\epsilon \leq \epsilon_0 = \frac{1-H}{\tilde{c}_0(|\eta_0|_{X^{1+n}_{\epsilon^{2+n}}} + |\mathbf{v}_0|_{X^{1+n}_{\epsilon^{1+n}}})}$, *there exists* $T > 0$ *independent of* ϵ, *such that* (1.3) *has a unique solution* $(\eta, \mathbf{v})$ *with* $\eta \in C([0, T/\epsilon]; X^{1+n}_{\epsilon^{2+n}}(\mathbb{R}^n))$ *and* $\mathbf{v} \in C([0, T/\epsilon]; X^{1+n}_{\epsilon^{1+n}}(\mathbb{R}^n))$. *Moreover,*

$$\begin{aligned} &\sup_{t \in (0, T/\epsilon]} \left(|\eta|_{X^2_{\epsilon^3}} + |\eta_t|_{X^1_{\epsilon^2}} + |\eta_{tt}|_{X^0_{\epsilon}} + |v|_{X^2_{\epsilon^2}} + |v_t|_{X^1_{\epsilon}} + |v_{tt}|_{L^2} \right) \\ &\leq C(|\eta_0|_{X^2_{\epsilon^3}} + |v_0|_{X^2_{\epsilon^2}}), \quad \textit{for } n = 1, \end{aligned} \tag{2.31}$$

and

$$\begin{aligned} &\sup_{t \in (0, T/\epsilon]} \left(|\eta|_{X^3_{\epsilon^4}} + |\eta_t|_{X^2_{\epsilon^3}} + |\eta_{tt}|_{X^1_{\epsilon^2}} + |\eta_{ttt}|_{X^0_{\epsilon}} \right. \\ &\left. + |\mathbf{v}|_{X^3_{\epsilon^3}} + |\mathbf{v}_t|_{X^2_{\epsilon^2}} + |\mathbf{v}_{tt}|_{X^1_{\epsilon}} + |\mathbf{v}_{ttt}|_{L^2} \right) \\ &\leq C(|\eta_0|_{X^3_{\epsilon^4}} + |\mathbf{v}_0|_{X^3_{\epsilon^3}}), \quad \textit{for } n = 2. \end{aligned} \tag{2.32}$$

Remark 12 An alternate proof of Theorem 3 is provided in [10], Theorem 2.2.

Remark 13 (1.3) can also be viewed from a physical viewpoint as the Saint-Venant (shallow-water) system with capillarity.

We mention here an interesting Boussinesq system, corresponding to $a = b = c = 0, d > 0$ in (1.2), namely

$$\begin{cases} \eta_t + \nabla \cdot \mathbf{v} + \epsilon \nabla \cdot (\eta \mathbf{v}) = 0 \\ \mathbf{v}_t + \nabla \eta + \epsilon [\frac{1}{2} \nabla |\mathbf{v}|^2 - \Delta \mathbf{v}_t] = 0, \end{cases} \tag{2.33}$$

Equation (2.33) is the "BBM" version of the (ill-posed) Boussinesq system

$$\begin{cases} \eta_t + \nabla \cdot \mathbf{v} + \epsilon \nabla \cdot (\eta \mathbf{v}) = 0 \\ \mathbf{v}_t + \nabla \eta + \epsilon [\frac{1}{2} \nabla |\mathbf{v}|^2 + \nabla \Delta \eta] = 0, \end{cases} \tag{2.34}$$

which is obtained by expanding the Dirichlet-Neumann operator arising in the Zakharov-Craig-Sulem formulation of the water waves system in the Boussinesq regime (see [15, 44]), and from which all the abcd Boussinesq systems are deduced.

It is worth noticing however that (1.3) and (2.33) behave quite differently. Actually (2.33) can be viewed mathematically as a perturbation of the Saint-Venant system and it keeps some "hyperbolic" features. This has been proved in the one dimensional case in the pioneering works [3, 61], see also [49] for an improved presentation.

Roughly speaking, under a non-cavitation condition that insures that the underlying Saint-Venant system is hyperbolic, the *positive convex entropy* of the Saint-Venant system implies an apriori global bound on the solutions of (2.33) leading to the global existence and uniqueness of solutions without size restrictions.[11]

The best known result is obtained in [49].

Theorem 4 ([49]) *Let $s > 1/2$. For $(\eta_0, v_0) \in H^s(\mathbb{R}) \times H^{s+1}(\mathbb{R})$ such that $1 + \epsilon\eta_0 > 0$, (2.33) has a unique solution $(\eta, v) \in C(\mathbb{R}^+; H^s(\mathbb{R})) \times H^{s+1}(\mathbb{R})) \cap C^1(\mathbb{R}^+, H^{s-1}(\mathbb{R}) \times H^s(\mathbb{R}))$.*

One also finds in [49] the existence of low regularity entropic solutions of (2.33) emanating from $(\eta_0 \in \Lambda_{\sigma_0}, v_0 \in H^1(\mathbb{R}))$ where Λ_{σ_0} is the Orlicz class

$$\Lambda_{\sigma_0} = \{\eta \text{ measurable} \mid \int_{\mathbb{R}} (\eta \ln \eta - \eta + 1)dx < +\infty\}.$$

Those results exploit the hyperbolic nature of the underlying Saint-Venant system. We do not know of similar results in the two-dimensional case.

Remark 14 The one dimensional system (2.33) is the only one in the family of abcd systems for the Cauchy problem is globally well-posed for arbitrary large initial data.

3 Euler-Korteweg

We will here revisit system (1.3) using the fact (unnoticed in [55]) that it is a particular case of a Euler-Korteweg system.

The (general) Euler-Korteweg system is a dispersive perturbation of the compressible Euler system. It writes

$$\begin{cases} \partial_t \rho + \nabla \cdot (\rho \mathbf{u}) = 0 \\ \partial_t \mathbf{u} + (\mathbf{u} \cdot \nabla)\mathbf{u} + \nabla(g(\rho)) = \nabla(K(\rho)\Delta\rho + \frac{1}{2}K'(\rho)|\nabla\rho|^2), \end{cases} \tag{3.1}$$

where ρ is the fluid density, $\mathbf{u}$ the velocity field, $K(\rho)$ is the capillarity coefficient (smooth and positive), $g(\rho)$ the pressure law.

When K is zero, (3.1) reduces to the standard Euler equations with velocity of sound $\sqrt{\rho g'(\rho)}$. The $K(\rho) = 1/(4\rho)$, corresponds to quantum fluids and (3.1) can then be deduced from the generalized Schrödinger equation via the Madelung transform, see [22], the classical Gross-Pitaevskii equation corresponding to $g(\rho) = \rho - 1$.

When $\mathbf{u} = \nabla\phi$ is potential, the second equation in (3.1) rewrites

$$\partial_t \phi + \frac{|\nabla\phi|^2}{2} + g(\rho) = K(\rho)\Delta\rho + \frac{1}{2}K'(\rho)|\nabla\rho|^2.$$

[11] This connexion to the Saint-Venant system apparently does not hold for the other Boussinesq systems.

Defining the energy (Hamiltonian)

$$E(\rho, \phi) = \int_{\mathbb{R}^d} \{\frac{K(\rho)|\nabla\rho|^2 + \rho|\nabla\phi|^2}{2} + G(\rho)\}dx, \tag{3.2}$$

where G is a primitive of g, leads in the irrotational case to the Hamiltonian structure:

$$\begin{cases} \partial_t\rho - \frac{\delta E}{\delta\phi} = 0 \\ \partial_t\phi + \frac{\delta E}{\delta\rho} = 0, \end{cases} \tag{3.3}$$

where $\delta\rho$ denotes the variational derivative of ρ.

Remark 15 As recalled in [9], (3.1) belongs to the more general family of systems

$$\begin{cases} \partial_t\rho + \nabla\cdot(\rho\mathbf{u}) = 0 \\ \partial_t\mathbf{u} + (\mathbf{u}\cdot\nabla)\mathbf{u} + \nabla(\delta\mathcal{E}) = 0, \end{cases} \tag{3.4}$$

where $\mathcal{E} = \mathcal{E}(\rho, \nabla\rho)$ is the internal energy and $\delta\mathcal{E}$ denotes the variational derivative of $\mathcal{E}$. (3.1) corresponds to the choice $\mathcal{E}(\rho, \nabla\rho) = F(\rho) + \frac{1}{2}K(\rho)||\nabla\rho||^2,\ g := \frac{dF}{d\rho}$.

The Boussinesq system (1.3) corresponds to a particular case of (3.1) by taking $\rho = 1 + \epsilon\eta$, $\mathbf{u} = \epsilon\mathbf{v}$, $K(\rho) = \epsilon$, $g(\rho) = \rho - 1$.

Remark 16 Note that (3.1) is physically relevant in spatial dimensions 1, 2 and 3 while (1.3) appears only in dimensions one and two.

Remark 17 As was already noticed (1.3) is referred to in [55] as a *Schrödinger type* Boussinesq system since the diagonalization of its linear part is given (see [55] Sect. 3) by $\partial_t \pm i[(-\Delta)^{1/2}(I - \epsilon\Delta)^{1/2}]$ in two-dimension and by $\partial_t \pm [\epsilon^{1/2}H\partial_x^2 + R_\epsilon]$ where H is the Hilbert transform and R_ϵ an order zero operator in dimension one.

On the other hand, many works have been recently devoted to the Euler-Korteweg system(3.1), see for instance [5, 8–12] and the goal of this section is to focus on those that apply to the particular version (1.3). Note that some papers on the Euler-Korteweg system (*e.g.* [6]) consider only the three-dimensional case which is not relevant for the Boussinesq systems.

One also finds in [10], Sect. 6.1 a *weakly transverse* Euler-Korteweg system that, in the particular case of the (abcd) system with $a = b = d = 0, c < 0$, is a *weakly transverse Boussinesq system* in the sense of [45]. It corresponds to the KP regime where the wave lengths in y are much larger than the x ones. As noticed in Remark 4 of [10] it would be interesting to prove a long time existence for this system and similar ones for the other members of the (abcd) family.

3.1 The Cauchy Problem

The Cauchy problem for (3.1) is studied in [11] in the one -dimensional case, in [12] in several space dimensions, and we now examine the implications of those results for (1.3) starting with the one-dimensional case studied in [11].

As aforementioned, (1.3) can be written in the framework of (3.1) by setting, in one spatial dimension and with $\epsilon = 1, \eta = \rho - 1, v = u, K(\rho) = 1, g(\rho) = \rho - 1$, so that it becomes

$$\begin{cases} \partial_t \rho + (\rho u)_x = 0 \\ \partial_t u + \rho_x + u u_x - \rho_{xxx} = 0, \end{cases} \tag{3.5}$$

which is in the framework of system (1.6) in [11]. The idea in [11] is to use a Lagrangian setting leading to the new formulation

$$\begin{cases} \partial_t v - \partial_y u = 0 \\ \partial_t u - \partial_y (\alpha(v) \partial_y w) = q(v) w, \\ \partial_t w + \partial_y (\alpha(v) \partial_y u) = 0, \end{cases} \tag{3.6}$$

where y is the mass Lagrangian coordinate, so that $dy = \rho dx - \rho u dt$, $v = \frac{1}{\rho}$, $\kappa(v) = (1/v)^5$, $\alpha = \sqrt{\kappa}$, $w = -\alpha \partial_y v$, $q(v) = -v^{-1/2}$.

Introducing the complex function $z = u + iw$, it is noticed in [11] that (3.6) can be viewed as the coupling of the variable Schrödinger equation

$$\partial_t z + i \partial_y (\alpha(v) \partial_y z) = q(v) \text{Im } z \tag{3.7}$$

with the compatibility condition

$$\partial_t v - \partial_y \text{Re } z = 0.$$

By the general Theorem 2.1 and Theorem 2.2 in [11], (3.6) is locally well-posed in $H^2(\mathbb{R})^3$.

More generally the following theorem displays what the local well-posedness result to the Euler-Korteweg system established in [12] implies to the context of (1.3) with $\epsilon = 1$ in both dimensions one and two.

Theorem 5 *For* $s > 1 + n/2$, $n = 1, 2$, *let* $(\eta_0, \mathbf{v}_0) \in H^{s+1}(\mathbb{R}^n) \times H^s(\mathbb{R}^n)$, $1 + \eta_0 > 0$ *and curl* $\mathbf{v}_0 = 0$ *when* $n = 2$. *Then there exist* $T > 0$ *and a unique solution* $(\eta, \mathbf{v}) \in C([0, T]; H^{s+1}(\mathbb{R}^n) \times H^s(\mathbb{R}^n)) \cap C^1([0, T]; H^{s-1}(\mathbb{R}^n) \times H^{s-2}(\mathbb{R}^n))$. *Moreover,* $1 + \eta > 0$ *on* $[0, T] \times \mathbb{R}^n$.

Remark 18 With respect to Theorem 3, Theorem 5 weakens the regularity of the data but it does not provide a long time existence result when the small parameter ϵ is introduced.

Remark 19 Enforcing the links between the Euler-Korteweg systems and Schrödinger equations, Audiard [7] proved a *Kato local smoothing effect* similar to the one established in [26] for general dispersive equations and in [31] for Schrödinger equations with variable coefficients. Applied to the particular case of the Boussinesq system (1.3), Theorem 3.1 in [7] implies that for (η_0, v_0) such that $\eta_0 \in L^\infty(\mathbb{R})$, $((\eta_0)_x, v_0) \in H^s(\mathbb{R})$, $s > 3/2$, the corresponding local solution (η, v) defined on $[0, T)$ satisfies for any $\epsilon > 0$

$$(\eta_x, v) \in L^2(0, T; H^{s+1/2}_{\text{loc}}(\mathbb{R})).$$

One also finds a blow-up criterion for the solutions of (3.1) in [12] which in the case of (1.3) reads:

Blow-up criterion: The local solution in Theorem 5 can be continued beyond $[0, T]$ provided
1. For $(x, t) \in \mathbb{R}^n \times [0, T]$, $\eta(x, t) \in J \subset \mathbb{R}^{+*}$, J compact.
2. $\int_0^T (||\Delta\eta(t)||_{L^\infty(\mathbb{R}^n)} + ||\text{div}\, \mathbf{v}(t)||_{L^\infty(\mathbb{R}^n)})dt < \infty$.

The proof of those results in [12] relies on energy estimates for an extended system and following [8] we give some details in the particular case of (1.3).

Setting $L = 2\sqrt{1+\eta}$, $\mathbf{w} = \sqrt{1/(1+\eta)}\, \nabla\eta = \nabla L$, $a = \sqrt{1+\eta}$, then one checks that $(L, \mathbf{v}, \mathbf{w})$ satisfies the system

$$\begin{cases} \partial_t L + \mathbf{v} \cdot \nabla L + a\text{div}\, \mathbf{v} = 0 \\ \partial \mathbf{v} + \mathbf{v} \cdot \nabla \mathbf{v} - \mathbf{w} \cdot \nabla \mathbf{w} - \nabla(a\text{div}\, \mathbf{w}) = -\nabla\eta, \\ \partial_t \mathbf{w} + \nabla(\mathbf{v} \cdot \mathbf{w}) + \nabla(a\text{div}\, \mathbf{w}) = 0. \end{cases} \tag{3.8}$$

Equivalently, setting[12] $\mathbf{z} = \mathbf{v} + i\mathbf{w}$, one gets

$$\begin{cases} \partial_t L + \mathbf{v} \cdot \nabla L + a\text{div}\, \mathbf{v} = 0 \\ \partial_t \mathbf{z} + \mathbf{z} \cdot \nabla \mathbf{z} + i\nabla(a\text{div}\mathbf{z}) = \nabla(\frac{L}{2})^2. \end{cases} \tag{3.9}$$

This again clarifies the dispersive structure of the system since the second equation in (3.9) is a degenerate quasilinear Schrödinger equation.

In [10] the authors consider several asymptotic regimes for the EK system, regarding the length and amplitude of the waves in particular regimes described by two counter propagating waves governed by the KdV or KP equations, generalizing previous similar results obtained for the Gross-Pitaevskii equation.

[12] We recall that we assume $1 + \eta > 0$.

4 Traveling Waves

Traveling wave solutions of (3.1) (resp. (1.3)), are solutions of type $(\rho(x-ct), u(x-ct)$, (resp. $(\eta(x-ct)), v(x-ct))$.

We are not aware of existence (or nonexistence) results proved directly on (1.3) but many works have been devoted to traveling wave solutions for the "general" Euler-Korteweg system (3.1), see for instance [13] and we describe now the consequences of results in this paper when applied to (1.3).

A traveling wave solution of (1.3) should thus satisfy the system

$$\begin{cases} [-c\eta + v(1+\epsilon\eta)]' = 0 \\ [-cv + \eta + \frac{\epsilon}{2}v^2 - \epsilon\eta'']' = 0, \end{cases} \tag{4.1}$$

where $'$ denotes the derivative with respect to $\xi = x - ct$.

In the context of Boussinesq systems it is natural to assume that η and v vanish at infinity so that (4.1) implies

$$\begin{cases} -c\eta + v(1+\epsilon\eta) = 0 \\ -cv + \eta + \frac{\epsilon}{2}v^2 - \epsilon\eta'' = 0, \end{cases} \tag{4.2}$$

Remark 20 This implies that contrary to (2.4), the system (1.3) does not possess non trivial one dimensional steady solutions satisfying $1 + \epsilon\eta > 0$, $x \in \mathbb{R}$ and $\eta \to 0$ as $x \to \pm\infty$.

Remark 21 A solitary wave (η, v) can also be viewed as a critical point of the Hamiltonian $H(\eta, v)$ under the constraint $M(\eta, v) = const$, the velocity c appearing as a Lagrange multplier.

Remark 22 Note that in the context of the *solitary waves* studied for the Euler-Korteweg system (3.1) in [13] is slightly different since then the system satisfied by a solution of the form $(\rho, u) = (\rho(x-ct), u(x-ct))$ is

$$\begin{cases} \big(\rho(u-c)\big)' = 0 \\ (u-c)u' = \big(K(\rho)\rho'' + \frac{1}{2}K(\rho)'|\rho'|^2 - g_0(\rho)\big)', \end{cases} \tag{4.3}$$

where $'$ denotes the derivative with respect to $\xi = x - ct$.

The first equation implies that $\rho(u-c) \equiv$ const and the constant should be equal to $j := \rho_\pm(u_\pm - c)$ where

$$\lim_{x\to\pm\infty} (\rho, u)(x) = (\rho_\pm, u_\pm).$$

The case of different limits at $\pm\infty$ would corresponds to the *kink* solutions in [13]. As aforementioned, in the particular case of the Boussinesq system one has $K(\rho) = \epsilon$, $g_0(\rho) = \rho$, $\rho = 1 + \epsilon\eta$, $\mathbf{u} = \epsilon\mathbf{v}$. Moreover, $u_\pm = 0$, $\rho_\pm = 1$.

Eliminating v in (4.2) leads to a single equation for η:

$$-\epsilon\eta'' + \eta\left(1 - \frac{c^2}{1+\epsilon\eta}\right) + \frac{\epsilon}{2}\frac{c^2\eta^2}{(1+\epsilon\eta)^2} = 0. \tag{4.4}$$

We now prove the existence of non trivial solutions of (4.4). A simple computation on (4.2) shows that

$$\epsilon(\eta')^2 = R_1(\eta) = \eta^2\left(1 - \frac{c^2}{1+\epsilon\eta}\right), \tag{4.5}$$

proving that no non trivial solitary waves satisfying $\eta \to 0$ as $x \to \pm\infty$ exist when $c > 1$.

When $0 < c < 1$ the existence of a solitary wave η amounts to finding an homoclinic orbit about the origin of (4.4). To do so, setting

$$\Phi = (1 - \frac{c^2}{1+\epsilon\eta})^{1/2},$$

we deduce from (4.5) that Φ satisfies

$$\frac{4\epsilon(1-c^2)^2}{c^4}|\Phi'|^2 = (1 - \frac{1}{1-c^2}\Phi^2)^2(1-\Phi^2)^2 \tag{4.6}$$

and $\lim_{\xi\to\pm\infty}\Phi(\xi) = \sqrt{1-c^2}$. By simple calculation, we get

$$4\epsilon\left|\left(\frac{1}{\sqrt{1-c^2}}\tanh^{-1}(\frac{1}{\sqrt{1-c^2}}\Phi) - \tanh^{-1}\Phi\right)'\right|^2 = 1,$$

where $\tanh^{-1}(x) = \frac{1}{2}\ln(\frac{1+x}{1-x})$ is the inverse function of $\tanh x$. Then there exists an unique (up to a translation) even solution $\Phi(\xi) \in (0, \sqrt{1-c^2})$ to (4.6) over $[0, \infty)$ such that $\lim_{\xi\to+\infty}\Phi(\xi) = \sqrt{1-c^2}$ and

$$F(\Phi) := 2\sqrt{\epsilon}\left(\frac{1}{\sqrt{1-c^2}}\tanh^{-1}(\frac{1}{\sqrt{1-c^2}}\Phi) - \tanh^{-1}\Phi\right) = \xi, \tag{4.7}$$

i.e.,

$$\Phi(\xi) = F^{-1}(\xi), \quad \text{for } \xi \in [0, \infty).$$

Then we obtain that (4.5) has an unique (up to a translation) even solution

$$\eta(\xi) = \frac{1}{\epsilon}\left(\frac{c^2}{1-\Phi^2(\xi)} - 1\right).$$

Remark 23 In [5], Theorem 1.1, C. Audiard constructs a family of localized traveling waves for (3.1) near the constant state $(\rho_0, 0)$ under the following hypothesis

1. $K : \mathbb{R}_+^* \to \mathbb{R}_+^*$ is a smooth function.
2. $g : \mathbb{R}_+^* \to R$ is a smooth function such that $g(\rho_0) = 0,\ g'(\rho_0) > 0$.
3. The number $\Gamma := 3 + \frac{\rho_0 g''(\rho_0)}{g'(\rho_0)}$ is not zero.

However those hypothesis are never satisfied by (1.3) since then $g(\rho_0) = \rho_0 = 1 + \eta_0 > 0$.

5 Final Remarks

We focussed in this paper on two particular cases of the so-called abcd Boussinesq systems, one of them being a particular case of an Euler-Korteweg system. As for the other members of the abcd family of Boussinesq systems many questions remain open concerning the long time behavior of solutions of the Cauchy problem and we emphasize two of them in the following conjectures:

1. Global existence (and scattering ?) of small solutions.
2. When the small parameter ϵ is introduced, one has the alternative: global well-posedness or finite time blow-up, at a time of order $O(1/\epsilon)$.
3. For the specific system (2.4) it would be desirable to use inverse scattering techniques to derive a rigorous theory of resolution of the Cauchy problem, providing information on the structure of the solutions.

Acknowledgements We thank the referees for very helpful remarks.
This work of the first author was partially supported by the ANR project ANuI (ANR-17-CE40-0035-02). This work of the second author was partially supported by NSFC under Grant 11671383 and 12171019.

References

1. A. Arlen, T. Fromcke and P. Nabelek, *On Periodic and Finite Genus Solutions to the Integrable Kaup-Broer System for Capillary Waves*, preprint (2022)
2. D.M. Ambrose, J.L. Bona, T. Milgram, Global solutions and ill-posedness for the Kaup system and related systems. Indiana Math. J. **68**(4), 1173–1198 (2019)
3. C.J. Amick, Regularity and uniqueness of solutions for the Boussinesq system of equations. J. Differ. Eq. **54**(1), 231–247 (1984)
4. J. Angulo Pava, On the Cauchy problem for a Boussinesq-type system. Adv. Diff. Eq. **4**, 457–492 (1999)
5. C. Audiard, Small energy traveling waves for the Euler-Korteweg system. Nonlinearity **30**, 3362–3399 (2017)

6. C. Audiard, On the time of existence of solutions of the Euler-Kortewed system. Ann. Fac. Sci. Toulouse Math. (6) **30**(2), 7–16 (2021)
7. C. Audiard, Dispersive smoothing for the Euler-Korteweg system. SIAM J. Math. Anal. **44**(4), 3018–3040 (2012)
8. C. Audiard, B. Haspot, Global well-posedness of the Euler-Korteweg system for small irrotational data. Commun. Math. Phys. **351**, 201–247 (2017)
9. S. Benzoni-Gavage, Planar traveling waves in capillary fluids. Differ. Integral Eq. **26**, 439–485 (2013)
10. S. Benzoni-Gavage, D. Chiron, Long wave asymptotics for the Euler-Korteweg system. Rev. Mat. Iberoam. **34**(21), 245–304 (2018)
11. S. Benzoni-Gavage, R. Danchin, S. Descombes, Well-posedness of one dimensional Korteweg models. Electron. J. Differ. Equ. **56**, 1–35 (2006)
12. S. Benzoni-Gavage, R. Danchin, S. Descombes, On the well-posedness for the Euler-Korteweg model in several dimensions. Indiana Math. J. **56**, 1499–1579 (2007)
13. S. Benzoni-Gavage, R. Danchin, S. Descombes, D. Jamet, Structure of Korteweg models ans stability of diffuse interfaces. Interfaces Free Bound. **7** (2005)- 371-414
14. H. Berestycki, P.-L. Lions, Nonlinear scalar field equation I. Arch. Rat. Mech. Anal. **32**, 313–346 (1983)
15. J.L. Bona, T. Colin, D. Lannes, Long-wave approximation for water waves. Arch. Ration. Mech. Anal. **178**, 373–410 (2005)
16. J.L. Bona, M. Chen, J.-C. Saut, Boussinesq equations and other systems for small-amplitude long waves in nonlinear dispersive media I: Derivation and the linear theory. J. Nonlinear Sci. **12**, 283–318 (2002)
17. J.L. Bona, M. Chen, J.-C. Saut, Boussinesq equations and other systems for small-amplitude long waves in nonlinear dispersive media II: the nonlinear theory. Nonlinearity **17**, 925–952 (2004)
18. J. Boussinesq, Théorie des ondes et des remous qui se propagent le long d'un canal rectangulaire horizontal, en communicant au liquide contenu dans ce canal des vitesses sensiblement pareilles de la surface au fond. J. Math. Pures Appl. **7**(2), 55–108 (1872)
19. L.F.J. Broer, Approximate equations for long water waves. Appl. Sci. Res. **31**, 377–395 (1975)
20. C. Burtea, New long time existence results for a class of Boussinesq-type systems. J. Math. Pures Appl. **106**(2), 203–236 (2016)
21. C. Burtea, Long time existence results for bore-type initial data for BBM-Boussinesq systems. J. Diff. Equ. **261**, 4825–4860 (2016)
22. R. Carles, R. Danchin, J.-C. Saut, Madelung, Gross-Pitaevskii and Korteweg. Nonlinearity **25**, 2843–2873 (2012)
23. Min Chen, Solitary-wave and multi-pulsed traveling-wave solutions of Boussinesq systems. Appl. Anal. **75**(1–2), 213–240 (2000)
24. T. Colin, M.I. Weinstein, On the ground states of vector nonlinear Schrödinger equations. Annales de lI'I.H.P. Sect. A, **65**(1), 57–79 (1996)
25. T. Congy, S.K. Ivanov, A.M. Kamchatnov, N. Pavloff, Evolution of initial discontinuities in the Riemann problem for the Kaup-Boussinesq equation with positive dispersion. Chaos **27**(8), 083107 (2017)
26. P. Constantin, J.-C. Saut, Local smoothing properties of dispersive equations. J. Amer. Math. Soc. **1**, 413–439 (1988)
27. O. Darrigol, *Worlds of Flow, A Hydrodynamics from the Bernoullis to Prandtl* (Oxford University Press, Oxford, 2005)
28. O. Darrigol, The spirited horse, the engineer, and the mathematician?: water waves in nineteenth-century hydrodynamics. Arch. Hist. Exact Sci. **58**, 21–95 (2003)
29. O. Darrigol, The legacy of Joseph Bussinesq's legacy in fluid mechanics. C.R. Mech. **345**, 427–445 (2017)
30. M.W. Dingemans, *Water Wave Propagation Over Uneven Bottoms. Part 2- Non-linear Wave Propagation*, Advances Series on Ocean Engineering, vol. 13 (World Scientific, 2000)

31. S.I. Doï, Remarks on the Cauchy problem for Schrödinger-type equations. Comm. Partial Diff. Eq. **21**, 163–178 (1996)
32. V.A. Dougalis, A. Duran, L. Saridaki, On solitary-wave solutions of Boussinesq/Boussinesq systems for internal waves. Phys. D (2021)
33. G.A. El, R.H.J. Grimshaw, M.V. Pavlov, Integrable shallow-water equations and undular bores. Studies in Appl. Math. **106**, 157–186 (2001)
34. M. Grillakis, J. Shatah, W.A. Strauss, Stability theory of solitary waves in the presence of symmetry. I. J. Funct. Anal. **74**, 160–197 (1987)
35. M.D. Groves, *Hamiltonian Long-wave Approximations for Water Waves in a Uniform Channel Systems*, ed. by L. Debnath, N.D. Wave (World Scientific, 1992), pp. 99–125
36. Hu. Junqi, Global well-posedness of the BCL system with viscosity. Chin. App. Math. Ser. B **30**(2), 153–172 (2009)
37. D.J. Kaup, A higher-order water-wave equation and the method for solving it. Progr. Theoret. Phys. **54**, 396–408 (1975)
38. C.E. Kenig, K.D. Koenig, On the local well-posedness of the Benjamin-Ono and modified Benjamin-Ono equations. Math. Res. Let. **10**, 879–895 (2003)
39. C.E. Kenig, G. Ponce, L. Vega, Smoothing effects and local existence theory for the generalized nonlinear Schrödinger equations. Invent. Math. **134**(3), 489–545 (1998)
40. N. Kita, J. Segata, Well-posedness for the Boussinesq-type system related to the water wave. Funkcial. Ekvac. **47**, 329–350 (2004)
41. B.A. Kuperschmidt, Mathematics of dispersive water waves. Commun. Math. Phys. **99**, 51–73 (1985)
42. C. Kwak, C. Munoz, F. Poblete, J.C. Pozo, The scattering problem for Hamiltonian ABCD Boussinesq systems in the energy space. J. Math. Pures et Appl. **127**, 121–159 (2019)
43. C. Kwak, C. Munoz, Asymptotic dynamics for the small data weakly dispersive one-dimensional Hamiltonian abcd systems. Trans. Am; Math. Soc. **373**(12), 1043–1107 (2019)
44. D. Lannes, *Water Waves?: Mathematical Theory and Asymptotics, Mathematical Surveys and Monographs*, vol. 188 (AMS, Providence, 2013)
45. D. Lannes, J.-C. Saut, Weakly transverse Boussinesq systems and the KP approximation. Nonlinearity **19**, 2853–2875 (2006)
46. F. Linares, D. Pilod, J.-C. Saut, Well-posedness of strongly dispersive two-dimensional surface waves Boussinesq systems. SIAM J. Math. Anal. **44**(6), 4195–4221 (2012)
47. V.B. Matveev, M.I. Yavor, Solutions presque périodiques et N-solitons de l'équation hydrodynamique non linéaire de Kaup. Annales de l'IHP, Section A **31**(1), 25–41 (1979)
48. M. Ming, J.-C. Saut, P. Zhang, Long time existence of solutions to Boussinesq systems. SIAM. J. Math. Anal. **44**(6), 4078–4100 (2012)
49. L. Molinet, R. Talhouk, I. Zaiter, The classical Boussinesq system revisited. Nonlinearity **34**(2), 744–775 (2021)
50. L. Molinet, S. Vento, Improvement of the energy method for strongly nonresonant dispersive equations and applications. Anal. PDE **8**(6), 1455–1495 (2015)
51. P.V. Nabelek, *Private Communication* (2022)
52. P.V. Nabelek, V.E. Zakharov, Solutions to the Kaup-Broer system and its (2+1) dimensional integrable generalization via a dressing method. Phys. D **409**, 132478 (2020)
53. D.H. Peregrine, Long waves in a uniform channel of arbitrary cross-section. J. Fluid Mech. **32**, 353–365 (1968)
54. G. Ponce, On the global well-posedness of the Benjamin-Ono equation. Diff. Int. Equ. **4**(3), 527–542 (1991)
55. J.C. Saut, C. Wang, L. Xu, The Cauchy problem on large time for surface waves Boussinesq systems II. SIAM J. Math. Anal. **49**(4), 2321–2386 (2017)
56. J.-C. Saut, L. Xu, The Cauchy problem on large time for surface waves Boussinesq systems. J. Math. Pures Appl. **97**, 635–662 (2012)
57. J.-C. Saut, L. Xu, Well-posedness on large time for a modified full dispersion system of surface waves. J. Math. Phys. **53**, 115606 (2012)

58. J.-C. Saut, L. Xu, Long time existence for a strongly dispersive Boussinesq system. SIAM J. Math. Anal. **52**(3), 2803–2848 (2020)
59. J.-C. Saut, L. Xu, Long time existence for a two-dimensional strongly dispersive Boussinesq system. Commun. PDE **46**(11), 2057–2087 (2021)
60. T. Tao, Global well-posedness of the Benjamin-Ono equation in $H^1(\mathbb{R})$. J. Hyperbolic Differ. Equ. **1**, 27–49 (2004)
61. M.E. Schonbek, Existence of solutions to the Boussinesq system of equations. J. Differ. Eq. **42**, 325–352 (1981)
62. C. Valls, Stability of some waves in the Boussinesq system. Commun. Pure Appl. Anal. **5** (4), 929-939
63. J. Zhu, X. Wang, Broer-Kaup system revisit?: inelastic interaction and blow-up. J. Math. Anal. Appl. **496**, 124791 (2021)

Representations of Pauli–Fierz Type Models by Path Measures

Fumio Hiroshima

Abstract Functional integral representations of the semigroups generated by Pauli–Fierz type Hamiltonians in quantum field theory are reviewed. Firstly we introduce functional integral representations for Schrödinger type operators. Secondly those for Pauli–Fierz type Hamiltonians are shown. Finally inequalities derived from functional integral representations are shown.

Keywords Pauli-Fierz model · Nonrelativistic QED · Feynman-Kac formula · Semigroup

1 Introduction

Congratulations on Prof. Toru Ozawa's sixtieth birthday. It is my great pleasure to be able to publish a paper on this occasion. I was a bit surprised to hear that he was "already" 60 years old, as he always seems to be young and full of mathematical power. On the other hand, Prof. Ozawa has been well known since he was quite young and has trained many excellent students, so I was also surprised to hear that he was "only" 60 years old. I first met Prof. Ozawa around 1994 or 1995 at Hokkaido University, when he was already well known for his scattering theory for nonlinear Schrödinger operators, and it is astonishing to think that he was only 32 or 33 years old at the time.

I also thank Prof. Ozawa for always giving me words of encouragement whenever we were together in conferences, and for inviting me to be a lecturer at SGU workshop to be held at Waseda University in 2020. I was also excited to be invited to Prof. Ozawa's sixtieth birthday conference in 2021, but both were cancelled due to COVID 19, which I regret. In 2022, after COVID19 subsided somewhat, I was grateful to be again invited to be a lecturer at SGU workshop of Waseda University.

Dedicated to Professor Toru Ozawa on his sixtieth birthday.

F. Hiroshima (✉)
Faculty of Mathematics, Kyushu University, Fukuoka, Japan
e-mail: hiroshima.fumio.965@m.kyushu-u.ac.jp

S. Machihara (ed.), *Mathematical Physics and Its Interactions*, Springer Proceedings in Mathematics & Statistics 451, https://doi.org/10.1007/978-981-97-0364-7_10

As an impressive memory of Prof. Ozawa, he gave me a question on the regularity of an integral kernel appearing in Theorem 3.3 in this paper at the Ph.D. dissertation defense of mine in 1996 in Sapporo. In one fourth of a century since then, the research of mine has been fortunately progressed considerably and a massive book [74] has been published in 2020.

I wish that Prof. Ozawa will continue to take good care of himself and work hard on his research for ever.

Our motivation of the investigation of functional integral representations of the semigroups generated by self-adjoint Hamiltonians was to solve problems in the statistical mechanics. Our interest has been then shifted to the construction of functional integral representations themselves, not only for Schrödinger operators, but also for relativistic Schrödinger operators and Schrödinger operators with spin $1/2$, etc. In this period we found papers [28–30] in physics discussed functional integral representations for Schrödinger type operators, which were very useful.

Furthermore, we were able to obtain functional integral representations for models coupled to quantum fields. Fortunately, around 2014, we could prove that functional integral representations were useful tools for non-perturbative analysis of Hamiltonians in quantum field theory, and our research were accelerated since then.

The following is a concrete explanation. The analysis of point spectra embedded in continuous spectrum in quantum field theory had been investigated, and the research had turned to analyzing the properties of their eigenvalues and eigenvectors. In particular a special attention was payed to ground states. We refer to see [67] and references therein for the development of the investigation of eigenvectors and ground states of models in quantum field theory. However, since the corresponding eigenvalues are embedded eigenvalues, there was no established method to analyze their eigenvectors.

From functional integral representations, we can define a measure on a path space, which is called the Gibbs measure [74, pp. 195, 378, 256, 491]. The Gibbs measure μ_{Gibbs} can be used to express the expectation value of a certain observable with respect to a ground state in a non-perturbative way:

$$(\Psi_{\mathrm{g}}, \mathcal{O}\Psi_{\mathrm{g}}) = \int F_{\mathcal{O}}(q)\mathrm{d}\mu_{\mathrm{Gibbs}}(q),$$

where $\mathcal{O}$ denotes an observable and Ψ_{g} a ground state of a model. The basic models of quantum field theory we are concerned are the Nelson model [111–113], the spin-boson model, and the Pauli–Fierz model in non-relativistic QED [117]. The Pauli–Fierz model is the main object in this paper. The Nelson model is an interaction model of charged particles linearly coupled with a scalar bose field and has been greatly analyzed. The analysis of the renormalized Nelson model has been also completed in a non-perturbative manner by using functional integral representations in [16, 49, 75]. The spin-boson model describes a linear interaction between spin (or two excited states) and a scalar bose field. The spectral properties of the spin-boson model has been studied in [59] in terms of a functional integration. On the other hand, for the Pauli–Fierz model, although a functional integral representation exists, we feel that

Table 1 Pauli–Fierz type operators

	Pauli–Fierz type operator	Hilbert space
H	$\frac{1}{2}(-\mathrm{i}\nabla - A)^2 + V + H_{\mathrm{f}}$	$L^2(\mathbb{R}^3) \otimes L^2(\mathrm{Q})$
$H(p)$	$\frac{1}{2}(p - \mathrm{P_f} - A(0))^2 + H_{\mathrm{f}}$	$L^2(\mathrm{Q})$
H_{R}	$((-\mathrm{i}\nabla - A)^2 + m^2)^{1/2} - m + H_{\mathrm{f}} + V$	$L^2(\mathbb{R}^3) \otimes L^2(\mathrm{Q})$
$H_{\mathrm{R}}(p)$	$\left((p - \mathrm{P_f} - A(0))^2 + m^2\right)^{1/2} - m + H_{\mathrm{f}}$	$L^2(\mathrm{Q})$
H_{S}	$\frac{1}{2}(-\mathrm{i}\nabla - A)^2 + V + H_{\mathrm{f}} - \frac{1}{2}\sigma \cdot B$	$L^2(\mathbb{R}^3 \times \mathbb{Z}_2) \otimes L^2(\mathrm{Q})$
$H_{\mathrm{S}}(p)$	$\frac{1}{2}(p - \mathrm{P_f} - A(0))^2 + H_{\mathrm{f}} - \frac{1}{2}\sigma \cdot B(0)$	$\mathbb{Z}_2 \otimes L^2(\mathrm{Q})$

its application is still insufficient. We hope that some young researchers will read this paper and become interested in this research.

In this article, we will introduce functional integral representations for the Pauli–Fierz type models [65, 66, 73]. See Table 1 for the list of models we consider in this paper.

A formal integration on the set of continuous paths due to Richard Feynman played an important role in the modern quantum physics. It has already existed due to Wiener's work on Brownian motion initiated in 1923 when the quantum mechanics was not yet found!, and it was Kac in 1949 who first showed a suitable framework, however, not for the path integral directly, see [83, 84]. In contrast to Feynman integrals, a Feynman–Kac formula offers an integral representation of semigroup e^{-tH} instead of unitary group $e^{-\mathrm{i}tH}$ on a path space endowed with a probability measure. Here we show below the reason why this improves the situation. Let

$$A(q, \dot{q}; 0, t) = \int_0^t \left(\frac{1}{2}\dot{q}^2(s) - V(q(s))\right) \mathrm{d}s,$$

where $\frac{1}{2}\dot{q}^2 - V(q)$ is a Lagrangian derived from the Legendre transform of Hamiltonian

$$h = -\frac{1}{2}\Delta + V.$$

The formal expression below

$$(e^{-\mathrm{i}tH})(x, y) = C \int_{\mathfrak{C}_{xy}} e^{\mathrm{i}A(q,\dot{q};0,t)} \prod_{0 \le s \le t} \mathrm{d}q(s) \tag{1.1}$$

gives an integral kernel of the operator $e^{-\mathrm{i}tH}$ for all t. Here $\mathrm{d}q(s)$ is Lebesgue measure for each s, and $\mathfrak{C}_{xy}$ the set of continuous paths $q(\cdot) : [0, t] \to \mathbb{R}^3$ such that $q(0) = x$ and $q(t) = y$. The right-hand side of (1.1) is Feynman's integral, and $e^{\mathrm{i}A(q,\dot{q};0,t)}$ is a phase factor introduced by Feynman. See Feynman's paper [38, 39]. Also we refer to see e.g., [2, 41, 46, 50, 81] for Feynman's integrals. Analytic continuation $s \to -is$, $ds \to -ids$ and the replacement $\dot{q}(s)^2 \to -\dot{q}(s)^2$ in (1.1) lead to the kernel

$$(e^{-tH})(x, y) = C \int_{\mathfrak{C}_{xy}} e^{-\int_0^t V(q(s))ds} e^{-\frac{1}{2}\int_0^t \dot{q}(s)^2 ds} \prod_{0 \le s \le t} dq(s). \tag{1.2}$$

It is possible to define a mathematically meaningful measure $d\mathcal{W}^{x,y}_{[0,t]}(q)$ whose formal expression is given by

$$\exp\left(-\int_0^t \frac{1}{2}\dot{q}(s)^2 ds\right) \prod_{0 \le s \le t} dq(s).$$

The paths of a random process $(B_t)_{t \ge 0}$ called Brownian motion just have the required properties and the Feynman–Kac formula

$$(e^{-tH})(x, y) = \int_{\mathfrak{C}} e^{-\int_0^t V(B_s(w))ds} d\mathcal{W}^{x,y}_{[0,t]}(w), \quad \forall t \ge 0, \forall x, y \in \mathbb{R}^3 \tag{1.3}$$

with $\mathfrak{C} = C([0, \infty), \mathbb{R}^3)$ rigorously holds. In particular, there is a measure supported on the space $C([0, \infty), \mathbb{R}^3)$ of continuous functions $[0, \infty) \to \mathbb{R}^3$, and it can be identified as Wiener measure $\mathcal{W}^{x,y}_{[0,t]}$ conditional on paths leaving from x at time 0 and ending in y at time t. Kac has actually proved [82] that the heat equation with an initial data $\phi(x)$:

$$\begin{cases} -\dfrac{\partial f}{\partial t} = -\dfrac{1}{2}\Delta f + Vf, \\ f(x, 0) = \phi(x), \end{cases} \tag{1.4}$$

is solved by the function

$$f(x, t) = \int_{\mathfrak{C}} e^{-\int_0^t V(B_s(w))ds} \phi(B_t(w)) d\mathcal{W}^x(w). \tag{1.5}$$

Here $\mathcal{W}^x$ is Wiener measure starting from x at $t = 0$. Equation (1.4) is actually the same as the imaginary time Schrödinger equation: See e.g., [18, 111]. The Feynman–Kac formula for the Schrödinger operator h derived from the heat equation can be written as (1.5):

$$e^{-th} f(x) = f(x, t) = (1.5).$$

We then construct functional integral representations for Schrödinger type operators [71, 72].

This paper includes a short review of [74, Chap. 3] and we add some new results. We can construct functional integral representations of other operators, derived from quantum field theory [49, 54, 61, 65, 66, 73]. Both functional integral representations are related each other. A Schrödinger operator with a vector potential $a = a(x) = (a_1(x), a_2(x), a_3(x))$ is given by

$$\frac{1}{2}(\sigma \cdot (-\mathrm{i}\nabla - a))^2 + V, \tag{1.6}$$

where $\sigma = (\sigma_1, \sigma_2, \sigma_3)$ is the 2×2 Pauli matrices given by

$$\sigma_1 = \begin{pmatrix} 0 & 1 \\ 1 & 0 \end{pmatrix}, \quad \sigma_2 = \begin{pmatrix} 0 & -\mathrm{i} \\ \mathrm{i} & 0 \end{pmatrix}, \quad \sigma_3 = \begin{pmatrix} 1 & 0 \\ 0 & -1 \end{pmatrix}.$$

On the other hand the Pauli–Fierz Hamiltonian with spin $1/2$ is given by

$$\frac{1}{2}(\sigma \cdot (-\mathrm{i}\nabla - A))^2 + V + H_{\mathrm{f}}, \tag{1.7}$$

where the quantized radiation field $A = A(x) = (A_1(x), A_2(x), A_3(x))$ and the free Hamiltonian H_{f} are defined by

$$A_\mu(x) = \frac{1}{\sqrt{2}} \sum_{j=\pm 1} \int_{\mathbb{R}^3} e_\mu^j(k) \left(\frac{\hat{\varphi}(k)}{\sqrt{|k|}} e^{-\mathrm{i}kx} a^*(k, j) + \frac{\hat{\varphi}(-k)}{\sqrt{|k|}} e^{+\mathrm{i}kx} a(k, j) \right) \mathrm{d}k$$

and

$$H_{\mathrm{f}} = \sum_{j=\pm 1} \int |k| a^*(k, j) a(k, j) \mathrm{d}k.$$

Here $a^*(k, j)$ and $a(k, j)$ are the creation and annihilation operators, respectively. They formally satisfy $[a(k, j), a^*(k', j')] = \delta(k - k')\delta_{jj'}$. Physically $\hat{\varphi}$ denotes the Fourier transform of the charge distribution φ. If $\varphi(x) = \delta(x)$, then

$$\hat{\varphi} = \text{constant}.$$

Since (1.6) and (1.7) are of similar forms, it would be easy to predict that the functional integral representations would also be similar. On the other hand, we can know the effect of quantum field by comparing a functional integral representation for (1.6) with that of (1.7).

This paper is organized as follows. In Sect. 2 we prepare stochastic tools and introduce functional integral representations for Schrödinger type operators. See the list of operators in Table 2. Section 3 is devoted to investigating the Pauli–Fierz Hamiltonian and its translation invariant version. Section 4 is devoted to investigating the relativistic Pauli–Fierz Hamiltonian and its translation invariant version. In Sect. 5 the Pauli–Fierz Hamiltonian with spin $1/2$ is studied. This section is rather complicated and several statements are shown in Appendices B and C. In Sect. 6 we derive

Table 2 Schrödinger operators, a: vector field, b: magnetic field, V: potential, σ: spin

	Schrödinger operator
h	$-\frac{1}{2}\Delta + V$
$h(a)$	$\frac{1}{2}(-\mathrm{i}\nabla - a)^2 + V$
$h_{\mathrm{R}}(a)$	$((-\mathrm{i}\nabla - a)^2 + m^2)^{1/2} - m + V$
$h_{\mathrm{S}}(a, b)$	$\frac{1}{2}(-\mathrm{i}\nabla - a)^2 + V - \frac{1}{2}\sigma \cdot b$
$h_{\mathrm{SR}}(a, b)$	$((-\mathrm{i}\nabla - a)^2 - \sigma \cdot b + m^2)^{1/2} - m + V$

energy comparison inequalities for the Pauli–Fierz type Hamiltonians. Section 7 is the concluding remarks.

2 Schrödinger Operators by Path Measures

In this section we introduce functional integral representations of Schrödinger type operators.

2.1 Stochastic Preparations

In order to construct functional integral representations of the semigroups generated by Schrödinger type operators we will need several independent stochastic processes $(B_t)_{t\geq 0}$, $(N_t)_{t\geq 0}$ and $(T_t)_{t\geq 0}$. See Table 3. Here we explain properties of these stochastic processes. We denote the expectation with respect to path measure W^x starting at x by

Table 3 Stochastic processes

	Process	Space	Measure
Brownian motion	$(B_t)_{t\geq 0}$	$\mathscr{X}$	$\mathcal{W}$
Poisson process	$(N_t)_{t\geq 0}$	$\mathscr{X}_\mu$	μ
Spin process	$(\theta_t)_{t\geq 0}$	$\mathscr{X}_\mu$	μ
Subordinator	$(T_t)_{t\geq 0}$	$\mathscr{X}_\nu$	ν
Subordinate brownian motion	$(B_{T_t})_{t\geq 0}$	$\mathscr{X} \times \mathscr{X}_\nu$	$\mathcal{W} \times \nu$
Subordinate spin process	$(\theta_{T_t})_{t\geq 0}$	$\mathscr{X}_\mu \times \mathscr{X}_\nu$	$\mu \times \nu$

$$\int_{\substack{\text{path}\\ \text{space}}} f \mathrm{d}W^x = \mathbb{E}_W^x[f].$$

(Brownian motion) We give the definition of Brownian motion. $(B_t)_{t\geq 0} = (B_t^1, B_t^2, B_t^3)_{t\geq 0}$ is 3D-Brownian motion on a probability space $(\mathscr{X}, \mathcal{B}, \mathcal{W}^x)$ iff

(1) $\mathcal{W}^x(B_0 = x) = 1$, i.e., $B_0(w) = x$ for almost surely $w \in \mathscr{X}$;
(2) the increments $(B_{t_i} - B_{t_{i-1}})_{1\leq i\leq n}$ are independent Gaussian random variables for every collection $0 = t_0 < t_1 \cdots < t_n$ with the mean and covariance are given by

$$\mathbb{E}_{\mathcal{W}}^0[B_t^\mu] = 0,$$
$$\mathbb{E}_{\mathcal{W}}^0[B_s^\mu B_t^\nu] = \delta_{\mu\nu}(s \wedge t),$$

respectively;
(3) the map $t \mapsto B_t(w)$ is continuous almost surely $w \in \mathscr{X}$.

By the definition of Brownian motion, $B_t - B_s$ and B_{t-s} have the same distribution on $\mathbb{R}^3$, and $B_t - B_s$ and $B_u - B_v$ for $v \leq u \leq s \leq t$ are independent. The distribution of $B_t - B_s$ for $t > s$ is given by

$$\Pi_{t-s}(x) = (2\pi(t-s))^{-3/2} e^{-|x|^2/2(t-s)}.$$

Thus it follows that

$$\mathbb{E}_{\mathcal{W}}^x[f(B_{t_1}, \ldots, B_{t_n})] = \int_{\mathbb{R}^{3N}} f(y_1, \ldots, y_n) \prod_{j=1}^n \Pi_{t_j - t_{j-1}}(y_j - y_{j-1}) \mathrm{d}y_1 \ldots . \mathrm{d}y_n \tag{2.1}$$

with $y_0 = x$ and $t_0 = 0$. Brownian motion is also Markov and then

$$\mathbb{E}_{\mathcal{W}}^x[f(B_{t+s}) \mid \mathscr{F}_s] = \mathbb{E}_{\mathcal{W}}^x[f(B_{t+s}) \mid \sigma(B_s)] = \mathbb{E}_{\mathcal{W}}^{B_s}[f(B_t)]$$

for any $x \in \mathbb{R}^3$, where $\mathbb{E}_{\mathcal{W}}^{B_s}[\ldots]$ is $\mathbb{E}_{\mathcal{W}}^y[\ldots]$ evaluated $y = B_s$, $\mathscr{F}_s = \sigma(B_u \mid 0 \leq u \leq s)$ is the minimal sigma-filed generated by $\{B_u \mid 0 \leq u \leq s\}$, and $\sigma(B_s)$ by B_s.

Remark 2.1 One concrete realisation of Brownian motion is as follows. There exists the so-called Wiener measure $\mathcal{W}^x$ on $C([0, \infty); \mathbb{R}^3)$ such that $B_t(w) = w(t)$ for $w \in C([0, \infty); \mathbb{R}^3)$ becomes Brownian motion under $\mathcal{W}^x$. In this paper we do not choose any special Brownian motion unless otherwise stated.

By (2.1) Brownian motion satisfies that

$$\mathbb{E}_{\mathcal{W}}^x[f(B_t)] = e^{\frac{t}{2}\Delta} f(x) \quad a.e.\ x \in \mathbb{R}^3$$

and hence the generator of Brownian motion is

$$\frac{1}{2}\Delta.$$

Let $M(0, t)$ be the set of measurable functions $f : [0, \infty) \times \mathscr{X} \to \mathbb{R}^3$ such that

(1) $f(t, \cdot)$ is $\mathbb{C}^3$-valued $\mathscr{F}_t$-measurable for each t,
(2) $\mathbb{E}^0_{\mathcal{W}}[\int_0^t |f(s, w)|^2 ds] < \infty$.

Then for $f \in M(0, t)$ one can define the stochastic integral X_t presented as

$$X_t = \int_0^t f(s, w) \cdot dB_s.$$

X_t satisfies that $\mathbb{E}^x[X_t] = 0$ and Itô-isometry $\mathbb{E}^x[|X_t|^2] = \int_0^t \mathbb{E}^x[|f(s, \cdot)|^2]ds$. The stochastic integral plays an important role through this article.

Now we define the so-called Lévy process. We refer to see e.g., [4, 121] and [100, Sect. 3.1]. A stochastic process $(X_t)_{t\geq 0}$ on a probability space $(\mathscr{Y}, \mathcal{B}, \mathcal{W})$ is a Lévy process iff

(1) $\mathcal{W}(X_0 = 0) = 1$;
(2) the increments $(X_{t_i} - X_{t_{i-1}})_{1\leq i\leq n}$ are independent random variables for every collection $0 = t_0 < t_1 \cdots < t_n$;
(3) $X_t - X_s$ and X_{t-s} have the same distribution for every $0 \leq s \leq t$;
(4) The map $t \mapsto X_t(w)$ is stochastically continuous, i.e.,

$$\lim_{s\to t} \mathcal{W}(|X_s - X_t| > \varepsilon) = 0$$

for all $t \geq 0$ and $\varepsilon > 0$.

Brownian motion is a Lévy process. Let $(X_t)_{t\geq 0}$ be a Lévy process on $(\mathscr{Y}, \mathcal{B}, \mu)$. It is known that there exists a Lévy process $(Y_t)_{t\geq 0}$ such that $Y_t(w) = X_t(w)$ for $w \in \mathscr{Y} \setminus N_t$, where $\mu(N_t) = 0$, and $t \mapsto Y_t(w)$ is right-continuous and has left-limits for almost surely w. See e.g., [100, Corollary 3.5]. $(Y_t)_{t\geq 0}$ is called the càdlàg version of $(X_t)_{t\geq 0}$. In what follows we assume that Lévy processes are right-continuous and have left-limits for almost surely w. Note also that a Lévy process is a Markov process.

We introduce two Lévy processes below.

(Poisson process) A nonnegative integer-valued Lévy process $(N_t)_{t\geq 0}$ on a probability space $(\mathscr{X}_\mu, \mathcal{B}_\mu, \mu)$ is called a Poisson process iff $\mathbb{E}_\mu[e^{iuN_t}] = e^{t(e^{iu}-1)}$ for $u \in \mathbb{R}$.

The distribution of $N_t - N_s$ on $\mathbb{R}$ is given by $\rho(u) = e^{(t-s)(e^{iu}-1)}$, $u \in \mathbb{R}$. It holds that

$$\mu(N_t = n) = \frac{t^n}{n!}e^{-t}, \quad n \in \mathbb{N} \cup \{0\}.$$

Since $\sum_{n=0}^{\infty} \mu(N_t = n) = 1$, we have

$$\mathbb{E}^0_\mu[f(N_t)] = \sum_{n=0}^\infty \frac{t^n}{n!} f(n) e^{-t}.$$

The poisson process $(N_t)_{t\geq 0}$ is also Markov and its generator is L given by

$$Lf(x) = f(x+1) - f(x).$$

We define integrals with respect to $(N_t)_{t\geq 0}$ in terms of the sum of evaluations at jumping times, i.e., for g we write

$$\int_a^b g(s, N_s)\mathrm{d}N_s = \sum_{\substack{a\leq r\leq b \\ N_{r+}\neq N_{r-}}} g(r, N_r).$$

Here $N_{r+} = \lim_{s\downarrow r} N_s$ and $N_{r-} = \lim_{s\uparrow r} N_s$. Then

$$\int_a^{b+} g(s, N_{s-})\mathrm{d}N_s = \begin{cases} \sum_{\substack{a\leq r<b \\ N_{r+}\neq N_{r-}}} g(r, N_{r-}), & N_{b+} = N_{b-}, \\ \sum_{\substack{a\leq r<b \\ N_{r+}\neq N_{r-}}} g(r, N_{r-}) + g(b, N_b), & N_{b+} \neq N_{b-}. \end{cases}$$

The expectation of $\int_a^{b+} g(s, N_{s-})\mathrm{d}N_s$ satisfies that

$$\mathbb{E}\left[\int_a^{b+} g(s, N_{s-})\mathrm{d}N_s\right] = \mathbb{E}\left[\int_a^{b+} g(s, N_{s-})\mathrm{d}s\right].$$

Let

$$\mathbb{Z}_2 = \{-1, +1\}. \tag{2.2}$$

By the Poisson process we also define a $\mathbb{Z}_2$-valued stochastic process $(\theta_t)_{t\geq 0}$ on $(\mathscr{X}_\mu, \mathcal{B}_\mu, \mu)$ by

$$\theta_t = (-1)^{N_t}. \tag{2.3}$$

This is called the spin process in this paper. Stochastic process $(\theta_t)_{t\geq 0}$ is useful to study functional integral representations for Schrödinger operators with spin $1/2$.

(Subordinator) A 1D-Lévy process $(T_t)_{t\geq 0}$ is called the subordinator iff whenever $s \leq t$ implies that $T_s \leq T_t$ almost surely.

Let $(T_t)_{t\geq 0}$ be a subordinator on a probability space $(\mathscr{X}_\nu, \mathcal{B}_\nu, \nu)$. $(T_t)_{t\geq 0}$ is non-negative and non-decreasing. Thus it seems to be "random time". Let us consider a subordinator $(T_t)_{t\geq 0}$ such that its Laplace transform is given by

$$\mathbb{E}_\nu^0[e^{-uT_t}] = e^{-t((2u+m^2)^{1/2}-m)}, \quad u \ge 0, \tag{2.4}$$

where $m \ge 0$ plays a role of the mass of quanta in physics. The existence of the subordinator (2.4) is established. See e.g., [100, Example 3.90]. Comparing $\mathbb{E}_{\mathcal{W}}^x[f(B_t)] = e^{\frac{t}{2}\Delta} f(x)$ and (2.4), we can see that

$$\mathbb{E}_\nu^0 \mathbb{E}_{\mathcal{W}}^x[f(B_{T_t})] = e^{-t((-\Delta+m^2)^{1/2}-m)} f(x) \quad a.e.\ x \in \mathbb{R}^3.$$

Hence the generator of the compound process $(B_{T_t})_{t\ge0}$ is the free relativistic Schrödinger operator:

$$-(-\Delta + m^2)^{1/2} + m.$$

The distribution of T_t on $\mathbb{R}$ is given by

$$\rho(r,t) = \frac{t}{\sqrt{2\pi r^3}} \exp\left(-\frac{1}{2}\left(\frac{t^2}{r} + m^2 r\right) + mt\right) 1\!\!1_{[0,\infty)}(r).$$

Furthermore the subordinate spin process $(\theta_{T_t})_{t\ge0}$ is used for constructing functional integral representations for relativistic Schrödinger operators with spin $1/2$.

We use the shorthand: $\mathbb{E}_{\mathcal{W}}^x \mathbb{E}_\mu^\alpha \mathbb{E}_\nu^0 = \mathbb{E}^{x,\alpha,0}$, $\mathbb{E}_{\mathcal{W}}^x \mathbb{E}_\mu^\alpha = \mathbb{E}^{x,\alpha}$ and $\mathbb{E}_{\mathcal{W}}^x = \mathbb{E}^x$ etc. The role of these three stochastic processes is as follows. Clearly, Schrödinger operator $-\frac{1}{2}\Delta + V$ can be described by $(B_t)_{t\ge0}$ under V. The spin process $(\theta_t)_{t\ge0}$ results from Schrödinger operators with spin $1/2$. Finally, the subordinator $(T_t)_{t\ge0}$ appears in relativistic Schrödinger operators. A particular combination of these three independent stochastic processes then yields functional integral representations of $(f, e^{-tK} g)$, where $K = h, h(a), h_{\rm R}(a), h_{\rm S}(a,b), h_{\rm SR}(a,b)$. See Table 2. For each K we shall show that

$$(f, e^{-tK} g) = \int_{\text{space}} \mathbb{E}_W^y[\bar{f}(\xi_0) g(\xi_t) e^{Z_t}] d\rho(y),$$

where $(\xi_t)_{t\ge0}$ is a stochastic process, W a path measure and e^{Z_t} an integral kernel. The generator $-G$ of ξ_t satisfies that

$$(f, e^{-tG} g) = \int_{\text{space}} \mathbb{E}_W^y[\bar{f}(\xi_0) g(\xi_t)] d\rho(y).$$

As is mentioned above, we then have examples of generators below:

$$(f, e^{\frac{t}{2}\Delta} g) = \int_{\mathbb{R}^3} \mathbb{E}^x[\bar{f}(B_0) g(B_t)] dx,$$

$$(f, e^{-t((-\Delta+m^2)^{1/2}-m)} g) = \int_{\mathbb{R}^3} \mathbb{E}^{x,0}[\bar{f}(B_0) g(B_{T_t})] dx.$$

Here we assumed that the space dimension is three, it can be however straightforwardly extended to any dimension.

2.2 Schrödinger Operators $h(a)$

We begin with showing function integral representations of $e^{-th(a)}$ with Schrödinger operators $h(a)$ without proofs as an introduction. We refer to see e.g., [1, 111, 128, 129] of functional integral representations of Schrödinger operators and its applications. Let $V : \mathbb{R}^3 \to \mathbb{R}$ be a measurable function and a Schrödinger operator is defined by

$$h = -\frac{1}{2}\Delta + V$$

acting in $L^2(\mathbb{R}^3)$. The following proposition is basic for functional integral representations.

Proposition 2.1 *Let $V \in L^\infty(\mathbb{R}^3)$. Then for $f, g \in L^2(\mathbb{R}^3)$,*

$$(f, e^{-th}g) = \int_{\mathbb{R}^3} \mathbb{E}^x[\overline{f(B_0)}g(B_t)e^{-\int_0^t V(B_s)\mathrm{d}s}]\mathrm{d}x. \tag{2.5}$$

In particular,

$$(e^{-th}g)(x) = \mathbb{E}^x[e^{-\int_0^t V(B_s)\mathrm{d}s}g(B_t)]. \tag{2.6}$$

Several proofs of this proposition are known. It involves (1) application of Itô formula [80]:

$$\begin{aligned}
&\mathrm{d}e^{-\int_0^t V(B_s)\mathrm{d}s}f(B_t)\\
&= \mathrm{d}e^{-\int_0^t V(B_s)\mathrm{d}s} \cdot f(B_t) + e^{-\int_0^t V(B_s)\mathrm{d}s} \cdot \mathrm{d}f(B_t) + \mathrm{d}e^{-\int_0^t V(B_s)\mathrm{d}s} \cdot \mathrm{d}f(B_t)\\
&= e^{-\int_0^t V(B_s)\mathrm{d}s}\left(-V(B_t)f(B_t) + \frac{1}{2}\Delta f(B_t)\right)\mathrm{d}t + e^{-\int_0^t V(B_s)\mathrm{d}s}\nabla f(B_t) \cdot \mathrm{d}B_t,
\end{aligned}$$

and (2) application of the Trotter product formula [22, 87–89, 134]:

$$e^{-th} = \lim_{n\to\infty}\left(e^{-\frac{t}{n}(-\frac{1}{2}\Delta)}e^{-\frac{t}{n}V}\right)^n.$$

Here $\mathrm{d}X_t$ denotes $\mathrm{d}X_t = X_t - X_0$. Thus

$$\mathrm{d}e^{-\int_0^t V(B_s)\mathrm{d}s}f(B_t) = e^{-\int_0^t V(B_s)\mathrm{d}s}f(B_t) - f(B_0).$$

This formula can be extended for general potentials V. Furthermore introducing a vector potential $a = (a_1, a_2, a_3) : \mathbb{R}^3 \to \mathbb{R}^3$, we define Schrödinger operators with a vector potential a by

$$h(a) = \frac{1}{2}(-\mathrm{i}\nabla - a)^2 + V.$$

Proposition 2.2 *Suppose that* $a \in (C_b^2(\mathbb{R}^3))^3$ *and* $V \in L^\infty(\mathbb{R}^3)$. *Then for* $f, g \in L^2(\mathbb{R}^3)$,

$$(f, e^{-th(a)}g) = \int_{\mathbb{R}^3} \mathbb{E}^x[\overline{f(B_0)}g(B_t)e^{Z_t}]\mathrm{d}x. \tag{2.7}$$

Here

$$Z_t = -\mathrm{i}\int_0^t a(B_s) \circ \mathrm{d}B_s - \int_0^t V(B_s)\mathrm{d}s \tag{2.8}$$

with

$$\int_0^t a(B_s) \circ \mathrm{d}B_s = \int_0^t a(B_s) \cdot \mathrm{d}B_s + \frac{1}{2}\int_0^t \nabla \cdot a(B_s)\mathrm{d}s.$$

In particular,

$$(e^{-th(a)}f)(x) = \mathbb{E}^x[e^{Z_t}f(B_t)]. \tag{2.9}$$

We give a comment on this formula. Since $h(a)$ is self-adjoint, it follows that

$$(f, e^{-th(a)}g) = \overline{(g, e^{-th(a)}f)}.$$

Integral kernel Z_t however has the purely imaginary part $-\mathrm{i}\int_0^t a(B_s) \circ \mathrm{d}B_s$. For a moment, we may feel odd. We are often asked a question on it by probabilists. We can see that

$$\begin{aligned}\overline{(g, e^{-th(a)}f)} &= \int_{\mathbb{R}^3} \mathbb{E}^x[g(B_0)\bar{f}(B_t)e^{\bar{Z}_t}]\mathrm{d}x = \int_{\mathbb{R}^3} \mathbb{E}^x[g(\dot{B}_0)\bar{f}(\dot{B}_t)e^{\dot{\bar{Z}}_t}]\mathrm{d}x \\ &= \int_{\mathbb{R}^3} \mathbb{E}^0[g(x)\bar{f}(\dot{B}_t + x)e^{\dot{\bar{Z}}_t(x)}]\mathrm{d}x,\end{aligned}$$

where $(\dot{B}_s)_{0\le s\le t} = (B_{t-s} - B_t)_{0\le s\le t}$ is also Brownian motion, and

$$\dot{\bar{Z}}_t(x) = +\mathrm{i}\int_0^t a(\dot{B}_s + x) \circ \mathrm{d}\dot{B}_s - \int_0^t V(\dot{B}_s + x)\mathrm{d}s.$$

Chang variables as $x \to y - \dot{B}_t = y + B_t$. Hence we can compute as

$$-\int_0^t V(y - \dot{B}_t + \dot{B}_s)\mathrm{d}s = -\int_0^t V(y + B_u)\mathrm{d}u,$$

while that of the stochastic integral is

$$+\mathrm{i}\int_0^t a(x + \dot{B}_s) \circ \mathrm{d}\dot{B}_s = +\mathrm{i}\int_0^t a(y - \dot{B}_t + \dot{B}_s) \circ \mathrm{d}\dot{B}_s = -\mathrm{i}\int_0^t a(y + B_s) \circ \mathrm{d}B_s.$$

Note that signature + is changed to signature −, because of the fact

$$\int_0^t a(B_s)\circ \mathrm{d}\dot{B}_s = \lim_{n\to\infty}\sum_{j=1}^{2^n}\frac{1}{2}\left(a(B_{\frac{jt}{2^n}})+a(B_{\frac{(j-1)t}{2^n}})\right)\left(B_{\frac{(j-1)t}{2^n}}-B_{\frac{jt}{2^n}}\right) = -\int_0^t a(B_s)\circ \mathrm{d}B_s.$$

We conclude that

$$\overline{(g, e^{-th(a)}f)} = \int_{\mathbb{R}^3}\mathbb{E}^y[g(y+B_t)\bar{f}(y)e^{Z_t}]\mathrm{d}y = (f, e^{-th(a)}g).$$

The self-adjointness of $h(a)$ and the existence of the purely imaginary part $-\mathrm{i}\int_0^t a(B_s)\circ \mathrm{d}B_s$ become compatible.

2.3 *Schrödinger Operators with Kato-Class Potentials*

We consider Schrödinger operators with singular potentials. Kato-class is a class of singular potentials but we can define Schrödinger operators with Kato-class potentials as self-adjoint operators by functional integral representations.

Definition 2.1 (1) $V:\mathbb{R}^d\to\mathbb{R}$ is called a Kato-class potential whenever

$$\lim_{r\to 0}\sup_{x\in\mathbb{R}^d}\int_{B_r(x)}|g(x-y)V(y)|\mathrm{d}y = 0 \tag{2.10}$$

holds, where $B_r(x)$ is the closed ball of radius r centered at x, and $g(x)$ is the function given by

$$g(x) = \begin{cases}|x|, & d=1,\\ -\log|x|, & d=2,\\ |x|^{2-d}, & d\geq 3.\end{cases} \tag{2.11}$$

We denote the set of Kato-class potentials by $\mathcal{K}(\mathbb{R}^d)$.

(2) V is a local Kato-class potential whenever $V\lceil_K \in \mathcal{K}(\mathbb{R}^d)$ for any compact set $K\subset\mathbb{R}^d$. We denote the set of local Kato-class potentials by $\mathcal{K}_{\mathrm{loc}}(\mathbb{R}^d)$.

(3) V is Kato-decomposable whenever $V=V_+-V_-$ with $V_+\in\mathcal{K}_{\mathrm{loc}}(\mathbb{R}^d)$ and $V_-\in\mathcal{K}(\mathbb{R}^d)$.

We give non-trivial examples of Kato-class potentials for $d=3$. Notice that

$$\left|\int_{B_r(x)}\frac{1}{|x-y|}|V(y)|\mathrm{d}y\right| \leq \left(\int_{B_r(x)}\frac{1}{|x-y|^p}\mathrm{d}y\right)^{1/p}\left(\int_{B_r(x)}|V(y)|^q\mathrm{d}y\right)^{1/q},$$

with $1/p + 1/q = 1$. When $p < 3$, the first factor goes to zero as $r \to 0$. Thus V is Kato-class if $\sup_{x\in\mathbb{R}^3}\int_{B_r(x)}|V(y)|^q dy < \infty$ for some $r > 0$ and some $q > 3/2$. Thus

$$V(x) = \frac{1}{|x|^{2-\varepsilon}}, \quad \varepsilon > 0$$

is Kato-class.

We are in the position to prove exponential integrability of integrals over some potentials. When $V_- \in \mathcal{K}(\mathbb{R}^3)$, it can be seen that the exponent $e^{\int_0^t V(B_s)ds}$ is integrable with respect to Wiener measure so that $\mathbb{E}^x[e^{\int_0^t V(B_s)ds}]$ is finite for all x.

Lemma 2.3 ([74, Lemma 4.105], [19]) *Let $V \in \mathcal{K}(\mathbb{R}^d)$ and $V(x) \geq 0$ a.e. $x \in \mathbb{R}^d$. Then there exist $\beta, \gamma > 0$ such that*

$$\sup_{x\in\mathbb{R}^d} \mathbb{E}^x[e^{\int_0^t V(B_s)ds}] < \gamma e^{\beta t}. \tag{2.12}$$

Furthermore, if $V \in L^p(\mathbb{R}^d)$ with $p > d/2$ and $1 \leq p < \infty$, then $\beta \leq c(p)^{1/\varepsilon}\Gamma(\varepsilon)^{1/\varepsilon}\|V\|_p^{1/\varepsilon}$, where $\varepsilon = 1 - \frac{d}{2p}$ and $c(p) = \begin{cases} (2\pi)^{-d/2} & p = 1, \\ (2\pi)^{-d/2p}q^{d/(2q)} & p > 1 \end{cases}$ with $\frac{1}{p} + \frac{1}{q} = 1$. In particular $L^p(\mathbb{R}^d) \subset \mathcal{K}(\mathbb{R}^d)$ for $p > d/2$ and $1 \leq p < \infty$.

Let V be Kato-class. Define the map K_t on $L^2(\mathbb{R}^3)$ by

$$(K_t f)(x) = \mathbb{E}^x[e^{-\int_0^t V(B_s)ds} f(B_t)].$$

Lemma 2.4 *$\{K_t : t \geq 0\}$ is a symmetric C_0-semigroup on $L^2(\mathbb{R}^3)$.*

Proof The boundedness of K_t follows from Lemma 2.3. Define $\tilde{B}_s = B_{t-s} - B_t$, $0 \leq s \leq t$, for a fixed $t > 0$. We see that $\tilde{B}_s$ is also a Brownian motion. Thus

$$(f, K_t g) = \mathbb{E}^0\left[\int_{\mathbb{R}^3} \overline{f(x)} e^{-\int_0^t V(\tilde{B}_s + x)ds} g(\tilde{B}_t + x)dx\right].$$

Changing the variable x to $y = \tilde{B}_t + x$, we obtain

$$\begin{aligned}(f, K_t g) &= \mathbb{E}^0\left[\int_{\mathbb{R}^3} \overline{f(y - \tilde{B}_t)} e^{-\int_0^t V(\tilde{B}_s - \tilde{B}_t + y)ds} g(y)dy\right] \\ &= \int_{\mathbb{R}^3} \mathbb{E}^0[\overline{f(y + B_t)} e^{-\int_0^t V(B_{t-s}+y)ds} g(y)]dy = (K_t f, g),\end{aligned} \tag{2.13}$$

i.e., K_t is symmetric. Write now $Z_t = e^{-\int_0^t V(B_s)ds}$. The semigroup property follows directly from the Markov property of Brownian motion:

$$\begin{aligned}(K_sK_tf)(x) &= \mathbb{E}^x[Z_s\mathbb{E}^{B_s}(Z_tf(B_t))]\\ &= \mathbb{E}^x[\mathbb{E}^x[Z_se^{-\int_0^t V(B_{s+u})\mathrm{d}u}f(B_{s+t})|\mathcal{F}_s]]\\ &= \mathbb{E}^x\left[Z_{s+t}f(B_{s+t})\right] = K_{s+t}f(x).\end{aligned}$$

The strong continuity of $t \mapsto K_t$ is implied by

$$\|K_tf - f\| \le \mathbb{E}^0[\|e^{-\int_0^t V(\cdot+B_s)\mathrm{d}s}f(\cdot + B_t) - f\|] \to 0$$

as $t \to 0$, by the Lebesgue dominated convergence theorem.

QED

By the Stone's theorem for semigroups there exists a unique self-adjoint operator K such that

$$K_t = e^{-tK}$$

for any $t \ge 0$. For example we can define a self-adjoint operator $K = -\frac{1}{2}\Delta - \frac{1}{|x|^\alpha}$ for $0 \le \alpha < 2$, since $\frac{1}{|x|^\alpha} \in \mathcal{K}(\mathbb{R}^3)$ for $0 \le \alpha < 2$.

$-\frac{1}{2}\Delta$ is defined as a self-adjoint operator in the Hilbert space $L^2(\mathbb{R}^3)$ and $e^{\frac{t}{2}\Delta}$ by the spectral measure associated with $-\Delta$. Hence $e^{\frac{t}{2}\Delta}f$ is well-defined for any $f \in L^2(\mathbb{R}^3)$, while it can be represented as $e^{\frac{t}{2}\Delta}f(x) = \mathbb{E}^x[f(B_t)]$ for each $x \in \mathbb{R}^3$. We note that the right-hand side $\mathbb{E}^x[f(B_t)]$ can be defined not only for $f \in L^2(\mathbb{R}^3)$ but also $f \in L^p(\mathbb{R}^3)$ for $1 \le p \le \infty$. More precisely $e^{\frac{t}{2}\Delta}f$ for $f \in L^p(\mathbb{R}^3)$ is defined by $\mathbb{E}^x[f(B_t)]$. We can show that $e^{\frac{t}{2}\Delta}$ maps $L^p(\mathbb{R}^3)$ to $L^q(\mathbb{R}^3)$ for every $1 \le p \le q \le \infty$. For $-\frac{1}{2}\Delta + V$ with Kato-decomposable V we can show the statements below:

(1) Let $V = V_+ - V_-$. Suppose that $V_+ \in L^1_{loc}(\mathbb{R}^3)$ and $V_- \in \mathcal{K}(\mathbb{R}^3)$. Then for each $x \in \mathbb{R}^3$, $\mathbb{E}^x[e^{-\int_0^t V(B_u)\mathrm{d}u}f(B_t)]$ is well-defined. We can however say more strong statements for kato-decomposable V. Let V be Kato-decomposable and $f \in L^p(\mathbb{R}^3)$ for $1 \le p \le \infty$. Then for every $t > 0$, $x \mapsto K_tf(x)$ is continuous [1].
(2) Let V be Kato-decomposable. Then for every $1 \le p \le q \le \infty$, K_t maps $L^p(\mathbb{R}^3)$ to $L^q(\mathbb{R}^3)$ as bounded operators, i.e., $\|K_tf\|_{L^q(\mathbb{R}^3)} \le C\|f\|_{L^p(\mathbb{R}^3)}$.

2.4 Schrödinger Operators with Spin 1/2 $h_S(a, b)$

In this section we discuss functional integral representations for Schrödinger operators with spin 1/2. We refer to see [29–31, 42, 71, 72] for functional integral representations of Schrödinger operators with spin 1/2. Since the Trotter product formula holds even when there is spin, it is easy to see that a functional integral representation can be obtained like Schrödinger operators, but an infinite product of matrices appears in the integral kernel. It may be possible to have an exact form of the infinite product of matrices, but we will introduce another way.

Let $\sigma = (\sigma_1, \sigma_2, \sigma_3)$ be the 2×2 Pauli matrices given by

$$\sigma_1 = \begin{pmatrix} 0 & 1 \\ 1 & 0 \end{pmatrix}, \quad \sigma_2 = \begin{pmatrix} 0 & -\mathrm{i} \\ \mathrm{i} & 0 \end{pmatrix}, \quad \sigma_3 = \begin{pmatrix} 1 & 0 \\ 0 & -1 \end{pmatrix}.$$

σ_js are traceless symmetric matrices. Consider the $\mathbb{R}^3 \times \mathbb{Z}_2$-valued joint Brownian and jump process

$$\mathscr{X} \times \mathscr{X}_\mu \ni (w, m) \mapsto X_t(w, m) = (B_t(w), \theta_t(m)) \in \mathbb{R}^3 \times \mathbb{Z}_2$$

with initial value X_0. The generator of $(X_t)_{t \geq 0}$ is

$$G_0 = \frac{1}{2}\Delta - \sigma_{\mathrm{F}},$$

where σ_{F} is the fermionic harmonic oscillator defined by

$$\sigma_{\mathrm{F}} = \frac{1}{2}(\sigma_3 + \mathrm{i}\sigma_2)(\sigma_3 - \mathrm{i}\sigma_2) = -\sigma_1 + \mathbb{1}_{2\times 2}.$$

Let $b = (b_1, b_2, b_3)$ be a magnetic potential, and

$$h_{\mathrm{S}}(a, b) = \frac{1}{2}(-\mathrm{i}\nabla - a)^2 - \frac{1}{2}\sigma \cdot b + V \tag{2.14}$$

be a Schrödinger operator with spin 1/2 defined in $\mathbb{C}^2 \otimes L^2(\mathbb{R}^3)$. In physics a Schrödinger operator with spin 1/2 is defined by

$$\frac{1}{2}(\sigma \cdot (-\mathrm{i}\nabla - a))^2 + V.$$

We can see that

$$\frac{1}{2}(\sigma \cdot (-\mathrm{i}\nabla - a))^2 + V = \frac{1}{2}(-\mathrm{i}\nabla - a)^2 - \frac{1}{2}\sigma \cdot \mathrm{rot}a + V$$

and $b = \mathrm{rot}a$. In this section however we assume that a and b are independent vectors in (2.14).

We identify $\mathbb{C}^2 \otimes L^2(\mathbb{R}^3) = L^2(\mathbb{R}^3 \times \mathbb{Z}_2)$ by $\mathbb{C}^2 \otimes L^2(\mathbb{R}^3) \ni \binom{f_1}{f_2} \mapsto f(x, \cdot) \in L^2(\mathbb{R}^3 \times \mathbb{Z}_2)$, where $f_j(x) = f(x, j)$. See Table 5 in Sect. 5. The proposition below is established.

Proposition 2.5 ([31, 71]) *Let $a \in (C_{\mathrm{b}}^2(\mathbb{R}^3))^3$, $b \in (L^\infty(\mathbb{R}^3))^3$ and $V \in L^\infty(\mathbb{R}^3)$. Suppose that*

$$\int_0^t \mathrm{d}s \int_{\mathbb{R}^3} \left| \log \frac{1}{2}\sqrt{b_1(y)^2 + b_2(y)^2} \right| \frac{e^{-|y-x|^2/(2s)}}{(2\pi s)^{3/2}} \mathrm{d}y < \infty$$

for all $(x, t) \in \mathbb{R}^3 \times \mathbb{R}^+$. Then

$$(f, e^{-th_S(a,b)}g) = e^t \sum_{\alpha \in \mathbb{Z}_2} \int_{\mathbb{R}^3} \mathbb{E}^{x,\alpha}[\overline{f(X_0)}g(X_t)e^{Z_t}]dx. \tag{2.15}$$

Here

$$\begin{aligned} Z_t = & -i \int_0^t a(B_s) \circ dB_s - \int_0^t V(B_s)ds \\ & + \int_0^t \frac{1}{2}\theta_{N_s} b_3(B_s)ds + \int_0^{t+} \log\left(\frac{1}{2}(b_1(B_s) - i\theta_{N_{s-}} b_2(B_s))\right) dN_s. \end{aligned} \tag{2.16}$$

Note that inserting $V = 0$, $a = (0, 0, 0)$ and $b = (2, 0, 0)$ yields that $h_S(a, b) = -\frac{1}{2}\Delta - \sigma_1$ and

$$Z_t = \int_0^{t+} \log 1 dN_s = 0.$$

Hence it follows that

$$e^t \sum_{\alpha \in \mathbb{Z}_2} \int_{\mathbb{R}^3} \mathbb{E}^{x,\alpha}[\overline{f(X_0)}g(X_t)e^{Z_t}]dx = (f, e^t e^{-t(-\frac{1}{2}\Delta + \sigma_F)}g) = (f, e^{-t(-\frac{1}{2}\Delta - \sigma_1)}g).$$

2.5 Relativistic Schrödinger Operators $h_R(a)$

We consider relativistic Schrödinger operators. We refer to see [20, 28, 71] for functional integral representations for relativistic Schrödinger operators. In the relativistic case, the subordinator explained above appears in addition. Let

$$h_R(a) = ((-i\nabla - a)^2 + m^2)^{1/2} - m + V.$$

When $V = 0$, we can see that

$$\mathbb{E}_\nu^0[e^{-T_t h(a)}] = e^{-th_R(a)}$$

by the definition of the subordinator $(T_t)_{t \geq 0}$. Hence by

$$(f, e^{-th(a)}g) = \int_{\mathbb{R}^3} \mathbb{E}^x[\overline{f(B_0)}g(B_t)e^{-i\int_0^t a(B_s)\circ dB_s}]dx,$$

we have

$$(f, e^{-th_R(a)}g) = \int_{\mathbb{R}^3} \mathbb{E}^{x,0}[\overline{f(B_0)}g(B_{T_t})e^{-i\int_0^{T_t} a(B_s)\circ dB_s}]dx. \tag{2.17}$$

We have the proposition below.

Proposition 2.6 ([28, 71]) *Suppose that* $V \in L^\infty(\mathbb{R}^3)$, $a \in (L^2_{\rm loc}(\mathbb{R}^3))^3$ *and* $\nabla \cdot a \in L^1_{\rm loc}(\mathbb{R}^3)$. *Then*

$$(f, e^{-th_{\rm R}(a)}g) = \int_{\mathbb{R}^3} \mathbb{E}^{x,0}[\overline{f(B_0)}g(B_{T_t})e^{Z_t}]\mathrm{d}x, \tag{2.18}$$

where

$$Z_t = -\mathrm{i}\int_0^{T_t} a(B_s)\circ \mathrm{d}B_s - \int_0^t V(B_{T_s})\mathrm{d}s. \tag{2.19}$$

Proof The proof is an application of the Trotter product formula and (2.17). QED

2.6 Relativistic Schrödinger Operators with Spin 1/2 $h_{\rm SR}(a,b)$

We consider functional integral representations for relativistic Schrödinger operators with spin $1/2$. Combining the relativistic case and the spin case well, we can get functional integral representations for relativistic Schrödinger operators with spin $1/2$. We refer to see [32, 71, 72] for functional integral representations for relativistic Schrödinger operators with spin $1/2$. We define the subordinate process $(q_t)_{t\geq 0}$ in terms of the $\mathbb{R}^3\times\mathbb{Z}_2$-valued stochastic process $(X_t)_{t\geq 0} = ((B_t,\theta_t))_{t\geq 0}$:

$$\mathscr{X}\times\mathscr{X}_\nu\times\mathscr{X}_\nu \ni (w,m,w_2)\mapsto q_t(w,m,w_2) = (B_{T_t(w_2)}(w), \theta_{T_t(w_2)}(m)) \in \mathbb{R}^3\times\mathbb{Z}_2.$$

In a similar manner to $(X_t)_{t\geq 0}$, we can identify the generator of $(q_t)_{t\geq 0}$. The generator of $(q_t)_{t\geq 0}$ is

$$G = -(-\Delta + 2\sigma_{\rm F} + m^2)^{1/2} + m. \tag{2.20}$$

This is obtained through the equalities

$$\begin{aligned}\sum_{\alpha\in\mathbb{Z}_2}\int_{\mathbb{R}^3}\mathbb{E}^{x,\alpha,0}\left[\overline{f(q_0)}g(q_t)\right]\mathrm{d}x &= \mathbb{E}^0_\nu\left[\sum_{\alpha\in\mathbb{Z}_2}\int_{\mathbb{R}^3}\mathbb{E}^{x,\alpha}_{\mathcal{W}\times\mu}\left[\overline{f(q_0)}g(q_t)\right]\mathrm{d}x\right]\\ &= \mathbb{E}^0_\nu[(f, e^{-T_t(-\frac{1}{2}\Delta+\sigma_F)}g)] = (f, e^{-tG}g).\end{aligned}$$

Hence it follows that (2.20) is the generator of $(q_t)_{t\geq 0}$.

Let

$$h_{\rm SR}(a,b) = ((-\mathrm{i}\nabla - a)^2 - \sigma\cdot b + m^2)^{1/2} - m + V$$

Table 4 Schrödinger operators

Schrödinger operator	Space	Path space	Process ξ_t	Generator G	Measure W	Kernel Z_t
$h(a)$	$\mathbb{R}^3$	$\mathscr{X}$	B_t	$\frac{1}{2}\Delta$	$\mathcal{W}$	(2.8)
$h_{\mathrm{R}}(a)$	$\mathbb{R}^3$	$\mathscr{X}\times\mathscr{X}_\nu$	B_{T_t}	$-(-\Delta+m^2)^{1/2}+m$	$\mathcal{W}\times\nu$	(2.19)
$h_{\mathrm{S}}(a,b)$	$\mathbb{R}^3\times\mathbb{Z}_2$	$\mathscr{X}\times\mathscr{X}_\mu$	$X_t=(B_t,\theta_t)$	$\frac{1}{2}\Delta-\sigma_F$	$\mathcal{W}\times\mu$	(2.16)
$h_{\mathrm{SR}}(a,b)$	$\mathbb{R}^3\times\mathbb{Z}_2$	$\mathscr{X}\times\mathscr{X}_\mu\times\mathscr{X}_\nu$	$q_t=(B_{T_t},\theta_{T_t})$	$-(-\Delta+2\sigma_F+m^2)^{1/2}+m$	$\mathcal{W}\times\mu\times\nu$	(2.22)

be a relativistic Schrödinger operator with vector potential a, magnetic potential b and spin 1/2 defined on $\mathbb{C}^2\otimes L^2(\mathbb{R}^3)$. Let $\rho(r,t)$ be the distribution of T_t on $\mathbb{R}$.

Proposition 2.7 ([32, 71]) *Let $V\in L^\infty(\mathbb{R}^3)$, $a\in(L^2_{\mathrm{loc}}(\mathbb{R}^3))^3$ and $\nabla\cdot a\in L^1_{\mathrm{loc}}(\mathbb{R}^3)$. Suppose that $b\in(L^\infty(\mathbb{R}^3))^3$ and*

$$\int_0^\infty \mathrm{d}r\rho(r,t)\int_0^r \mathrm{d}s\int_{\mathbb{R}^3}\left|\log\left(\frac{1}{2}\sqrt{b_1(y)^2+b_2(y)^2}\right)\right|\frac{e^{-|y-x|^2/2s}}{(2\pi s)^{3/2}}\mathrm{d}y<\infty$$

for all $(x,t)\in\mathbb{R}^3\times\mathbb{R}^+$. Then

$$(f,e^{-th_{\mathrm{SR}}(a,b)}g)=\sum_{\alpha=0,1}\int_{\mathbb{R}^3}\mathbb{E}^{x,\alpha,0}[e^{T_t}\overline{f(q_0)}g(q_t)e^{Z_t}]\mathrm{d}x, \qquad (2.21)$$

where

$$Z_t=-\mathrm{i}\int_0^{T_t}a(B_s)\circ\mathrm{d}B_s-\int_0^t V(B_{T_s})\mathrm{d}s+\int_0^{T_t}\frac{1}{2}\theta_{N_s}b_3(B_s)\mathrm{d}s$$
$$+\int_0^{T_t+}\log\left(\frac{1}{2}(b_1(B_s)-\mathrm{i}\theta_{N_{s-}}b_2(B_s))\right)\mathrm{d}N_s. \qquad (2.22)$$

Note that $V=0$, $a=(0,0,0)$ and $b=(2,0,0)$. Then $h_{\mathrm{SR}}(a,b)=(-\Delta-2\sigma_1+m^2)^{1/2}-m$ and $Z_t=\int_0^{T_t+}\log 1\mathrm{d}N_s=0$. Hence it follows that

$$\sum_{\alpha=0,1}\int_{\mathbb{R}^3}\mathbb{E}^{x,\alpha,0}[e^{T_t}\overline{f(q_0)}g(q_t)e^{Z_t}]\mathrm{d}x=\mathbb{E}^0[e^{T_t}(f,e^{-T_t(-\frac{1}{2}\Delta+\sigma_F)}g)]$$
$$=(f,e^{-t((-\Delta-2\sigma_1+m^2)^{1/2}-m)}g).$$

We exhibit all the results mentioned above in Table 4.

2.7 Brief Summaries of Applications

There are many applications of functional integral representations to spectral analysis of Schrödinger operators. We refer to see e.g., [128]. Here are some of them.

(Singular potentials) By functional integral representations we can define a Schrödinger operator with a singular potential V. E.g., Kato-class potentials (see Definition 2.1). We define $\mathrm{K}_t f(x) = \mathbb{E}^x[e^{-\int_0^t V(B_s)\mathrm{d}s} f(B_t)]$ for a Kato-class potential V. One can show that K_t defines a symmetric C_0-semigroup on $L^2(\mathbb{R}^3)$. Thus there exists a unique self-adjoint operator K such that $\mathrm{K}_t = e^{-tK}$ by the Stone's theorem.

Functional integral representations of e^{-th} are also useful to study properties of Schrödinger operators with singular potentials. It includes Klauder phenomena which yields

$$\lim_{\varepsilon \to 0} -\frac{1}{2}\Delta + \varepsilon V \neq -\frac{1}{2}\Delta$$

for some singular V. See [33, 34, 125].

(Ergodic properties) [36, 37, 118] From the formula

$$(f, e^{-th}g) = \int_{\mathbb{R}^3} \mathbb{E}^x[\ldots]\mathrm{d}x,$$

one can immediately see that $(f, e^{-th}g) > 0$ for any $f, g \geq 0$. This derives that e^{-th} is positivity improving or ergodic.

(Spatial decay) [1, 19, 20, 72] If h admits an eigenfunction ϕ at eigenvalue E, then $e^{-th}\phi = e^{-Et}\phi$ and hence the identity

$$\phi = e^{-t(h-E)}\phi \tag{2.23}$$

follows for any $t \geq 0$. Thus this eigenfunction can be represented in terms of an average over the paths of a stochastic process. This makes possible to obtain information on the spectral properties of h by probabilistic means. Let $h = -\frac{1}{2}\Delta + V$. Then we have

$$\phi(x) = e^{tE}\mathbb{E}^x[e^{-\int_0^t V(B_s)\mathrm{d}s}\phi(B_t)]. \tag{2.24}$$

From this we can see e.g., the spatial decay of $\phi(x)$ from both upper and lower. While we define

$$X_t(x) = e^{tE}e^{-\int_0^t V(B_s+x)\mathrm{d}s}\phi(B_t + x).$$

It can be shown that $(X(x)_t)_{t\geq 0}$ is a martingale and then

$$\mathbb{E}^0[X_{t\wedge \tau}(x)] = \mathbb{E}^0[X_0(x)] = \phi(x)$$

for any stopping time τ. Choosing a suitable stopping time τ, we can also estimate the spatial decay of $\phi(x)$ from both upper and lower. We refer to see e.g., [20].

(Hypercontractivity) [25–27] It is not hard to establish that $e^{\Delta/2} : L^p(\mathbb{R}^3) \to L^q(\mathbb{R}^3)$ for every $1 \le p \le q \le \infty$. We can extend this to e^{-th}. From (2.24) and the Riesz–Thorin interpolation [74, Lemma 4.108] we can also show that e^{-th} is hypercontractive. I.e., $e^{-th} : L^p(\mathbb{R}^3) \to L^q(\mathbb{R}^3)$ for every $1 \le p \le q \le \infty$.

(Smoothing effect) [1] For suitable V we can see that

$$x \mapsto \mathbb{E}^x[e^{-\int_0^t V(B_s)\mathrm{d}s}\phi(B_t)]$$

is continuous. E.g., $V = V_+ - V_-$ with Kato-class V_+ and local Kato-class V_-. From this the spatial continuity of $x \mapsto \phi(x)$ can be obtained by (2.23).

(Lieb–Thirring inequality) [24, 94–96, 98, 99] The number of non-positive eigenvalues

$$N = \#\{e \mid e \in \sigma_p(h), e \le 0\}$$

of h can be also estimated by using the functional integral representation of e^{-th}, which is known as Lieb–Thirring inequality:

$$N \le a_3 \int_{\mathbb{R}^3} |V_-(x)|^{3/2}\mathrm{d}x.$$

This inequality can be extended to general Schrödinger type operators. In general for

$$h^\Psi = \Psi(-\Delta) + V$$

with a Bernstein function Ψ [122], it can be seen under some conditions that

$$N \le a_{3,\Psi} \int_{\mathbb{R}^3} \Psi^{-1}(|V_-(x)|)^{3/2}\mathrm{d}x.$$

Here Ψ^{-1} is the inverse function of Ψ. The examples of Bernstein functions include $\Psi(x) = x^\alpha$ for $0 < \alpha < 1$ and $\Psi(x) = 1 - e^{-\beta x}$ for $\beta > 0$. Let $\Psi(x) = \sqrt{x}$. Then $N \le a_{3,\Psi} \int_{\mathbb{R}^3} |V_-(x)|^3\mathrm{d}x$ follows. See [24].

(Non-relativistic limit) [68] Let c be the velocity of the light. Let $T_t(c)$ be the subordinator such that

$$\mathbb{E}_\nu^0[e^{-uT_t(c)}] = e^{-t(\sqrt{2c^2u+m^2c^4}-mc^2)}.$$

Thus one can show that

$$\lim_{c\to\infty} \mathbb{E}_\nu^0[f(T_t(c))] = f\left(\frac{t}{m}\right)$$

for any bounded continuous function f. The right-hand side $f(\frac{t}{m})$ is deterministic. Let

$$h_c = \sqrt{-\Delta + m^2c^4} - mc^2 + V.$$

Thus by (2.18) one can see that

$$\begin{aligned}\lim_{c\to\infty}(f, e^{-th_c}g) &= \lim_{c\to\infty}\int_{\mathbb{R}^3}\mathbb{E}^{x,0}[\overline{f(B_0)}g(B_{T_t(c)})e^{-\int_0^t V(B_{T_s(c)})\mathrm{d}s}]\mathrm{d}x \\ &= \int_{\mathbb{R}^3}\mathbb{E}^x[\overline{f(B_0)}g(B_{\frac{t}{m}})e^{-\int_0^t V(B_{\frac{s}{m}})\mathrm{d}s}]\mathrm{d}x = (f, e^{-t(-\frac{1}{2m}\Delta+V)}g).\end{aligned}$$

Applications mentioned above are available not only for h but also other Schrödinger operators $h(a)$, $h_{\mathrm{R}}(a)$, $h_S(a,b)$ and $h_{\mathrm{SR}}(a,b)$.

(Rabi-model) [58] The Rabi Hamiltonian is a self-adjoint operator in $\mathbb{C}^2 \otimes L^2(\mathbb{R})$ which is defined by

$$H_{\mathrm{Rabi}} = \omega a^*a + g\sigma_1(a + a^*) - \Delta\sigma_3.$$

Here $a = \frac{1}{\sqrt{2}}(x + \frac{d}{dx})$ and $a^* = \frac{1}{\sqrt{2}}(x - \frac{d}{dx})$ are the annihilation operator and the creation operator, both $\omega > 0$ and $\Delta \geq 0$ are constants. The rotation group in $\mathbb{R}^3$ has an adjoint representation on $SU(2)$. Let $n \in \mathbb{R}^3$ be a unit vector and $\theta \in [0, 2\pi)$. Thus $e^{(i/2)\theta n\cdot\sigma}$ satisfies that

$$e^{(i/2)\theta n\cdot\sigma}\sigma_\mu e^{-(i/2)\theta n\cdot\sigma} = (R\sigma)_\mu,$$

where R denotes 3×3 matrix representing the rotation around n with angle θ. In particular for $n = (0, 1, 0)$ and $\theta = \pi/2$, we have

$$\begin{aligned}e^{(i/2)\theta n\cdot\sigma}\sigma_1 e^{-(i/2)\theta n\cdot\sigma} &= \sigma_3, \\ e^{(i/2)\theta n\cdot\sigma}\sigma_3 e^{-(i/2)\theta n\cdot\sigma} &= -\sigma_1.\end{aligned}$$

Set $U = e^{(i\pi/4)\sigma_2}$. Then

$$UH_{\mathrm{Rabi}}U^{-1} = \omega a^*a + g\sigma_3(a + a^*) - \Delta\sigma_1.$$

Since $\Psi_{\mathrm{g}} = \pi^{-1/4}e^{-|x|^2/2}$ is strictly positive, we can define the unitary operator

$$U_g : L^2(\mathbb{R}) \to L^2(\mathbb{R}, \Psi_{\mathrm{g}}^2\mathrm{d}x)$$

by $U_g f = \Psi_{\mathrm{g}}^{-1}f$. We set the probability measure $\Psi_{\mathrm{g}}^2\mathrm{d}x$ on $\mathbb{R}$ by $\mathrm{d}\mu$. Thus $UH_{\mathrm{Rabi}}U^{-1}$ is transformed to the operator:

$$H = U_g U H_{\text{Rabi}} U^{-1} U_g^{-1} = \frac{1}{2}\left(-\frac{d^2}{dx^2} + \omega x \frac{d}{dx}\right) + g\sigma_3\sqrt{2\omega}x - \Delta\sigma_1$$

in $\mathbb{C}^2 \otimes L^2(\mathbb{R}, \mathrm{d}\mu)$. Let

$$h = \frac{1}{2}\left(-\frac{d^2}{dx^2} + \omega x \frac{d}{dx}\right)$$

and $(X_t)_{t\geq 0}$ be the Ornstein–Uhrenbeck process on a probability space $(\mathscr{X}, \mathcal{B}, P^x)$. We have $P^x(X_0 = x) = 1$

$$\int_{\mathbb{R}} \mathbb{E}_{P^x}[X_t]\,\mathrm{d}\mu(x) = 0, \quad \int_{\mathbb{R}} \mathbb{E}_{P^x}[X_t X_s]\,\mathrm{d}\mu(x) = \frac{e^{-|t-s|\omega}}{2\omega}.$$

The generator of X_t is given by $-h$ and

$$(f, e^{-th}g) = \int_{\mathbb{R}} \mathbb{E}_{P^x}\left[\overline{f(X_0)}g(X_t)\right]\mathrm{d}\mu(x).$$

The distribution $\rho_t(x, y)$ of X_t under P^x is given by

$$\rho_t(x, y) = \Psi_{\mathrm{g}}(x)^{-1} K_t(x, y)\Psi_{\mathrm{g}}(y),$$

where $K_t(x, y)$ denotes the Mehler kernel:

$$K_t(x, y) = \frac{1}{\sqrt{\pi(1 - e^{-2t})}} \exp\left(\frac{4xye^{-t} - (x^2 + y^2)(1 + e^{-2t})}{2(1 - e^{-2t})}\right).$$

Let $\sum_{\sigma\in\mathbb{Z}_2} \int_{\mathbb{R}} \mathbb{E}_{P^x}\mathbb{E}_\nu[\cdots]\,\mathrm{d}\mu(x) = \mathbb{E}[\cdots]$. It follows that

$$(\Delta > 0) \quad (f, e^{-tH}g)_{\mathcal{H}} = e^t\mathbb{E}\left[\overline{f(X_0, \sigma_0)}g(X_t, \sigma_t)e^{-g\sqrt{2\omega}\int_0^t \sigma_s X_s \mathrm{d}s}\Delta^{N_t}\right],$$

$$(\Delta = 0) \quad (f, e^{-tH}g)_{\mathcal{H}} = e^t\mathbb{E}\left[\mathbb{1}_{N_t=0}\overline{f(X_0, \sigma)}g(X_t, \sigma)e^{-g\sigma\sqrt{2\omega}\int_0^t X_s \mathrm{d}s}\right].$$

The proof is as follows. Let $\Delta > 0$. We can see that

$$(f, e^{-tH}g)_{\mathcal{H}} = e^t\mathbb{E}\left[\overline{f(X_0, \sigma_0)}g(X_t, \sigma_t)e^{-g\sqrt{2\omega}\int_0^t \sigma_s X_s \mathrm{d}s}e^{\int_0^t \log\Delta \mathrm{d}N_s}\right].$$

Since $e^{\int_0^t \log\Delta \mathrm{d}N_s} = e^{\log\Delta^{N_t}} = \Delta^{N_t}$, the first statement follows. In the case of $\Delta = 0$ only the set $N_t = 0$ contributes to the path integral. Then the second statement follows.

3 Pauli–Fierz Model

3.1 Newton–Maxwell Equation

Consider a single classical particles in the configuration space $\mathbb{R}^3$, interacting with an electromagnetic field. The particle has an assigned mass $m = 1$ and a charge distribution $\hat{\varphi}$, and their dynamics is completely characterized by their momentum $p \in \mathbb{R}^3$ and position $q \in \mathbb{R}^3$. On the other hand, it is convenient for our purposes to describe the electromagnetic field in the Coulomb gauge by a complex vector field:

$$\alpha = (\alpha_\lambda)_{\lambda=1,2} : \mathbb{R}^3 \longrightarrow \mathbb{C}.$$

Thus the Hamiltonian of the Newton–Maxwell system takes the form

$$H(p,q,\alpha) = \frac{1}{2}(p - A(q,\alpha))^2 + V(q) + \sum_{\lambda=1,2} \int_{\mathbb{R}^3} \bar{\alpha}_\lambda(k)|k|\alpha_\lambda(k)\mathrm{d}k.$$

This leads to the Newton–Maxwell equations:

$$\begin{cases} \nabla_t p = -\dfrac{\partial H}{\partial q}, \\ \nabla_t q = \dfrac{\partial H}{\partial p}, \\ \mathrm{i}\nabla_t \alpha_\lambda(k) = \dfrac{\partial H}{\partial \bar{\alpha}(k)}. \end{cases}$$

Hence we obtain

$$(\mathrm{NM}) \begin{cases} \nabla_t p = (p - A(q,\alpha)) \cdot \nabla_q A(q,\alpha) - \nabla_q V(q), \\ \nabla_t q = p - A(q,\alpha), \\ \mathrm{i}\nabla_t \alpha_\lambda(k) = |k|\alpha_\lambda(k) - \dfrac{\hat{\varphi}(k)}{\sqrt{2|k|}}(p - A(q,\alpha)) \cdot e_\lambda(k) e^{-ikq}. \end{cases}$$

The model considered in this paper is the so-called Pauli–Fierz model [117], which is the quantized version of $H(p,q,\alpha)$. In the classical limit, $\hbar \to 0$, of the Pauli–Fierz model leads to the Newton–Maxwell equations. We refer to see [3, 132].

3.2 Pauli–Fierz Hamiltonian

The Hamiltonian of non-relativistic QED is defined as a self-adjoint operator on a Hilbert space. The Pauli–Fierz Hamiltonian describes the minimal interaction between electrons and a quantized radiation field, where electrons are treated as

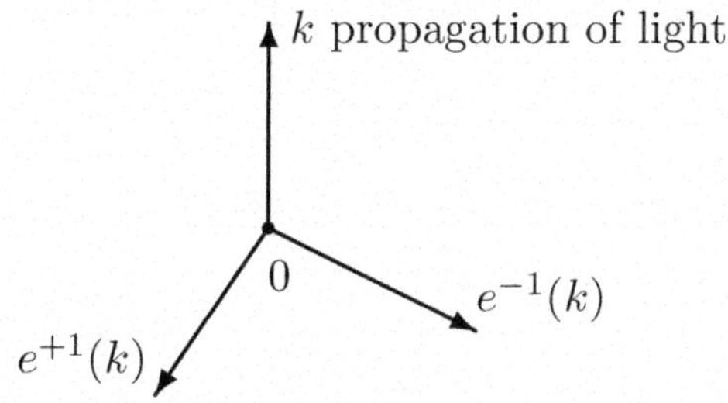

Fig. 1 Polarization vector

quantum mechanical matters and the number of electrons is fixed at one. Hence the Pauli–Fierz Hamiltonian can be interpreted as a Schrödinger operator of one electron coupled with a quantized radiation field. We refer to see [7, 44, 126] for tools of quantum field theory. Let

$$\mathcal{H} = L^2(\mathbb{R}^3) \otimes \mathcal{F}$$

be the Hilbert space describing the joint electron-photon state vectors. Here

$$\mathcal{F} = \mathcal{F}(L^2(\mathbb{R}^3 \times \mathbb{Z}_2))$$

is the boson Fock space over $L^2(\mathbb{R}^3 \times \mathbb{Z}_2)$. Here the boson Fock space $\mathcal{F}(\mathcal{K})$ over $\mathcal{K}$ is defined by

$$\mathcal{F}(\mathcal{K}) = \bigoplus_{n=0}^{\infty} [\otimes_s^n \mathcal{K}],$$

where $\otimes_s^n$ denotes the n-fold symmetric tensor product. See Appendix A for the fundamental facts on boson Fock spaces. The elements of the set $\mathbb{Z}_2$ account for the fact that a photon is a transversal wave perpendicular to the direction of its propagation, thus it has two components. See Fig. 1. The Fock vacuum in $\mathcal{F}$ is defined by

$$\Omega = 1\!\!1 \oplus 0 \oplus 0 \cdots .$$

We identify $\mathcal{H}$ as the set of $\mathcal{F}$-valued L^2-functions on $\mathbb{R}^3$:

$$\mathcal{H} \cong \int_{\mathbb{R}^3}^{\oplus} \mathcal{F} \mathrm{d}x. \tag{3.1}$$

This will be used without further notice in what follows. Let $a(f)$ and $a^*(f)$ be the annihilation operator and the creation operator on $\mathcal{F}$ smeared by $f \in L^2(\mathbb{R}^3 \times \mathbb{Z}_2)$, respectively. Then

$$[a(f), a^*(g)] = \sum_{j=\pm 1} (\bar{f}_j, g_j),$$

$$[a(f), a(g)] = 0 = [a^*(f), a^*(g)].$$

We use the identification $L^2(\mathbb{R}^3 \times \mathbb{Z}_2) \cong L^2(\mathbb{R}^3) \oplus L^2(\mathbb{R}^3)$ and set

$$a^\sharp(f, +1) = a^\sharp(f \oplus 0), \quad a^\sharp(f, -1) = a^\sharp(0 \oplus f),$$

where $a^\sharp$ stands for either operator. The finite particle subspace of $\mathcal{F}$ is given by

$$\mathcal{F}_{\text{fin}} = \{\Psi = \{\Psi^{(n)}\}_{n=0}^\infty | \Psi^{(m)} = 0 \text{ for all } m > N \text{ with some } N\}.$$

Next we define the quantized radiation field with a cutoff function $\hat{\varphi}$. Put

$$\wp_\mu^j(x) = \frac{\hat{\varphi}(k)}{\sqrt{\omega(k)}} e_\mu^j(k) e^{-ikx} \in L^2(\mathbb{R}_k^3),$$
$$\tilde{\wp}_\mu^j(x) = \frac{\hat{\varphi}(-k)}{\sqrt{\omega(k)}} e_\mu^j(k) e^{ikx} \in L^2(\mathbb{R}_k^3),$$

for $x \in \mathbb{R}^3$, $j = \pm 1$ and $\mu = 1, 2, 3$, where ω is the dispersion relation defined by

$$\omega(k) = |k|.$$

Here $\hat{\varphi}$ is the Fourier transform of the charge distribution φ. The vectors $e^{+1}(k)$ and $e^{-1}(k)$ are called polarization vectors, that is, $e^{+1}(k)$, $e^{-1}(k)$ and $k/|k|$ form a right-hand system at $k \in \mathbb{R}^3$;

$$e^i(k) \cdot e^j(k) = \delta_{ij}, \quad e^j(k) \cdot k = 0, \quad e^{+1}(k) \times e^{-1}(k) = k/|k|.$$

The quantized radiation field with cutoff function $\hat{\varphi}$ is defined by

$$A_\mu(x) = \frac{1}{\sqrt{2}} \sum_{j=\pm 1} (a^*(\wp_\mu^j(x), j) + a(\tilde{\wp}_\mu^j(x), j)), \quad \mu = 1, 2, 3.$$

In the case of $\hat{\varphi}/\sqrt{\omega} \in L^2(\mathbb{R}^3)$ and $\overline{\hat{\varphi}(k)} = \hat{\varphi}(-k)$, $A_\mu(x)$ is symmetric, and moreover essentially self-adjoint on $\mathcal{F}_{\text{fin}}$ for each $x \in \mathbb{R}^3$. We denote the closure of $A_\mu(x)\lceil_{\mathcal{F}_{\text{fin}}}$ by the same symbol. A_μ is a self-adjoint operator on

$$D(A_\mu) = \left\{ F \in \mathcal{H} \middle| F(x) \in D(A_\mu(x)) \text{ a.e. } x \text{ and } \int_{\mathbb{R}^3} \|A_\mu(x)F(x)\|_{\mathcal{F}}^2 \mathrm{d}x < \infty \right\}$$

and acts as $(A_\mu F)(x) = A_\mu(x)F(x)$ for $F \in D(A_\mu)$. Since $k \cdot e^j(k) = 0$, it implies the Coulomb gauge condition

$$\nabla_x \cdot A = 0.$$

This in turn yields $\sum_{\mu=1}^3 [\nabla_\mu, A_\mu] = 0$. The formally it is written as

$$A_\mu(x) = \frac{1}{\sqrt{2}} \sum_{j=\pm 1} \int_{\mathbb{R}^3} e_\mu^j(k) \left(\frac{\hat{\varphi}(k)}{\sqrt{|k|}} e^{-ikx} a^*(k,j) + \frac{\hat{\varphi}(-k)}{\sqrt{|k|}} e^{+ikx} a(k,j) \right) dk.$$

Let us explain the term $\frac{\hat{\varphi}(k)}{\sqrt{|k|}}dk$ in $A_\mu(x)$. Let

$$H_m = \{k \in \mathbb{R}^4 \mid k \cdot \tilde{k} = m^2, k_0 > 0\}$$

be called mass hyperboloids. Here $\tilde{k} = (k_0, -k_1, -k_2, -k_3)$. The Lorentz group $\mathscr{L}$ is the set of linear transformations on $\mathbb{R}^4$ that preserve $k \cdot \tilde{l}$, i.e.,

$$\Lambda \in \mathscr{L} \text{ iff } k \cdot \tilde{l} = \Lambda k \cdot \widetilde{\Lambda l} \quad \forall k, l \in \mathbb{R}^4.$$

The restricted Lorentz group $\mathscr{L}_+^\uparrow$ is the subgroup of $\mathscr{L}$ such that $\Lambda = (\Lambda_{\mu\nu})_{0 \le \mu,\nu \le 3} \in \mathscr{L}_+^\uparrow$ iff $\det\Lambda = 1$ and $\Lambda_{00} > 0$. The mass hyperboloids is invariant under the restricted Lorentz group $\mathscr{L}_+^\uparrow$. Let $V_m : H_m \to \mathbb{R}^3$ for $m > 0$ and $V_0 : H_0 \to \mathbb{R}^3 \setminus \{0\}$ for $m = 0$ be the homeomorphisms defined by $V_m(k_0, k_1, k_2, k_3) = (k_1, k_2, k_3)$. Define

$$\rho(E) = \int_{V_m(E)} \frac{1}{\sqrt{k_1^2 + k_2^2 + k_3^2 + m^2}} dk_1 dk_2 dk_3, \quad E \subset H_m.$$

Then ρ is a measure on H_m. The measure ρ is invariant with respect to all $\Lambda \in \mathscr{L}_+^\uparrow$, i.e., $\rho(\Lambda E) = \rho(E)$. From this the measure

$$\frac{1}{\sqrt{|k|}} dk$$

on $\mathbb{R}^3$ is called the relativistically covariant measure for $m = 0$. If $\hat{\varphi} \neq 1$, then

$$\frac{\hat{\varphi}(k)}{\sqrt{|k|}} dk$$

breaks relativistic covariance.

Next we introduce the free field Hamiltonian on $\mathcal{F}$. Let us consider the massless and free relativistic Schrödinger operator $\sqrt{-\Delta}$ on $L^2(\mathbb{R}^3)$. Then the N-body massless and free relativistic Schrödinger operator is given by

$$\sum_{j=1}^N \sqrt{-\Delta_j} \tag{3.2}$$

on $L^2(\mathbb{R}^{3N}) \cong \otimes^N L^2(\mathbb{R}^3)$. The free field Hamiltonian H_f on $\mathcal{F}$ is the direct sum of (3.2), and then given in terms of the second quantization of $\omega(k)$:

$$H_{\rm f} = {\rm d}\Gamma(\omega).$$

It leaves the n-particle subspace $\mathcal{F}^{(n)} = \otimes_s^n L^2(\mathbb{R}^3 \times \mathbb{Z}_2)$ invariant, i.e., for $f \in \mathcal{F}^{(n)}$,

$$H_{\rm f} f(k_1, \ldots, k_n, j_1, \ldots, j_n) = \left(\sum_{j=1}^N \omega(k_j)\right) f(k_1, \ldots, k_n, j_1, \ldots, j_n).$$

See Appendix A.2 for second quantizations.

The Pauli–Fierz model describes the minimal interaction between an electron and the quantized radiation field. The electron is described by the Schrödinger operator

$$H_{\rm p} = -\frac{1}{2}\Delta + V$$

in $L^2(\mathbb{R}^3)$. The Hamiltonian for the electron decoupled with the quantized radiation field is given by

$$H_0 = -\frac{1}{2}\Delta \otimes 1\!\!1 + 1\!\!1 \otimes H_{\rm f}$$

with domain

$$D(H_0) = D\left(-\frac{1}{2}\Delta \otimes 1\!\!1\right) \cap D(1\!\!1 \otimes H_{\rm f}).$$

The interaction is obtained by the minimal coupling $-{\rm i}\nabla_\mu \otimes 1\!\!1 \mapsto -{\rm i}\nabla_\mu \otimes 1\!\!1 - A_\mu$. Then the Pauli–Fierz Hamiltonian is defined by

$$H = \frac{1}{2}(-{\rm i}\nabla \otimes 1\!\!1 - A)^2 + V \otimes 1\!\!1 + 1\!\!1 \otimes H_{\rm f}. \tag{3.3}$$

We give a remark on the definition of H. When defining the sum of operators, each operator should be defined on the same Hilbert space. For the Pauli–Fierz Hamiltonian, however, $-{\rm i}\nabla \otimes 1\!\!1$, $V \otimes 1\!\!1$ and $1\!\!1 \otimes H_{\rm f}$ are defined on Hilbert space $L^2(\mathbb{R}^3) \otimes \mathcal{F}$, but A on $\int_{\mathbb{R}^3}^{\oplus} \mathcal{F} {\rm d}x$. It confuses readers who see it for the first time. However we respect and follow the traditional definition of the Pauli–Fierz Hamiltonian under the identification of $L^2(\mathbb{R}^3) \otimes \mathcal{F}$ and $\int_{\mathbb{R}^3}^{\oplus} \mathcal{F} {\rm d}x$ in (3.1).

It is also known that Hamiltonians defined through different polarizations are unitary equivalent each others. Thus we may fix polarization vectors as they are the most convenient.

In what follows we neglect tensor notation $\otimes$ unless confusions may arise. Then H is simply written as

$$H = \frac{1}{2}(-{\rm i}\nabla - A)^2 + V + H_{\rm f}.$$

We say $V \in \mathcal{R}_{\rm Kato}$ iff $D(\Delta) \subset D(V)$ and there exist $0 \leq a < 1$ and $0 \leq b$ such that

$$\|Vf\| \le a \left\| -\frac{1}{2}\Delta f \right\| + b\|f\|$$

for $f \in D(\Delta)$. Moreover we say that $\varphi \in \mathcal{U}$ iff $\varphi \in \mathscr{S}'(\mathbb{R}^3)$ satisfies that $\hat{\varphi}$ is a local L^1-function, $\hat{\varphi}(-k) = \overline{\hat{\varphi}(k)}$ and

$$\sqrt{\omega}\hat{\varphi} \in L^2(\mathbb{R}^3), \ \hat{\varphi}/\sqrt{\omega} \in L^2(\mathbb{R}^3), \ \hat{\varphi}/\omega \in L^2(\mathbb{R}^3).$$

We give comments on the class $\mathcal{U}$. $\hat{\varphi}(-k) = \overline{\hat{\varphi}(k)}$ and $\hat{\varphi}/\sqrt{\omega} \in L^2(\mathbb{R}^3)$ ensure that A is well-defined and symmetric. $\hat{\varphi}/\omega \in L^2(\mathbb{R}^3)$ implies that A is relatively bounded with respect to $H_{\rm f}^{1/2}$, and $\sqrt{\omega}\hat{\varphi} \in L^2(\mathbb{R}^3)$ yields $\nabla_\mu A_\nu$ is well-defined.

In this paper unless otherwise stated we suppose the following assumptions.

Assumption 3.1 $\varphi \in \mathcal{U}$ and $V \in \mathcal{R}_{\rm Kato}$.

Under this assumption on V, by the Kato–Rellich theorem [85] one can see that $H_{\rm p}$ is self-adjoint on $D(\Delta)$.

Proposition 3.1 ([35, 52, 62, 64, 106]) *Suppose that Assumption 3.1 holds. Then H is self-adjoint on $D(H_0)$ and essentially self-adjoint on any core of H_0.*

Note that condition for the self-adjointness is weakened in [106].

3.3 Pauli–Fierz Hamiltonian in Schrödinger Representation

We introduce a Q-space associated with the quantized radiation field and reformulate the Pauli–Fierz Hamiltonian on $L^2(\mathbb{R}^3) \otimes L^2(\mathrm{Q})$ instead of $\mathcal{H}$. This is called Schrödinger representation. Furthermore, we introduce Euclidean quantum fields associated with the Pauli–Fierz Hamiltonian to derive a functional integral representation of e^{-tH}. We refer to see [123, 124] for a Q-space representation of a Fock space and [115, 126] for Euclidean fields.

The following setting is taken from [62]. For a real-valued $f \in L^2(\mathbb{R}^3)$ we set

$$A_\mu(f) = \frac{1}{\sqrt{2}} \sum_{j=\pm 1} \int_{\mathbb{R}^3} e_\mu^j(k)(\hat{f}(k)a^*(k,j) + \hat{f}(-k)a(k,j))\mathrm{d}k.$$

With this notation we write $A_\mu(x) = A_\mu(\tilde{\varphi}(\cdot - x))$, where $\tilde{\varphi} = (\hat{\varphi}/\sqrt{\omega})^\vee$. The relations

$$(\Omega, A_\mu(f)\Omega) = 0, \tag{3.4}$$

$$(\Omega, A_\mu(f)A_\nu(g)\Omega) = \frac{1}{2}(\hat{f}, \delta_{\mu\nu}^\perp \hat{g}) \tag{3.5}$$

are immediate. Here $\delta_{\mu\nu}^\perp(k)$ is the transversal delta function defined by

$$\delta^{\perp}_{\mu\nu}(k) = \delta_{\mu\nu} - \frac{k_\mu k_\nu}{|k|^2}, \quad \mu, \nu = 1, 2, 3,$$

and we used the identity:

$$\sum_{j=1,2} e^j_\mu(k) e^j_\nu(k) + \frac{k_\mu k_\nu}{|k|^2} = \delta_{\mu\nu}.$$

Let $\delta^\perp = (\delta^\perp_{\mu\nu})_{1\le\mu,\nu\le 3}$ be the 3×3 matrix:

$$\delta^\perp(k) = \begin{pmatrix} 1 - \frac{k_1^2}{|k|^2} & -\frac{k_1 k_2}{|k|^2} & -\frac{k_1 k_3}{|k|^2} \\ -\frac{k_2 k_1}{|k|^2} & 1 - \frac{k_2^2}{|k|^2} & -\frac{k_2 k_3}{|k|^2} \\ -\frac{k_3 k_1}{|k|^2} & -\frac{k_3 k_2}{|k|^2} & 1 - \frac{k_3^2}{|k|^2} \end{pmatrix}.$$

In order to construct the functional integral representation of $(F, e^{-tH} G)$ we prepare probability spaces $(\mathrm{Q}_\#, \Sigma_\#, \mu_\#)$, $\# = 0, 1$, and Gaussian random variables $\hat{A}_\#(f)$ indexed by $f \in \bigoplus^3 L^2_{\mathbb{R}}(\mathbb{R}^{3+\#})$ with mean zero:

$$\mathbb{E}_{\mu_\#}[\hat{A}_\#(f)] = 0,$$

and the covariance:

$$\mathbb{E}_{\mu_\#}[\hat{A}_\#(f)\hat{A}_\#(g)] = \mathrm{q}_\#(f, g).$$

Here the bilinear forms $\mathrm{q}_\#$ on $(\bigoplus^3 L^2_{\mathbb{R}}(\mathbb{R}^{3+\#})) \times (\bigoplus^3 L^2_{\mathbb{R}}(\mathbb{R}^{3+\#}))$ are defined by

$$\mathrm{q}_0(f, g) = \frac{1}{2}(\hat{f}, \delta^\perp \hat{g}), \tag{3.6}$$

$$\mathrm{q}_1(F, G) = \frac{1}{2}(\hat{F}, \delta^\perp \hat{G}). \tag{3.7}$$

Note that $\hat{F} = \hat{F}(k_0, k)$, $\hat{G} = \hat{G}(k_0, k)$ but $\delta^\perp = \delta^\perp(k)$ for $(k_0, k) \in \mathbb{R} \times \mathbb{R}^3$. The definitions of (3.6) and (3.7) are motivated by (3.4) and (3.5). It is established that there exist a probability space $(\mathrm{Q}_\#, \Sigma_\#, \mu_\#)$ and a family of Gaussian random variables $(\hat{A}_\#(f), f \in \bigoplus^3 L_{\mathbb{R}}(\mathbb{R}^{3+\#}))$ such that the mean is zero and the covariance is given by $\mathrm{q}_\#$. Define the μth component of $\hat{A}_\#$ by

$$\hat{A}_{\#,\mu}(f) = \hat{A}_\#\left(\bigoplus_{\nu=1}^3 \delta_{\mu\nu} f\right), \quad f \in L^2(\mathbb{R}^{3+\#}).$$

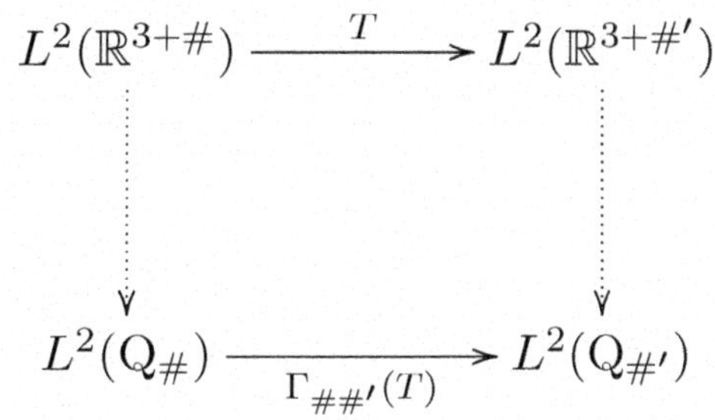

Fig. 2 Functor $\Gamma_{\#\#'}$

We shall set $\hat{A} = \hat{A}_0$, $\mathrm{q} = \mathrm{q}_0$, $\mathrm{Q} = \mathrm{Q}_0$, $\mu_M = \mu_0$, and $\mathbb{A} = \hat{A}_1$, $\mathrm{q}_\mathrm{E} = \mathrm{q}_1$, $\mathrm{Q}_\mathrm{E} = \mathrm{Q}_1$, $\mu_\mathrm{E} = \mu_1$.

It is known that

$$\mathrm{LH}\left\{ :\prod_{j=1}^{n} \hat{A}_\#(f_j):, \mathbb{1} \;\middle|\; f_j \in \bigoplus^3 L^2(\mathbb{R}^{3+\#}), j = 1, \ldots, n, n \in \mathbb{N} \right\} \tag{3.8}$$

is dense in $L^2(\mathrm{Q}_\#)$ Here $:\ldots:$ denotes the Wick product. Thus

$$\int_{\mathrm{Q}_\#} :\prod_{j=1}^{n} \hat{A}_\#(f_j)::\prod_{i=1}^{m} \hat{A}_\#(g_i):\mathrm{d}\mu_\# = 0, \quad n \neq m.$$

We set (3.8) as $L^2_{\mathrm{fin}}(\mathrm{Q}_\#)$. We define the second quantization $\Gamma_{\#\#'}(T)$. Here and in what follows for the operator $S : L^2(\mathbb{R}^{3+\#}) \to L^2(\mathbb{R}^{3+\#'})$ we use the same notation S as for the operator $\bigoplus^3 S : \bigoplus^3 L^2(\mathbb{R}^{3+\#}) \to \bigoplus^3 L^2(\mathbb{R}^{3+\#'})$, $(f_1, f_2, f_3) \mapsto (Sf_1, Sf_2, Sf_3)$ for notational convenience, and we write simply $\hat{A}_\#(Tf)$ for $\hat{A}_\#((\bigoplus^3 T)f)$ etc. Let $\Gamma_{\#\#'}(T)\mathbb{1} = \mathbb{1}$ and

$$\Gamma_{\#\#'}(T):\prod_{i=1}^{n} \hat{A}_\#(f_i): = :\prod_{i=1}^{n} \hat{A}_{\#'}(Tf_i):.$$

If T is a contraction operator, then so is $\Gamma_{\#\#'}(T)$. Thus $\Gamma_{\#\#'}$ is a functor between sets of contraction operators $\{T : L^2(\mathbb{R}^{3+\#}) \to L^2(\mathbb{R}^{3+\#'})\}$. We write $\Gamma_{00} = \Gamma : L^2(\mathrm{Q}) \to L^2(\mathrm{Q})$, $\Gamma_{11} = \Gamma_\mathrm{E} : L^2(\mathrm{Q}_\mathrm{E}) \to L^2(\mathrm{Q}_\mathrm{E})$ and $\Gamma_{01} = \Gamma_{\mathrm{Int}} : L^2(\mathrm{Q}) \to L^2(\mathrm{Q}_\mathrm{E})$ (Fig. 2).

We introduce a family of isometries connecting the Minkowski quantum field and the Euclidean quantum field. Let $\dot{H}_\nu(\mathbb{R}^n)$ be the homogeneous Sobolev space:

$$\dot{H}_\nu(\mathbb{R}^n) = \{f \in \mathscr{S}'(\mathbb{R}^3) \mid \hat{f}(k)|k|^{\nu/2} \in L^2(\mathbb{R}^3)\}.$$

Thus

$$\tau_t : \dot{H}_{-1/2}(\mathbb{R}^3) \to \dot{H}_{-1}(\mathbb{R}^4)$$

is defined by $\tau_t f = \delta_t \otimes f$. Then $\|\tau_t f\|_{\dot{H}_{-1}(\mathbb{R}^4)} = \|f\|_{\dot{H}_{-1/2}(\mathbb{R}^3)}$. Let $i_{-1/2} : L^2(\mathbb{R}^3) \to \dot{H}_{-1/2}(\mathbb{R}^3)$ and $i_{-1} : L^2(\mathbb{R}^4) \to \dot{H}_{-1}(\mathbb{R}^4)$ be given by

$$\widehat{i_{-1/2} f}(k) = \sqrt{\omega(k)} \hat{f}(k),$$
$$\widehat{i_{-1} f}(k_0, k) = \sqrt{\omega(k)^2 + |k_0|^2} \hat{f}(k_0, k).$$

In [62] we define the family of isometries $\mathrm{j}_t : L^2(\mathbb{R}^3) \to L^2(\mathbb{R}^4)$ by

$$\mathrm{j}_t = (i_{-1})^{-1} \circ \tau_t \circ i_{-1/2}, \quad t \in \mathbb{R}.$$

See Fig. 3. We see that

$$\mathrm{j}_s^* \mathrm{j}_t = e^{-|s-t|\hat{\omega}},$$

where $\hat{\omega} = \omega(-\mathrm{i}\nabla)$. See Fig. 4.

Define the family of isometries $\mathrm{J}_t : L^2(\mathrm{Q}) \to L^2(\mathrm{Q_E})$ by the second quantization of j_t, i.e.,

$$\mathrm{J}_t = \Gamma_{\mathrm{Int}}(\mathrm{j}_t).$$

Let K be a self-adjoint operator in $L^2(\mathbb{R}^3)$. Then $\{\Gamma(e^{\mathrm{i}tK}) : t \in \mathbb{R}\}$ turns to be the strongly continuous one-parameter unitary group. Then the Stone's theorem tells us that there exists a unique self-adjoint operator $\mathrm{d}\Gamma(K)$ such that

Fig. 3 Isometries j_t between $L^2(\mathbb{R}^3)$ and $L^2(\mathbb{R}^4)$

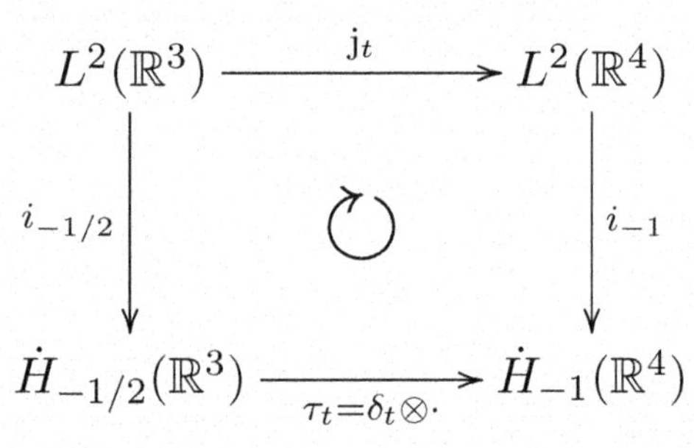

Fig. 4 Decomposition of $e^{-|s-t|\hat{\omega}}$

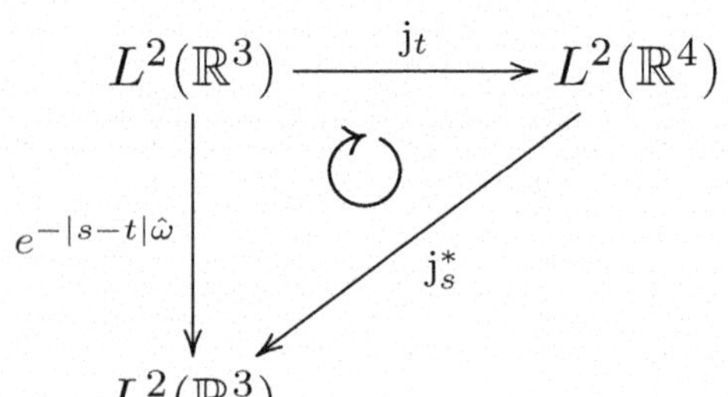

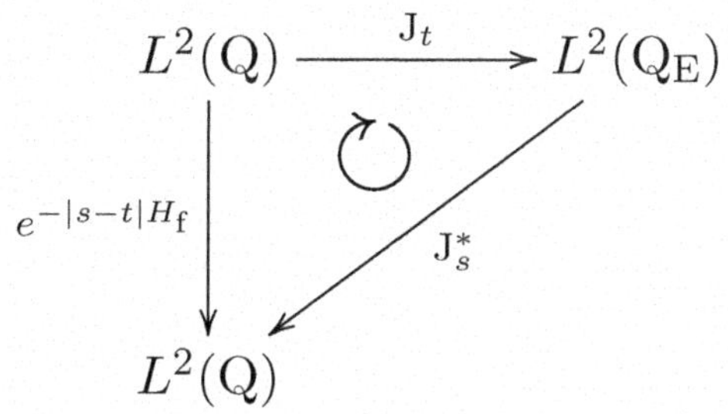

Fig. 5 Decomposition of $e^{-|s-t|H_\mathrm{f}}$

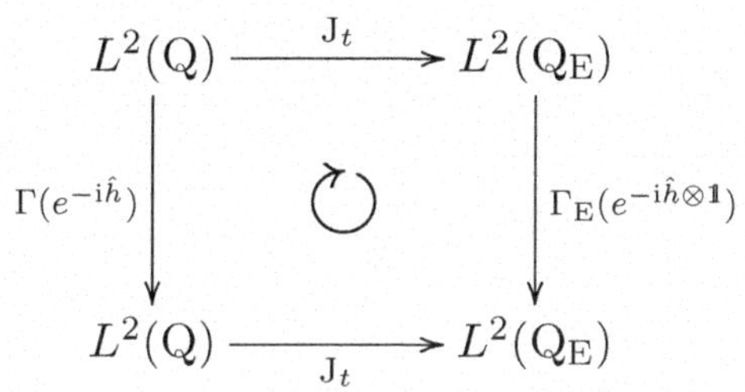

Fig. 6 Intertwining properties

$$\Gamma(e^{\mathrm{i}tK}) = e^{\mathrm{i}t\mathrm{d}\Gamma(K)}, \quad t \in \mathbb{R}.$$

Let

$$\mathrm{d}\Gamma(\hat{\omega}) = \hat{H}_\mathrm{f}.$$

Since $e^{-tH_\mathrm{f}} = \Gamma(e^{-t\hat{\omega}})$, $e^{-t\hat{H}_\mathrm{f}}$ can be decomposed as

$$\mathrm{J}_t^*\mathrm{J}_s = e^{-|s-t|\hat{H}_\mathrm{f}}.$$

See Fig. 5. Isometry J_t plays an important role in functional integral representations, which connects Minkowskian quantum fields with Euclidean quantum fields. Let $\hat{h} = h(-\mathrm{i}\nabla)$ be the self-adjoint operator with a real-valued symbol h. Then $\mathrm{J}_t\Gamma(e^{-\mathrm{i}\hat{h}}) = \Gamma_\mathrm{E}(e^{-\mathrm{i}\hat{h}\otimes\mathbb{1}})\mathrm{J}_t$ and $\mathrm{J}_t\mathrm{d}\Gamma(\hat{h}) = \mathrm{d}\Gamma_\mathrm{E}(\hat{h}\otimes\mathbb{1})\mathrm{J}_t$ hold. See Fig. 6.

3.4 Wiener–Itô–Segal Isomorphism Between $\mathcal{F}$ and $L^2(\mathrm{Q})$

The Wiener–Itô–Segal isomorphism U is an isomorphism between $L^2(\mathrm{Q})$ and $\mathcal{F}$. Let us define $U : \mathcal{F} \to L^2(\mathrm{Q})$ by

$$U\Omega = \mathbb{1},$$

$$U{:}\prod_{i=1}^n A(f_i){:}\Omega = {:}\prod_{i=1}^n \hat{A}(f_i){:}.$$

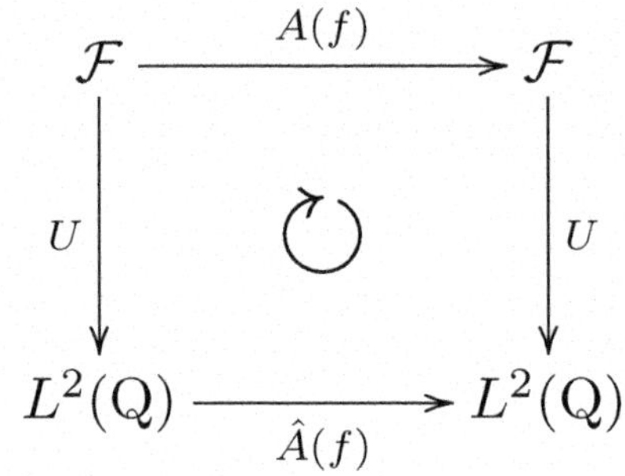

Fig. 7 Wiener–Itô–Segal isomorphism U

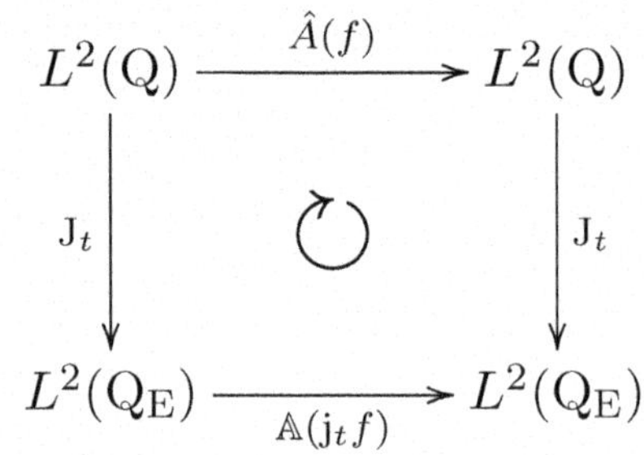

Fig. 8 $\hat{A}(f)$ and $\mathbb{A}(f)$

Thus U becomes a unitary operator from $\mathcal{F}$ to $L^2(\mathrm{Q})$. We denote $\mathbb{1} \otimes U : \mathcal{H} \to L^2(\mathbb{R}^3) \otimes L^2(\mathrm{Q})$ by U for simplicity. Note that the inverse Fourier transform of $g(k, x) = e^{-ikx}\hat{\varphi}(k)/\sqrt{\omega(k)}$ equals $\check{g}(y, x) = \tilde{\varphi}(y - x)$, where $\tilde{\varphi} = (\hat{\varphi}/\sqrt{\omega})\check{} \in L^2(\mathbb{R}^3)$. Notice that the test function of $A_\mu(f)$ is $\hat{f}$ but not f (Fig. 7).

As seen above, the isometry J_t connects $L^2(\mathrm{Q})$ with $L^2(\mathrm{Q}_\mathrm{E})$. We can also see the intertwining properties:

$$\mathrm{J}_t \hat{A}(f) = \mathbb{A}(\mathrm{j}_t f)\mathrm{J}_t.$$

See Fig. 8. This property is very important to construct functional integral representations for Pauli–Fierz type Hamiltonians.

Now we define the Pauli–Fierz Hamiltonian in the Schrödinger representation. The relations

$$\begin{aligned} U A_\mu(x) U^{-1} &= \hat{A}_\mu(\tilde{\varphi}(\cdot - x)), \\ U H_\mathrm{f} U^{-1} &= \hat{H}_\mathrm{f}, \end{aligned}$$

follow directly. As a result

$$U H U^{-1} = \frac{1}{2}(-\mathrm{i}\nabla - \hat{A})^2 + V + \hat{H}_\mathrm{f},$$

with $\hat{A} = (\hat{A}_1, \hat{A}_2, \hat{A}_3)$. In what follows we use notation H for UHU^{-1} and H_f for $\hat{H}_\mathrm{f}$. The Pauli–Fierz Hamiltonian in Schrödinger representation is defined by

$$H = \frac{1}{2}(-\mathrm{i}\nabla - \hat{A})^2 + V + H_{\mathrm{f}}$$

in $L^2(\mathbb{R}^3) \otimes L^2(\mathrm{Q})$. Unless confusions arise we also denote $L^2(\mathbb{R}^3) \otimes L^2(\mathrm{Q})$ by $\mathcal{H}$, and we set $\mathcal{H}_{\mathrm{E}} = L^2(\mathbb{R}^3) \otimes L^2(\mathrm{Q}_{\mathrm{E}})$.

3.5 Hilbert Space-Valued Stochastic Integrals

We define a Hilbert space-valued stochastic integral. It will be first explained in some generality and then applied to the Pauli–Fierz Hamiltonian. Let $\mathcal{K}$ be a Hilbert space and define

$$C^n(\mathbb{R}^3; \mathcal{K}) = \left\{ f : \mathbb{R}^3 \to \mathcal{K} \,|\, f \text{ is } n \text{ times strongly continuously differentiable} \right\}$$

and

$$C^n_{\mathrm{b}}(\mathbb{R}^3; \mathcal{K}) = \left\{ f \in C^n(\mathbb{R}^3; \mathcal{K}) \middle| \sup_{|z| \le n, x \in \mathbb{R}^3} \|\nabla^z f(x)\|_{\mathcal{K}} < \infty \right\},$$

where $|z| = z_1 + z_2 + z_3$ for $z = (z_1, z_2, z_3)$ and $\nabla^z = \nabla^{z_1}_{x_1} \nabla^{z_2}_{x_2} \nabla^{z_3}_{x_3}$ denotes strong derivative. We set $L^2(\mathscr{X}) = L^2(\mathscr{X}, \mathrm{d}\mathcal{W}^x)$. The proof of the following lemma is straightforward and similar to the case of real-valued processes. Let $f \in C^1_{\mathrm{b}}(\mathbb{R} \times \mathbb{R}^3; \mathcal{K})$. The sequence defined by

$$\mathrm{J}^\mu_n(f) = \sum_{j=1}^{2^n} f\left(\frac{j-1}{2^n} t, B_{\frac{j-1}{2^n} t}\right) \left(B^\mu_{\frac{j}{2^n} t} - B^\mu_{\frac{j-1}{2^n} t}\right)$$

is a Cauchy sequence in $L^2(\mathscr{X}) \otimes \mathcal{K}$. For $f \in C^1_{\mathrm{b}}(\mathbb{R} \times \mathbb{R}^3; \mathcal{K})$ the limit

$$\int_0^t f(s, B_s) \mathrm{d}B^\mu_s = s - \lim_{n \to \infty} \mathrm{J}^\mu_n(f)$$

defines a $\mathcal{K}$-valued stochastic integral. By the above definition

$$\mathbb{E}\left[\left(\int_0^t f(s, B_s) \mathrm{d}B^\mu_s, \int_0^t g(s, B_s) \mathrm{d}B^\nu_s \right)_{\mathcal{K}} \right] = \delta_{\mu\nu} \mathbb{E}\left[\int_0^t (f(s, B_s), g(s, B_s))_{\mathcal{K}} \mathrm{d}s \right]$$

holds. $\mathrm{J}^\mu_n(f)$ is defined by the sum of f evaluated at the left endpoints. Then $f(\frac{j-1}{2^n} t, B_{\frac{j-1}{2^n} t})$ and $B^\mu_{\frac{j}{2^n} t} - B^\mu_{\frac{j-1}{2^n} t}$ are independent, while we define $S^\mu_n(f)$ below by evaluating at both of the left and right endpoints. Let $f \in C^2_{\mathrm{b}}(\mathbb{R}^3; \mathcal{K})$ and

$$S_n^\mu(f) = \sum_{j=1}^{2^n} \left(f(B_{\frac{j}{2^n}t}) + f(B_{\frac{j-1}{2^n}t})\right)\left(B^\mu_{\frac{j}{2^n}t} - B^\mu_{\frac{j-1}{2^n}t}\right).$$

Similarly to the case of real-valued processes we can see that

$$s-\lim_{n\to\infty} S_n^\mu(f) = \int_0^t f(B_s)\mathrm{d}B_s^\mu + \frac{1}{2}\int_0^t \nabla_\mu f(B_s)\mathrm{d}s$$

holds in $L^2(\mathscr{X}) \otimes \mathcal{K}$. Thus extra term $\frac{1}{2}\int_0^t \nabla_\mu f(B_s)\mathrm{d}s$ appears.

We will construct the functional integral through the Euclidean quantum field, and will use an $L^2(\mathbb{R}^4)$-valued stochastic integral of the form

$$\int_0^t \mathrm{j}_s\tilde{\varphi}(\cdot - B_s)\mathrm{d}B_s^\mu. \tag{3.9}$$

However, since

$$\frac{\|\mathrm{j}_s f - \mathrm{j}_t f\|^2}{|s-t|^2} = 2\left(\hat{f}, \frac{(1-e^{-|s-t|\omega})}{|s-t|}\hat{f}\right)\frac{1}{|s-t|}$$

diverges as $t \to s$ for $\hat{f} \in D(\omega)$, $\mathbb{R}\times\mathbb{R}^3 \ni (s,x) \mapsto \mathrm{j}_s\tilde{\varphi}(\cdot - x) \in L^2(\mathbb{R}^4)$ is not strongly differentiable in $s \in \mathbb{R}$. Then $\mathrm{j}_s\tilde{\varphi}(\cdot - x) \notin C_\mathrm{b}^1(\mathbb{R}_s \times \mathbb{R}^3_x; L^2(\mathbb{R}^4))$. Therefore we need to give a proper definition of (3.9).

Lemma 3.2 *If $\hat{\lambda}, \omega\hat{\lambda} \in L^2(\mathbb{R}^4)$, then for each $\mu = 1, 2, 3$,*

$$S_n^\mu(\lambda) = \sum_{j=1}^{2^n} \mathrm{j}_{\frac{(j-1)t}{2^n}}\lambda\left(\cdot - B_{\frac{(j-1)t}{2^n}}\right)\left(B^\mu_{\frac{jt}{2^n}} - B^\mu_{\frac{(j-1)t}{2^n}}\right), \quad n = 1, 2, 3, \ldots,$$

is a Cauchy sequence in $L^2(\mathscr{X}) \otimes L^2(\mathbb{R}^4)$.

Proof Fix an μ. Write $S_n = S_n^\mu(\lambda)$ and $\eta_* = \mathrm{j}_*\lambda(\cdot - B_*)$. Then

$$S_{n+1} - S_n = \sum_{m=1}^{2^n}\left(\eta_{\frac{(2m-1)t}{2^{n+1}}} - \eta_{\frac{(2m-2)t}{2^{n+1}}}\right)\left(B^\mu_{\frac{2mt}{2^{n+1}}} - B^\mu_{\frac{(2m-1)t}{2^{n+1}}}\right).$$

Hence

$$\mathbb{E}^x[\|S_{n+1} - S_n\|^2] = \sum_{m=1}^{2^n}\mathbb{E}^x\left[\left\|\eta_{\frac{(2m-1)t}{2^{n+1}}} - \eta_{\frac{(2m-2)t}{2^{n+1}}}\right\|^2\right]\frac{t}{2^{n+1}}.$$

We have

$$\left\|\eta_{\frac{(2m-1)t}{2^{n+1}}} - \eta_{\frac{(2m-2)t}{2^{n+1}}}\right\|^2 \le \|\omega\hat{\lambda}\|^2\left|B^\mu_{\frac{(2m-1)t}{2^{n+1}}} - B^\mu_{\frac{(2m-2)t}{2^{n+1}}}\right|^2 + \frac{t}{2^n}\|\lambda\|\|\omega\hat{\lambda}\|.$$

We conclude that

$$\left(\mathbb{E}^x[\|S_m - S_n\|^2]\right)^{1/2} \le \left(\frac{2\|\lambda\|\|\omega\hat{\lambda}\| + \|\omega\hat{\lambda}\|^2}{2}\right)^{1/2} \sum_{j=n+1}^{m} \frac{t}{2^{(j+1)/2}},$$

thus $\{S_n\}_{n=1}^{\infty}$ is a Cauchy sequence.

QED

Let $\hat{\lambda}, \omega\hat{\lambda} \in L^2(\mathbb{R}^4)$. Then

$$\int_0^t \mathrm{j}_s\lambda(\cdot - B_s)\mathrm{d}B_s^{\mu} = s-\lim_{n\to\infty} S_n^{\mu}, \quad \mu = 1, 2, 3,$$

defines an $L^2(\mathbb{R}^4)$-valued stochastic integral, where the strong limit is in the strong topology of $L^2(\mathscr{X}) \otimes L^2(\mathbb{R}^4)$. By the definition it is seen that

$$\mathbb{E}^x\left[\left(\int_0^t \mathrm{j}_s\lambda(\cdot - B_s)\mathrm{d}B_s^{\mu}, \int_0^t \mathrm{j}_s\rho(\cdot - B_s)\mathrm{d}B_s^{\nu}\right)\right] = t\delta_{\mu\nu}(\lambda, \rho).$$

3.6 Functional Integral Representations for the Pauli–Fierz Hamiltonian

Using the Trotter product formula and the factorization formula $e^{-|s-t|H_{\mathrm{f}}} = \mathrm{J}_t^*\mathrm{J}_s$, we derive a functional integral representation of e^{-tH}. A combination of the functional integral representation of $e^{-tH_{\mathrm{p}}}$ and the equality $e^{-tH_{\mathrm{f}}} = \mathrm{J}_0^*\mathrm{J}_t$ gives

$$(F, e^{-tH_0}G)_{\mathcal{H}} = (\mathrm{J}_0F, e^{-tH_{\mathrm{p}}}\mathrm{J}_tG)_{\mathcal{H}_{\mathrm{E}}} = \int_{\mathbb{R}^3} \mathbb{E}^x[e^{-\int_0^t V(B_s)\mathrm{d}s}(\mathrm{J}_0F(B_0), \mathrm{J}_tG(B_t))_{L^2(\mathrm{Q_E})}]\mathrm{d}x.$$

The next theorem is due to [61]. See also [40, 105, 131] for great extensions.

Theorem 3.3 *We suppose that $V \in L^{\infty}(\mathbb{R}^3)$. Then*

$$(F, e^{-tH}G) = \int_{\mathbb{R}^3} \mathbb{E}^x\left[e^{-\int_0^t V(B_s)\mathrm{d}s}(\mathrm{J}_0F(B_0), e^{-\mathrm{i}\mathbb{A}(K_t)}\mathrm{J}_tG(B_t))_{L^2(\mathrm{Q_E})}\right]\mathrm{d}x. \quad (3.10)$$

Here K_t denotes the $\bigoplus^3 L^2(\mathbb{R}^3)$-valued stochastic integral given by

$$K_t = \bigoplus_{\mu=1}^{3} \int_0^t \mathrm{j}_s\tilde{\varphi}(\cdot - B_s)\mathrm{d}B_s^{\mu}.$$

Proof Firstly we assume that $V = 0$, and we write $\mathbb{A}_s(x) = \mathbb{A}(\bigoplus^3 \mathrm{j}_s\tilde{\varphi}(\cdot - x))$ in this proof. Define the family of symmetric contraction operators $P_s : \mathcal{H} \to \mathcal{H}$ by

$$P_s F(x) = \int_{\mathbb{R}^3} \Pi_s(x-y) e^{\mathrm{i}H(x,y)} F(y) \mathrm{d}y, \quad s > 0,$$

with $P_0 F = F$, where $\Pi_s(x) = \frac{1}{(2\pi s)^{3/2}} \exp(-|x|^2/2s)$ is the heat kernel and

$$H(x, y) = \frac{1}{2}(\hat{A}(x) + \hat{A}(y)) \cdot (x - y).$$

By a direct computation

$$(F, (P_{\frac{t}{2^n}})^{2^n} G) = \int_{\mathbb{R}^3} \mathrm{d}x \int_{(\mathbb{R}^3)^{2^n}} \overline{F(x)} e^{\mathrm{i}\sum_{j=1}^{2^n} H_j} G(x_{2^n}) \left(\prod_{j=1}^{2^n} \Pi_{\frac{t}{2^n}}(x_{j-1} - x_j) \right) \prod_{j=1}^{2^n} \mathrm{d}x_j$$

with $x = x_0$, $H_j = H(x_{j-1}, x_j)$. This can be expressed by using Brownian motion as

$$(F, (P_{\frac{t}{2^n}})^{2^n} G) = \int_{\mathbb{R}^3} \mathbb{E}^x[(F(B_0), e^{-\mathrm{i}\hat{A}(L_n)} G(B_t))] \mathrm{d}x,$$

where $L_n = \bigoplus_{\mu=1}^3 \sum_{m=0}^{2^n} (\tilde{\varphi}(\cdot - B_{t\frac{m}{2^n}}) + \tilde{\varphi}(\cdot - B_{t\frac{m-1}{2^n}}))(B^\mu_{t\frac{m}{2^n}} - B^\mu_{t\frac{m-1}{2^n}})$. It is seen that

$$L_n \to L_t = \bigoplus_{\mu=1}^3 \int_0^t \tilde{\varphi}(\cdot - B_s) \mathrm{d}B^\mu_s$$

as $n \to \infty$ strongly in $\bigoplus^3 \left(L^2(\mathscr{X}) \otimes L^2(\mathbb{R}^3)\right)$. Here we used the Coulomb gauge condition. This implies that for $F, G \in \mathcal{H}_{\mathrm{E}}$,

$$\lim_{n\to\infty} (F, (P_{\frac{t}{2^n}})^{2^n} G) = \int_{\mathbb{R}^3} \mathbb{E}^x[(F(B_0), e^{-\mathrm{i}\hat{A}(L_t)} G(B_t))] \mathrm{d}x. \tag{3.11}$$

From (3.11) it follows that $|\lim_{n\to\infty} (F, (P_{\frac{t}{2^n}})^{2^n} G)| \le \|F\| \|G\|$. Hence for each $t \ge 0$ there exists a symmetric bounded operator S_t such that

$$\lim_{n\to\infty} (F, (P_{\frac{t}{2^n}})^{2^n} G) = (F, S_t G).$$

Since $(P_{\frac{t}{2^n}})^{2^n}$ is uniformly bounded as $\|(P_{\frac{t}{2^n}})^{2^n}\| \le 1$, the above weak convergence improves to

$$s-\lim_{n\to\infty} (P_{\frac{t}{2^n}})^{2^n} = S_t, \quad t \ge 0. \tag{3.12}$$

Furthermore, by (3.11)

$$(F, S_t G) = \int_{\mathbb{R}^3} \mathbb{E}^x[(F(B_0), e^{-i\hat{A}(L_t)} G(B_t))] dx. \tag{3.13}$$

Putting these together we can show that $\{S_t : t \geq 0\}$ is a symmetric C_0-semigroup, thus there exists a unique self-adjoint operator K such that $S_t = e^{-tK}$, $t \geq 0$. Let

$$H(\hat{A}) = \frac{1}{2}(-i\nabla - \hat{A})^2.$$

We have

$$\lim_{t\to\infty} (F, t^{-1}(\mathbb{1} - P_t)G) = (F, H(\hat{A})G) \tag{3.14}$$

for $F, G \in C_0^\infty(\mathbb{R}^3) \otimes L^2_{\rm fin}(Q)$ in Lemma 3.4 below. This leads to

$$(t^{-1}(e^{-tK} - \mathbb{1})F, G) = \lim_{n\to\infty} (t^{-1}((P_{\frac{t}{2^n}})^{2^n} - \mathbb{1})F, G) = -\int_0^1 (H(\hat{A})F, e^{-tsH(\hat{A})}G) ds. \tag{3.15}$$

In the second equality above we used (3.14). Equation (3.15) can be immediately extended to vectors $F \in D(H_0)$ and $G \in D(K)$. Take $t \downarrow 0$ on both sides of (3.15). Then it holds that $(F, KG) = (H(\hat{A})F, G)$ for $F \in D(H_0)$ and $G \in D(K)$, which implies that $K \supset H(\hat{A})\lceil_{D(H_0)}$ as $H(\hat{A})$ is self-adjoint. Define

$$\hat{H} = K \dotplus H_{\rm f},$$

where $\dotplus$ denotes the quadratic form sum [86]. The Trotter product formula for quadratic form sums [88, 89] and the factorization $e^{-\frac{t}{n}H_{\rm f}} = J^*_{\frac{kt}{n}} J_{\frac{(k+1)t}{n}}$ yield that

$$(F, e^{-t\hat{H}}G) = \lim_{n\to\infty} \left(J_0 F, \left(\prod_{i=0}^{n-1} R_i \right) J_t G \right),$$

where $R_j = J_{\frac{jt}{n}} e^{-\frac{t}{n}K} J^*_{\frac{jt}{n}}$. Using the definition of e^{-tK} we get

$$J_s e^{-tK} J_s^* G(x) = s-\lim_{n\to\infty} \int_{(\mathbb{R}^3)^{2^n}} J_s e^{i\sum_{j=1}^{2^n} H_j} J_s^* G(x_{2^n}) \left(\prod_{j=1}^{2^n} \Pi_{\frac{t}{2^n}}(x_{j-1} - x_j) \right) \prod_{j=1}^{2^n} dx_j$$

with $x = x_0$. Write $\delta_j = \delta_j(n, t, n_j) = t/(n2^{n_j})$ for $j = 0, 1, \ldots, n-1$ and define $P_{s,j} : \mathcal{H}_{\rm E} \to \mathcal{H}_{\rm E}$ by P_s with $H(x, y)$ replaced by the Euclidean version $\mathbb{H}_{\frac{jt}{n}}(x, y)$ given by

$$\mathbb{H}_s(x, y) = \frac{1}{2}(\mathbb{A}_s(x) + \mathbb{A}_s(y)) \cdot (x - y)$$

with $\mathbb{A}_s(x) = \mathbb{A}(\mathrm{j}_s\tilde{\varphi}(\cdot - x))$ as

$$P_{s,j}F(x) = \int_{\mathbb{R}^3} \Pi_s(x-y)e^{\mathrm{i}\mathbb{H}_{\frac{jt}{n}}(x,y)}F(y)\mathrm{d}y.$$

We have

$$\left(\mathrm{J}_0F, \left(\prod_{i=0}^{n-1} R_i\right)\mathrm{J}_tG\right) = \lim_{n_0\to\infty}\cdots\lim_{n_{n-1}\to\infty}\left(\mathrm{J}_0F, \left(\prod_{i=0}^{n-1}(P_{\delta_i,i})^{2^{n_i}}\right)\mathrm{J}_tG\right).$$

Here we used the Markov property [114, 116, 126] of the projection $E_s = \mathrm{J}_s\mathrm{J}_s^*$ (see Fig. 9). As a result we have

$$\left(\mathrm{J}_0F, \left(\prod_{i=0}^{n-1}R_i\right)\mathrm{J}_tG\right) = \lim_{n_0\to\infty}\cdots\lim_{n_{n-1}\to\infty}\int_{\mathbb{R}^3}\mathbb{E}^x[(\mathrm{J}_0F(B_0), e^{-\mathrm{i}\mathbb{A}(K)}\mathrm{J}_tG(B_t))]\mathrm{d}x$$

with $K = K(n_0, n_1, \ldots, n_{n-1}, n)$ given by

$$K = \bigoplus_{\mu=1}^{3}\sum_{j=0}^{n-1}\sum_{m=1}^{2^{n_j}}\mathrm{j}_{\frac{jt}{n}}\left(\tilde{\varphi}\left(\cdot - B_{\frac{t}{n}\left(\frac{m}{2^{n_j}}+j\right)}\right) + \tilde{\varphi}\left(\cdot - B_{\frac{t}{n}\left(\frac{m-1}{2^{n_j}}+j\right)}\right)\right)\left(B^{\mu}_{\frac{t}{n}\left(\frac{m}{2^{n_j}}+j\right)} - B^{\mu}_{\frac{t}{n}\left(\frac{m-1}{2^{n_j}}+j\right)}\right).$$

Note that

$$K \to \bigoplus_{\mu=1}^{3}\sum_{j=0}^{n-1}\int_{\frac{jt}{n}}^{t(j+1)/n}\mathrm{j}_{\frac{jt}{n}}\tilde{\varphi}(\cdot - B_s)\mathrm{d}B_s^{\mu}$$

as $n_0, n_1, \ldots, n_{n-1} \to \infty$ in $\bigoplus^3(L^2(\mathscr{X}) \otimes L^2(\mathbb{R}^3))$. Finally as $n \to \infty$ we have

$$(F, e^{-t\hat{H}}G) = \int_{\mathbb{R}^3}\mathbb{E}^x[(\mathrm{J}_0F(B_0), e^{-\mathrm{i}\mathbb{A}(K_t)}\mathrm{J}_tG(B_t))]\mathrm{d}x. \tag{3.16}$$

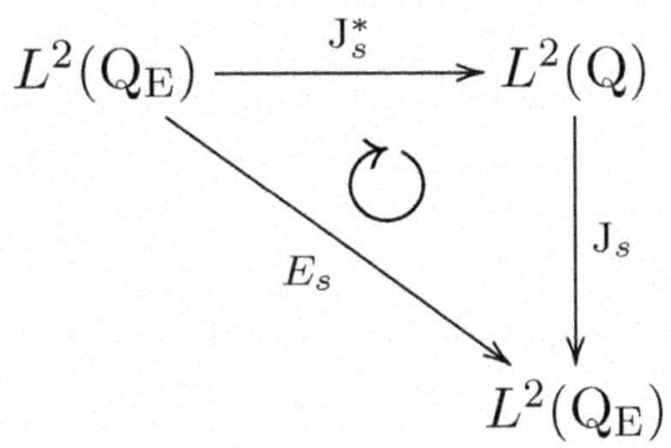

Fig. 9 Projection E_s

By the construction of $\hat{H}$,

$$\hat{H} \supset \frac{1}{2}(-\mathrm{i}\nabla - \hat{A})^2 + H_{\mathrm{f}}\lceil_{D(H_0)}, \tag{3.17}$$

but Proposition 3.1 yields that

$$\hat{H} = \frac{1}{2}(-\mathrm{i}\nabla - \hat{A})^2 + H_{\mathrm{f}}$$

follows, and $\hat{H}$ is self-adjoint on $D(H_0)$. A functional integral representation of $e^{-t\hat{H}}$ including non-zero V can be obtained by the Trotter product formula

$$(F, e^{-t\hat{H}}G) = \lim_{n\to\infty}(F, (e^{-\frac{t}{n}K}e^{-\frac{t}{n}V}e^{-\frac{t}{n}H_{\mathrm{f}}})^n G).$$

This completes the proof.

QED

It remains to show (3.14).

Lemma 3.4 *It follows (3.14), i.e.,*

$$\lim_{t\to\infty}(F, t^{-1}(\mathbb{1} - P_t)G) = (F, H(\hat{A})G)$$

for $F, G \in C_0^\infty(\mathbb{R}^3) \otimes L^2_{\mathrm{fin}}(\mathrm{Q})$.

Proof It is directly seen that

$$\frac{d}{ds}(F, P_s G) = \int_{\mathbb{R}^3} \mathrm{d}x \int_{\mathbb{R}^3} \Pi_s(x-y)\left(F(x), \frac{1}{2}\Delta_y e^{\mathrm{i}H(x,y)}G(y)\right)\mathrm{d}y.$$

We have

$$(F(x), \Delta_{y_\mu} e^{\mathrm{i}H(x,y)}G(y)) = \mathrm{I}^+_\mu(x,y) + \mathrm{I}^-_\mu(x,y),$$

where

$$\mathrm{I}^+_\mu(x,y) = \left(F(x), e^{\mathrm{i}H(x,y)}\left(\hat{A}^\mu(y)\mathrm{i}\nabla_{y_\mu} + \frac{1}{2}\mathrm{i}\hat{A}^{\mu\mu}(y) + \frac{1}{2}\hat{A}^\mu(y)(\hat{A}_\mu(x) + \hat{A}_\mu(y))\right)G(y)\right)(x-y)$$
$$-\frac{1}{4}(F(x), e^{\mathrm{i}H(x,y)}(\hat{A}^\mu(y)(x-y))^2 G(y)),$$

$$\mathrm{I}^-_\mu(x,y) = \left(F(x), e^{\mathrm{i}H(x,y)}\left(\Delta_{y_\mu} - (\hat{A}_\mu(x) + \hat{A}_\mu(y))\mathrm{i}\nabla_{y_\mu} - \frac{1}{4}(\hat{A}_\mu(x) + \hat{A}_\mu(y))^2\right)G(y)\right).$$

Here $\hat{A}^\mu(y) = \nabla_{y_\mu}\hat{A}(y)$ and $\hat{A}^{\mu\mu}(y) = \nabla^2_{y_\mu}\hat{A}(y)$. We have

$$\lim_{t\to 0}\int_{\mathbb{R}^3}\mathrm{d}x\int_{\mathbb{R}^3}\Pi_t(x-y)\mathrm{I}^+_\mu(x,y)\mathrm{d}y = 0.$$

It can be also seen that the map $y \mapsto \mathrm{I}_\mu^-(x,y)$ is continuous for each $x \in \mathbb{R}^3$ and it follows that $\lim_{y\to x} \mathrm{I}_\mu^-(x,y) = -(F(x), (-\mathrm{i}\nabla_{x_\mu} - A_\mu(x))^2 G(x))$. Hence we have

$$\lim_{t\to 0}\int_{\mathbb{R}^3} \mathrm{d}x \int_{\mathbb{R}^3} \Pi_t(x-y)\mathrm{I}_\mu^-(x,y)\mathrm{d}y = -\int_{\mathbb{R}^3} (F(x), (-\mathrm{i}\nabla_{x_\mu} - A_\mu)^2 G(x))\mathrm{d}x.$$

Together with them we have

$$\lim_{t\to\infty} \frac{1}{t}(F, (\mathbb{1} - P_t)G) = -\lim_{t\to 0+} \frac{d}{dt}(F, P_t G) = (F, H(\hat{A})G).$$

Thus the lemma follows.

QED

In Theorem 3.3 we assumed boundedness of external potentials. We can offer an extension of the functional integral representation to a wider potential class.

Theorem 3.5 ([61]) *Take*

$$H = \frac{1}{2}(-\mathrm{i}\nabla - \hat{A})^2 + H_{\mathrm{f}} \dotplus V_+ \dot{-} V_-$$

for $V_+ \in L^1_{\mathrm{loc}}(\mathbb{R}^3)$ *and* V_- *which is* $-\frac{1}{2}\Delta$ *form bounded with relative bound* < 1. *Then* (3.10) *holds for* H.

Proof The proof is a minor modification of [128, Theorem 6.2]. The following proof is taken from [61]. Let

$$V_{+n}(x) = \begin{cases} V_+(x), & V_+(x) < n, \\ n, & V_+(x) \geq n, \end{cases} \qquad V_{-m}(x) = \begin{cases} V_-(x), & V_-(x) < m, \\ m, & V_-(x) \geq m. \end{cases}$$

Set $V_{n,m} = V_{+n} - V_{-m}$ and $h = \frac{1}{2}(-\mathrm{i}\nabla - \hat{A})^2 + H_{\mathrm{f}}$. Then

$$(F, e^{-t(H+V_{n,m})}G) = \int_{\mathbb{R}^3} \mathbb{E}^x \left[e^{-\int_0^t V_{n,m}(B_s)\mathrm{d}s} \left(F(B_0), \mathrm{J}_0^* e^{-\mathrm{i}\mathbb{A}(K_t)} \mathrm{J}_t G(B_t)\right)\right] \mathrm{d}x. \tag{3.18}$$

Define the closed quadratic forms

$$\mathfrak{q}_{n,m}(F,F) = (h^{1/2}F, h^{1/2}F) + (V_{+n}^{1/2}F, V_{+n}^{1/2}F) - (V_{-m}^{1/2}F, V_{-m}^{1/2}F),$$
$$\mathfrak{q}_{n,\infty}(F,F) = (h^{1/2}F, h^{1/2}F) + (V_{+n}^{1/2}F, V_{+n}^{1/2}F) - (V_{-}^{1/2}F, V_{-}^{1/2}F),$$
$$\mathfrak{q}_{\infty,\infty}(F,F) = (h^{1/2}F, h^{1/2}F) + (V_{+}^{1/2}F, V_{+}^{1/2}F) - (V_{-}^{1/2}F, V_{-}^{1/2}F),$$

whose form domains are $Q(\mathfrak{q}_{n,m}) = Q(h)$, $Q(\mathfrak{q}_{n,\infty}) = Q(h)$ and $Q(\mathfrak{q}_{\infty,\infty}) = Q(h) \cap Q(V_+)$. Note that

$$\mathfrak{q}_{n,m} \downarrow \mathfrak{q}_{n,\infty}, \quad m \uparrow \infty$$

and

$$\mathfrak{q}_{n,\infty} \uparrow \mathfrak{q}_{\infty,\infty}, \quad n \uparrow \infty.$$

By [86, VIII. Theorem 3.11], $\mathfrak{q}_{n,\infty}$ is a closed quadratic form, and we can conclude that for all $t \geq 0$,

$$\exp\left(-t\left(h + V_{+n} - V_{-m}\right)\right) \to \exp\left(-t\left(h + V_{+n} \dot{-} V_{-}\right)\right) \tag{3.19}$$

as $m \to \infty$. By [86, VIII. Theorem 3.13] and [127], $\mathfrak{q}_{\infty,\infty}$ is also a closed quadratic form, and

$$\exp(-t(h \dot{+} V_{+n} \dot{-} V_{-})) \to \exp(-t(h \dot{+} V_{+} \dot{-} V_{-})) \tag{3.20}$$

as $n \to \infty$ strongly. By taking first $n \to \infty$ and then $m \to \infty$ it can be proven that both sides of (3.18) converge. I.e., the left-hand side of (3.18) converges by (3.19) and (3.20). We have the inequality

$$\begin{aligned}
&\int_{\mathbb{R}^3} \mathbb{E}^x \left[e^{-\int_0^t V_{n,m}(B_s)\mathrm{d}s} \left| \left(F(B_0), \mathrm{J}_0^* e^{-\mathrm{i}\mathbb{A}(K_t)} \mathrm{J}_t G(B_t) \right) \right| \right] \mathrm{d}x \\
&\leq \int_{\mathbb{R}^3} \mathbb{E}^x \left[e^{-\int_0^t V_{n,\infty}(B_s)\mathrm{d}s} (|F(B_0)|, e^{-tH_\mathrm{f}} |G(B_t)|) \right] \mathrm{d}x < \infty.
\end{aligned}$$

Here the finiteness of the second term can be derived. Since

$$|e^{-\int_0^t V_{n,m}(B_s)\mathrm{d}s} (F(B_0), \mathrm{J}_0^* e^{-\mathrm{i}\mathbb{A}(K_t)} \mathrm{J}_t G(B_t))| \leq e^{-\int_0^t V_{n,\infty}(B_s)\mathrm{d}s} (|F(B_0)|, e^{-tH_\mathrm{f}} |G(B_t)|),$$

the dominated convergence theorem yields that

$$\begin{aligned}
&\lim_{m \to \infty} \int_{\mathbb{R}^3} \mathbb{E}^x \left[e^{-\int_0^t V_{n,m}(B_s)\mathrm{d}s} \left(F(B_0), \mathrm{J}_0^* e^{-\mathrm{i}\mathbb{A}(K_t)} \mathrm{J}_t G(B_t) \right) \right] \mathrm{d}x \\
&= \int_{\mathbb{R}^3} \mathbb{E}^x \left[e^{-\int_0^t V_{n,\infty}(B_s)\mathrm{d}s} (F(B_0), \mathrm{J}_0^* e^{-\mathrm{i}\mathbb{A}(K_t)} \mathrm{J}_t G(B_t)) \right] \mathrm{d}x.
\end{aligned}$$

Furthermore since

$$\begin{aligned}
&\int_{\mathbb{R}^3} \mathbb{E}^x \left[e^{-\int_0^t V_{n,\infty}(B_s)\mathrm{d}s} \left| \left(F(B_0), \mathrm{J}_0^* e^{-\mathrm{i}\mathbb{A}(K_t)} \mathrm{J}_t G(B_t) \right) \right| \right] \mathrm{d}x \\
&\leq \int_{\mathbb{R}^3} \mathbb{E}^x \left[e^{-\int_0^t V_{n,\infty}(B_s)\mathrm{d}s} (|F(B_0)|, e^{-tH_\mathrm{f}} |G(B_t)|) \right] \mathrm{d}x < \infty
\end{aligned}$$

and

$$\left| e^{-\int_0^t V_{n',\infty}(B_s)\mathrm{d}s} \left(F(B_0), \mathrm{J}_0^* e^{-\mathrm{i}\mathbb{A}(K_t)} \mathrm{J}_t G(B_t) \right) \right| \leq e^{-\int_0^t V_{n,\infty}(B_s)\mathrm{d}s} (|F(B_0)|, e^{-tH_\mathrm{f}} |G(B_t)|)$$

for $n \leq n'$, the dominated convergence theorem again yields that

$$\lim_{n\to\infty}\int_{\mathbb{R}^3} \mathbb{E}^x\left[e^{-\int_0^t V_{n,\infty}(B_s)ds}\left(F(B_0), J_0^* e^{-i\mathbb{A}(K_t)} J_t G(B_t)\right)\right] dx$$
$$= \int_{\mathbb{R}^3} \mathbb{E}^x\left[e^{-\int_0^t V(B_s)ds}(F(B_0), J_0^* e^{-i\mathbb{A}(K_t)} J_t G(B_t))\right] dx.$$

Together with them the right-hand side of (3.18) converges to

$$\int_{\mathbb{R}^3} \mathbb{E}^x[e^{-\int_0^t V(B_s)ds}(F(B_0), J_0^* e^{-i\mathbb{A}(K_t)} J_t G(B_t))]dx$$

for any $F, G \in D(H_0)$ as first $m \to \infty$ and then $n \to \infty$. Then the proof is complete. QED

3.7 Pauli–Fierz Hamiltonian with Kato-Class Potential

We consider the Pauli–Fierz Hamiltonian with Kato-class potential V. This section is due to [54]. We introduce the assumption below.

Assumption 3.2 $\varphi \in \mathcal{U}$ and V is Kato-decomposable.

Firstly we are interested in defining H with a Kato-class potential as a self-adjoint operator. This will be done through the functional integral representation established in the previous section. We have

$$(e^{-tH} G)(x) = \mathbb{E}^x[e^{-\int_0^t V(B_s)ds} J_0^* e^{-i\mathbb{A}(K_t)} J_t G(B_t)]. \tag{3.21}$$

Conversely, we shall show that a sufficient condition to define the right-hand side of (3.21) is that V is of Kato-class. The idea used for Schrödinger operators with Kato-class potentials [23] can be extended to the Pauli–Fierz Hamiltonian.

Let V be a Kato-decomposable potential and define the family of operators

$$(K_t F)(x) = \mathbb{E}^x[e^{-\int_0^t V(B_r)dr} J_0^* e^{-i\mathbb{A}(K_t)} J_t F(B_t)].$$

Lemma 3.6 *Suppose Assumption 3.2. Then* K_t *is bounded on* $\mathcal{H}$.

Proof Let $F \in \mathcal{H}$. By Schwarz inequality we have

$$\|K_t F\|_{\mathcal{H}}^2 \leq \int_{\mathbb{R}^3} \mathbb{E}^0[e^{-2\int_0^t V(B_r+x)dr}]\mathbb{E}^0[\|F(B_t+x)\|^2]dx.$$

Since V is of Kato-class, we have $\sup_{x\in\mathbb{R}^3} \mathbb{E}^0[e^{-2\int_0^t V(B_r+x)dr}] = C < \infty$, and thus

$$\|\mathrm{K}_t F\|_{\mathcal{H}}^2 \leq C\|F\|_{\mathcal{H}}^2.$$

QED

We shall show that $\{\mathrm{K}_t : t \geq 0\}$ is a symmetric C_0-semigroup. To do that we introduce a time shift operator u_t on $L^2(\mathbb{R}^4)$ by

$$u_t f(x_0, x) = f(x_0 - t, x), \quad x = (x_0, x) \in \mathbb{R} \times \mathbb{R}^3.$$

It is straightforward that $u_t^* = u_{-t}$ and $u_t^* u_t = 1$. We denote the second quantization of u_t by $U_t = \Gamma_{\mathrm{E}}(u_t)$ which acts on $L^2(\mathrm{Q_E})$ and is unitary. It follows that $u_t \mathrm{j}_s = \mathrm{j}_{s+t}$ for every $t, s \in \mathbb{R}$. It derives the formula $U_t \mathrm{J}_s = \mathrm{J}_{s+t}$ (Fig. 10).

Lemma 3.7 *Suppose Assumption 3.2. Then* $\mathrm{K}_s\mathrm{K}_t = \mathrm{K}_{s+t}$ *holds true for* $s, t \geq 0$. *Moreover* $t \mapsto \mathrm{K}_t$ *is strongly continuous and* $\mathrm{K}_0 = \mathbb{1}$.

Proof By the definition of K_t we have

$$\mathrm{K}_s\mathrm{K}_t F = \mathbb{E}^x\left[e^{-\int_0^s V(B_r)\mathrm{d}r}\mathrm{J}_0^* e^{-\mathrm{i}\mathbb{A}(K_s)}\mathrm{J}_s \mathbb{E}^{B_s}[e^{-\int_0^t V(B_r)\mathrm{d}r}\mathrm{J}_0^* e^{-\mathrm{i}\mathbb{A}(K_t)}\mathrm{J}_t F(B_t)]\right]. \quad (3.22)$$

By the formulae $\mathrm{J}_s\mathrm{J}_0^* = \mathrm{E}_s U_{-s}^*$ (Fig. 11) and $\mathrm{J}_t = U_{-s}\mathrm{J}_{t+s}$, we see that (3.22) is equal to

$$\mathbb{E}^x\left[e^{-\int_0^s V(B_r)\mathrm{d}r}\mathrm{J}_0^* e^{-\mathrm{i}\mathbb{A}(K_s)}\mathrm{E}_s \mathbb{E}^{B_s}[e^{-\int_0^t V(B_r)\mathrm{d}r} U_{-s}^* e^{-\mathrm{i}\mathbb{A}(K_t)} U_{-s}\mathrm{J}_{t+s} F(B_t)]\right]. \quad (3.23)$$

Since U_s is unitary, we have

$$U_{-s}^* e^{-\mathrm{i}\mathbb{A}(K_t)} U_{-s} = e^{-\mathrm{i}\mathbb{A}(u_{-s}^* K_t)}$$

as an operator. The test function of the exponent $u_{-s}^* K_t$ is given by

$$u_{-s}^* K_t = \bigoplus_{\mu=1}^3 \int_0^t \mathrm{j}_{r+s}\tilde{\varphi}(\cdot - B_r)\mathrm{d}B_r^\mu.$$

Moreover by the Markov property of E_t, $t \in \mathbb{R}$, we can neglect E_s in (3.23), and by the Markov property of $(B_t)_{t\geq 0}$ we have

$$\begin{aligned}(\mathrm{K}_s\mathrm{K}_t F)(x) &= \mathbb{E}^x[e^{-\int_0^s V(B_r)\mathrm{d}r}\mathrm{J}_0^* e^{-\mathrm{i}\mathbb{A}(K_s)}\mathbb{E}^x[e^{-\int_s^{s+t} V(B_r)\mathrm{d}r} e^{-\mathrm{i}\mathbb{A}(K_s^{s+t})}\mathrm{J}_{s+t} F(B_{s+t})|\mathscr{F}_s]] \\ &= \mathbb{E}^x[e^{-\int_0^{s+t} V(B_r)\mathrm{d}r}\mathrm{J}_0^* e^{-\mathrm{i}\mathbb{A}(K_{s+t})}\mathrm{J}_{s+t} F(B_{s+t})] = K_{s+t}F,\end{aligned}$$

where we recall that $(\mathscr{F}_t)_{t\geq 0}$ denotes the natural filtration of $(B_t)_{t\geq 0}$. The second statement can be proven immediately.

QED

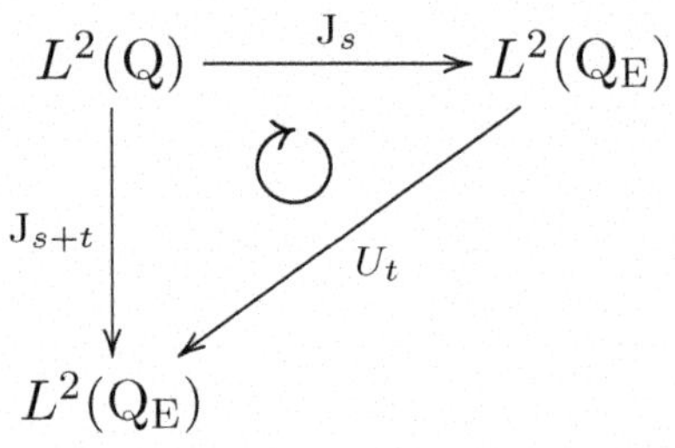

Fig. 10 Shift by U_t

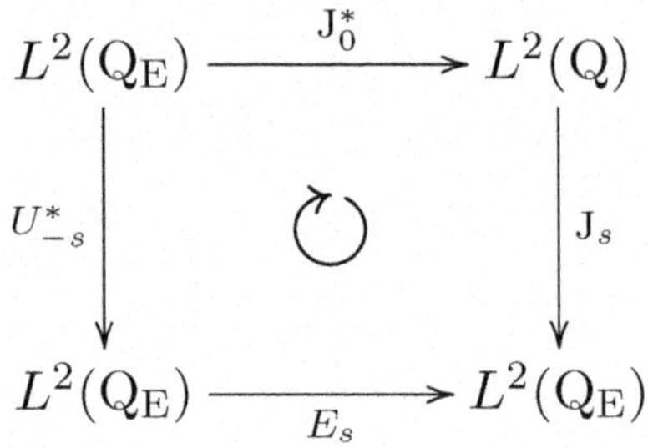

Fig. 11 E_s and U_s

Theorem 3.8 *Suppose Assumption 3.2. Then* $\{\mathrm{K}_t : t \geq 0\}$ *is a symmetric* C_0*-semigroup.*

Proof It was shown that $\{\mathrm{K}_t : t \geq 0\}$ is a C_0-semigroup. Hence it is enough to show that K_t is symmetric for each $t \geq 0$. Recall that $\mathrm{R} = \Gamma(r)$ is the second quantization of the reflection r, where $r : L^2(\mathbb{R}^4) \to L^2(\mathbb{R}^4)$ such that $rf(x_0, x) = f(-x_0, x)$ for $(x_0, x) \in \mathbb{R} \times \mathbb{R}^3$. We have

$$(F, \mathrm{K}_t G) = (U_t \mathrm{R} F, U_t \mathrm{R} \mathrm{K}_t G) = \int_{\mathbb{R}^3} \mathbb{E}^x \left[e^{-\int_0^t V(B_s)\mathrm{d}s} \left(\mathrm{J}_t F(B_0), e^{-\mathrm{i}\mathbb{A}(u_t r K_t)} \mathrm{J}_0 G(B_t) \right) \right] \mathrm{d}x.$$

Notice that $u_t r K_t = \int_0^t \mathrm{j}_{t-s} \tilde{\varphi}(\cdot - B_s) \mathrm{d}B_s^\mu$. Note that $\dot{B}_s = B_{t-s} - B_t$, $0 \leq s \leq t$ is also Brownian motion. Exchanging integrals $\int_{\mathscr{X}} \mathrm{d}\mathcal{W}^0$ and $\int_{\mathbb{R}^3} \mathrm{d}x$ and changing the variable x to $y + B_t$, we can see

$$(F, \mathrm{K}_t G) = \mathbb{E}^0 \left[\int_{\mathbb{R}^3} e^{-\int_0^t V(B_{t-s}+y)\mathrm{d}s} \left(\mathrm{J}_t F(B_t + y), e^{+\mathrm{i}\mathbb{A}(K_t)} \mathrm{J}_0 G(y) \right) \right] \mathrm{d}y.$$

Then

$$(F, \mathrm{K}_t G) = \int_{\mathbb{R}^3} \mathbb{E}^y \left[e^{-\int_0^t V(B_s)\mathrm{d}s} \left(\mathrm{J}_0^* e^{-\mathrm{i}\mathbb{A}(K_t)} \mathrm{J}_t F(y + B_t), G(y) \right) \right] \mathrm{d}y = (\mathrm{K}_t F, G)$$

and K_t is symmetric. Then the proof is complete.

QED

By Theorem 3.8 and the Stone's theorem for semigroups there exists a self-adjoint operator K such that

$$\mathrm{K}_t = e^{-t\mathrm{K}}, \quad t \geq 0.$$

3.8 Positivity Improving

It is known that second quantizations and positivity improving are deeply related. We refer to see [36, 43, 48, 101, 102, 109, 110, 125, 126, 130] for positivity improving, and this section is due to [63]. Let T be a contraction operator on $L^2(\mathbb{R}^3)$. Then it is known that $\Gamma(T)$ is positivity preserving in $L^2(\mathrm{Q})$. I.e.,

$$(\Phi, \Gamma(T)\Psi) \geq 0, \quad \Phi \geq 0, \ \Psi \geq 0.$$

Furthermore let $f : \mathbb{R}^3 \to \mathbb{R}$ be a non-negative function such that the Lebesgue measure of $\{k \in \mathbb{R}^3 \mid f(k) = 0\}$ is zero. Then $\Gamma(e^{-f(-\mathrm{i}\nabla)})$ is positivity improving on $L^2(\mathrm{Q})$. I.e.,

$$(\Phi, \Gamma(e^{-f(-\mathrm{i}\nabla)})\Psi) \gneq 0, \quad \Phi \geq 0, \ \Psi \geq 0.$$

In particular $\mathrm{J}_0^*\mathrm{J}_t = e^{-tH_\mathrm{f}} = \Gamma(e^{-t\hat{\omega}})$ is positivity improving on $L^2(\mathrm{Q})$. See Fig. 12. We are interested in asking if $\mathrm{J}_0^* X \mathrm{J}_t$ is positivity improving or not for some X.

Let $\mathfrak{S}_\# = \exp(-\mathrm{i}\frac{\pi}{2}N_\#)$, where N denotes the Number operator on $L^2(\mathrm{Q})$ and N_E that on $L^2(\mathrm{Q}_\mathrm{E})$. $\mathfrak{S}$ (resp. $\mathfrak{S}_\mathrm{E}$) is a unitary operator on $L^2(\mathrm{Q})$ (resp. $L^2(\mathrm{Q}_\mathrm{E})$). Then we can show that $\mathfrak{S}_\mathrm{E}^{-1} e^{-\mathrm{i}\mathbb{A}(f)}\mathfrak{S}_\mathrm{E}$ is a shift operator on $L^2(\mathrm{Q}_\mathrm{E})$. In particular, it is positivity preserving on $L^2(\mathrm{Q}_\mathrm{E})$. Since J_t and J_0^* are positivity preserving, we see that $\mathrm{J}_0^*\mathfrak{S}_\mathrm{E}^{-1} e^{-\mathrm{i}\mathbb{A}(f)}\mathfrak{S}_\mathrm{E}\mathrm{J}_t$ is positivity preserving on $L^2(\mathrm{Q})$. We can show a stronger statement below. See Fig. 13.

Proposition 3.9 ([63]) $\mathrm{J}_0^*\mathfrak{S}_\mathrm{E}^{-1} e^{-\mathrm{i}\mathbb{A}(f)}\mathfrak{S}_\mathrm{E}\mathrm{J}_t$ *is positivity improving in* $L^2(\mathrm{Q}_\mathrm{E})$.

Corollary 3.10 ([63]) $\mathfrak{S}^{-1}e^{-tH}\mathfrak{S}$ *is positivity improving.*

Proof Let $F, G \in \mathcal{H}$ be non-negative but not identically zero. We have

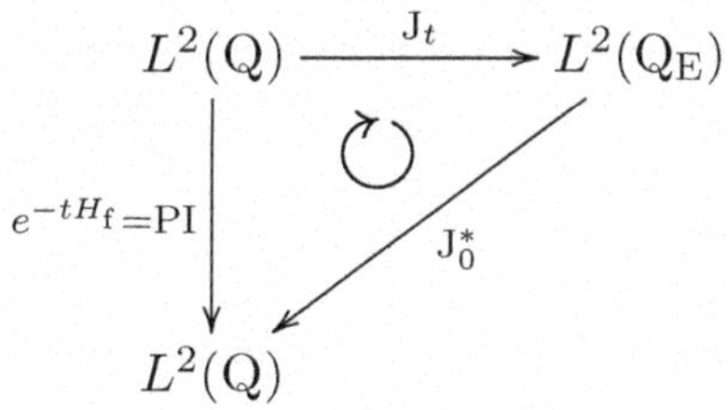

Fig. 12 Positivity improving I

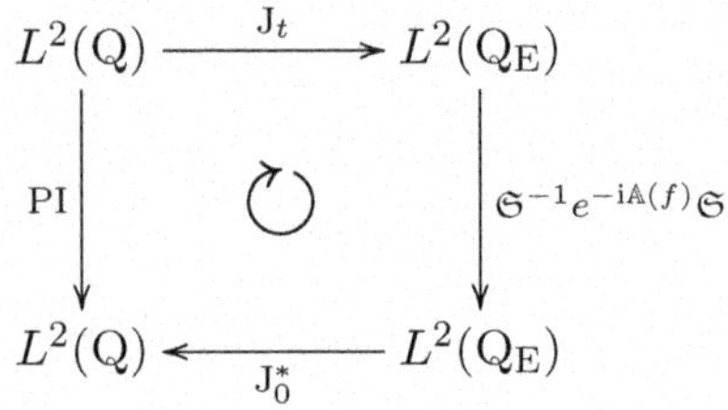

Fig. 13 Positivity improving II

$$
\begin{aligned}
(F, \mathfrak{S}^{-1}e^{-tH}\mathfrak{S}G) &= \int_{\mathbb{R}^3} \mathbb{E}^x[e^{-\int_0^t V(B_s)\mathrm{d}s}(F(B_0), \mathrm{J}_0^*\mathfrak{S}_\mathrm{E}^{-1}e^{-\mathrm{i}\mathbb{A}(K_t)}\mathfrak{S}_\mathrm{E}\mathrm{J}_t G(B_t))]\mathrm{d}x \\
&= \int_{\mathbb{R}^3} \mathbb{E}^0[e^{-\int_0^t V(B_s+x)\mathrm{d}s}(F(x), \mathrm{J}_0^*\mathfrak{S}_\mathrm{E}^{-1}e^{-\mathrm{i}\mathbb{A}(K_t(x))}\mathfrak{S}_\mathrm{E}\mathrm{J}_t G(B_t+x))]\mathrm{d}x,
\end{aligned}
$$

where $K_t(x) = \int_0^t \mathrm{j}_s\tilde{\varphi}(\cdot - B_s - x)\mathrm{d}B_s$.

$$
f(x, w) = e^{-\int_0^t V(B_s)\mathrm{d}s}(F(x), \mathrm{J}_0^*\mathfrak{S}_\mathrm{E}^{-1}e^{-\mathrm{i}\mathbb{A}(K_t(x))}\mathfrak{S}_\mathrm{E}\mathrm{J}_t G(B_t+x))
$$

is a function with respect to $(x, w) \in \mathbb{R}^3 \times \mathscr{X}$. We can see that there exists a measurable set $M \times B \subset \mathbb{R}^3 \times \mathscr{X}$ such that $\int_{M\times B} \mathrm{d}\mathcal{W}\mathrm{d}x > 0$ and $f(x, w) > 0$ for $(x, w) \in M \times B$ from Proposition 3.9. Hence $(F, \mathfrak{S}^{-1}e^{-tH}\mathfrak{S}G) = \int_{\mathbb{R}^3} \mathbb{E}^0[f]\mathrm{d}x \geq \int_{M\times B} f\mathrm{d}\mathcal{W}\mathrm{d}x > 0$ and the corollary is proven.

QED

Corollary 3.11 ([63]) *If H has a ground state Ψ_g. Then $\mathfrak{S}^{-1}\Psi_\mathrm{g}$ is strictly positive.*

Proof This follows from Corollary 3.10 and the Perron Frobenius theorem.

QED

3.9 Baker–Campbell–Hausdorff Formula and Fock Representation

In Sects. 3.6 and 3.7 we obtain the functional integral representation of the semigroup generated by H in the Schrödinger representation. In this section we show a functional integral representation of e^{-tH} in the Fock representation by applying Baker–Campbell–Hausdorff formula. Let $a_\mathrm{E}^\sharp$ be the annihilation operator and the creation operator acting in $\mathcal{F}(L^2(\mathbb{R}^4) \oplus L^2(\mathbb{R}^4))$. Let

$$
\tilde{\varphi}_j = (\tilde{\varphi}_{1j}, \tilde{\varphi}_{2j}, \tilde{\varphi}_{3j}), \quad j = \pm 1,
$$

where $\tilde{\varphi}_{\mu j}$ denotes the inverse Fourier transform of $e_\mu^j \frac{\hat{\varphi}}{\sqrt{\omega}}$. Let

$$
\hat{\mathrm{j}}_s = F^{-1}\mathrm{j}_s F, \tag{3.24}
$$

where F denotes the Fourier transform on $L^2(\mathbb{R}^3)$. We note that following identification:

$$\hat{A}(K_t) \cong \frac{1}{\sqrt{2}}\left(a_{\mathrm{E}}^*(M_t) + a_{\mathrm{E}}(\tilde{M}_t)\right),$$

where

$$M_t = \bigoplus_{j=\pm 1} \int_0^t \hat{\mathrm{j}}_s \frac{\hat{\varphi}}{\sqrt{\omega}} e^{-\mathrm{i}kB_s} e^j \cdot \mathrm{d}B_s,$$

$$\tilde{M}_t = \bigoplus_{j=\pm 1} \int_0^t \hat{\mathrm{j}}_s \frac{\tilde{\hat{\varphi}}}{\sqrt{\omega}} e^{+\mathrm{i}kB_s} e^j \cdot \mathrm{d}B_s.$$

We have the theorem below.

Theorem 3.12 *Equation (3.10) can be represented as*

$$(F, e^{-tH}G) = \int_{\mathbb{R}^3} \mathbb{E}^x \left[e^{-\int_0^t V(B_s)\mathrm{d}s} e^{-\mathfrak{K}_t} (F(B_0), e^{a^*(L_t)} e^{-tH_{\mathrm{f}}} e^{a(\tilde{L}_t)} G(B_t))_{L^2(\mathrm{Q})}\right] \mathrm{d}x. \tag{3.25}$$

Here

$$L_t = -\frac{\mathrm{i}}{\sqrt{2}} \bigoplus_{j=\pm 1} \int_0^t e^{-|t-s|\omega(k)} \frac{\hat{\varphi}(k)}{\sqrt{\omega(k)}} e^{-\mathrm{i}kB_s} e^j(k) \cdot \mathrm{d}B_s, \tag{3.26}$$

$$\tilde{L}_t = -\frac{\mathrm{i}}{\sqrt{2}} \bigoplus_{j=\pm 1} \int_0^t e^{-|s|\omega(k)} \frac{\hat{\varphi}(-k)}{\sqrt{\omega(k)}} e^{+\mathrm{i}kB_s} e^j(k) \cdot \mathrm{d}B_s \tag{3.27}$$

and

$$\mathfrak{K}_t = \frac{1}{4} \sum_{j=\pm 1} \left\| \int_0^t \mathrm{j}_s \tilde{\varphi}_j(\cdot - B_s) \cdot \mathrm{d}B_s \right\|^2_{L^2(\mathbb{R}^4)} = \frac{1}{2} \mathrm{q}_{\mathrm{E}}(K_t, K_t). \tag{3.28}$$

Proof By Baker–Campbell–Hausdorff formula we can see that

$$\mathrm{J}_0^* e^{-\mathrm{i}\mathbb{A}(K_t)} \mathrm{J}_t = \mathrm{J}_0^* e^{-\mathfrak{K}_t} e^{-\mathrm{i}a_{\mathrm{E}}^*(M_t)} e^{-\mathrm{i}a_{\mathrm{E}}(\tilde{M}_t)} \mathrm{J}_t = e^{-\mathfrak{K}_t} e^{-\mathrm{i}a^*(\mathrm{j}_0^* M_t)} e^{-tH_{\mathrm{f}}} e^{-\mathrm{i}a(\mathrm{j}_t^* \tilde{M}_t)}.$$

It can be seen that $-\mathrm{i}\mathrm{j}_t^* M_t = L_t$ and $-\mathrm{i}\mathrm{j}_0^* \tilde{M}_t = \tilde{L}_t$. Together with them we obtain the desired results.

QED

The exponent $\mathfrak{K}_t$ is formally written as

$$\mathfrak{K}_t \stackrel{formal}{=} \frac{1}{4} \sum_{1 \le \mu, \nu \le 3} \int_0^t \mathrm{d}B_s^\mu \int_0^t \mathrm{d}B_r^\nu \int_{\mathbb{R}^3} \frac{|\hat{\varphi}(k)|^2}{\omega(k)} e^{-|s-r|\omega(k)} e^{-\mathrm{i}k(B_s - B_r)} \left(\delta_{\mu\nu} - \frac{k_\mu k_\nu}{|k|^2}\right) \mathrm{d}k.$$

This expression is formal. The double stochastic integral $\int_0^t\int_0^t \ldots \mathrm{d}B_s^\mu \mathrm{d}B_r^\nu$ is delicate. We discuss this in [15]. In fact it is derived in [15, Proposition 3.1] that

$$\mathfrak{K}_t = \frac{1}{4}\sum_{1\le\mu,\nu\le 3}\int_{\mathbb{R}^3}\mathrm{d}k\frac{|\hat{\varphi}(k)|^2}{\omega(k)}\int_0^t e^{-\mathrm{i}kB_s}\mathrm{d}B_s^\mu\left(\delta_{\mu\nu}-\frac{k_\mu k_\nu}{|k|^2}\right)\int_0^s e^{-(s-r)\omega(k)}e^{\mathrm{i}kB_r}\mathrm{d}B_r^\nu.$$

3.10 Translation Invariant Pauli–Fierz Hamiltonian

We consider the translation invariant Pauli–Fierz Hamiltonian, which is obtained by setting the external potential V identically zero, resulting in the fact that H commutes with the total momentum operator. This section is due to [65].

We begin with defining a fiber Hamiltonian. As said, a standing assumption throughout this section is Assumption 3.2 but $V = 0$. Put

$$\mathrm{P_f}_\mu = \mathrm{d}\Gamma(k_\mu), \quad \mu = 1, 2, 3,$$

which describes the field momentum. The total momentum operator P^T on $\mathcal{H}$ is defined by the sum of the momentum operator for the particle and that of field:

$$\mathrm{P}_\mu^\mathrm{T} = -\mathrm{i}\nabla_\mu + \mathrm{P_f}_\mu, \quad \mu = 1, 2, 3.$$

It follows that

$$[H, \mathrm{P}_\mu^\mathrm{T}] = 0, \quad \mu = 1, 2, 3. \tag{3.29}$$

Remark 3.1 We give a comment on (3.29). Both H and P^T are unbounded, and then (3.29) holds on $D(H\mathrm{P}^\mathrm{T}) \cap D(H\mathrm{P}^\mathrm{T})$. Precisely we can show that $[e^{-tH}, e^{-\mathrm{i}s\mathrm{P}^\mathrm{T}}] = 0$ for $t \ge 0$ and $s \in \mathbb{R}$ on the whole Hilbert space by a functional integral representation.

This leads to a decomposition of H with respect to the spectrum of the total momentum operator $\sigma(\mathrm{P}_\mu^\mathrm{T}) = \mathbb{R}$. The Pauli–Fierz Hamiltonian with total momentum $p \in \mathbb{R}^3$ is defined by

$$H(p) = \frac{1}{2}(p - \mathrm{P_f} - A(0))^2 + H_\mathrm{f}, \quad p \in \mathbb{R}^3,$$

with domain $D(H(p)) = D(H_\mathrm{f}) \cap D(\mathrm{P}_\mathrm{f}^2)$, where $A_\mu(0) = A_\mu(x = 0)$. We give a relationship between H and $H(p)$. Define the unitary operator

$$\mathscr{T} : L^2(\mathbb{R}_x) \otimes \mathcal{F} \to L^2(\mathbb{R}_p) \otimes \mathcal{F}$$

by

$$\mathscr{T} = (\hat{F} \otimes 1\!\!1) \int_{\mathbb{R}^3}^{\oplus} \exp(ix\mathrm{P_f})\mathrm{d}x, \tag{3.30}$$

with $\hat{F}$ denoting the Fourier transformation from $L^2(\mathbb{R}_x)$ to $L^2(\mathbb{R}_p)$. For $\Psi \in \mathcal{H}$,

$$(\mathscr{T}\Psi)(p) = \frac{1}{(2\pi)^{3/2}} \int_{\mathbb{R}^3} e^{-\mathrm{i}xp} e^{\mathrm{i}x\mathrm{P_f}} \Psi(x)\mathrm{d}x.$$

Theorem 3.13 *For each $p \in \mathbb{R}^3$, $H(p)$ is a non-negative self-adjoint operator, and*

$$\int_{\mathbb{R}^3}^{\oplus} \mathcal{F}\mathrm{d}p \cong \mathcal{H}, \quad \int_{\mathbb{R}^3}^{\oplus} H(p)\mathrm{d}p \cong H.$$

Here the unitary equivalence is implemented by $\mathscr{T}$.

The self-adjointness of $H(p)$ ensures that $\{e^{-tH(p)} : t \geq 0\}$ is a symmetric C_0-semigroup. As in the previous section, we transform $H(p)$ from the Fock representation to the Schrödinger representation in order to construct a functional integral representation. Then $H(p)$ becomes

$$H(p) = \frac{1}{2}(p - \mathrm{P_f} - \hat{A}(0))^2 + H_\mathrm{f}$$

on $L^2(\mathrm{Q})$, where $\hat{A}(0) = \hat{A}(x = 0)$. We use the same notations $H(p)$, H_f and $\mathrm{P_f}$ in both the Fock representation and the Schrödinger representation. Recall that $\mathrm{P_f} = \mathrm{d}\Gamma(-\mathrm{i}\nabla)$ and $H_\mathrm{f} = \mathrm{d}\Gamma(\omega(-\mathrm{i}\nabla))$. The functional integral representation of $e^{-tH(p)}$ can be also constructed as an application of that of e^{-tH}.

Theorem 3.14 ([65]) *Let $\Psi, \Phi \in L^2(\mathrm{Q})$. Then*

$$(\Psi, e^{-tH(p)}\Phi) = \mathbb{E}^0[(\mathrm{J}_0\Psi, e^{-\mathrm{i}\mathbb{A}(K_t)}\mathrm{J}_t e^{+\mathrm{i}(p-\mathrm{P_f})B_t}\Phi)_{L^2(\mathrm{Q_E})}]. \tag{3.31}$$

In particular

$$e^{-tH(p)}\Phi = \mathbb{E}^0[\mathrm{J}_0^* e^{-\mathrm{i}\mathbb{A}(K_t)}\mathrm{J}_t e^{+\mathrm{i}(p-\mathrm{P_f})B_t}\Phi]. \tag{3.32}$$

Proof Write $F_s = \Pi_s \otimes \Psi$ and $G_r = \Pi_r \otimes \Phi$, where Π_s is the 3D-heat kernel:

$$\Pi_s(x) = \frac{1}{(2\pi s)^{3/2}} \exp(-|x|^2/2s).$$

Due to the fact that $H = \mathscr{T}(\int_{\mathbb{R}^3}^{\oplus} H(p)\mathrm{d}p)\mathscr{T}^{-1}$, we have the key identity:

$$(F_s, e^{-tH}e^{-\mathrm{i}\xi\mathrm{P}^\mathrm{T}}G_r) = \int_{\mathbb{R}^3} e^{-\mathrm{i}\xi p}((\mathscr{T}F_s)(p), e^{-tH(p)}(\mathscr{T}G_r)(p))\mathrm{d}p$$

for any $\xi \in \mathbb{R}^3$. Note that $\lim_{s\to 0}(\mathscr{T}F_s)(p) = \frac{1}{(2\pi)^{3/2}}\Psi$ strongly in $L^2(\mathrm{Q})$ for each $p \in \mathbb{R}^3$. Hence

$$\lim_{s\to 0}(F_s, e^{-tH}e^{-\mathrm{i}\xi \mathrm{P}^{\mathrm{T}}}G_r) = \frac{1}{(2\pi)^{3/2}}\int_{\mathbb{R}^3}(\Psi, e^{-tH(p)}e^{-\mathrm{i}\xi p}(\mathscr{T}G_r)(p))\mathrm{d}p. \tag{3.33}$$

By using the fact $e^{\mathrm{i}\xi \mathrm{P}^{\mathrm{T}}}G_r(x) = \Pi_r(x-\xi)e^{-\mathrm{i}\xi \mathrm{P}_{\mathrm{f}}}\Phi$, we obtain by the functional integral representation in Theorem 3.3 that

$$(F_s, e^{-tH}e^{-\mathrm{i}\xi \mathrm{P}^{\mathrm{T}}}G_r) = \int_{\mathbb{R}^3}\mathbb{E}^x[\Pi_s(x)\Pi_r(B_t-\xi)(\mathrm{J}_0\Psi, e^{-\mathrm{i}\mathbb{A}(K_t)}\mathrm{J}_t e^{-\mathrm{i}\xi \mathrm{P}_{\mathrm{f}}}\Phi)]\mathrm{d}x.$$

Then it follows from $\Pi_s(x) \to \delta(x)$ as $s \to 0$ that

$$\lim_{s\to 0}(F_s, e^{-tH}e^{-\mathrm{i}\xi \mathrm{P}^{\mathrm{T}}}G_r) = \mathbb{E}^0[\Pi_r(B_t-\xi)(\mathrm{J}_0\Psi, e^{-\mathrm{i}\mathbb{A}(K_t)}\mathrm{J}_t e^{-\mathrm{i}\xi \mathrm{P}_{\mathrm{f}}}\Phi)]. \tag{3.34}$$

Combining (3.33) and (3.34) leads to

$$\frac{1}{(2\pi)^{3/2}}\int_{\mathbb{R}^3}e^{-\mathrm{i}\xi p}(\Psi, e^{-tH(p)}(\mathscr{T}G_r)(p))\mathrm{d}p = \mathbb{E}^0[\Pi_r(B_t-\xi)(\mathrm{J}_0\Psi, e^{-\mathrm{i}\mathbb{A}(K_t)}\mathrm{J}_t e^{-\mathrm{i}\xi \mathrm{P}_{\mathrm{f}}}\Phi)]. \tag{3.35}$$

Taking the inverse Fourier transform on both sides of (3.35) with respect to p gives

$$(\Psi, e^{-tH(p)}(\mathscr{T}G_r)(p)) = \frac{1}{(2\pi)^{3/2}}\mathbb{E}^0\left[\int_{\mathbb{R}^3}e^{\mathrm{i}\xi p}\Pi_r(B_t-\xi)(\mathrm{J}_0\Psi, e^{-\mathrm{i}\mathbb{A}(K_t)}\mathrm{J}_t e^{-\mathrm{i}\xi \mathrm{P}_{\mathrm{f}}}\Phi)d\xi\right] \tag{3.36}$$

for almost every $p \in \mathbb{R}^3$. Since both sides of (3.36) are continuous in p, the equality stays valid for all $p \in \mathbb{R}^3$. After taking $r \to 0$ on both sides we arrive at (3.31).

QED

By Baker–Campbell–Hausdorff formula we can also have a functional integral representation in the Fock representation. Then the proof is similar to that of Theorem 3.12.

Corollary 3.15 *Let* $\Psi, \Phi \in L^2(\mathrm{Q})$. *Then*

$$(\Psi, e^{-tH(p)}\Phi) = \mathbb{E}^0[e^{-\mathfrak{K}_t}(\Psi, e^{a^*(L_t)}e^{-tH_{\mathrm{f}}}e^{a(\tilde{L}_t)}e^{+\mathrm{i}(p-\mathrm{P}_{\mathrm{f}})B_t}\Phi)]. \tag{3.37}$$

In particular

$$e^{-tH(p)}\Phi = \mathbb{E}^0[e^{-\mathfrak{K}_t}e^{a^*(L_t)}e^{-tH_{\mathrm{f}}}e^{a(\tilde{L}_t)}e^{+\mathrm{i}(p-\mathrm{P}_{\mathrm{f}})B_t}\Phi]. \tag{3.38}$$

Here L_t, $\tilde{L}_t$ *and* $\mathfrak{K}_t$ *are given by (3.26), (3.27) and (3.28), respectively.*

3.11 Pauli–Fierz Hamiltonian with the Dipole Approximation

The Pauli–Fierz Hamiltonian with the dipole approximation is defined by H with $A = \int_{\mathbb{R}^3}^{\oplus} A(x)\mathrm{d}x$ replaced by $\mathbb{1} \otimes A(0)$. This implies that collisions between an electron and photons are ignored. Hence it is of the form

$$H_{\text{dip}} = \frac{1}{2}(-\mathrm{i}\nabla \otimes \mathbb{1} - \mathbb{1} \otimes A(0)) + V \otimes \mathbb{1} + \mathbb{1} \otimes H_{\mathrm{f}}.$$

By introducing the dipole approximation the model can be solvable. More precisely it can be diagonalized when $V = 0$.

3.11.1 Bogoliubov Transformation

In this section we suppose that

(1) $\hat{\varphi}/\omega^{3/2} \in L^2(\mathbb{R}^3)$, $\overline{\hat{\varphi}(k)} = \hat{\varphi}(-k)$, and $\hat{\varphi}(k) = \hat{\varphi}(|k|)$.
(2) $\hat{\varphi} \in C_0^\infty(\mathbb{R}^3)$.

Note that we can weaken these conditions, but for simplicity we assume (1) and (2). A weaker conditions are given in [77, Assumption 3.4].

Let K be a Hilbert-Schmidt operator on $L^2(\mathbb{R}^3) \oplus L^2(\mathbb{R}^3)$ such that

$$Kf = \sum_{n=0}^{\infty} \lambda_n (\psi_n, f)\phi_n.$$

Here $\{\psi_n\}$ and $\{\phi_n\}$ are orthogonal vectors in $L^2(\mathbb{R}^3) \oplus L^2(\mathbb{R}^3)$, and $\sum_{n=1}^{\infty} \lambda_n^2 < \infty$. Note that $\{\psi_n\}$ and $\{\phi_n\}$ are not necessarily orthonormal systems in $L^2(\mathbb{R}^3) \oplus L^2(\mathbb{R}^3)$. Then we define

$$\Delta_K^* = \lim_{M \to \infty} \sum_{n=1}^{M} \lambda_n a^*(\bar{\psi}_n) a^*(\phi_n),$$

$$\Delta_K = \lim_{M \to \infty} \sum_{n=1}^{M} \lambda_n a(\bar{\psi}_n) a(\phi_n).$$

It is established in [6] that Δ_K^* and Δ_K are densely defined closed operators in $\mathcal{F}$. Moreover let $\{e_n\}$ be a orthonormal system in $L^2(\mathbb{R}^3) \oplus L^2(\mathbb{R}^3)$, and S a bounded operator on $L^2(\mathbb{R}^3) \oplus L^2(\mathbb{R}^3)$. We define

$$N_S = \lim_{M \to \infty} \sum_{n=1}^{M} a^*(e_n) a(\overline{S^* e_n}).$$

This is also a densely defined closed operator. In particular choosing $S = \mathbb{1}$, we see that $N_{\mathbb{1}}$ is the number operator. These are studied in [6, 119, 120]. Let

$$W_{\pm} = (W_{\pm ij})_{1\le i,j\le 2}$$

be the operators defined by

$$W_{+ij} = \frac{1}{2} \sum_{1\le\mu,\nu\le 3} e_\mu^i \left(\frac{1}{\sqrt{\omega}} T_{\mu\nu}^* \sqrt{\omega} + \sqrt{\omega} T_{\mu\nu}^* \frac{1}{\sqrt{\omega}} \right) e_\nu^j,$$

$$W_{-ij} = \frac{1}{2} \sum_{1\le\mu,\nu\le 3} e_\mu^i \left(\frac{1}{\sqrt{\omega}} T_{\mu\nu}^* \sqrt{\omega} - \sqrt{\omega} T_{\mu\nu}^* \frac{1}{\sqrt{\omega}} \right) T e_\nu^j.$$

Here T is defined by $Tf(k) = f(-k)$ and $T_{\mu\nu}$ by

$$T_{\mu\nu} f = \delta_{\mu\nu} f + \frac{\hat{\varphi}}{D_+(\omega^2)} \sqrt{\omega} G \sqrt{\omega} \delta_{\mu\nu}^\perp \hat{\varphi} f$$

with

$$Gf(k) = \lim_{\varepsilon\downarrow 0} \int_{\mathbb{R}^3} \frac{f(k')}{(\omega(k)^2 - \omega(k')^2 + \mathrm{i}\varepsilon)\sqrt{\omega(k)\omega(k')}} \mathrm{d}k'$$

and

$$D_+(s) = 1 - \frac{4\pi}{3} \left(\lim_{\varepsilon\downarrow 0} \int_{|s-x|>\varepsilon} \frac{\hat{\varphi}(\sqrt{x})|^2 \sqrt{|x|}}{s-x} \mathrm{d}x - \pi \mathrm{i} |\hat{\varphi}(\sqrt{s})|^2 \sqrt{s} \right).$$

Let

$$S_p = \exp(-\mathrm{i}\Pi_p),$$

where

$$\Pi_p = \frac{\mathrm{i}}{\sqrt{2}} \frac{1}{m_{\mathrm{eff}}} \sum_{j=\pm 1} \left(a^* \left(p \cdot e^j \frac{\hat{\varphi}}{\omega^{3/2}}, j \right) - a \left(p \cdot e^j \frac{\hat{\varphi}}{\omega^{3/2}}, j \right) \right)$$

with

$$m_{\mathrm{eff}} = 1 + \frac{2}{3} \left\| \frac{\hat{\varphi}}{\omega} \right\|^2.$$

Let

$$U = C \exp(-\frac{1}{2} \Delta^*_{W_- W_+^{-1}}) : \exp(-N_{\mathbb{1}-\overline{(W_+^{-1})^*}}) : \exp(-\frac{1}{2} \Delta_{-W_+^{-1}\overline{W_-}}),$$

where

$$C = \det(\mathbb{1} - (W_- W_+^{-1})^* (W_- W_+^{-1}))^{1/4}.$$

Note that $\|U\mathbb{1}\| = 1$. Define

$$\mathcal{U} = \int_{\mathbb{R}^3}^{\oplus} S_p U e^{\mathrm{i}\frac{\pi}{2}N} \mathrm{d}p.$$

Thus we can see that $\mathcal{U}$ is the unitary operator on $\mathcal{H}$. See [5] and [77, Sect. 3.4].

Proposition 3.16 ([5, 78]) *It follows that*

$$\mathcal{U}^{-1} H_{\mathrm{dip}} \mathcal{U} = -\frac{1}{2m_{\mathrm{eff}}} \Delta \otimes \mathbb{1} + \mathbb{1} \otimes H_{\mathrm{f}} + g. \tag{3.39}$$

Here

$$g = \frac{3}{2\pi} \int_{-\infty}^{\infty} \frac{\frac{2}{3} \left\| \frac{t}{\sqrt{t^2+\omega^2}} \frac{\hat{\varphi}}{\sqrt{t^2+\omega^2}} \right\|^2}{1 + \frac{2}{3} \left\| \frac{\hat{\varphi}}{\sqrt{t^2+\omega^2}} \right\|^2} \mathrm{d}t.$$

A functional integral representation of $e^{-tH_{\mathrm{dip}}}$ is a minor modification of the full Hamiltonian. The stochastic integral $\int_0^t \mathrm{j}_s \tilde{\varphi}(\cdot - B_s) \mathrm{dB}_{\mathrm{s}}^{\mu}$ appearing in the functional integral representation of e^{-tH} is replaced by $\int_0^t \mathrm{j}_s \tilde{\varphi} \mathrm{dB}_{\mathrm{s}}^{\mu}$.

Proposition 3.17 *It follows that*

$$(F, e^{-tH_{\mathrm{dip}}} G) = \mathbb{E}^x [e^{-\int_0^t V(B_s)\mathrm{d}s} (F(B_0), \mathrm{J}_0^* e^{-\mathrm{i}\mathbb{A}(K_t^{\mathrm{dip}})} \mathrm{J}_t G(B_t))] \tag{3.40}$$

where $K_t^{\mathrm{dip}} = \bigoplus_{\mu=1}^3 \int_0^t \mathrm{j}_s \tilde{\varphi} \mathrm{d}B_s^{\mu}$.

Let $V = 0$. Then together with Propositions 3.16 and 3.17 we have the identity:

$$(F, e^{-tH_{\mathrm{dip}}} G) = e^{-tg} \int_{\mathbb{R}^3} \mathbb{E}^x [(\mathcal{U}^{-1} F(B_0), e^{-tH_{\mathrm{f}}} \mathcal{U}^{-1} G(B_{t/m_{\mathrm{eff}}})] \mathrm{d}x. \tag{3.41}$$

3.11.2 Translation Invariant Pauli–Fierz Hamiltonian with the Dipole Approximation

Let $V = 0$. Then

$$[H_{\mathrm{dip}}, p_\mu] = 0 \quad \mu = 1, 2, 3.$$

We have

$$H_{\mathrm{dip}} = \int_{\mathbb{R}^3}^{\oplus} H_{\mathrm{dip}}(p) \mathrm{d}p,$$

where

$$H_{\mathrm{dip}}(p) = \frac{1}{2}(p - A(0)) + H_{\mathrm{f}}.$$

Thus we have the proposition below.

Proposition 3.18 *It follows that*

$$(\Psi, e^{-tH_{\mathrm{dip}}(p)}\Phi) = \mathbb{E}^0[(\Psi, \mathrm{J}_0^* e^{-\mathrm{i}\mathbb{A}(K_t^{\mathrm{dip}})}\mathrm{J}_t e^{+\mathrm{i}(p-\mathrm{P}_{\mathrm{f}})B_t}\Phi)]. \tag{3.42}$$

Proof The proof is a minor modification of that of Theorem 3.14.

QED

Let $p = 0$ and $\Psi = \Phi = \mathbb{1}$. Thus we have

$$(\mathbb{1}, e^{-tH_{\mathrm{dip}}(p)}\mathbb{1}) = \mathbb{E}^0[e^{-\frac{1}{2}\mathrm{q}_{\mathrm{E}}(K_t^{\mathrm{dip}}, K_t^{\mathrm{dip}})}]. \tag{3.43}$$

Here we can see that

$$\frac{1}{2}\mathrm{q}_{\mathrm{E}}(K_t^{\mathrm{dip}}, K_t^{\mathrm{dip}}) \stackrel{formal}{=} \frac{1}{4}\sum_{1\leq\mu,\nu\leq 3}\int_0^t \mathrm{d}B_s^\mu \int_0^t \mathrm{d}B_r^\nu \int_{\mathbb{R}^3} \frac{|\hat{\varphi}(k)|^2}{\omega(k)} e^{-|s-r|\omega(k)}\left(\delta_{\mu\nu} - \frac{k_\mu k_\nu}{|k|^2}\right)\mathrm{d}k.$$

This expression is formal. In a similar manner to [15, Proposition 3.1] it is derived that

$$\frac{1}{2}\mathrm{q}_{\mathrm{E}}(K_t^{\mathrm{dip}}, K_t^{\mathrm{dip}}) = \frac{1}{4}\sum_{1\leq\mu,\nu\leq 3}\int_{\mathbb{R}^3}\mathrm{d}k\frac{|\hat{\varphi}(k)|^2}{\omega(k)}\int_0^t \mathrm{d}B_s^\mu\left(\delta_{\mu\nu} - \frac{k_\mu k_\nu}{|k|^2}\right)\int_0^s e^{-(s-r)\omega(k)}\mathrm{d}B_r^\nu.$$

Corollary 3.19 *It follows that*

$$\mathbb{E}^0[e^{-\frac{1}{2}\mathrm{q}_{\mathrm{E}}(K_t^{\mathrm{dip}}, K_t^{\mathrm{dip}})}] = e^{-tg}\det(\mathbb{1} - (W_- W_+^{-1})^*(W_- W_+^{-1}))^{1/2}(e^{-\frac{1}{2}\Delta_{K_0}^*}\mathbb{1}, e^{-\frac{1}{2}\Delta_{K_t}^*}\mathbb{1}),$$

where $K_t = -\mathrm{j}_t^*(W_+^{-1}\overline{W_-})^*\mathrm{j}_t$ *and* $K_0 = -\mathrm{j}_0^*(W_+^{-1}\overline{W_-})^*\mathrm{j}_0$.

Proof We have

$$U(p)^{-1}H_{\mathrm{dip}}(p)U(p) = \frac{1}{2m_{\mathrm{eff}}}p^2 + H_{\mathrm{f}} + g$$

for every $p \in \mathbb{R}^3$. Here

$$U(p) = S_p U e^{\mathrm{i}\frac{\pi}{2}N}.$$

Then

$$(\mathbb{1}, e^{-tH_{\mathrm{dip}}(p)}\mathbb{1}) = e^{-tg}\mathbb{E}^0[(U(0)^{-1}\mathbb{1}, e^{-tH_{\mathrm{f}}}U(0)^{-1}\mathbb{1})] = e^{-tg}\mathbb{E}^0[(\mathrm{J}_0 U(0)^{-1}\mathbb{1}, \mathrm{J}_t U(0)^{-1}\mathbb{1})].$$

Note that

$$U(0)^{-1}\mathbb{1} = \det(\mathbb{1} - (W_- W_+^{-1})^*(W_- W_+^{-1}))^{1/4} e^{-\frac{1}{2}\Delta^*_{-(W_+^{-1}\overline{W_-})^*}} \mathbb{1}.$$

Since $\mathrm{j}_\# e^{-\frac{1}{2}\Delta^*_{-(W_+^{-1}\overline{W_-})^*}} \mathbb{1} = e^{-\frac{1}{2}\Delta^*_{K_\#}} \mathbb{1}$ for $\# = 0, t$. Then the corollary follows.

QED

4 Relativistic Pauli–Fierz Model

4.1 Relativistic Pauli–Fierz Hamiltonian

In quantum mechanics a relativistic Schrödinger operator with a vector potential a is defined by $H_{\mathrm{R}}(a) = ((-\mathrm{i}\nabla - a)^2 + m^2)^{1/2} - m + V$, and a functional integral representation of $e^{-tH_{\mathrm{R}}(a)}$ is shown in Sect. 2.2. A key element in the construction of the functional integral representation of $e^{-tH_{\mathrm{R}}(a)}$ is to use the subordinator $(T_t)_{t\geq 0}$. In this section the analogue version of the Pauli–Fierz model is defined and its functional integral representation is given. This section is due to [55, 66]. We also refer to see [56, 57, 90–93, 108]. We say that $V \in \mathcal{R}_{\mathrm{RKato}}$ iff $D(\sqrt{-\Delta}) \subset D(V)$ and there exist $0 \leq a < 1$ and $0 \leq b$ such that $\|Vf\| \leq a\|\sqrt{-\Delta}f\| + b\|f\|$ for $f \in D(\sqrt{-\Delta})$ with $0 \leq a < 1$ and $0 \leq b$. Instead of Assumption 3.1 throughout this section we suppose the assumption below unless otherwise stated.

Assumption 4.1 $\varphi \in \mathcal{U}$, $\omega^{3/2}\hat{\varphi} \in L^2(\mathbb{R}^3)$, and $V \in \mathcal{R}_{\mathrm{RKato}}$.

In Assumptions 4.1 we add the extra condition $\omega^{3/2}\hat{\varphi} \in L^2(\mathbb{R}^3)$ to Assumption 3.1, and instead of $V \in \mathcal{R}_{\mathrm{Kato}}$, we suppose that $V \in \mathcal{R}_{\mathrm{RKato}}$. Let $\mathscr{D} = D(\Delta) \cap C^\infty(N)$ and

$$H(\hat{A}) = \frac{1}{2}(-\mathrm{i}\nabla - \hat{A})^2 \lceil_{\mathscr{D}}.$$

To define the relativistic Pauli–Fierz Hamiltonian $(2H(\hat{A}) + m^2)^{1/2} - m + H_{\mathrm{f}} + V$, we have to define $(2H(\hat{A}) + m^2)^{1/2}$ as a self-adjoint operator. It is however not trivial to choose a self-adjoint extension of $H(\hat{A})$. We have the lemma below.

Proposition 4.1 ([62]) *Suppose that* $\omega^{3/2}\hat{\varphi} \in L^2(\mathbb{R}^3)$, $\hat{\varphi}/\sqrt{\omega} \in L^2(\mathbb{R}^3)$ *and* $\hat{\varphi}(-k) = \overline{\hat{\varphi}(k)}$. *Then* $H(\hat{A})$ *is essentially self-adjoint.*

We keep denoting the unique self-adjoint extension of $H(\hat{A})$ by the same symbol $H(\hat{A})$ for simplicity, and we define $(2H(\hat{A}) + m^2)^{1/2}$ by the spectral resolution of $H(\hat{A})$. The relativistic Pauli–Fierz Hamiltonian is defined by

$$H_{\mathrm{R}} = (2H(\hat{A}) + m^2)^{1/2} - m \dotplus H_{\mathrm{f}} \dotplus V.$$

Here $m \geq 0$ is a parameter, but it describes the mass of an electron in physics.

4.2 Functional Integral Representations for the Relativistic Pauli–Fierz Hamiltonian

We will construct a functional integral representation of $e^{-tH_{\mathrm{R}}}$ through the Trotter product formula. We set

$$T_{\mathrm{k}} = (2H(\hat{A}) + m^2)^{1/2} - m. \tag{4.1}$$

By the Trotter product formula for quadratic form sums we see that

$$(F, e^{-tH_{\mathrm{R}}}G) = \lim_{n\to\infty} \left(F, \left(e^{-\frac{t}{2^n}T_{\mathrm{k}}} e^{-\frac{t}{2^n}H_{\mathrm{f}}} e^{-\frac{t}{2^n}V} \right)^{2^n} G \right).$$

Suppose $V \in C_0^\infty(\mathbb{R}^3)$. Then

$$\left(F, \left(e^{-\frac{t}{2^n}T_{\mathrm{k}}} e^{-\frac{t}{2^n}H_{\mathrm{f}}} e^{-\frac{t}{2^n}V} \right)^{2^n} G \right)$$
$$= \int_{\mathbb{R}^3} \mathbb{E}^{x,0}\left[\left(\mathrm{J}_0 F(B_{T_0}), e^{-\mathrm{i}\mathbb{A}(K_t^{\mathrm{R}}(n))} \mathrm{J}_t G(B_{T_t}) \right) e^{-\sum_{j=0}^{2^n} \frac{t}{2^n} V(B_{T_{t_j}})} \right] \mathrm{d}x,$$

where

$$K_t^{\mathrm{R}}(n) = \bigoplus_{\mu=1}^{3} \sum_{j=1}^{2^n} \int_{T_{t_{j-1}}}^{T_{t_j}} \mathrm{j}_{t_{j-1}} \tilde{\varphi}(\cdot - B_s) \mathrm{d}B_s^\mu$$

with $t_j = \frac{jt}{2^n}$, and $\int_{T_{t_{j-1}}}^{T_{t_j}} \mathrm{j}_{t_{j-1}} \tilde{\varphi}(\cdot - B_s) \mathrm{d}B_s^\mu$ denotes an $L^2(\mathbb{R}^4)$-valued stochastic integral

$$\int_T^S \mathrm{j}_{t_{j-1}} \tilde{\varphi}(\cdot - B_s) \mathrm{d}B_s^\mu$$

evaluated at $T = T_{t_{j-1}}$ and $S = T_{t_j}$. $(K_t^{\mathrm{R}}(n))_{t\geq 0}$ can be regarded as a sequence of $\bigoplus^3 L^2(\mathbb{R}^4)$-valued random processes on the product probability space $(\mathscr{X} \times \mathscr{X}_\nu, \mathcal{B} \times \mathcal{B}_\nu, \mathcal{W}^x \otimes \nu)$. Let $N_\nu \in \mathcal{B}_\nu$ be a null set, i.e., $\nu(N_\nu) = 0$, such that for arbitrary $w \in \mathscr{X}_\nu \setminus N_\nu$, the path $t \mapsto T_t(w)$ is nondecreasing and right-continuous, and has the left-limit. We have the lemma below.

Lemma 4.2 ([66]) *For each $w \in \mathscr{X}_\nu \setminus N_\nu$ and each $t \geq 0$, $K_t^{\mathrm{R}}(n)$ strongly converges in $L^2(\mathscr{X}) \otimes (\bigoplus^3 L^2(\mathbb{R}^4))$ as $n \to \infty$, i.e., there exists*

$$K_t^{\mathrm{R}} \in L^2(\mathscr{X}) \otimes (\bigoplus^3 L^2(\mathbb{R}^4))$$

such that $\lim_{n\to\infty} \mathbb{E}^x \left[\| K_t^{\mathrm{R}}(n) - K_t^{\mathrm{R}} \|^2 \right] = 0.$

By Lemma 4.2 and a limiting argument it follows that

$$(F, e^{-t(T_{\mathrm{k}} \dot{+} H_{\mathrm{f}})} G) = \int_{\mathbb{R}^3} \mathbb{E}^{x,0}\left[\left(\mathrm{J}_0 F(B_{T_0}), e^{-\mathrm{i}\mathbb{A}(K_t^{\mathrm{R}})} \mathrm{J}_t G(B_{T_t})\right)\right] \mathrm{d}x.$$

The immediate consequence is bounds of $e^{-t(T_{\mathrm{k}} \dot{+} H_{\mathrm{f}})}$. Let $F, G \in \mathcal{H}$. Then it follows that

$$|(F, e^{-t(T_{\mathrm{k}} \dot{+} H_{\mathrm{f}})} G)| \le (|F|, e^{-t((-\Delta+m^2)^{1/2}-m+H_{\mathrm{f}})} |G|)_{\mathcal{H}},$$
$$|(F, e^{-t(T_{\mathrm{k}} \dot{+} H_{\mathrm{f}})} G)| \le (\|F\|_{L^2(\mathrm{Q})}, e^{-t((-\Delta+m^2)^{1/2}-m)} \|G\|_{L^2(\mathrm{Q})})_{L^2(\mathbb{R}^3)}.$$

From these bounds we can conclude a relative boundedness with respect to V.

Lemma 4.3 *(1) and (2) follow.*

(1) *If V is relatively form bounded with respect to $(-\Delta + m^2)^{1/2} - m$ with a relative bound a, then $|V|$ is also relatively form bounded with respect to $(T_{\mathrm{k}} \dot{+} H_{\mathrm{f}})$ with a relative bound smaller than a.*

(2) *If V is relatively bounded with respect to $(-\Delta + m^2)^{1/2} - m$ with a relative bound a, then V is also relatively bounded with respect to $T_{\mathrm{k}} \dot{+} H_{\mathrm{f}}$ with a relative bound a.*

Suppose Assumption 4.1. Then $(T_{\mathrm{k}} \dot{+} H_{\mathrm{f}}) + V$ is self-adjoint on $D(T_{\mathrm{k}} \dot{+} H_{\mathrm{f}})$ by (1) of Lemma 4.3.

Theorem 4.4 *Suppose Assumption 4.1. Then*

$$(F, e^{-tH_{\mathrm{R}}} G) = \int_{\mathbb{R}^3} \mathbb{E}^{x,0}\left[e^{-\int_0^t V(B_{T_s})\mathrm{d}s} (\mathrm{J}_0 F(B_0), e^{-\mathrm{i}\mathbb{A}(K_t^{\mathrm{R}})} \mathrm{J}_t G(B_{T_t}))\right] \mathrm{d}x. \quad (4.2)$$

Proof When V is bounded and continuous, the theorem can be proven by the Trotter formula. Furthermore, it can be extended to a general V in the same way as that of Theorem 3.5.

QED

We have the corollary. Let $\mathfrak{S}_{\#} = \exp(-\mathrm{i}\frac{\pi}{2} N_{\#})$.

Corollary 4.5 ([66]) *(1) $\mathfrak{S}^{-1} e^{-tH_R} \mathfrak{S}$ is positivity improving. (2) If H_{R} has a ground state Ψ_{g}. Then $\mathfrak{S}^{-1}\Psi_{\mathrm{g}}$ is strictly positive.*

Proof Let $F, G \in \mathcal{H}$ be non-negative but not identically zero. We have

$$(F, \mathfrak{S}^{-1} e^{-tH_R} \mathfrak{S} G) = \int_{\mathbb{R}^3} \mathbb{E}^{x,0}\left[e^{-\int_0^t V(B_s)\mathrm{d}s} (F(B_0), \mathrm{J}_0^* \mathfrak{S}_{\mathrm{E}}^{-1} e^{-\mathrm{i}\mathbb{A}(K_t^{\mathrm{R}})} \mathfrak{S}_{\mathrm{E}} \mathrm{J}_t G(B_{T_t}))\right] \mathrm{d}x.$$

Since $\mathrm{J}_0^* \mathfrak{S}_{\mathrm{E}}^{-1} e^{-\mathrm{i}\mathbb{A}(K_t^{\mathrm{R}})} \mathfrak{S}_{\mathrm{E}} \mathrm{J}_t$ is positivity improving, the corollary can be proven in a similar manner to Corollary 3.10.

QED

4.3 Invariant Domain and Self-adjointness

In this section by using the functional integral representation derived in Theorem 4.4 we show the self-adjointness of H_R. The fundamental lemma is as follows.

Lemma 4.6 *Let K be a non-negative self-adjoint operator. Suppose that there exists a dense domain D such that $D \subset D(K)$ and $e^{-tK}D \subset D$ for all $t \geq 0$. Then $K\lceil_D$ is essentially self-adjoint.*

Proof It suffices to show that for some $\lambda > 0$, $\mathrm{Ran}((\lambda + K)\lceil_D)$ is dense. Suppose the contrary. Then there exists nonzero f such that $(f, (\lambda + K)\psi) = 0$ for all $\psi \in D$. We have

$$\frac{d}{dt}(f, e^{-tK}\psi) = (f, -Ke^{-tK}\psi) = \lambda(f, e^{-tK}\psi).$$

Thus $(f, e^{-tK}\psi) = e^{\lambda t}(f, \psi)$. If $(f, \psi) \neq 0$, then $\lim_{t\to\infty}|(f, e^{-tK}\psi)| = \infty$, contradicting the fact that e^{-tK} is a contraction. Hence $(f, \psi) = 0$ for all $\psi \in D(K)$, but (f, ψ) can not equal zero for all $\psi \in D(K)$, since $D(K)$ is dense. Hence we conclude that $\mathrm{Ran}((\lambda + K)\lceil_D)$ is dense.

QED

To prove the essential self-adjointness of H_R we find an invariant domain D so that

$$D \subset D(H_R), \quad e^{-tH_R}D \subset D.$$

Let $\hat{\omega} = \omega(-i\nabla) \otimes 1\!\!1$ under $L^2(\mathbb{R}^4) \cong L^2(\mathbb{R}^3) \otimes L^2(\mathbb{R})$. Let $\alpha > 0$. We can estimate the forth moment $\mathbb{E}^{x,0}\left[\|\hat{\omega}^{\alpha/2}K_t^R\|^4_{\bigoplus^3 L^2(\mathbb{R}^4)}\right]$. Suppose that $\hat{\varphi}/\sqrt{\omega}, \omega^{(\alpha-1)/2}\hat{\varphi} \in L^2(\mathbb{R}^3)$. Then the Burkholder–Davis–Gundy (BDG) type inequality [62, 66] holds:

$$\mathbb{E}^{x,0}\left[\|\hat{\omega}^{\alpha/2}K_t^R\|^4_{\bigoplus^3 L^2(\mathbb{R}^4)}\right] \leq C\|\omega^{(\alpha-1)/2}\hat{\varphi}\|^4_{L^2(\mathbb{R}^3)},$$

where C is a constant.

Lemma 4.7 *Let $V = 0$. Then (1) and (2) hold true.*
(1) For $F \in D(-i\nabla_\mu)$ and $G \in D(-i\nabla_\mu) \cap D(H_f^{1/2})$ it follows that

$$(-i\nabla_\mu F, e^{-tH_R}G) \leq C\left((\|\sqrt{\omega}\hat{\varphi}\| + \|\hat{\varphi}\|)\|(H_f + 1\!\!1)^{1/2}G\| + \|-i\nabla_\mu G\|\right)\|F\|.$$

In particular $e^{-tH_R}D(\sqrt{-\Delta}) \cap D(H_f^{1/2}) \subset D(\sqrt{-\Delta})$ for $t \geq 0$.
(2) For $F, G \in D(H_f)$ it follows that

$$(H_f F, e^{-tH_R}G) \leq \left(\|H_f G\| + (\|\sqrt{\omega}\hat{\varphi}\| + \|\hat{\varphi}\|)\|(H_f + 1\!\!1)^{1/2}G\| + \|\hat{\varphi}/\sqrt{\omega}\|^2\|G\|\right)\|F\|.$$

In particular $e^{-tH_R}D(H_f) \subset D(H_f)$ for $t \geq 0$.

Proof We show the outline of a proof. Notice that

$$(e^{is(-i\nabla_\mu)}F, e^{-tH_R}G) = \int_{\mathbb{R}^3} \mathbb{E}^{x,0}\Big[\Big(J_0 F(B_{T_0}), e^{+isP_{f\mu}}e^{-i\mathbb{A}(K_t^R)}e^{-isP_{f\mu}}J_t e^{-is(-i\nabla_\mu)}G(B_{T_t})\Big)\Big]dx. \tag{4.3}$$

Here we used

$$e^{is(-i\nabla_\mu+P_{f\mu})}e^{-tH_R} = e^{-tH_R}e^{is(-i\nabla_\mu+P_{f\mu})}.$$

We see that

$$e^{+isP_{f\mu}}e^{-i\mathbb{A}(K_t^R)}e^{-isP_{f\mu}} = e^{-i\mathbb{A}(e^{is(-i\nabla_\mu\otimes 1)}K_t^R)}.$$

Take the derivative at $s = 0$ on both sides of (4.3). We have

$$\begin{aligned}(\nabla_\mu F, e^{-tH_R}G) = &\int_{\mathbb{R}^3} \mathbb{E}^{x,0}\Big[\Big(J_0 F(B_{T_0}), -i\mathbb{A}_\mu(\nabla_\mu K_t^R)e^{-i\mathbb{A}(K_t^R)}J_t G(B_{T_t})\Big)\Big]dx\\ &+\int_{\mathbb{R}^3} \mathbb{E}^{x,0}\Big[\Big(J_0 F(B_{T_0}), e^{-i\mathbb{A}(K_t^R)}J_t(-\nabla_\mu G)(B_{T_t})\Big)\Big]dx.\end{aligned} \tag{4.4}$$

It is trivial to see that

$$\left|\int_{\mathbb{R}^3} \mathbb{E}^{x,0}\Big[\Big(J_0 F(B_{T_0}), e^{-i\mathbb{A}(K_t^R)}J_t(-\nabla_\mu G)(B_{T_t})\Big)\Big]dx\right| \le \|F\|\|\nabla_\mu G\|.$$

We can estimate the first term on the right-hand side of (4.4) by BDG inequality as

$$\begin{aligned}&\left|\int_{\mathbb{R}^3} \mathbb{E}^{x,0}\Big[\Big(J_0 F(B_{T_0}), \mathbb{A}_\mu(\nabla_\mu K_t^R)e^{-i\mathbb{A}_\mu(K_t^R)}J_t G(B_{T_t})\Big)\Big]dx\right|\\ &\le C(\|\omega^{1/2}\hat{\varphi}\| + \|\hat{\varphi}\|)\|(H_f + \mathbb{1})^{1/2}F\|\|G\|.\end{aligned}$$

Then (1) follows. We have

$$(H_f F, e^{-tH_R}G) = \int_{\mathbb{R}^3} \mathbb{E}^{x,0}\Big[\Big(J_0 F(B_{T_0}), e^{-i\mathbb{A}(K_t^R)}SJ_t G(B_{T_t})\Big)\Big]dx,$$

where $S = e^{i\mathbb{A}(K_t^R)}H_f e^{-i\mathbb{A}(K_t^R)} = H_f - i[H_f, \mathbb{A}(K_t^R)] + g$ with $g = q(K_t^R, K_t^R)$. It is trivial to see that

$$\left|\int_{\mathbb{R}^3} \mathbb{E}^{x,0}\Big[\Big(J_0 F(B_{T_0}), e^{-i\mathbb{A}(K_t^R)}H_f J_t G(B_{T_t})\Big)\Big]dx\right| \le \|F\|\|H_f G\|.$$

In the same way as the estimate of the first term of the right-hand side of (4.4) we can see that

$$\left|\int_{\mathbb{R}^3} \mathbb{E}^{x,0}\Big[\Big(\mathrm{J}_0 F(B_{T_0}), e^{-\mathrm{i}\mathbb{A}(K_t^{\mathrm{R}})}[H_{\mathrm{f}}, \mathbb{A}(K_t^{\mathrm{R}})]\mathrm{J}_t G(B_{T_t})\Big)\Big] \mathrm{d}x\right|$$
$$\leq C(\|\sqrt{\omega}\hat{\varphi}\|+\|\hat{\varphi}\|)\|F\|\|(H_{\mathrm{f}}+\mathbb{1})^{1/2}G\|$$

with some constant $C > 0$. Here we used BDG inequality. Finally we see that $g \leq C\|K_t^{\mathrm{R}}\|^2$ and by BDG inequality again,

$$\left|\int_{\mathbb{R}^3} \mathbb{E}^{x,0}\Big[\Big(\mathrm{J}_0 F(B_{T_0}), e^{-\mathrm{i}\mathbb{A}(K_t^{\mathrm{R}})} g \mathrm{J}_t G(B_{T_t})\Big)\Big] \mathrm{d}x\right| \leq C\|\hat{\varphi}/\sqrt{\omega}\|^2\|F\|\|G\|.$$

Then (2) follows.

QED

Theorem 4.8 ([66]) *Suppose Assumption 4.1 and* $m > 0$. *Then* H_{R} *is essentially self-adjoint on* $D(\sqrt{-\Delta}) \cap D(H_{\mathrm{f}})$.

Proof Suppose $V = 0$. Let $D_\infty = C_0^\infty(\mathbb{R}^3) \otimes L^2_{\mathrm{fin}}(\mathrm{Q})$. Let $F \in D_\infty$. We see that

$$\|(T_{\mathrm{k}} \dot{+} H_{\mathrm{f}})F\|^2 \leq C_1\|\sqrt{-\Delta}F\|^2 + C_2\|H_{\mathrm{f}}F\|^2 + C_3\|F\|^2 \tag{4.5}$$

with some constants C_1, C_2 and C_3. Since D_∞ is a core of $\sqrt{-\Delta} + H_{\mathrm{f}}$,

$$D(\sqrt{-\Delta}) \cap D(H_{\mathrm{f}}) \subset D(T_{\mathrm{k}} \dot{+} H_{\mathrm{f}}) \tag{4.6}$$

follows from a limiting argument. By Lemmas 4.7, we also see that

$$e^{-t(T_{\mathrm{k}} \dot{+} H_{\mathrm{f}})}\Big(D(\sqrt{-\Delta}) \cap D(H_{\mathrm{f}})\Big) \subset \Big(D(\sqrt{-\Delta}) \cap D(H_{\mathrm{f}})\Big). \tag{4.7}$$

Equations (4.6) and (4.7) imply that $T_{\mathrm{k}} \dot{+} H_{\mathrm{f}}$ is essentially self-adjoint on $D(\sqrt{-\Delta}) \cap D(H_{\mathrm{f}})$ by Lemma 4.6. Next we suppose that V satisfies Assumption 4.1. Then V is also relatively bounded with respect to $T_{\mathrm{k}} \dot{+} H_{\mathrm{f}}$ with a relative bound strictly smaller than one. Then the theorem follows by the Kato–Rellich theorem.

QED

Furthermore we can establish the self-adjointness of H_{R}. The key inequality to show the self-adjointness of H_{R} on $D(\sqrt{-\Delta}) \cap D(H_{\mathrm{f}})$ is the following inequality.

Lemma 4.9 ([55]) *Suppose Assumption 4.1 and that* $m \geq 0$ *and* $V = 0$. *Then there exists a constant* C *such that*

$$\|\sqrt{-\Delta}F\|^2 + \|H_{\mathrm{f}}F\|^2 \leq C\|(T_{\mathrm{k}} \dot{+} H_{\mathrm{f}} + \mathbb{1})F\|^2 \tag{4.8}$$

for all $F \in D(\sqrt{-\Delta}) \cap D(H_{\mathrm{f}})$.

Theorem 4.10 ([55]) *Suppose Assumption 4.1. Then* H_{R} *is self-adjoint on* $D(\sqrt{-\Delta}) \cap D(H_{\mathrm{f}})$.

Proof Suppose that $V = 0$. We write $H^{(m)}$ for H_{R} to emphasize m-dependence. Let $m > 0$. Then $H^{(m)}$ is essentially self-adjoint on $D(\sqrt{-\Delta}) \cap D(H_{\mathrm{f}})$. While by (4.8), $H^{(m)}\lceil_{D(\sqrt{-\Delta})\cap D(H_{\mathrm{f}})}$ is closed. Then $H^{(m)}$ is self-adjoint on $D(\sqrt{-\Delta}) \cap D(H_{\mathrm{f}})$. Note that

$$H^{(0)} = H^{(m)} + (H^{(0)} - H^{(m)})$$

and $H^{(0)} - H^{(m)}$ is bounded. Then $H^{(0)}$ is also self-adjoint on $D(\sqrt{-\Delta}) \cap D(H_{\mathrm{f}})$ for $V = 0$. Finally let V be potential satisfying Assumption 4.1. Then V is also relatively bounded with respect to $H^{(m)}$ with a relative bound strictly smaller than one. Then the theorem follows from the Kato–Rellich theorem.

QED

A spinless hydrogen like atom is defined by introducing the Coulomb potential

$$V(x) = -\frac{g}{|x|}, \quad g > 0,$$

which is relatively form bounded with respect to $(-\Delta + m^2)^{1/2}$ with a relative bound strictly smaller than one if $g \leq 2/\pi$ by [53]. Furthermore if $g < 1/2$, V is relatively bounded with respect to $(-\Delta + m^2)^{1/2}$ with a relative bound strictly smaller than one. Let $\hat{A}_\Lambda$ be the quantized radiation field with

$$\hat{\varphi}(k) = \frac{1}{(2\pi)^{3/2}} 1\!\!1_{|k|\leq\Lambda}(k).$$

By Theorem 4.10 when $g < 1/2$, H_{R} is self-adjoint on $D(\sqrt{-\Delta}) \cap D(H_{\mathrm{f}})$.

4.4 *Non-relativistic Limit*

Schrödinger operators do not contain the velocity of the light in the definition. On the other hand Dirac operators

$$H(c) = c\alpha \cdot \left(-\mathrm{i}\nabla - \frac{1}{c}a\right) + \beta mc^2 + V$$

can be regarded as relativistic versions of Schrödinger operators. Here c denotes the velocity of the light, and α and β are 4×4 Dirac matrices given by

$$\alpha = \begin{pmatrix} 0 & \sigma_i \\ \sigma_i & 0 \end{pmatrix}, i = 1, 2, 3, \quad \beta = \begin{pmatrix} 1\!\!1_{2\times2} & 0 \\ 0 & -1\!\!1_{2\times2} \end{pmatrix}$$

with 2×2 Pauli matrices σ_1, σ_2 and σ_3. This was introduced by P. A. M. Dirac to construct a relativistically covariant quantum theory. Hence non-relativistic limits of Dirac operators have been studied so far. We refer to see [133, Sect. 6] for non-relativistic limits of Dirac operators and [103, 104] for those of non-linear Dirac and Klein-Gordon equations. Formally it can be shown that

$$\lim_{c\to\infty}(H(c) - mc^2 - z)^{-1} = \begin{pmatrix} \left(\frac{1}{2m}(-\mathrm{i}\nabla - a)^2 + V - z\right)^{-1} & 0 \\ 0 & 0 \end{pmatrix}.$$

On the other hand we are concerned with semi-relativistic Schrödinger operators in this section. Since we have

$$\sqrt{c^2|p|^2 + m^2c^4} - mc^2 = \frac{1}{2m}|p|^2 + \mathcal{O}\left(\frac{|p|^4}{m^3c^2}\right),$$

intuitively we have

$$\exp\left(-t\left(\sqrt{c^2(-\Delta) + m^2c^4} - mc^2\right)\right) \to \exp\left(\frac{t}{2m}\Delta\right)$$

as $c \to \infty$. This intuition becomes substantial by means of the so-called non-relativistic limit of semi-relativistic Schrödinger operators. Define

$$H_c = \sqrt{-c^2\Delta + m^2c^4} - mc^2 + V.$$

By using a functional integral representation we can show that $H_c \to H_\infty$ as $c \to \infty$ in a specific sense, and the limit operator is the Schrödinger operator

$$H_\infty = -\frac{1}{2m}\Delta + V.$$

For every $c > 0$ consider the subordinator $(T_t^c)_{t\geq 0}$ with parameter c such that

$$\mathbb{E}[e^{-uT_t^c}] = e^{-t(\sqrt{2c^2u + m^2c^4} - mc^2)}, \quad u \geq 0.$$

Proposition 4.11 *Let f be a bounded continuous function on $\mathbb{R}$. Then*

$$\lim_{c\to\infty}\mathbb{E}_P[f(T_t^c)] = f\left(\frac{t}{m}\right).$$

Proof Let $f \in \mathscr{S}(\mathbb{R})$. We have

$$\mathbb{E}[f(T_t^c)] = \frac{1}{\sqrt{2\pi}}\int_{\mathbb{R}} \check{f}(k)\mathbb{E}[e^{\mathrm{i}kT_t^c}]\mathrm{d}k.$$

We see that

$$\mathbb{E}[e^{\mathrm{i}kT_t^c}] = \int_0^\infty \frac{ct}{\sqrt{2\pi}} e^{mc^2t} s^{-3/2} \exp\left(-\frac{1}{2}\left(\frac{c^2t^2}{s} + (2k\mathrm{i} + m^2c^2)s\right)\right) \mathrm{d}s$$

and then

$$\mathbb{E}_\mu[e^{\mathrm{i}kT_t^c}] = e^{-(\sqrt{2k\mathrm{i}+m^2c^2}-mc)ct}.$$

Furthermore, we have

$$c(\sqrt{2k\mathrm{i} + m^2c^2} - mc) = c((4k^2 + m^4c^4)^{1/4} - mc)e^{\mathrm{i}\theta/2} + mc^2(e^{\mathrm{i}\theta/2} - 1),$$

where $\tan\theta = 2k/m^2c^2$ with $|\theta| < \pi/2$. The first term converges to zero as $c \to \infty$. For the second term it can be seen that $mc^2(e^{\mathrm{i}\theta/2} - 1) = mc^2(\cos(\theta/2) - 1 + \mathrm{i}\sin(\theta/2))$. Since

$$mc^2(\cos(\theta/2) - 1) \sim mc^2(\theta/2)^2/2 \sim mc^2(k/m^2c^2)^2/2 \sim 0$$

and

$$imc^2 \sin(\theta/2) \sim imc^2\theta/2 \sim \mathrm{i}k/m$$

as $c \to \infty$, we get

$$\lim_{c\to\infty} \mathbb{E}_P[e^{\mathrm{i}kT_t^c}] = e^{\mathrm{i}\frac{tk}{m}}$$

as $c \to \infty$, and the proposition follows. When f is chosen to be a bounded continuous function, it can be uniformly approximated by functions in $\mathscr{S}(\mathbb{R})$ and the proof is completed by a simple limiting argument.

QED

We derive the non-relativistic limit of e^{-tH_c}.

Corollary 4.12 *Let V be a bounded continuous function. Then*

$$s - \lim_{c\to\infty} e^{-tH_c} = e^{-tH_\infty}.$$

Proof We suppose that V is non-negative without loss of generality. It is enough to show the weak limit

$$\lim_{c\to\infty} (f, e^{-tH_c}g) = (f, e^{-tH_\infty}g). \tag{4.9}$$

Since $H_c \geq 0$ for every $c > 0$, $\|e^{-tH_c}\| \leq 1$ uniformly with respect to $c > 0$. It is also sufficient to show (4.9) for arbitrary $f, g \in \mathscr{S}(\mathbb{R})$ by a simple limiting argument. Note that by Proposition 4.11 it can be seen that

$$(f, e^{-t(\sqrt{-\Delta+m^2c^4}-mc^2+V)}g) = \int_{\mathbb{R}^3} \mathbb{E}^{x,0}\left[\bar{f}(x)g(B_{T_t^c})e^{-\int_0^t V(B_{T_s^c})\mathrm{d}s}\right]\mathrm{d}x$$

$$\to \int_{\mathbb{R}^3} \mathbb{E}^x\left[\bar{f}(x)g(B_{\frac{t}{m}})e^{-\int_0^t V(B_{\frac{s}{m}})\mathrm{d}s}\right]\mathrm{d}x = (f, e^{-t(-\frac{1}{2m}\Delta+V)}g)$$

as $c \to \infty$.

QED

$K_t^{\mathrm{R}}(c)$ is defined by K_t^{R} with T_t replaced by T_t^c.

Lemma 4.13 *It follows that* $\lim_{c\to\infty} K_t^{\mathrm{R}}(c) = K_t$ *strongly in* $L^2(\mathscr{X} \times \mathscr{X}_\nu) \otimes (\bigoplus^3 L^2(\mathbb{R}^4))$.

Proof Let

$$\mathrm{I}_n^c = \bigoplus_{\mu=1}^3 \sum_{j=1}^{2^n} \int_{T_{t_{j-1}}^c}^{T_{t_j}^c} \mathrm{j}_{t_{j-1}}\tilde{\varphi}(\cdot - B_s)\mathrm{d}B_s^\mu,$$

$$\mathrm{I}_n = \bigoplus_{\mu=1}^3 \sum_{j=1}^{2^n} \int_{\frac{t_{j-1}}{m}}^{\frac{t_j}{m}} \mathrm{j}_{t_{j-1}}\tilde{\varphi}(\cdot - B_s)\mathrm{d}B_s^\mu.$$

We have seen that $\mathrm{I}_n^c \to K_t^{\mathrm{R}}(c)$ and $\mathrm{I}_n \to K_t$ as $n \to \infty$ strongly in $L^2(\mathscr{X} \times \mathscr{X}_\nu) \otimes (\bigoplus^3 L^2(\mathbb{R}^4))$. We have

$$\|K_t^{\mathrm{R}}(c) - K_t\| \le \|K_t^{\mathrm{R}}(c) - \mathrm{I}_n^c\| + \|\mathrm{I}_n^c - \mathrm{I}_n\| + \|\mathrm{I}_n - K_t\|$$

and

$$\mathbb{E}^x[\|\mathrm{I}_n^c - \mathrm{I}_k^c\|^2] \le 3T_t^c\|\hat{\varphi}/\sqrt{\omega}\|^2\left(\sum_{j=n+1}^k 2^{-j/2}\right)^2.$$

From this we have

$$\mathbb{E}^{x,0}[\|\mathrm{I}_n^c - K_t^{\mathrm{R}}(c)\|^2] \le 3\mathbb{E}^0[T_t^c]\|\hat{\varphi}/\sqrt{\omega}\|^2\left(\sum_{j=n+1}^\infty 2^{-j/2}\right)^2.$$

Since $\mathbb{E}^0[T_t^c] = \frac{t}{m}$ which is independent of $c > 0$, we obtain that

$$\mathbb{E}^{x,0}[\|\mathrm{I}_n^c - K_t^{\mathrm{R}}(c)\|^2] \le \frac{3t}{m}\|\hat{\varphi}/\sqrt{\omega}\|^2\left(\sum_{j=n+1}^\infty 2^{-j/2}\right)^2$$

and we conclude that

$$\mathbb{E}^{x,0}[\|\mathrm{I}_n^c - K_t^{\mathrm{R}}(c)\|^2] \to 0 \tag{4.10}$$

as $n \to \infty$ uniformly in c. Let $\varepsilon > 0$ be arbitrary. There exists n_0 such that for all $n > n_0$ $\mathbb{E}^{x,0}[\|K_t^{\mathrm{R}}(c) - \mathrm{I}_n^c\|^2] < \varepsilon^2$ and $\mathbb{E}^{x,0}[\|\mathrm{I}_n - K_t\|^2] < \varepsilon^2$ uniformly in c. Now we estimate $\|\mathrm{I}_n^c - \mathrm{I}_n\|$. We have

$$\mathrm{I}_n^c - \mathrm{I}_n = \bigoplus_{\mu=1}^{3} \sum_{j=1}^{2^n} \left(\int_{T^c_{t_{j-1}}}^{T^c_{t_j}} \mathrm{j}_{t_{j-1}} \tilde{\varphi}(\cdot - B_s) \mathrm{d}B_s^\mu - \int_{\frac{t_{j-1}}{m}}^{\frac{t_j}{m}} \mathrm{j}_{t_{j-1}} \tilde{\varphi}(\cdot - B_s) \mathrm{d}B_s^\mu \right).$$

We note that $s \to \int_a^s \mathrm{j}_{t_{j-1}} \tilde{\varphi}(\cdot - B_s) \mathrm{d}B_s^\mu$ and $s \to \int_s^b \mathrm{j}_{t_{j-1}} \tilde{\varphi}(\cdot - B_s) \mathrm{d}B_s^\mu$ are almost surely continuous. Hene

$$(S, T) \to \mathbb{E}^x \left[\left(\int_S^T \mathrm{j}_{t_{j-1}} \tilde{\varphi}(\cdot - B_s) \mathrm{d}B_s^\mu, \int_{\frac{t_{j-1}}{m}}^{\frac{t_j}{m}} \mathrm{j}_{t_{j-1}} \tilde{\varphi}(\cdot - B_s) \mathrm{d}B_s^\mu \right) \right]$$

is continuous. This implies that for every j,

$$\begin{aligned} &\mathbb{E}^{x,0} \left[\left(\int_{T^c_{t_{j-1}}}^{T^c_{t_j}} \mathrm{j}_{t_{j-1}} \tilde{\varphi}(\cdot - B_s) \mathrm{d}B_s^\mu, \int_{\frac{t_{j-1}}{m}}^{\frac{t_j}{m}} \mathrm{j}_{t_{j-1}} \tilde{\varphi}(\cdot - B_s) \mathrm{d}B_s^\mu \right) \right] \\ &\to \mathbb{E}^x \left[\left(\int_{\frac{t_{j-1}}{m}}^{\frac{t_j}{m}} \mathrm{j}_{t_{j-1}} \tilde{\varphi}(\cdot - B_s) \mathrm{d}B_s^\mu, \int_{\frac{t_{j-1}}{m}}^{\frac{t_j}{m}} \mathrm{j}_{t_{j-1}} \tilde{\varphi}(\cdot - B_s) \mathrm{d}B_s^\mu \right) \right] = \frac{(t_j - t_{j-1})}{m} \|\hat{\varphi}/\sqrt{\omega}\|^2 \end{aligned} \tag{4.11}$$

as $c \to \infty$. We have

$$\mathbb{E}^{x,0}[\|\mathrm{I}_n^c - \mathrm{I}_n\|^2] = 3 \sum_{j=1}^{2^n} \mathbb{E}^{x,0} \left[\left\| \int_{T^c_{t_{j-1}}}^{T^c_{t_j}} \mathrm{j}_{t_{j-1}} \tilde{\varphi}(\cdot - B_s) \mathrm{d}B_s^\mu - \int_{\frac{t_{j-1}}{m}}^{\frac{t_j}{m}} \mathrm{j}_{t_{j-1}} \tilde{\varphi}(\cdot - B_s) \mathrm{d}B_s^\mu \right\|^2 \right],$$

and

$$\begin{aligned} &\mathbb{E}^{x,0} \left[\left\| \int_{T^c_{t_{j-1}}}^{T^c_{t_j}} \mathrm{j}_{t_{j-1}} \tilde{\varphi}(\cdot - B_s) \mathrm{d}B_s^\mu - \int_{\frac{t_{j-1}}{m}}^{\frac{t_j}{m}} \mathrm{j}_{t_{j-1}} \tilde{\varphi}(\cdot - B_s) \mathrm{d}B_s^\mu \right\|^2 \right] \\ &= \mathbb{E}^{x,0} \left[\left\| \int_{T^c_{t_{j-1}}}^{T^c_{t_j}} \mathrm{j}_{t_{j-1}} \tilde{\varphi}(\cdot - B_s) \mathrm{d}B_s^\mu \right\|^2 \right] + \mathbb{E}^{x,0} \left[\left\| \int_{\frac{t_{j-1}}{m}}^{\frac{t_j}{m}} \mathrm{j}_{t_{j-1}} \tilde{\varphi}(\cdot - B_s) \mathrm{d}B_s^\mu \right\|^2 \right] \\ &\quad - 2\mathbb{E}^{x,0} \left[\left(\int_{T^c_{t_{j-1}}}^{T^c_{t_j}} \mathrm{j}_{t_{j-1}} \tilde{\varphi}(\cdot - B_s) \mathrm{d}B_s^\mu, \int_{\frac{t_{j-1}}{m}}^{\frac{t_j}{m}} \mathrm{j}_{t_{j-1}} \tilde{\varphi}(\cdot - B_s) \mathrm{d}B_s^\mu \right) \right] \\ &= \frac{1}{m} \|\hat{\varphi}/\sqrt{\omega}\|^2 (\mathbb{E}^0[T^c_{t_j} - T^c_{t_{j-1}}] + t_j - t_{j-1}) \\ &\quad - 2\mathbb{E}^{x,0} \left[\left(\int_{T^c_{t_{j-1}}}^{T^c_{t_j}} \mathrm{j}_{t_{j-1}} \tilde{\varphi}(\cdot - B_s) \mathrm{d}B_s^\mu, \int_{\frac{t_{j-1}}{m}}^{\frac{t_j}{m}} \mathrm{j}_{t_{j-1}} \tilde{\varphi}(\cdot - B_s) \mathrm{d}B_s^\mu \right) \right]. \end{aligned}$$

Note that $\mathbb{E}^0[T^c_{t_j} - T^c_{t_{j-1}}] = t_j - t_{j-1}$ and (4.11). We can see that $\mathbb{E}^{x,0}[\|\mathrm{I}^c_n - \mathrm{I}_n\|^2] \to 0$ as $c \to \infty$. We have

$$\lim_{c\to\infty}(\mathbb{E}^{x,0}[\|K^{\mathrm{R}}_t(c) - K_t\|]^2)^{1/2} \le 2\varepsilon + \lim_{c\to\infty}(\mathbb{E}^{x,0}[\|\mathrm{I}^c_n - \mathrm{I}_n\|])^{1/2} = 2\varepsilon.$$

Thus the lemma is proven.

QED

Now we show a non-relativistic limit of the relativistic Pauli–Fierz Hamiltonian.

Theorem 4.14 ([68]) *Suppose that V is bounded and continuous. Then for every $t \ge 0$ it follows that*

$$\mathrm{s}-\lim_{c\to\infty} e^{-tH_{\mathrm{R}}} = e^{-tH}.$$

Proof Suppose that $F, G \in C_0^\infty(\mathbb{R}^3) \otimes L^2(\mathrm{Q})$. We have

$$(F, e^{-tH_{\mathrm{R}}}G) = \int_{\mathbb{R}^3} \mathbb{E}^{x,0}\left[e^{-\int_0^t V(B_{T^c_s})\mathrm{d}s}(\mathrm{J}_0F(x), e^{-\mathrm{i}\mathbb{A}(K^{\mathrm{R}}_t(c))}\mathrm{J}_tG(B_{T^c_t}))\right]\mathrm{d}x.$$

It follows that

$$\lim_{c\to\infty}(F, e^{-tH_{\mathrm{R}}}G) = \int_{\mathbb{R}^3} \mathbb{E}^x\left[e^{-\int_0^t V(B_{\frac{s}{m}})\mathrm{d}s}(\mathrm{J}_0F(x), e^{-\mathrm{i}\mathbb{A}(K_t)}\mathrm{J}_tG(B_{\frac{t}{m}}))\right]\mathrm{d}x = (F, e^{-tH}G).$$

Since $H_{\mathrm{R}} \ge \inf_{x\in\mathbb{R}^3} V(x) = g > -\infty$, $e^{-tH_{\mathrm{R}}} \le e^{-tg}$. Let $F, G \in \mathcal{H}$. There exists $F_n, G_n \in C_0^\infty(\mathbb{R}^3) \otimes L^2(\mathrm{Q})$ such that $F_n \to F$ and $G_n \to G$ strongly as $n \to \infty$. By the uniform bound $e^{-tH_{\mathrm{R}}} \le e^{-tg}$, we can show $\lim_{c\to\infty}(F, e^{-tH_{\mathrm{R}}}G) = (F, e^{-tH}G)$. Finally since the weak convergence of $e^{-tH_{\mathrm{R}}}$ implies the strong convergence, the theorem follows.

QED

4.5 Translation Invariant Relativistic Pauli–Fierz Hamiltonian

For the relativistic Pauli–Fierz Hamiltonian with $V = 0$, as well as H, it can be seen that $[H_{\mathrm{R}}, \mathrm{P}^{\mathrm{T}}_\mu] = 0$. This allows that there exists a self-adjoint operator $H_{\mathrm{R}}(p)$ in $\mathcal{F}$ such that

$$H_{\mathrm{R}} = \int_{\mathbb{R}^3}^{\oplus} H_{\mathrm{R}}(p)\mathrm{d}p.$$

The self-adjoint operator $H_{\mathrm{R}}(p)$, $p \in \mathbb{R}^3$, is called the relativistic Pauli–Fierz Hamiltonian with a total momentum p. We can show the similar results to those of $H(p)$ by using the functional integral representation of $e^{-tH_{\mathrm{R}}}$. The theorem below can be proven in a similar manner to that of Theorem 3.14.

Theorem 4.15 ([66]) *Suppose Assumption 4.1 with* $V = 0$. *Let* $\Psi, \Phi \in L^2(\mathrm{Q})$. *Then*

$$(\Psi, e^{-tH_{\mathrm{R}}(p)}\Phi) = \mathbb{E}^{0,0}\left[\left(\mathrm{J}_0\Psi, e^{-\mathrm{i}\mathbb{A}(K_t^{\mathrm{R}})}\mathrm{J}_t e^{+\mathrm{i}(p-\mathrm{P_f})B_{T_t}}\Phi\right)\right].$$

We can see the explicit form of the fiber Hamiltonian $H_{\mathrm{R}}(p)$. Let

$$K(p) = (p - \mathrm{P_f} - \hat{A}(0))^2 + m^2.$$

Then we have

$$(\Psi, e^{-tK(p)}\Phi) = e^{-tm^2}\mathbb{E}^{0,0}\left[e^{\mathrm{i}pB_t}\left(\Psi, e^{-\mathrm{i}\hat{A}(L_t)}e^{-\mathrm{i}\mathrm{P_f}B_t}\Phi\right)\right].$$

Let $D = D(\mathrm{P_f^2}) \cap D(H_{\mathrm{f}})$. Set $\bar{K}(p) = \overline{K(p)\lceil_D}$. We define $L_{\mathrm{R}}(p)$ by

$$L_{\mathrm{R}}(p) = \bar{K}(p)^{1/2} \dot{+} H_{\mathrm{f}}, \quad p \in \mathbb{R}^3.$$

Theorem 4.16 ([66]) *Suppose Assumption 4.1 with* $V = 0$. *Then*

$$H_{\mathrm{R}} \cong \int_{\mathbb{R}^3}^{\oplus} L_{\mathrm{R}}(p)\mathrm{d}p.$$

In particular $H_{\mathrm{R}}(p) = L_{\mathrm{R}}(p)$.

5 Pauli–Fierz Model with Spin 1/2

5.1 Pauli–Fierz Hamiltonian with Spin 1/2

In this section we are concerned with the Pauli–Fierz Hamiltonian with spin 1/2. This section is due to [73], and in this section we assume Assumption 3.1 unless otherwise stated. The Hilbert space consisting of state vectors of the Pauli–Fierz Hamiltonian with spin 1/2 is

$$\mathcal{H}_S = L^2(\mathbb{R}^3; \mathbb{C}^2) \otimes L^2(\mathrm{Q}).$$

The Pauli–Fierz Hamiltonian with spin 1/2 is formally given by

$$\frac{1}{2}\left(\sigma \cdot (-\mathrm{i}\nabla - \hat{A})\right)^2 + V + H_{\mathrm{f}} = \frac{1}{2}(-\mathrm{i}\nabla - \hat{A})^2 + V + H_{\mathrm{f}} - \frac{1}{2}\sigma \cdot \hat{B}. \tag{5.1}$$

Here $\sigma = (\sigma_1, \sigma_2, \sigma_3)$ are the 2×2 Pauli matrices, and the quantized magnetic field $\hat{B}$ is defined by the curl of $\hat{A}$ as usual:

$$\hat{B}(x) = (\hat{B}_1(x), \hat{B}_2(x), \hat{B}_3(x)) = \mathrm{curl}_x \hat{A}(x).$$

Both sides of (5.1) are formally identical. It is straightforward to see that

$$\hat{B}_\mu(x) = \sum_{\lambda,\alpha,\nu=1}^{3} \hat{A}_\lambda(\delta_{\lambda\nu}\varepsilon^{\mu\alpha\nu}\nabla_{x_\alpha}\tilde{\varphi}(\cdot - x)), \tag{5.2}$$

where $\varepsilon^{\alpha\beta\gamma}$ denotes the antisymmetric tensor defined by

$$\varepsilon^{\alpha\beta\gamma} = \begin{cases} 1 & \alpha\beta\gamma \text{ is an even permutation of } 123, \\ -1 & \alpha\beta\gamma \text{ is an odd permutation of } 123, \\ 0 & \text{otherwise.} \end{cases}$$

Then

$$\hat{B}_\mu = \frac{1}{\sqrt{2}}\sum_{j=\pm 1}\left\{a^*\left(-\mathrm{i}(k\times e^j)_\mu \frac{\hat{\varphi}}{\sqrt{\omega}}e^{-\mathrm{i}kx}, j\right) + a^*\left(\mathrm{i}(k\times e^j)_\mu \frac{\tilde{\hat{\varphi}}}{\sqrt{\omega}}e^{\mathrm{i}kx}, j\right)\right\}.$$

In this paper the right-hand side of (5.1) is adopted as the definition of the Pauli–Fierz Hamiltonian with spin 1/2:

$$H_\mathrm{S} = \frac{1}{2}(-\mathrm{i}\nabla - \hat{A})^2 + V + H_\mathrm{f} - \frac{1}{2}\sigma\cdot\hat{B}. \tag{5.3}$$

H_S is self-adjoint on $D(-\Delta)\cap D(H_\mathrm{f})$ and bounded from below. Moreover, it is essentially self-adjoint on any core of $H_0 = H_\mathrm{p} + H_\mathrm{f}$.

5.2 *Scalar Representations*

As in the classical case in order to construct a functional integral representation of $(F, e^{-tH_\mathrm{S}}G)$ with a scalar integrand we introduce a two-valued spin variable $\theta \in \mathbb{Z}_2$ and redefine the Pauli–Fierz Hamiltonian with spin 1/2. We identify $\mathcal{H}_\mathrm{S}$ with $L^2(\mathbb{R}^3\times\mathbb{Z}_2)\otimes L^2(\mathrm{Q})$ by

$$\mathcal{H}_\mathrm{S} \ni F = \begin{pmatrix} F(\cdot,+1) \\ F(\cdot,-1)\end{pmatrix} \cong F(\cdot,\theta) \in L^2(\mathbb{R}^3\times\mathbb{Z}_2)\otimes L^2(\mathrm{Q}).$$

Since

$$H_\mathrm{S} = \frac{1}{2}(-\mathrm{i}\nabla - \hat{A})^2 + V + H_\mathrm{f} - \frac{1}{2}\begin{pmatrix} \hat{B}_3 & \hat{B}_1 - \mathrm{i}\hat{B}_2 \\ \hat{B}_1 + \mathrm{i}\hat{B}_2 & -\hat{B}_3 \end{pmatrix},$$

H_S can be regarded as an operator $H_{\mathbb{Z}_2}$ acting on $L^2(\mathbb{R}^3\times\mathbb{Z}_2)\otimes L^2(\mathrm{Q})$:

Table 5 Correspondence between spin 1/2 and $\mathbb{Z}_2$-actions

Spin	$\sigma_1 = \begin{pmatrix} 0 & 1 \\ 1 & 0 \end{pmatrix}$	$\sigma_2 = \begin{pmatrix} 0 & -\mathrm{i} \\ \mathrm{i} & 0 \end{pmatrix}$	$\sigma_3 = \begin{pmatrix} 1 & 0 \\ 0 & -1 \end{pmatrix}$
$\mathbb{Z}_2$-action	$f(\theta) \to f(-\theta)$	$f(\theta) \to -\mathrm{i}\theta f(-\theta)$	$f(\theta) \to \theta f(\theta)$

$$(H_{\mathbb{Z}_2} F)(\theta) = \left(\frac{1}{2}(-\mathrm{i}\nabla - \hat{A})^2 + V + H_{\mathrm{f}} + \hat{H}_{\mathrm{d}}(\theta) \right) F(\theta) + \hat{H}_{\mathrm{od}}(-\theta) F(-\theta).$$

See Table 5.

Here $\hat{H}_{\mathrm{d}}$ and $\hat{H}_{\mathrm{od}}$ denote the diagonal part and the off-diagonal part, respectively, which are explicitly given by

$$\hat{H}_{\mathrm{d}}(x, \theta) = -\frac{1}{2}\theta \hat{B}_3(x), \quad \hat{H}_{\mathrm{od}}(x, -\theta) = -\frac{1}{2}\left(\hat{B}_1(x) - i\theta \hat{B}_2(x) \right), \quad \theta \in \mathbb{Z}_2.$$

Write

$$\mathcal{K} = L^2(\mathbb{R}^3 \times \mathbb{Z}_2) \otimes L^2(\mathrm{Q}).$$

We define the Pauli–Fierz Hamiltonian with spin 1/2 on $\mathcal{K}$ by $H_{\mathbb{Z}_2}$, and consider a functional integral representation of $(F, e^{-tH_{\mathbb{Z}_2}} G)$ for $(F, e^{-tH_{\mathrm{S}}} G)$.

5.3 Functional Integral Representations for the Pauli–Fierz Hamiltonian with Spin 1/2

The idea of constructing a functional integral representation of $e^{-tH_{\mathbb{Z}_2}}$ is to use the identification:

$$\mathcal{K} \cong \int_{\mathrm{Q}}^{\oplus} L^2(\mathbb{R}^3 \times \mathbb{Z}_2) \mathrm{d}\mu(\phi). \tag{5.4}$$

In other words, we regard $\mathcal{K}$ as the set of $L^2(\mathbb{R}^3 \times \mathbb{Z}_2)$-valued L^2-functions on Q. We make the decomposition

$$H_{\mathbb{Z}_2} = \int_{\mathrm{Q}}^{\oplus} K(\phi) \mathrm{d}\mu(\phi) + H_{\mathrm{f}},$$

where $K(\phi)$ is a self-adjoint operator on $L^2(\mathbb{R}^3 \times \mathbb{Z}_2)$ for each $\phi \in \mathrm{Q}$. For each $\phi \in \mathrm{Q}$, we define the Hamiltonian $K(\phi)$ on $L^2(\mathbb{R}^3 \times \mathbb{Z}_2)$ by

$$(K(\phi)F)(x, \theta) = \left(\frac{1}{2}(-\mathrm{i}\nabla - \hat{A}(\phi))^2 + V + \hat{H}_{\mathrm{d}}(\theta, \phi) \right) F(x, \theta) + \hat{H}_{\mathrm{od}}(-\theta, \phi) F(x, -\theta). \tag{5.5}$$

We construct the functional integral representation of $e^{-tH_{\mathbb{Z}_2}}$ through functional integral representations of both $e^{-tK(\phi)}$ and e^{-tH_f}, and the Trotter product formula. In order to do that we will use the identity

$$(F, e^{-t\int_Q^\oplus K(\phi)d\mu}G)_{\mathcal{K}} = \int_Q (F(\phi), e^{-tK(\phi)}G(\phi))_{L^2(\mathbb{R}^3\times\mathbb{Z}_2)} d\mu(\phi),$$

while we have already done that of $(F(\phi), e^{-tK(\phi)}G(\phi))$, $\phi \in Q$, in the classical case $(f, e^{-th_S(a,b)}g)$ introduced in Sect. 2.1. To prevent the off-diagonal part $\hat{H}_{od}$ vanishes we introduce a regularization $H_{\mathbb{Z}_2,\varepsilon}$ of $H_{\mathbb{Z}_2}$ by

$$H_{\mathbb{Z}_2,\varepsilon}F(\theta) = \left(\frac{1}{2}(-i\nabla - \hat{A})^2 + V + H_f + \hat{H}_d(\theta)\right)F(\theta) + \Psi_\varepsilon(\hat{H}_{od}(-\theta))F(-\theta),$$

where $\Psi_\varepsilon(X) = X + \varepsilon\psi_\varepsilon(|X|)$ and $\psi_\varepsilon \in C_0^\infty(\mathbb{R})$ is given by

$$\psi_\varepsilon(z) = \psi_\varepsilon(|z|) = \begin{cases} 1, & |z| < \varepsilon/2, \\ \leq 1, & \varepsilon/2 \leq |z| \leq \varepsilon, \\ 0, & |z| > \varepsilon. \end{cases}$$

Also, let $K_\varepsilon(\phi)$ be the counterpart of $K(\phi)$ with $\hat{H}_{od}(\phi)$ replaced by $\Psi_\varepsilon(\hat{H}_{od}(\phi))$, i.e.,

$$(K_\varepsilon(\phi)F)(x,\theta) = \left(\frac{1}{2}(-i\nabla - \hat{A}(\phi))^2 + V + \hat{H}_d(\theta,\phi)\right)F(x,\theta) + \Psi_\varepsilon(\hat{H}_{od}(-\theta,\phi))F(x,-\theta).$$

Recall that $\theta_t = (-1)^{N_t}$ and $(X_t)_{t\geq 0} = (B_t, \theta_t)_{t\geq 0}$ is the $(\mathbb{R}^3 \times \mathbb{Z}_2)$-valued random process on $\mathscr{X} \times \mathscr{X}_\mu$. If $\tilde{\varphi} \in C_0^\infty(\mathbb{R}^3)$, then for every $\phi \in Q$, $K_\varepsilon(\phi)$ is self-adjoint on $D(-\Delta) \otimes \mathbb{Z}_2$ and for $g \in L^2(\mathbb{R}^3 \times \mathbb{Z}_2)$,

$$(e^{-tK_\varepsilon(\phi)}g)(x,\alpha) = e^t\mathbb{E}^{x,\alpha}[e^{-\int_0^t V(B_s)ds}e^{Z_t(\phi,\varepsilon)}g(X_t)], \quad (x,\alpha) \in \mathbb{R}^3 \times \mathbb{Z}_2,$$

where

$$\begin{aligned} Z_t(\phi,\varepsilon) = & -i\int_0^t \hat{A}(\tilde{\varphi}(\cdot - B_s), \phi)dB_s \\ & - \int_0^t \hat{H}_d(B_s, \theta_{N_s}, \phi)ds + \int_0^{t+} \log\left(-\Psi_\varepsilon(\hat{H}_{od}(B_s, -\theta_{N_{s-}}, \phi))\right)dN_s. \end{aligned}$$

Next we define an operator $K_\varepsilon(\hat{A})$ on $\mathcal{K}$ through $K_\varepsilon(\phi)$ and the constant fiber direct integral representation (5.4) of $\mathcal{K}$. Take $\tilde{\varphi} \in C_0^\infty(\mathbb{R}^3)$ and define the self-adjoint operator $K_\varepsilon(\hat{A})$ on $\mathcal{K}$ by

$$K_\varepsilon(\hat{A}) = \int_Q^\oplus K_\varepsilon(\phi)\mathrm{d}\mu(\phi).$$

Then we can define the self-adjoint operator K_ε by

$$K_\varepsilon = K_\varepsilon(\hat{A}) \dot{+} H_\mathrm{f}.$$

In what follows we construct a functional integral representation of e^{-tK_ε} and show that $e^{-tK_\varepsilon} = e^{-tH_{\mathbb{Z}_2,\varepsilon}}$. Let us define a dense subspace by $\mathcal{K}_\infty = C_0^\infty(\mathbb{R}^3 \times \mathbb{Z}_2)\hat{\otimes}L^2_{\mathrm{fin}}(\mathrm{Q})$.

Lemma 5.1 *Under Assumption 3.1 it follows that*

$$s - \lim_{\varepsilon\downarrow 0} e^{-tH_{\mathbb{Z}_2,\varepsilon}} = e^{-tH_{\mathbb{Z}_2}}. \tag{5.6}$$

Suppose that $\tilde{\varphi} \in C_0^\infty(\mathbb{R}^3)$. *Then* $H_{\mathbb{Z}_2,\varepsilon} = K_\varepsilon$ *and in particular it follows that*

$$(F, e^{-tH_{\mathbb{Z}_2}}G) = \lim_{\varepsilon\downarrow 0}(F, e^{-tK_\varepsilon}G). \tag{5.7}$$

Proof It is seen that $K_\varepsilon = H_{\mathbb{Z}_2,\varepsilon}$ on $\mathcal{K}_\infty$, implying that $K_\varepsilon = H_{\mathbb{Z}_2,\varepsilon}$ as a self-adjoint operator since $\mathcal{K}_\infty$ is a core of $H_{\mathbb{Z}_2,\varepsilon}$. Moreover, $H_{\mathbb{Z}_2,\varepsilon} \to H_{\mathbb{Z}_2}$ on $\mathcal{K}_\infty$ as $\varepsilon \to 0$ and $\mathcal{K}_\infty$ is a common core of the sequence $\{H_{\mathbb{Z}_2,\varepsilon}\}_{\varepsilon\geq 0}$. Thus (5.6) and (5.7) follow.

QED

By Lemma 5.1 to give the functional integral representation of $(F, e^{-tH_{\mathbb{Z}_2}}G)$ for any $\hat{\varphi}$ given in Assumption 3.1 it suffices to construct a functional integral representation of the right-hand side of (5.7) and to take an approximation argument on $\hat{\varphi}$. To obtain the functional integral representation of e^{-tK_ε}, we apply the Trotter product formula as usual, i.e.,

$$e^{-tK_\varepsilon} = s - \lim_{n\to\infty}(e^{-\frac{t}{n}K_\varepsilon(\hat{A})}e^{-\frac{t}{n}H_\mathrm{f}})^n = s - \lim_{n\to\infty} \mathrm{J}_0^*\Big(\prod_{i=0}^{n-1} \mathrm{J}_{\frac{it}{n}} e^{-\frac{t}{n}K_\varepsilon(\hat{A})}\mathrm{J}_{\frac{it}{n}}^*\Big)\mathrm{J}_t.$$

The Euclidean version of $\hat{B}_\mu(g)$ with test function $g \in L^2(\mathbb{R}^4)$ is defined by

$$\mathbb{B}_\mu(g) = \sum_{\lambda,\alpha,\nu=1}^{3} \mathbb{A}_\lambda(\delta_{\lambda\nu}\varepsilon^{\mu\alpha\nu}\nabla_{x_\alpha}g). \tag{5.8}$$

Here $\nabla_{x_\alpha}g = \nabla_{x_\alpha}g(\cdot - x)\lceil_{x=0}$, hence $\widehat{\nabla_{x_\alpha}g} = -ik_\alpha\hat{g}(k)$. Define the Euclidean versions of $\hat{H}_\mathrm{d}(x,\theta)$ and $\hat{H}_\mathrm{od}(x,-\theta)$ by

$$\mathbb{H}_{\mathrm{d}}(x,\theta,s) = -\frac{1}{2}\theta\mathbb{B}_3\left(\mathrm{j}_s\tilde{\varphi}(\cdot - x)\right), \tag{5.9}$$

$$\mathbb{H}_{\mathrm{od}}(x,-\theta,s) = -\frac{1}{2}(\mathbb{B}_1(\mathrm{j}_s\tilde{\varphi}(\cdot - x)) - \mathrm{i}\theta\mathbb{B}_2(\mathrm{j}_s\tilde{\varphi}(\cdot - x))), \tag{5.10}$$

respectively.

5.4 Technical Estimates

This section is a brief version of [74, Sects. 3.8.7 and 3.8.8]. To avoid complicated computations we introduce self-adjoint operators L and L_R for $R > 0$ which satisfy that

$$(|F|, e^{-tL_R}|G|)| \uparrow (|F|, e^{-tL}|G|) \quad \text{as } R \uparrow \infty$$

and diamagnetic type inequality:

$$|(F, e^{-tH_{\mathbb{Z}_2}}G)| \le (|F|, e^{-tL}|G|).$$

See Table 6. Applying this inequality we can avoid technical difficulties to construct the functional integral representation of $(F, e^{-tH_{\mathbb{Z}_2}}G)$. We introduce two cutoff functions for each $R > 0$. Let $\chi_R^- \in C^\infty(\mathbb{R})$ be defined by

(1) $\chi_R^-(x) = x$ for $x > -R+1$,
(2) $-R \le \chi_R^-(x) \le -R+1$ for $-R \le x \le -R+1$,
(3) $\chi_R^-(x) = -R$ for $x < -R$.

We also define $\chi_R^+ \in C^\infty(\mathbb{R})$ by

(1) $\chi_R^+(x) = R$ for $x > R$,
(2) $R-1 \le \chi_R^+(x) \le R$ for $R-1 \le x \le R$,
(3) $\chi_R^+(x) = x$ for $x < R-1$.

Furthermore note that $|\chi_R^\pm(x) - \chi_R^\pm(y)| \le c|x-y|$ with some c for any $x, y \in \mathbb{R}$. We define $\Psi_\varepsilon(|\hat{H}_{\mathrm{od}}(-\theta)|)_R$ and $\hat{H}_{\mathrm{d}}(\theta)_R$ by

Table 6 List of operators and exponents

Operator	Exponent
e^{-tM_R}	$L(t,\varepsilon,R) = -\int_0^t \hat{H}_{\mathrm{d}}(B_s,\theta_{N_s})_R\mathrm{d}s + \int_0^{t+}\log\Psi_\varepsilon(\lvert\hat{H}_{\mathrm{od}}(B_s,-\theta_{N_s})\rvert)_R\mathrm{d}N_s$
$\mathrm{J}_s e^{-tM_R}\mathrm{J}_s^*$	$L(t,s,\varepsilon,R) = -\int_0^t \mathbb{H}_{\mathrm{d}}(B_r,\theta_{N_r},s)_R\mathrm{d}r + \int_0^{t+}\log\Psi_\varepsilon(\lvert\mathbb{H}_{\mathrm{od}}(B_r,-\theta_{N_r},s)\rvert)_R\mathrm{d}N_r$
e^{-tL_R}	$Y_R = -\int_0^t \mathbb{H}_{\mathrm{d}}(B_s,\theta_{N_s},s)_R\mathrm{d}s + \int_0^{t+}\log\Psi_\varepsilon(\lvert\mathbb{H}_{\mathrm{od}}(B_s,-\theta_{N_s},s)\rvert)_R\mathrm{d}N_s$
e^{-tL}	$Y = -\int_0^t \mathbb{H}_{\mathrm{d}}(B_s,\theta_{N_s},s)\mathrm{d}s + \int_0^{t+}\log\Psi_\varepsilon(\lvert\mathbb{H}_{\mathrm{od}}(B_s,-\theta_{N_s},s)\rvert)\mathrm{d}N_s$
$e^{-tH_{\mathbb{Z}_2,\varepsilon}}$	$Z_t(\varepsilon) = -\mathrm{i}\mathbb{A}(K_t) - \int_0^t \mathbb{H}_{\mathrm{d}}(B_s,\theta_{N_s},s)\mathrm{d}s + \int_0^{t+}\log\Psi_\varepsilon(-\mathbb{H}_{\mathrm{od}}(B_s,-\theta_{N_s},s))\mathrm{d}N_s$
$e^{-tH_{\mathbb{Z}_2}}$	$Z_t = -\mathrm{i}\mathbb{A}(K_t) - \int_0^t \mathbb{H}_{\mathrm{d}}(B_s,\theta_{N_s},s)\mathrm{d}s + \int_0^{t+}\log\Psi_\varepsilon(-\mathbb{H}_{\mathrm{od}}(B_s,-\theta_{N_s},s))\mathrm{d}N_s$

$$\Psi_\varepsilon(|\hat{H}_{\rm od}(-\theta)|)_R = \chi_R^+\left(\Psi_\varepsilon(|\hat{H}_{\rm od}(-\theta)|)\right), \tag{5.11}$$

$$\hat{H}_{\rm d}(\theta)_R = \chi_R^-\left(\hat{H}_{\rm d}(\theta)\right). \tag{5.12}$$

Hence $-R \le \hat{H}_{\rm d}(\theta)_R$ and $\varepsilon \le \Psi_\varepsilon(|\hat{H}_{\rm od}(-\theta)|)_R \le R$. Let $\tilde{\varphi} \in C_0^\infty(\mathbb{R}^3)$. We set for each $\phi \in$ Q,

$$(M_R(\phi)F)(\theta) = (H_{\rm p} + \hat{H}_{\rm d}(\phi,\theta)_R)F(\theta) - \Psi_\varepsilon(|\hat{H}_{\rm od}(\phi,-\theta)|)_R F(-\theta)$$

for each $R > 0$, and

$$(M(\phi)F)(\theta) = (H_{\rm p} + \hat{H}_{\rm d}(\phi,\theta))F(\theta) - \Psi_\varepsilon(|\hat{H}_{\rm od}(\phi,-\theta)|)F(-\theta).$$

Then $M_R(\phi)$ and $M(\phi)$ are self-adjoint on $L^2(\mathbb{R}^3 \times \mathbb{Z}_2)$ and we define

$$M_R = \int_{\rm Q}^{\oplus} M_R(\phi) {\rm d}\mu(\phi), \quad M = \int_{\rm Q}^{\oplus} M(\phi) {\rm d}\mu(\phi).$$

We also define

$$L_R = M_R \dot{+} H_{\rm f}, \qquad L = M \dot{+} H_{\rm f}.$$

In the Fock representation L_R and L are given by

$$L_R = H_{\rm p} + \hat{H}_{\rm f} - \begin{pmatrix} (\frac{1}{2}\hat{B}_3)_R & \Psi_\varepsilon\left(\frac{1}{2}\sqrt{\hat{B}_1^2+\hat{B}_2^2}\right)_R \\ \Psi_\varepsilon\left(\frac{1}{2}\sqrt{\hat{B}_1^2+\hat{B}_2^2}\right)_R & -(\frac{1}{2}\hat{B}_3)_R \end{pmatrix},$$

$$L = H_{\rm p} + \hat{H}_{\rm f} - \begin{pmatrix} \frac{1}{2}\hat{B}_3 & \Psi_\varepsilon\left(\frac{1}{2}\sqrt{\hat{B}_1^2+\hat{B}_2^2}\right) \\ \Psi_\varepsilon\left(\frac{1}{2}\sqrt{\hat{B}_1^2+\hat{B}_2^2}\right) & -\frac{1}{2}\hat{B}_3 \end{pmatrix}.$$

In the Fock representation we have $|\hat{H}_{\rm od}(-\theta)| = \frac{1}{2}\sqrt{\hat{B}_1^2+\hat{B}_2^2}$ which leads that $|\hat{H}_{\rm od}(-\theta)|$ is independent of θ. The family of self-adjoint operators L_R, $R > 0$, have also a common core $\mathcal{K}_\infty$ and $\lim_{R\to\infty} L_R F = LF$ for $F \in \mathcal{K}_\infty$. Hence $e^{-tL_R} \to e^{-tL}$ strongly as $R \to \infty$. A functional integral representation of e^{-tL_R} can be done by the Trotter product formula.

5.4.1 Estimate of $\mathrm{J}_s e^{-tM_R}\mathrm{J}_s^*$

Let

$$L(t,s,\varepsilon,R) = -\int_0^t \mathbb{H}_{\mathrm{d}}(B_r,\theta_{N_r},s)_R \mathrm{d}r + \int_0^{t+} \log \Psi_\varepsilon(|\mathbb{H}_{\mathrm{od}}(B_r,-\theta_{N_r},s)|)_R \mathrm{d}N_r.$$

Here $\mathbb{H}_{\mathrm{d}}(x,\theta,s)$ and $\mathbb{H}_{\mathrm{od}}(x,-\theta,s)$ are given by (5.9) and (5.10), respectively, and truncated functions $\mathbb{H}_{\mathrm{d}}(B_r,\theta_{N_r},s)_R$ and $\Psi_\varepsilon(|\mathbb{H}_{\mathrm{od}}(x,-\theta,s)|)_R$ are defined in a similar way to (5.11) and (5.12), respectively. Let $\mathcal{K}_{\mathrm{E}} = L^2(\mathbb{R}^3)\otimes L^2(\mathrm{Q_E})$.

Lemma 5.2 *Assume that $V \in L^\infty(\mathbb{R}^3)$ and $\tilde{\varphi} \in C_0^\infty(\mathbb{R}^3)$. Let $F, G \in \mathcal{K}_{\mathrm{E}}$. Then*

$$\int_{\mathbb{R}^3} \sum_{\alpha\in\mathbb{Z}_2} \mathbb{E}^{x,\alpha}\left[e^{-\int_0^t V(B_r)\mathrm{d}r}(F(q_0), \mathrm{E}_s e^{L(t,s,\varepsilon,R)}\mathrm{E}_s G(q_t))\right]\mathrm{d}x$$

is finite and

$$(F, \mathrm{J}_s e^{-tM_R}\mathrm{J}_s^* G) = e^t \int_{\mathbb{R}^3} \sum_{\alpha\in\mathbb{Z}_2} \mathbb{E}^{x,\alpha}\left[e^{-\int_0^t V(B_r)\mathrm{d}r}(F(q_0), \mathrm{E}_s e^{L(t,s,\varepsilon,R)}\mathrm{E}_s G(q_t))\right]\mathrm{d}x. \tag{5.13}$$

Proof Let us set $V_\infty = \sup_{x\in\mathbb{R}^3} \mathbb{E}^x[e^{-2\int_0^t V(B_s)\mathrm{d}s}] < \infty$. Notice that the right-hand side of (5.13) is finite, since $e^{L(t,s,\varepsilon,R)}$ is a bounded function. For each $(x,\alpha,w,m) \in \mathbb{R}^3\times\mathbb{Z}_2\times\mathscr{X}\times\mathscr{X}_\mu$, we have

$$|(F(q_0), \mathrm{E}_s e^{L(t,s,\varepsilon,R)}\mathrm{E}_s G(q_t))| \le \|F(q_0)\|\|G(q_t)\|\|e^{L(t,s,\varepsilon,R)}\|.$$

Here $\|e^{L(t,s,\varepsilon,R)}\|$ is the operator norm of bounded operator $e^{L(t,s,\varepsilon,R)}$ on $L^2(\mathrm{Q_E})$. Then it follows that

$$|\mathrm{RHS}(5.13)| \le e^t \int_{\mathbb{R}^3} \sum_{\alpha\in\mathbb{Z}_2} \mathbb{E}^{x,\alpha}\left[e^{-\int_0^t V(B_r)\mathrm{d}r}\|F(q_0)\|\|G(q_t)\|\|e^{L(t,s,\varepsilon,R)}\|\right]\mathrm{d}x. \tag{5.14}$$

We will also prove in Lemma 5.3 below that there exists a random variable A_t on $(\mathscr{X}_\mu, \mathcal{B}_\mu, \mu)$ such that

(1) $\|e^{L(t,s,\varepsilon,R)}\| \le A_t$,
(2) A_t is independent of $(x,\alpha,w) \in \mathbb{R}^3\times\mathbb{Z}_2\times\mathscr{X}$,
(3) $\mathbb{E}_\mu[A_t^2] < \infty$.

By (1), (2) and (3) above and (5.14),

$$|\mathrm{RHS}\ (5.13)| \le \|G\|\|F\|V_\infty^{1/2}(\mathbb{E}_\mu[A_t^2])^{1/2} < \infty. \tag{5.15}$$

Next we prove the equality (5.13). Note that M_R is defined by a direct fiber integral representation we have

$$(\mathrm{J}_s^* F, e^{-tM_R}\mathrm{J}_s^* G) = e^t \int_{\mathbb{R}^3} \sum_{\alpha\in\mathbb{Z}_2} \mathbb{E}^{x,\alpha}\left[e^{-\int_0^t V(B_r)\mathrm{d}r}\mathbb{E}_{\mu_{\mathrm{E}}}\left[\overline{(\mathrm{J}_s^* F)(\phi, q_0)} e^{L_t(\phi,\varepsilon)}(\mathrm{J}_s^* G)(\phi, q_t)\right]\right]\mathrm{d}x.$$

Here we used Fubini's lemma and

$$L_t(\phi, \varepsilon) = -\int_0^t \hat{H}_{\mathrm{d}}(\phi, B_s, \theta_{N_s})_R \mathrm{d}s + \int_0^{t+} \log \Psi_\varepsilon(|\hat{H}_{\mathrm{od}}(\phi, B_s, -\theta_{N_s})|)_R \mathrm{d}N_s.$$

Let

$$L(t, \varepsilon, R) = -\int_0^t \hat{H}_{\mathrm{d}}(B_s, \theta_{N_s})_R \mathrm{d}s + \int_0^{t+} \log \Psi_\varepsilon(|\hat{H}_{\mathrm{od}}(B_s, -\theta_{N_s})|)_R \mathrm{d}N_s.$$

We rewrite as

$$(\mathrm{J}_s^* F, e^{-tM_R}\mathrm{J}_s^* G) = e^t \int_{\mathbb{R}^3} \sum_{\alpha\in\mathbb{Z}_2} \mathbb{E}^{x,\alpha}\left[e^{-\int_0^t V(B_r)\mathrm{d}r}(F(q_0), \mathrm{J}_s e^{L(t,\varepsilon,R)}\mathrm{J}_s^* G(q_t))\right]\mathrm{d}x.$$

We can see that $\mathrm{J}_s e^{L(t,\varepsilon,R)}\mathrm{J}_s^* = \mathrm{E}_s e^{L(t,s,\varepsilon,R)}\mathrm{E}_s$ by a limiting argument, leading to (5.13). Then the proof can be completed.

QED

Now we have to prove (1), (2) and (3) used in the proof of Lemma 5.2.

Lemma 5.3 *For each $t \geq 0$, operator $e^{L(t,s,\varepsilon,R)}$ is bounded, and there exists a random variable A_t on $(\mathscr{X}_\mu, \mathcal{B}_\mu, \mu)$ satisfying* (1)–(3) *in the proof of Lemma* 5.2.

Proof Since $R < \infty$, we have $|e^{L(t,s,\varepsilon,R)}| \leq e^{tR} e^{\int_0^{t+} \log R \mathrm{d}N_s}$. For each $m \in \mathscr{X}_\mu$, the number of jumps of map $s \mapsto N_s(m)$ for $0 \leq s \leq t$ is denoted by $N(m)$. Hence $\int_0^{t+} \log R\mathrm{d}N_s = \log R^{N(m)}$ and then $|e^{L(t,s,\varepsilon,R)}| \leq e^{tR} R^{N(m)}$. Set $A_t = A_t(m) = e^{tR} R^{N(m)}$. Then $\mathbb{E}_\mu[A_t^2] = e^{2tR} \sum_{N=0}^\infty \frac{t^N R^{2N}}{N!} e^{-t} = e^{t(R^2+2R-1)} < \infty$. Then A_t satisfies (1), (2) and (3).

QED

5.4.2 Estimate of e^{-tL_R}

In the next lemma we construct a functional integral representation of e^{-tL_R} for $R > 0$. Let

$$Y_R = -\int_0^t \mathbb{H}_{\mathrm{d}}(B_s, \theta_{N_s}, s)_R \mathrm{d}s + \int_0^{t+} \log \Psi_\varepsilon(|\mathbb{H}_{\mathrm{od}}(B_s, -\theta_{N_{s-}}, s)|)_R \mathrm{d}N_s.$$

Lemma 5.4 *Suppose that* $V \in L^\infty(\mathbb{R}^3)$ *and* $\tilde{\varphi} \in C_0^\infty(\mathbb{R}^3)$. *Then for every* $t \geq 0$ *and all* $F, G \in \mathcal{K}$ *it follows that* $\mathrm{J}_0^* e^{Y_R} \mathrm{J}_t$ *is a bounded operator on* $L^2(\mathrm{Q})$ *for every* $(x, \alpha, w, m) \in \mathbb{R}^3 \times \mathbb{Z}_2 \times \mathscr{X} \times \mathscr{X}_\mu$ *and*

$$(F, e^{-tL_R} G) = e^t \int_{\mathbb{R}^3} \sum_{\alpha \in \mathbb{Z}_2} \mathbb{E}^{x,\alpha} \left[e^{-\int_0^t V(B_s) \mathrm{d}s} (F(q_0), \mathrm{J}_0^* e^{Y_R} \mathrm{J}_t G(q_t)) \right] \mathrm{d}x. \quad (5.16)$$

Proof In a similar manner to the proof of Lemma 5.3 for each $m \in \mathscr{X}_\mu$, the number of jumps of map $s \mapsto N_s(m)$ for $0 \leq s \leq t$ is denoted by $N(m)$. Fix $m \in \mathscr{X}_\mu$. Hence by virtue of cutoff parameter $R > 0$ it can be seen that $\|\mathrm{J}_0^* e^{Y_R} \mathrm{J}_t \Phi\| \leq e^{tR} R^{N(m)} \|\Phi\|$, which implies that $\mathrm{J}_0^* e^{Y_R} \mathrm{J}_t$ is bounded for every $(x, \alpha, w, m) \in \mathbb{R}^3 \times \mathbb{Z}_2 \times \mathscr{X} \times \mathscr{X}_\mu$.

By the Trotter product formula, the Markov property of $\mathrm{J}_s \mathrm{J}_s^* = \mathrm{E}_s$ and Lemma 5.2 we can see that

$$(F, e^{-tL_R} G) = \lim_{n\to\infty} (F, \prod_{j=0}^{n} \left(e^{-\frac{t}{n} M_R} e^{-\frac{t}{n} H_\mathrm{f}} \right)^n G) = \lim_{n\to\infty} (F, \mathrm{J}_0^* \prod_{j=0}^{n-1} \left(\mathrm{J}_{\frac{jt}{n}} e^{-\frac{t}{n} M_R} \mathrm{J}_{\frac{jt}{n}}^* \right) \mathrm{J}_t G)$$

$$= \lim_{n\to\infty} \sum_{\alpha \in \mathbb{Z}_2} \int_{\mathbb{R}^3} \mathbb{E}^{x,\alpha} \left[e^{-\int_0^t V(B_s) \mathrm{d}s} (F(q_0), \mathrm{J}_0^* e^{Y_R(n)} \mathrm{J}_t G(q_t)) \right] \mathrm{d}x, \quad (5.17)$$

where

$$Y_R(n) = -\sum_{i=1}^{n} \int_{\frac{(i-1)t}{n}}^{\frac{it}{n}} \mathbb{H}_\mathrm{d} \left(B_s, \theta_{N_s}, \frac{(i-1)t}{n} \right)_R \mathrm{d}s$$
$$+ \sum_{i=1}^{n} \int_{\frac{(i-1)t}{n}}^{\frac{it}{n}+} \log \Psi_\varepsilon \left(\left| \mathbb{H}_\mathrm{od} \left(B_s, -\theta_{N_{s-}}, \frac{(i-1)t}{n} \right) \right| \right)_R \mathrm{d}N_s.$$

Then

$$\left| \sum_{\alpha \in \mathbb{Z}_2} \int_{\mathbb{R}^3} \mathbb{E}^{x,\alpha} \left[e^{-\int_0^t V(B_s) \mathrm{d}s} (F(q_0), \mathrm{J}_0^* e^{Y_R(n)} \mathrm{J}_t G(q_t)) \right] \mathrm{d}x \right|$$
$$\leq \sum_{\alpha \in \mathbb{Z}_2} \int_{\mathbb{R}^3} \mathbb{E}^{x,\alpha} \left[e^{-\int_0^t V(B_s) \mathrm{d}s} \|F(q_0)\| \|G(q_t))\| e^{tR} R^{N(\cdot)} \right] \mathrm{d}x \leq \|F\| \|G\| V_\infty^{1/2} e^{\frac{t}{2}(R^2+2R-1)} < \infty.$$
(5.18)

Since the right-hand side above is independent of n, it is enough to prove (5.16) for $F, G \in \mathcal{K}_\infty$, hence we suppose that $F, G \in \mathcal{K}_\infty$ in what follows in this proof. We can see that

$$\left\| \sum_{i=1}^{n} \int_{\frac{(i-1)t}{n}}^{\frac{it}{n}} \mathbb{H}_\mathrm{d} \left(B_s, \theta_{N_s}, \frac{(i-1)t}{n} \right)_R \mathrm{d}s F \right\| \leq tC \| |k| \hat{\varphi} / \sqrt{\omega} \| \| (N + \mathbb{1})^{1/2} F \|$$

and similarly

$$\left\| \int_0^t \mathbb{H}_{\rm d}\left(B_s, \theta_{N_s}, s\right)_R {\rm d}s F \right\| \le tC \| |k|\hat{\varphi}/\sqrt{\omega} \| \|(N+\mathbb{1})^{1/2} F\|$$

with some constant C. By $|\chi_R^-(x) - \chi_R^-(y)| \le c|x-y|$ with some constant c, we can also see that

$$\begin{aligned} &\left\| \left(\sum_{i=1}^n \int_{\frac{(i-1)t}{n}}^{\frac{it}{n}} \mathbb{H}_{\rm d}\left(B_s, \theta_{N_s}, \frac{(i-1)t}{n}\right)_R {\rm d}s - \int_0^t \mathbb{H}_{\rm d}\left(B_s, \theta_{N_s}, s\right)_R {\rm d}s \right) F \right\| \\ &\le c \sum_{i=1}^n \left(\int_{\frac{(i-1)t}{n}}^{\frac{it}{n}} \left\| |\mathbb{H}_{\rm d}\left(B_s, \theta_{N_s}, \frac{(i-1)t}{n}\right) - \mathbb{H}_{\rm d}(B_s, \theta_{N_s}, s)| F \right\|^2 {\rm d}s \right)^{1/2} \frac{\sqrt{t}}{\sqrt{n}}. \end{aligned}$$

It can be seen that

$$\left\| \left| \mathbb{H}_{\rm d}\left(B_s, \theta_{N_s}, \frac{(i-1)t}{n}\right) - \mathbb{H}_{\rm d}(B_s, \theta_{N_s}, s) \right| F \right\|^2 \le c' \frac{t}{n} \| |k|\hat{\varphi}\|^2 \|(N+\mathbb{1})^{1/2} F\|^2$$

with some constant c'. Then

$$\begin{aligned} &\left\| \left(\sum_{i=1}^n \int_{\frac{(i-1)t}{n}}^{\frac{it}{n}} \mathbb{H}_{\rm d}\left(B_s, \theta_{N_s}, \frac{(i-1)t}{n}\right)_R {\rm d}s - \int_0^t \mathbb{H}_{\rm d}(B_s, \theta_{N_s}, s)_R {\rm d}s \right) F \right\| \\ &\le cc' \| |k|\hat{\varphi}\| \frac{t\sqrt{t}}{\sqrt{n}} \|(N+\mathbb{1})^{1/2} F\| \to 0 \end{aligned}$$

as $n \to \infty$. Thus it can be straightforwardly seen that

$$\exp\left(-\sum_{i=1}^n \int_{\frac{(i-1)t}{n}}^{\frac{it}{n}} \mathbb{H}_{\rm d}\left(B_s, \theta_{N_s}, \frac{(i-1)t}{n}\right)_R {\rm d}s \right) F \to \exp\left(-\int_0^t \mathbb{H}_{\rm d}(B_s, \theta_{N_s}, s)_R {\rm d}s \right) F. \tag{5.19}$$

Next we consider

$$\exp\left(\sum_{i=1}^n \int_{\frac{(i-1)t}{n}}^{\frac{it}{n}+} \log \Psi_\varepsilon(|\mathbb{H}_{\rm od}\left(B_s, -\theta_{N_{s-}}, \frac{(i-1)t}{n})|\right)_R {\rm d}N_s \right).$$

Points of discontinuity of map $r \mapsto N_r(m)$ are denoted by $s_1 = s_1(m), \ldots, s_N = s_N(m) \in (0, \infty)$. For a sufficiently large n the number of discontinuous points in interval $(\frac{(i-1)t}{n}, \frac{it}{n}]$ is at most one. Then by taking n large enough and denoting $(n(s_i), n(s_i) + \frac{t}{n}]$ for the interval containing s_i, we get

$$\exp\left(\sum_{i=1}^n \int_{\frac{(i-1)t}{n}}^{\frac{it}{n}+} \log \Psi_\varepsilon\left(|\mathbb{H}_{\rm od}\left(B_s, -\theta_{N_{s-}}, \frac{(i-1)t}{n}\right)| \right)_R {\rm d}N_s \right) = \prod_{i=1}^N \chi_R^+(|\phi_i| + \varepsilon\psi_\varepsilon(|\phi_i|)),$$

where $\phi_i = \mathbb{H}_{\mathrm{od}}(B_{s_i}, -\theta_{N_{s_i-}}, n(s_i))$. By the Lipschitz continuity of χ_R^+ and ψ_ε we have

$$\left\| \left(\Psi_\varepsilon(|\mathbb{H}_{\mathrm{od}}(B_s, -\theta_{N_{s-}}, n(s_i))|)_R - \Psi_\varepsilon(|\mathbb{H}_{\mathrm{od}}(B_s, -\theta_{N_{s-}}, s_i)|)_R\right) F \right\|$$
$$\leq C \| \left(|\mathbb{H}_{\mathrm{od}}(B_s, -\theta_{N_{s-}}, n(s_i))| - |\mathbb{H}_{\mathrm{od}}(B_s, -\theta_{N_{s-}}, s_i)|\right) F \|.$$

Note that

$$\| \left(|\mathbb{H}_{\mathrm{od}}(B_s, -\theta_{N_{s-}}, n(s_i))| - |\mathbb{H}_{\mathrm{od}}(B_s, -\theta_{N_{s-}}, s_i)|\right) F \|^2$$
$$\leq C \| |k| \hat{\varphi} \|^2 |n(s_i) - s_i| \| (N + \mathbb{1})^{1/2} F \|^2$$

with some $C > 0$. Clearly, $n(s_i) \to s_i$ as $n \to \infty$. Then

$$\lim_{n\to\infty} \mathbb{H}_{\mathrm{od}}(B_{s_i}, -\theta_{N_{s_i-}}, n(s_i)) = \mathbb{H}_{\mathrm{od}}(B_{s_i}, -\theta_{N_{s_i-}}, s_i)$$

on $\mathcal{K}_\infty$ and we have

$$\lim_{n\to\infty} \exp\left(\sum_{i=1}^n \int_{\frac{(i-1)t}{n}}^{\frac{it}{n}+} \log \Psi_\varepsilon \left(|\mathbb{H}_{\mathrm{od}} \left(B_s, -\theta_{N_{s-}}, \frac{(i-1)t}{n} \right)| \right)_R \mathrm{d}N_s \right)$$
$$= \prod_{i=1}^N \Psi_\varepsilon(|\mathbb{H}_{\mathrm{od}}(B_{s_i}, -\theta_{N_{s_i-}}, s_i)|)_R = \exp\left(\int_0^{t+} \log \Psi_\varepsilon(|\mathbb{H}_{\mathrm{od}}(B_s, -\theta_{N_{s-}}, s|)_R \mathrm{d}N_s \right). \tag{5.20}$$

Then by (5.19) and (5.20) we can see that for each $(x, \alpha, w, m) \in \mathbb{R}^3 \times \mathbb{Z}_2 \times \mathscr{X} \times \mathscr{X}_\mu$,

$$\left(F(q_0), \mathrm{J}_0^* e^{Y_R(n)} \mathrm{J}_t G(q_t)\right) \to \left(F(q_0), \mathrm{J}_0^* e^{Y_R} \mathrm{J}_t G(q_t)\right)$$

as $n \to \infty$. Together with (5.18), the Lebesgue dominated convergence theorem yields that

$$\int_{\mathbb{R}^3} \sum_{\alpha\in\mathbb{Z}_2} \mathbb{E}\left[e^{-\int_0^t V(B_r)\mathrm{d}r} \left(F(q_0), \mathrm{J}_0^* e^{Y_R(n)} \mathrm{J}_t G(q_t)\right)\right] \mathrm{d}x$$
$$\to \int_{\mathbb{R}^3} \sum_{\alpha\in\mathbb{Z}_2} \mathbb{E}\left[e^{-\int_0^t V(B_r)\mathrm{d}r} \left(F(q_0), \mathrm{J}_0^* e^{Y_R} \mathrm{J}_t G(q_t)\right)\right] \mathrm{d}x$$

for $F, G \in \mathcal{K}_\infty$. Then the proof is complete.

QED

5.4.3 Estimate of e^{-tL}

In this section we estimate the integral kernel of e^{-tL} by using the Baker–Campbell–Hausdorff formula and Theorem B.8 in Appendix B. Let us define

$$Y = -\int_0^t \mathbb{H}_{\mathrm{d}}(B_s, \theta_{N_s}, s)\mathrm{d}s + \int_0^{t+} \log \Psi_\varepsilon(|\mathbb{H}_{\mathrm{od}}(B_s, -\theta_{N_{s-}}, s)|)\mathrm{d}N_s. \tag{5.21}$$

We have already seen that $\mathrm{J}_0^* e^{Y_R} \mathrm{J}_t$ is bounded for each $R < \infty$, but it is not trivial to see that $\mathrm{J}_0^* e^{Y} \mathrm{J}_t$ is bounded.

Proposition 5.5 ([74, Lemma 3.93], [73, Lemma 4.9]) *For each* $(x, \alpha, w, m) \in \mathbb{R}^3 \times \mathbb{Z}_2 \times \mathscr{X} \times \mathscr{X}_\mu$, *operator* $\mathrm{J}_0^* e^Y \mathrm{J}_t$ *is bounded in* $L^2(\mathrm{Q})$.

Proof Let $Y = Y_1 + Y_2$, where

$$Y_1 = -\int_0^t \mathbb{H}_{\mathrm{d}}(B_r, \theta_{N_r}, s)_R \mathrm{d}r,$$

$$Y_2 = \int_0^{t+} \log(\Psi_\varepsilon(|\mathbb{H}_{\mathrm{od}}(B_r, -\theta_{N_{r-}}, s)|))\mathrm{d}N_r.$$

Using Baker–Campbell–Hausdorff formula we expand $\mathrm{J}_0^* e^Y \mathrm{J}_t$. Thus formally we have to estimate

$$\mathrm{J}_0^* e^Y \mathrm{J}_t = \mathrm{J}_0^* e^{\frac{B^\dagger}{\sqrt{2}} + \frac{B}{\sqrt{2}}} \prod_{j=1}^N \left(\left| \frac{C_j^\dagger}{\sqrt{2}} + \frac{C_j}{\sqrt{2}} \right| + \varepsilon \psi_\varepsilon \left(\left| \frac{C_j^\dagger}{\sqrt{2}} + \frac{C_j}{\sqrt{2}} \right| \right) \right) \mathrm{J}_t, \tag{5.22}$$

where in the Fock representation $B^\sharp$ and $C_i^\sharp$ are defined by

$$B^\dagger = -a_{\mathrm{E}}^* \left(\bigoplus_{j=\pm 1} \int_0^t S_j(\theta_{N_r}) \hat{\mathrm{j}}_s \left(\frac{\hat{\varphi}}{\sqrt{\omega}} \right) e^{-\mathrm{i}kB_r} \mathrm{d}r \right),$$

$$B = -a_{\mathrm{E}} \left(\bigoplus_{j=\pm 1} \int_0^t S_j(\theta_{N_r}) \hat{\mathrm{j}}_s \left(\frac{\hat{\varphi}}{\sqrt{\omega}} \right) e^{\mathrm{i}kB_r} \mathrm{d}r \right),$$

$$C_i^\dagger = a_{\mathrm{E}}^* \left(\bigoplus_{j=\pm 1} T_j(\theta_{N_{s_i}}) \hat{\mathrm{j}}_s \left(\frac{\hat{\varphi}}{\sqrt{\omega}} \right) e^{-\mathrm{i}kB_{s_i}} \right),$$

$$C_i = a_{\mathrm{E}} \left(\bigoplus_{j=\pm 1} T_j(\theta_{N_{s_i}}) \hat{\mathrm{j}}_s \left(\frac{\hat{\varphi}}{\sqrt{\omega}} \right) e^{\mathrm{i}kB_{s_i}} \right).$$

Here

$$S_j(\theta) = -\frac{1}{2}\theta\eta_3^j,$$
$$T_j(\theta) = -\frac{1}{2}(\eta_1^j - \mathrm{i}\theta\eta_2^j)$$

with $\eta^j = -\mathrm{i}k \times e^j(k)$. Note that $\|\psi_\varepsilon(\left|\frac{C_j^\dagger}{\sqrt{2}} + \frac{C_j}{\sqrt{2}}\right|)\| \leq 1$ for any j. Now fix $(w, m) \in \mathscr{X} \times \mathscr{X}_\mu$. In a similar way to the case of $R < \infty$ there exist $N = N(m) \in \mathbb{N}$ and points $0 \leq s_1 < \cdots < s_N \leq t$, $s_j = s_j(m)$, $j = 1, \ldots, N$, depending on m such that $s \mapsto N_s(m)$ is not continuous. Then by taking n large enough and denoting $(n(s_i), n(s_i) + t/n]$ for the interval containing s_i, we get

$$e^{Y_2} = \prod_{i=1}^{N}(|\phi_i| + \varepsilon\psi_\varepsilon(|\phi_i|)),$$

where $\phi_i = \mathbb{H}_{\mathrm{od}}(B_{s_i}, -\theta_{N_{s_i-}}, n(s_i))$. We have

$$\prod_{i=1}^{N}(|\phi_i| + \varepsilon\psi_\varepsilon(|\phi_i|)) \leq 1 + \prod_{i=1}^{N}(\phi_i + \varepsilon\psi_\varepsilon(\phi_i))\overline{(\phi_i + \varepsilon\psi_\varepsilon(\phi_i))}$$

and $(\phi_i + \varepsilon\psi_\varepsilon(\phi_i))\overline{(\phi_i + \varepsilon\psi_\varepsilon(\phi_i))} \leq (1 + \varepsilon)\phi_i\overline{\phi_i} + \varepsilon^2$. Then

$$|e^{Y_2}| \leq 1 + \prod_{i=1}^{N}((1 + \varepsilon)\phi_i\overline{\phi_i} + \varepsilon^2).$$

Hence it follows that

$$|(F, \mathrm{J}_0^* e^Y \mathrm{J}_t G)| \leq \left(|F|, \mathrm{J}_0^* e^{Y_1}\left(1 + \prod_{i=1}^{N}((1 + \varepsilon)\phi_i\overline{\phi_i} + \varepsilon^2)\right)\mathrm{J}_t|G|\right). \tag{5.23}$$

Since Baker–Campbell–Hausdorff formula yields that

$$e^{Y_1} = e^{\beta} e^{\frac{B^\dagger}{\sqrt{2}}} e^{\frac{B}{\sqrt{2}}},$$

where $\beta = [B, B^\dagger]/4$. We have

$$\mathrm{J}_0^* e^{Y_1} \prod_{i=1}^{k}\phi_i \mathrm{J}_t = \mathrm{J}_0^* e^{\beta} e^{\frac{B^\dagger}{\sqrt{2}}} e^{\frac{B}{\sqrt{2}}} \prod_{i=1}^{k}\left(\frac{C_i^\dagger}{\sqrt{2}} + \frac{C_i}{\sqrt{2}}\right)\mathrm{J}_t. \tag{5.24}$$

We apply Wick's theorem to compute the commutator between $\prod_{i=1}^{k}\left(\frac{C_i^\dagger}{\sqrt{2}}+\frac{C_i}{\sqrt{2}}\right)$ and $e^{\frac{B^\dagger}{\sqrt{2}}}e^{\frac{B}{\sqrt{2}}}$. Set $j_i=i$ for simplicity. Hence

$$\prod_{i=1}^{k}\left(\frac{C_i^\dagger}{\sqrt{2}}+\frac{C_i}{\sqrt{2}}\right)=\sum_{p=0}^{[k/2]}\frac{1}{2^p}\sum_{i_1,\ldots,i_{2p}}^{\text{pair}}\left(\prod_{l=1}^{p}C_{i_{2l-1},i_{2l}}\right)c^k_{i_1,\ldots,i_{2k}}.$$

Here $C_{i,j}=[C_i,C_j^\dagger]$, and $\sum_{i_1,\ldots,i_{2p}}^{\text{pair}}$ denotes the summation over all p-pairs chosen from $\{1,\ldots,k\}$, and

$$c^k_{i_1,\ldots,i_{2p}}=2^{-(k-2p)/2}\sum_{l=0}^{k-2p}\sum_{\{n_1,\ldots,n_l\}\subset\{i_1,\ldots,i_{2p}\}^c}\prod_{i\in\{n_1,\ldots,n_l\}}C_i^\dagger\prod_{i\in\{n_1,\ldots,n_l\}^c\cap\{i_1,\ldots,i_{2p}\}^c}C_i.$$

Hence

$$e^{\frac{B^\dagger}{\sqrt{2}}}e^{\frac{B}{\sqrt{2}}}\prod_{i=1}^{k}\left(\frac{C_i^\dagger}{\sqrt{2}}+\frac{C_i}{\sqrt{2}}\right)=\sum_{p=0}^{[k/2]}\frac{1}{2^p}\sum_{i_1\ldots,i_{2p}}^{\text{pair}}\left(\prod_{l=1}^{p}C_{i_{2l-1},i_{2l}}\right)\sum_{l=0}^{k-2p}\sum_{\{n_1,\ldots,n_l\}\subset\{i_1,\ldots,i_{2p}\}^c}Xe^{\frac{B^\dagger}{\sqrt{2}}}e^{\frac{B}{\sqrt{2}}}Z,\tag{5.25}$$

where $X=\prod_{i\in\{n_1,\ldots,n_l\}}(C_i^*+y_i)$ and $Z=\prod_{i\in\{n_1,\ldots,n_l\}^c\cap\{i_1,\ldots,i_{2p}\}^c}C_i$ with $y_j=[B,C_j^\dagger]$. Let us define operators in $\mathcal{F}$ by

$$B^\dagger(l)=-a^*\left(\bigoplus_{j=\pm1}\int_0^t e^{-|s-l|\omega(k)}S_j(\theta_{N_r})\frac{\hat{\varphi}}{\sqrt{\omega}}e^{-\mathrm{i}kB_r(w)}\mathrm{d}r\right),$$

$$B(l)=-a\left(\bigoplus_{j=\pm1}\int_0^t e^{-|s-l|\omega(k)}S_j(\theta_{N_r})\frac{\hat{\varphi}}{\sqrt{\omega}}e^{\mathrm{i}kB_r(w)}\mathrm{d}r\right),$$

$$C_i^\dagger(l)=a^*\left(\bigoplus_{j=\pm1}e^{-|s-l|\omega(k)}T_j(\theta_{N_{s_i}})\frac{\hat{\varphi}}{\sqrt{\omega}}e^{-\mathrm{i}kB_{s_i}}\right),$$

$$C_i(l)=a\left(\bigoplus_{j=\pm1}e^{-|s-l|\omega(k)}T_j(\theta_{N_{s_i}})\frac{\hat{\varphi}}{\sqrt{\omega}}e^{\mathrm{i}kB_{s_i}}\right).$$

By intertwining properties $\mathrm{J}_0^*B^\dagger=B^\dagger(0)\mathrm{J}_0^*$, $\mathrm{J}_0B(0)=B\mathrm{J}_0$, $\mathrm{J}_0C_i^\dagger=C_i^\dagger(0)\mathrm{J}_0^*$ and $\mathrm{J}_0C_i(0)=C_i\mathrm{J}_0$, and factorization formula $\mathrm{J}_0^*\mathrm{J}_t=e^{-tH_{\mathrm{f}}}$, we see by (5.25) that

$$
\begin{aligned}
W(k) &= \mathrm{J}_0^* e^{\frac{B^\dagger}{\sqrt{2}}} e^{\frac{B}{\sqrt{2}}} \prod_{i=1}^{k} \left(\frac{C_i^\dagger}{\sqrt{2}} + \frac{C_i}{\sqrt{2}} \right) \mathrm{J}_t \\
&= \sum_{p=0}^{[k/2]} \frac{1}{2^p} \sum_{i_1,\ldots,i_{2p}}^{\text{pair}} \left(\prod_{l=1}^{p} C_{i_{2l-1},i_{2l}} \right) \sum_{l=0}^{k-2p} \sum_{\{n_1,\ldots,n_l\} \subset \{i_1,\ldots,i_{2p}\}^c} X(0) e^{\frac{B^\dagger(0)}{\sqrt{2}}} e^{-t\hat{H}_{\mathrm{f}}} e^{\frac{B(t)}{\sqrt{2}}} Z(t),
\end{aligned}
$$

where $X(0)$ and $Z(t)$ are defined by X and Z with C_j replaced by $C_j(0)$ and $C_j(t)$, respectively. Let $\alpha = \frac{1}{\sqrt{2}}(\| |k|\hat{\varphi}/\sqrt{\omega}\| + \| |k|\hat{\varphi}/\omega\|)$ and $\beta = \frac{1}{\sqrt{2}}\| |k|\hat{\varphi}/\sqrt{\omega}\|$. Then $\|[C_i(l), C_j^\dagger(l)]\| \le \beta^2$ for any i, j. Let

$$
\xi(l,m) = \sum_{j,j'=0}^{\infty} \frac{\gamma^{j+j''} \sqrt{(j+l)!}\sqrt{(j'+m)!}}{j!j'!}
$$

and

$$
\xi_r(l,m) = \sum_{j,j'=0}^{\infty} \frac{\gamma^{j+j''} \sqrt{(j+l)!}\sqrt{(j'+m)!}}{r^{(j+l)/2} r^{(j'+m)/2} j!j'!}.
$$

Here $\gamma = \| |k|\hat{\varphi}/\sqrt{\omega}\| + \| |k|\hat{\varphi}/\omega\|$. The following estimates (5.26) and (5.27) are proven in Theorem B.8.
Case ($t \ge 1$):

$$
\|W(k)\Psi\| \le w(k) \left\| e^{-\frac{1}{2}(t-1)\hat{H}_{\mathrm{f}}} \right\|^2 \|\Psi\|, \tag{5.26}
$$

where

$$
w(k) = k! \sum_{p=0}^{[k/2]} \sum_{l+m+2p=k} \frac{\left(\frac{\beta^2}{4}\right)^p (\alpha+y)^l \alpha^m}{p!l!m!} \xi(l,m)
$$

and $y = t\beta^2$.
Case ($t < 1$): For any $0 < r < t$, we have

$$
\|W(k)\Psi\| \le w(r,k) \left\| e^{-\frac{1}{2}(t-r)\hat{H}_{\mathrm{f}}} \right\|^2 \|\Psi\|, \tag{5.27}
$$

where

$$
w(r,k) = k! \sum_{p=0}^{[k/2]} \sum_{l+m+2p=k} \frac{\left(\frac{\beta^2}{4}\right)^p (\alpha+y)^l \alpha^m}{p!l!m!} \xi_r(l,m).
$$

Hence $W(k)$ is bounded. Since

$$J_0^* e^{Y_1}\left(1+\prod_{i=1}^N((1+\varepsilon)\phi_i\overline{\phi_i}+\varepsilon^2)\right)J_t$$

$$=J_0^* e^{Y_1}J_t+\sum_{k=0}^N\sum_{\{i_1,\ldots,i_k\}\subset\{1,\ldots,N\}}\varepsilon^{2(N-k)}(1+\varepsilon)^k J_0^* e^{Y_1}\left(\prod_{j=1}^k\phi_{i_j}\bar{\phi}_{i_j}\right)J_t,$$

in a similar manner to (5.26) and (5.27) we see the bound:

$$\left\|J_0^* e^{Y_1}\left(\prod_{j=1}^k\phi_{i_j}\bar{\phi}_{i_j}\right)J_t\right\|\le w(r,2k)\left\|e^{-\frac{1}{2}(t-r)\hat{H}_{\mathrm{f}}}\right\|^2$$

with $0<r<t$ for $t<1$ and $r=1$ for $t\ge 1$, we have

$$\left\|J_0^* e^{Y_1}\left(1+\prod_{i=1}^N((1+\varepsilon)\phi_i\overline{\phi_i}+\varepsilon^2)\right)J_t\right\|$$

$$\le\left(w(r,0)+\sum_{k=0}^N\frac{N!}{(N-k)!k!}\varepsilon^{2(N-k)}(1+\varepsilon)^k w(r,2k)\right)\left\|e^{-\frac{1}{2}(t-r)\hat{H}_{\mathrm{f}}}\right\|^2.$$

Then the proof is complete.

QED

Lemma 5.6 *Let $V\in L^\infty(\mathbb{R}^3)$ and $\tilde{\varphi}\in C_0^\infty(\mathbb{R}^3)$. Then for every $t\ge 0$ and all $F,G\in\mathcal{K}$ it follows that*

$$(F,e^{-tL}G)=e^t\int_{\mathbb{R}^3}\sum_{\alpha\in\mathbb{Z}_2}\mathbb{E}^{x,\alpha}[e^{-\int_0^t V(B_s)\mathrm{d}s}(F(q_0),J_0^* e^Y J_t G(q_t))]\mathrm{d}x. \tag{5.28}$$

Proof Let $F,G\in\mathcal{K}$ such that $F\ge 0$ and $G\ge 0$. By Lemma 5.4 we have

$$(F,e^{-tL_R}G)=e^t\int_{\mathbb{R}^3}\sum_{\alpha\in\mathbb{Z}_2}\mathbb{E}^{x,\alpha}[e^{-\int_0^t V(B_s)\mathrm{d}s}(F(q_0),J_0^* e^{Y_R}J_t G(q_t))]\mathrm{d}x. \tag{5.29}$$

Here $0\le g(R)=(F(q_0),J_0^* e^{Y_R}J_t G(q_t))$ and $g(R)$ is a monotonously increasing function in R. Hence $(F,e^{-tL_R}G)$ is also increasing in R and $(F,e^{-tL_R}G)\uparrow(F,e^{-tL}G)$ as $R\uparrow\infty$. For each $\phi\in\mathrm{Q}$, it follows that

$$J_0F(q_0)\cdot e^{Y_R}J_t G(q_t)\to J_0F(q_0)\cdot e^Y J_t G(q_t)$$

as $R\to\infty$. Then the monotone convergence theorem yields that the function on the right-hand side above $J_0F(q_0)\cdot e^Y J_t G(q_t)$ is finite for a.e. $(\phi,x,\alpha,w,m)\in\mathrm{Q_E}\times\mathbb{R}^3\times\mathbb{Z}_2\times\mathscr{X}\times\mathscr{X}_\mu$, and

$$
\begin{aligned}
(F, e^{-tL}G) &= \lim_{R\to\infty}(F, e^{-tL_R}G)\\
&= e^t\int_{\mathbb{R}^3}\sum_{\alpha\in\mathbb{Z}_2}\mathbb{E}^{x,\alpha}\left[\lim_{R\to\infty} e^{-\int_0^t V(B_s)\mathrm{d}s}(F(q_0), \mathrm{J}_0^* e^{Y_R}\mathrm{J}_t G(q_t))\right]\mathrm{d}x\\
&= e^t\int_{\mathbb{R}^3}\sum_{\alpha\in\mathbb{Z}_2}\mathbb{E}^{x,\alpha}\left[e^{-\int_0^t V(B_s)\mathrm{d}s}\mathbb{E}_{\mu_{\mathrm{E}}}[\mathrm{J}_0 F(q_0)\cdot e^Y\mathrm{J}_t G(q_t)]\right]\mathrm{d}x.
\end{aligned}
$$

For general $F, G \in \mathscr{K}$, the lemma is proven by decomposing F and G as a linear sum of positive functions: $F = \Re F_+ - \Re F_- + \mathrm{i}(\Im F_+ - \Im F_-)$ and $G = \Re G_+ - \Re G_- + \mathrm{i}(\Im G_+ - \Im G_-)$. Finally we show that $\mathrm{J}_0^* e^Y \mathrm{J}_t G(q_t) \in L^2(\mathrm{Q})$ for each $(x, \alpha, w, m) \in \mathbb{R}^3 \times \mathbb{Z}_2 \times \mathscr{X} \times \mathscr{X}_\mu$ in Proposition 5.5. Then $\mathbb{E}_{\mu_{\mathrm{E}}}[\mathrm{J}_0 F(q_0) \cdot e^Y \mathrm{J}_t G(q_t)] = (F(q_0), \mathrm{J}_0^* e^Y \mathrm{J}_t G(q_t))$ follows and the lemma is proven.

QED

5.4.4 Estimate of $e^{-tH_{\mathbb{Z}_2,\varepsilon}}$

Let $Z_t = Y_t(1) + Y_t(2) + Y_t(3)$, where

$$
\begin{aligned}
Y_t(1) &= -\mathrm{i}\mathbb{A}\left(\bigoplus_{\mu=1}^3\int_0^t \mathrm{j}_s\tilde{\varphi}(\cdot - B_s)\mathrm{d}B_s^\mu\right),\\
Y_t(2) &= -\int_0^t \mathbb{H}_{\mathrm{d}}(B_s, \theta_{N_s}, s)\mathrm{d}s,\\
Y_t(3) &= \int_0^{t+}\log(-\mathbb{H}_{\mathrm{od}}(B_s, -\theta_{N_{s-}}, s))\mathrm{d}N_s.
\end{aligned}
$$

We shall see that Z_t turns to be the exponent of an integral kernel of $e^{-tH_{\mathbb{Z}_2}}$. We furthermore define $Z_t(\varepsilon)$ by Z_t with $Y_t(3)$ replaced by $Y_t(3,\varepsilon)$, i.e., $Z_t(\varepsilon) = Y_t(1) + Y_t(2) + Y_t(3,\varepsilon)$, where

$$
Y_t(3,\varepsilon) = \int_0^{t+}\log(-\Psi_\varepsilon(\mathbb{H}_{\mathrm{od}}(B_s, -\theta_{N_{s-}}, s)))\mathrm{d}N_s.
$$

Lemma 5.7 *For each* $(x, \alpha, w, m) \in \mathbb{R}^3 \times \mathbb{Z}_2 \times \mathscr{X} \times \mathscr{X}_\mu$, $\mathrm{J}_0^* e^{Z_t(\varepsilon)}\mathrm{J}_t$ *is a bounded operator.*

Proof Let $\Phi, \Psi \in L^2(\mathrm{Q})$. By Y defined in (5.21) we see that

$$
|(\Phi, \mathrm{J}_0^* e^{Z_t(\varepsilon)}\mathrm{J}_t\Psi)| \le (|\Phi|, \mathrm{J}_0^* e^Y \mathrm{J}_t|\Psi|) \le \|\Phi\|\|\Psi\|\|\mathrm{J}_0^* e^Y\mathrm{J}_t\|
$$

follows from Proposition 5.5. Then the lemma is proven.

QED

Theorem 5.8 *Suppose Assumption 3.1. Then for every $t \geq 0$ and all $F, G \in \mathcal{K}$ it follows that* $(F, e^{-tH_{\mathbb{Z}_2}} G) = \lim_{\varepsilon\downarrow 0}(F, e^{-tH_{\mathbb{Z}_2,\varepsilon}} G)$ *and*

$$(F, e^{-tH_{\mathbb{Z}_2,\varepsilon}} G) = e^t \int_{\mathbb{R}^3} \sum_{\alpha\in\mathbb{Z}_2} \mathbb{E}[e^{-\int_0^t V(B_s)\mathrm{d}s}(F(q_0), \mathrm{J}_0^* e^{Z_t(\varepsilon)} \mathrm{J}_t G(q_t))]\mathrm{d}x. \quad (5.30)$$

Proof Since $e^{-tH_{\mathbb{Z}_2,\varepsilon}} \to e^{-tH^{\mathbb{Z}_2}}$ strongly as $\varepsilon \to 0$, $(F, e^{-tH_{\mathbb{Z}_2}} G) = \lim_{\varepsilon\downarrow 0}(F, e^{-tH_{\mathbb{Z}_2,\varepsilon}} G)$ follows. Now we turn to proving (5.30). Suppose that $\tilde{\varphi} \in C_0^\infty(\mathbb{R}^3)$ and $V \in L^\infty(\mathbb{R}^3)$. Write

$$\begin{aligned} X_{S,T}(\varepsilon, s) = &-\mathrm{i}\int_S^T \mathbb{A}(\mathrm{j}_s\tilde{\varphi}(\cdot - B_r))\mathrm{d}B_r \\ &- \int_S^T \mathbb{H}_\mathrm{d}(B_r, \theta_{N_r}, s)\mathrm{d}r + \int_S^{T+} \log(-\Psi_\varepsilon(\mathbb{H}_\mathrm{od}(B_r, -\theta_{N_{r-}}, s)))\mathrm{d}N_r. \end{aligned}$$

Define $S_{t,s}^\varepsilon : \mathcal{K}_\mathrm{E}^\infty \to \mathcal{K}_\mathrm{E}^\infty$ by

$$(S_{t,s}^\varepsilon G)(x, \theta_\alpha) = e^t \mathbb{E}^{x,\alpha}[e^{-\int_0^t V(B_r)\mathrm{d}r} e^{X_{0,t}(\varepsilon,s)} G(q_t)].$$

It can be seen that $S_{t,s}^\varepsilon$ has the property:

$$S_{t,s}^\varepsilon S_{t',s'}^\varepsilon G(x, \theta_\alpha) = e^{t+t'} \mathbb{E}^{x,\alpha}[e^{-\int_0^{t+t'} V(B_r)\mathrm{d}r} e^{X_{0,t}(\varepsilon,s)+X_{t,t+t'}(\varepsilon,s')} G(q_{t+t'})]. \quad (5.31)$$

Note that for $s_1 \leq \cdots \leq s_n$,

$$e^{X_{0,t_1}(\varepsilon,s_1)+X_{t_1,t_1+t_2}(\varepsilon,s_2)+\cdots+X_{t_1+\cdots+t_{n-1},t_1+\cdots+t_n}(\varepsilon,s_n)} \in \mathscr{E}_{[s_1,s_n]}. \quad (5.32)$$

Since $(F, e^{-tH_{\mathbb{Z}_2,\varepsilon}} G) = (F, e^{-tK_\varepsilon} G)$, by the Trotter product formula, (5.31), (5.32) and the Markov property of E_s, $s \in \mathbb{R}$, we obtain that

$$(F, e^{-tH_{\mathbb{Z}_2,\varepsilon}} G) = e^t \lim_{n\to\infty} \int_{\mathbb{R}^3} \sum_{\alpha\in\mathbb{Z}_2} \mathbb{E}^{x,\alpha}\left[e^{-\int_0^t V(B_r)\mathrm{d}r} \left(F(q_0), \mathrm{J}_0^* e^{Z_t^n(\varepsilon)} \mathrm{J}_t G(q_t)\right)\right]\mathrm{d}x, \quad (5.33)$$

where $Z_t^n(\varepsilon) = Y_t^n(1) + Y_t^n(2) + Y_t^n(3, \varepsilon)$ with

$$Y_t^n(1) = -\mathrm{i}\mathbb{A}\left(\bigoplus_{\mu=1}^3 \sum_{j=1}^n \int_{t_{j-1}}^{t_j} \mathrm{j}_{t_{j-1}}\tilde{\varphi}(\cdot - B_s)\mathrm{d}B_s^\mu\right),$$

$$Y_t^n(2) = -\sum_{j=1}^n \int_{t_{j-1}}^{t_j} \mathbb{H}_{\mathrm{d}}(B_s, \theta_{N_s}, t_{j-1})\mathrm{d}s,$$

$$Y_t^n(3,\varepsilon) = \sum_{j=1}^n \int_{t_{j-1}}^{t_j} \log(-\Psi_\varepsilon(\mathbb{H}_{\mathrm{od}}(B_s, -\theta_{N_{s-}}, t_{j-1})))\mathrm{d}N_s$$

and $t_j = jt/n$. Put

$$\langle F, T_n G\rangle = e^t \int_{\mathbb{R}^3} \sum_{\alpha\in\mathbb{Z}_2} \mathbb{E}^{x,\alpha}\left[e^{-\int_0^t V(B_r)\mathrm{d}r}\left(F(q_0), \mathrm{J}_0^* e^{Z_t^n(\varepsilon)}\mathrm{J}_t G(q_t)\right)\right]\mathrm{d}x,$$

$$\langle F, TG\rangle = e^t \int_{\mathbb{R}^3} \sum_{\alpha\in\mathbb{Z}_2} \mathbb{E}^{x,\alpha}\left[e^{-\int_0^t V(B_r)\mathrm{d}r}\left(F(q_0), \mathrm{J}_0^* e^{Z_t(\varepsilon)}\mathrm{J}_t G(q_t)\right)\right]\mathrm{d}x.$$

Notice that by the definition of $\langle F, T_n G\rangle$ we see that $|\langle F, T_n G\rangle| \le \|F\|\|G\|$. Suppose that $F_m \to F$ and $G_m \to G$ as $m \to \infty$. Then by a telescoping we can see that $|\langle F, T_n G\rangle - \langle F_m, T_n G_m\rangle| \le \|F - F_m\|\|G\| + \|G - G_m\|\|F_m\|$. By using facts that $|\mathrm{J}_t F| \le \mathrm{J}_t|F|$, $|e^{\pm Y_t(1)}F| = |F|$ and $|e^{Y_t(2)+Y_t(3,\varepsilon)}F| = e^Y|F|$, we can see that $|\langle F, TG\rangle| \le a\|F\|\|G\|$, where $a = e^{-t\inf\mathrm{Spec}(L)}$. Hence we have

$$|\langle F_m, TG_m\rangle - \langle F, TG\rangle| \le a\|F_m - F\|\|G_m\| + a\|F\|\|G - G_m\|.$$

Suppose that $\|F - F_m\| < \varepsilon$ and $\|G - G_m\| < \varepsilon$. Together with them we have

$$|\langle F, T_n G\rangle - \langle F, TG\rangle| \le \varepsilon(\|G\| + a\|G_m\| + \|F_m\| + a\|F\|) + |\langle F_m, T_n G_m\rangle - \langle F_m, TG_m\rangle|. \tag{5.34}$$

It can be concluded from (5.34) that it is enough to show the lemma for arbitrary F, G included in some dense domain. We shall show the lemma for $F, G \in \mathcal{K}_\infty$. We claim that

(1) For each $(x, \alpha, w, m) \in \mathbb{R}^3 \times \mathbb{Z}_2 \times \mathscr{X} \times \mathscr{X}_\mu$, there exists $C_t(m)$ such that

$$|(F, \mathrm{J}_0^* e^{Z_t^n(\varepsilon)}\mathrm{J}_t G)_{L^2(\mathrm{Q})}| \le C_t(m),$$

where $C_t(m)$ is independent of $(x, \alpha, w) \in \mathbb{R}^3 \times \mathbb{Z}_2 \times \mathscr{X}$, $\varepsilon > 0$ and n, and it is satisfied that $\mathbb{E}_\mu[C_t^2] < \infty$.

(2) For each $(x, \alpha, m) \in \mathbb{R}^3 \times \mathbb{Z}_2 \times \mathscr{X}_\mu$,

$$\lim_{n\to\infty} \mathbb{E}_{\mathcal{W}}^x[(F(q_0), \mathrm{J}_0^* e^{Z_t^n(\varepsilon)}\mathrm{J}_t G(q_t))] = \mathbb{E}_{\mathcal{W}}^x[(F(q_0), \mathrm{J}_0^* e^{Z_t(\varepsilon)}\mathrm{J}_t G(q_t))].$$

The proof of (1) and (2) will be given in Lemma C.9 and Lemma C.10 in Appendix C below, respectively. We set

$$\text{RHS (5.33)} = e^t \int_{\mathbb{R}^3} \sum_{\alpha\in\mathbb{Z}_2} \mathbb{E}^{x,\alpha}[\xi_n]\mathrm{d}x, \tag{5.35}$$

where $\xi_n = e^{-\int_0^t V(B_s)\mathrm{d}s}\left(F(q_0), \mathrm{J}_0^* e^{Z_t^n(\varepsilon)}\mathrm{J}_t G(q_t)\right)$. Thus we have

$$\mathbb{E}^x_{\mathcal{W}}[|\xi_n|] \le \mathbb{E}^x_{\mathcal{W}}[e^{-\int_0^t V(B_s)\mathrm{d}s} C_t(m)\|F(q_0)\|\|G(q_t)\|]$$

and

$$\int_{\mathbb{R}^3} \sum_{\alpha\in\mathbb{Z}_2} \mathbb{E}^{x,\alpha}[e^{-\int_0^t V(B_s)\mathrm{d}s} C_t(m)\|F(q_0)\|\|G(q_t)\|]\mathrm{d}x \le V_\infty^{1/2}(\mathbb{E}_\mu[C_t^2])^{1/2}\|F\|\|G\| < \infty.$$

Since $\mathbb{E}^x_{\mathcal{W}}[\xi_n] \to \mathbb{E}^x_{\mathcal{W}}[\xi_\infty]$ as $n \to \infty$ for each $(x, \alpha, m) \in \mathbb{R}^3 \times \mathbb{Z}_2 \times \mathscr{X}_\mu$, by the Lebesgue dominated convergence theorem,

$$\lim_{n\to\infty} \int_{\mathbb{R}^3} \sum_{\alpha\in\mathbb{Z}_2} \mathbb{E}^\alpha_\mu[\mathbb{E}^x_{\mathcal{W}}[\xi_n]]\mathrm{d}x = \int_{\mathbb{R}^3} \sum_{\alpha\in\mathbb{Z}_2} \mathbb{E}^\alpha_\mu[\mathbb{E}^x_{\mathcal{W}}[\xi_\infty]]\mathrm{d}x$$

follows. For $V \in \mathcal{R}_{\text{Kato}}$ by a limiting argument we can show the theorem. Finally for $\hat{\varphi}$ in Assumption 3.1, we can see (5.30) by an approximation, which is also shown in Lemma 5.9 below.

QED

Lemma 5.9 (5.30) *is valid for* $\hat{\varphi}$ *in Assumption 3.1.*

Proof It is enough to show (5.30) for $F, G \in \mathcal{K}_\infty$ by an approximation argument. Take a sequence $\hat{\varphi}_n \in C_0^\infty(\mathbb{R}^3)$ such that $|k|\hat{\varphi}_n/\sqrt{\omega} \to |k|\hat{\varphi}/\sqrt{\omega}$ as $n \to \infty$. Then (5.30) is valid for each $\hat{\varphi}_n$:

$$(F, e^{-tH_{\mathbb{Z}_2,\varepsilon}(n)}G) = \int_{\mathbb{R}^3} \sum_{\alpha\in\mathbb{Z}_2} \mathbb{E}[e^{-\int_0^t V(B_s)\mathrm{d}s}(F(q_0), \mathrm{J}_0^* e^{Z_t^n(\varepsilon)}\mathrm{J}_t G(q_t))]\mathrm{d}x.$$

Here $Z_t^n(\varepsilon)$ is defined by $Z_t(\varepsilon)$ with $\hat{\varphi}$ replaced by $\hat{\varphi}_n$, and $H_{\mathbb{Z}_2,\varepsilon}(n)$ by $H_{\mathbb{Z}_2,\varepsilon}$ with $\hat{\varphi}$ replaced by $\hat{\varphi}_n$. It can be seen that $H_{\mathbb{Z}_2,\varepsilon}(n) \to H_{\mathbb{Z}_2,\varepsilon}$ as $n \to \infty$ on a common core $\mathcal{K}_\infty$. Then $e^{-tH_{\mathbb{Z}_2,\varepsilon}(n)} \to e^{-tH_{\mathbb{Z}_2,\varepsilon}}$ strongly as $n \to \infty$. Let $Y_t^n(1)$, $Y_t^n(2)$ and $Y_t^n(3)$ be $Y_t(1)$, $Y_t(2)$ and $Y_t(3)$ with $\hat{\varphi}$ replaced by $\hat{\varphi}_n$, respectively. In a similar approximation argument to the proof of Lemma C.10 we can show that

(1) $\mathbb{E}^x_{\mathcal{W}}[\|(e^{Y_t^n(1)} - e^{Y_t(1)})F] \to 0$ as $\|(\hat{\varphi}_n - \hat{\varphi})/\sqrt{\omega}\| \to 0$,
(2) $\|(e^{Y_t^n(2)} - e^{Y_t(2)})F\| \to 0$ as $\||k|(\hat{\varphi}_n - \hat{\varphi})/\sqrt{\omega}\| \to 0$,
(3) $\|(e^{Y_t^n(3,\varepsilon)} - e^{Y_t(3,\varepsilon)})F\| \to 0$ as $\||k|(\hat{\varphi}_n - \hat{\varphi})/\sqrt{\omega}\| \to 0$

for $F \in L^2_{\rm fin}(\mathrm{Q})$. Hence

$$
\begin{aligned}
(F, e^{-tH_{\mathbb{Z}_2,\varepsilon}} G) &= \lim_{n\to\infty} \int_{\mathbb{R}^3} \sum_{\alpha\in\mathbb{Z}_2} \mathbb{E}[e^{-\int_0^t V(B_s)\mathrm{d}s} (F(q_0), \mathrm{J}_0^* e^{Z_t^n(\varepsilon)} \mathrm{J}_t G(q_t))]\mathrm{d}x \\
&= \int_{\mathbb{R}^3} \sum_{\alpha\in\mathbb{Z}_2} \mathbb{E}[e^{-\int_0^t V(B_s)\mathrm{d}s} (F(q_0), \mathrm{J}_0^* e^{Z_t(\varepsilon)} \mathrm{J}_t G(q_t))]\mathrm{d}x
\end{aligned}
$$

for $F, G \in \mathcal{K}_\infty$. Then the proof is complete.

QED

5.5 Functional Integral Representations of $e^{-tH_{\mathbb{Z}_2}}$

As was mentioned above we need the regularization $\Psi_\varepsilon(\hat{H}_{\rm od})$ of $\hat{H}_{\rm od}$ to prevent zeros of $\hat{H}_{\rm od}$. The zeros of $\hat{H}_{\rm od}$ produces the zeros of $e^{\int_0^{t+} \log(-\mathbb{H}_{\rm od}(B_s, -\theta_{N_{s-}}, s))\mathrm{d}N_s}$. In order to avoid $\lim_{\varepsilon\to 0}$ in the functional integral representation, instead of introducing regularization Ψ_ε we introduce a subset W of $\mathbb{R}^3 \times \mathscr{X} \times \mathscr{X}_N \times \mathrm{Q_E}$ by

$$
W = \left\{ \int_0^{t+} \log\left(\frac{1}{2} \sqrt{|b_1(s, x, B_s)|^2 + |b_2(s, x, B_s)|^2} \right) \mathrm{d}N_s > -\infty \right\}, \tag{5.36}
$$

where $b_\alpha(s, x, B_s) = \mathbb{B}_\alpha(\mathrm{j}_s \tilde{\varphi}(\cdot - B_s - x))$.

Theorem 5.10 *Suppose Assumption 3.1. Then for every $t \geq 0$ and all $F, G \in \mathcal{K}$ it follows that*

$$
(F, e^{-tH_{\mathbb{Z}_2}} G) = e^t \int_{\mathbb{R}^3} \sum_{\alpha\in\mathbb{Z}_2} \mathbb{E}^{x,\alpha}[e^{-\int_0^t V(B_s)\mathrm{d}s} (\mathrm{J}_0 F(q_0), e^{Z_t} \mathbb{1}_W \mathrm{J}_t G(q_t))]\mathrm{d}x. \tag{5.37}
$$

Here W is given by (5.36) *and the exponent Z_t is defined by*

$$
\begin{aligned}
Z_t = &-\mathrm{i} \int_0^t \mathbb{A}(\mathrm{j}_s \tilde{\varphi}(\cdot - B_s))\mathrm{d}B_s \\
&- \int_0^t \mathbb{H}_{\rm d}(B_s, \theta_{N_s}, s)\mathrm{d}s + \int_0^{t+} \log(-\mathbb{H}_{\rm od}(B_s, -\theta_{N_{s-}}, s))\mathrm{d}N_s
\end{aligned}
$$

and

$$
\int_0^{t+} \log(-\mathbb{H}_{\rm od}(B_s, -\theta_{N_{s-}}, s))\mathrm{d}N_s = \sum_{\substack{r\in[0,t] \\ N_{r+}\neq N_{r-}}} \log(-\mathbb{H}_{\rm od}(B_r, -\theta_{N_{r-}}, r))
$$

possibly takes infinity.

Proof By the Lebesgue dominated convergence theorem we have

$$\lim_{\varepsilon\to 0}\int_{\mathbb{R}^3}\sum_{\alpha\in\mathbb{Z}_2}\mathbb{E}^{x,\alpha}[e^{-\int_0^t V(B_s)\mathrm{d}s}(\mathrm{J}_0 F(q_0), e^{Z_t(\varepsilon)}\mathrm{J}_t G(q_t))]\mathrm{d}x$$
$$=\int_{\mathbb{R}^3}\sum_{\alpha\in\mathbb{Z}_2}\mathbb{E}^{x,\alpha}[e^{-\int_0^t V(B_s)\mathrm{d}s}\lim_{\varepsilon\to 0}(\mathrm{J}_0 F(q_0), e^{Z_t(\varepsilon)}\mathrm{J}_t G(q_t))]\mathrm{d}x.$$

We can also show that $\lim_{\varepsilon\to 0}(\mathrm{J}_0 F(q_0), \mathbb{1}_{W^c} e^{Z_t(\varepsilon)}\mathrm{J}_t G(q_t)) = 0$ and

$$\lim_{\varepsilon\to 0}(\mathrm{J}_0 F(q_0), e^{Z_t(\varepsilon)}\mathrm{J}_t G(q_t)) = \lim_{\varepsilon\to 0}(\mathrm{J}_0 F(q_0), \mathbb{1}_W e^{Z_t(\varepsilon)}\mathrm{J}_t G(q_t)) = (\mathrm{J}_0 F(q_0), \mathbb{1}_W e^{Z_t}\mathrm{J}_t G(q_t)).$$

Then the theorem follows.

QED

5.6 *Translation Invariant Pauli–Fierz Hamiltonian with Spin* 1/2

We study the translation invariant Pauli–Fierz Hamiltonian with spin $1/2$, that is, H_{S} with V identically zero. We suppose Assumption 3.1 with $V = 0$ in this section. The Pauli–Fierz Hamiltonian with spin $1/2$ and a fixed total momentum p is defined by

$$H_{\mathrm{S}}(p) = \frac{1}{2}(p - \mathrm{P_f} - \hat{A}(0))^2 + H_{\mathrm{f}} - \frac{1}{2}\sigma\cdot\hat{B}(0), \quad p\in\mathbb{R}^3,$$

with domain $D(H_{\mathrm{S}}(p)) = D(H_{\mathrm{f}})\cap D(\mathrm{P_f^2})$. $H_{\mathrm{S}}(p)$ is self-adjoint and essentially self-adjoint on any core of $H_{\mathrm{f}} + \mathrm{P_f^2}$.

As in the case of $H(p)$ and $H_{\mathrm{R}}(p)$ we have the unitary equivalence below:

$$L^2(\mathbb{R}^3;\mathbb{C}^2)\otimes L^2(\mathrm{Q}) \cong \int_{\mathbb{R}^3}^{\oplus}\mathbb{C}^2\otimes L^2(\mathrm{Q})\mathrm{d}p,$$
$$H_{\mathrm{S}} \cong \int_{\mathbb{R}^3}^{\oplus} H_{\mathrm{S}}(p)\mathrm{d}p,$$

which are implemented by $\mathscr{T}$ defined in (3.30). We now construct a functional integral representation for the translation invariant Pauli–Fierz Hamiltonian with spin $1/2$. As before, we transform $H_{\mathrm{S}}(p)$ on $\mathbb{C}^2\otimes L^2(\mathrm{Q})$ to $H_{\mathbb{Z}_2}(p)$ on $\ell^2(\mathbb{Z}_2)\otimes L^2(\mathrm{Q})$ which is defined by

$$(H_{\mathbb{Z}_2}(p)\Psi)(\theta) = \left(\frac{1}{2}(p - \mathrm{P_f} - \hat{A}(0))^2 + \hat{H}_{\mathrm{d}}(0,\theta) + H_{\mathrm{f}}\right)\Psi(\theta) + \hat{H}_{\mathrm{od}}(0,-\theta)\Psi(-\theta),$$

where

$$\hat{H}_{\rm d}(0,\theta) = -\frac{1}{2}\theta\hat{B}_3(0),$$
$$\hat{H}_{\rm od}(0,-\theta) = -\frac{1}{2}(\hat{B}_1(0) - {\rm i}\theta\hat{B}_2(0)).$$

The strategy of constructing a functional integral representation of $e^{-tH_{\mathbb{Z}_2}(p)}$ is similar to that of the spinless case.

Theorem 5.11 *Let* $\Phi, \Psi \in \ell^2(\mathbb{Z}_2) \otimes L^2({\rm Q})$. *Let* $W_0 \subset \mathscr{X} \times \mathscr{X}_\mu \times {\rm Q_E}$ *be defined by*

$$W_0 = \left\{(w, m, \phi) \in \mathscr{X} \times \mathscr{X}_\mu \times {\rm Q_E} \mid \int_0^{t+} \log\left(\frac{1}{2}\sqrt{|b_1(s,B_s)|^2 + |b_2(s,B_s)|^2}\right) {\rm d}N_s > -\infty\right\}.$$

Then

$$(\Phi, e^{-tH_{\rm S}(p)}\Psi) = e^t \sum_{\alpha\in\mathbb{Z}_2} \mathbb{E}^{0,\alpha}[({\rm J}_0\Phi(X_0), e^{Z_t} 1\!\!1_{W_0} {\rm J}_t e^{{\rm i}(p-{\rm P_f})B_t}\Psi(X_t))],$$

where the exponent Z_t *is given by*

$$Z_t = -{\rm i}\int_0^t \mathbb{A}({\rm j}_s\tilde{\varphi}(\cdot - B_s)){\rm d}B_s - \int_0^t \mathbb{H}_{\rm d}(B_s, \theta_{N_s}, s){\rm d}s + \int_0^{t+} \log(-\mathbb{H}_{\rm od}(B_s, -\theta_{N_{s-}}, s)){\rm d}N_s.$$

Proof In addition to the proof of Theorem 3.14 the theorem can be shown by mimicking the proof of Theorem 5.10.

QED

5.7 Symmetry and Non-degeneracy of Ground States

For the Pauli–Fierz Hamiltonian H and the relativistic Pauli–Fierz Hamiltonian H_R, we can show that the associated semigroups are positivity improving. Thus the ground state for H and $H_{\rm R}$ is unique if it exists. We are concerned with $e^{-tH_{\rm S}}$ in this section.

Before concerning with $H_{\rm S}$ we study a toy model $H(\varepsilon)$ defined by

$$H(\varepsilon) = \frac{1}{2}(-{\rm i}\nabla - A)^2 + V + H_{\rm f} - \varepsilon\sigma_1, \quad \varepsilon \in \mathbb{R}.$$

When $\varepsilon = 0$,

$$H(0) = \begin{pmatrix} \frac{1}{2}(-{\rm i}\nabla - A)^2 + V + H_{\rm f} & 0 \\ 0 & \frac{1}{2}(-{\rm i}\nabla - A)^2 + V + H_{\rm f} \end{pmatrix}.$$

Then all the eigenvalues of $H(0)$ are degenerate. Let us consider $\varepsilon \neq 0$. Since $H(\varepsilon) \cong H(-\varepsilon)$, we may assume $\varepsilon > 0$.

Proposition 5.12 *Let $\varepsilon > 0$. Then $\mathfrak{S}^{-1}e^{-H(\varepsilon)}\mathfrak{S}$ is positivity improving. In particular the ground state of $H(\varepsilon)$ is unique, if the ground state exists.*

Proof Let $F, G \in \mathcal{H}_\mathrm{S}$. It follows that

$$(F, e^{-tH(\varepsilon)}G) = e^t \sum_{\alpha\in\mathbb{Z}_2} \int_{\mathbb{R}^3} \mathbb{E}^{x,\alpha}[e^{-\int_0^t V(B_s)\mathrm{d}s} (\mathrm{J}_0 F(X_0), e^{Z_t}\mathrm{J}_t G(X_t))]\mathrm{d}x.$$

Here $Z_t = -\mathrm{i}\int_0^t \mathbb{A}(\mathrm{j}_s\tilde{\varphi}(\cdot - B_s))\mathrm{d}B_s + \int_0^{t+} \log\varepsilon \mathrm{d}N_s$. Let $F, G \geq 0$. Then

$$\begin{aligned}&(F, \mathfrak{S}^{-1}e^{-tH(\varepsilon)}\mathfrak{S}G)\\ &= e^t \sum_{\alpha\in\mathbb{Z}_2} \int_{\mathbb{R}^3} \mathbb{E}^{x,\alpha}\left[\varepsilon^{N_t} e^{-\int_0^t V(B_s)\mathrm{d}s} (\mathrm{J}_0 F(X_0), \mathfrak{S}_\mathrm{E}^{-1} e^{-\mathrm{i}\int_0^t \mathbb{A}(\mathrm{j}_s\tilde{\varphi}(\cdot-B_s))\mathrm{d}B_s} \mathfrak{S}_\mathrm{E}\mathrm{J}_t G(X_t))\right]\mathrm{d}x > 0\end{aligned}$$

and $\mathfrak{S}^{-1}e^{-tH(\varepsilon)}\mathfrak{S}$ is positivity improving.

QED

We study H_S from now on. Suppose that $B = 0$. Then similar to $H(0)$ mentioned above, all the eigenvalues of H_S are degenerate. However $B \neq 0$, unlike $H(\varepsilon)$ the ground state of H_S is also degenerate. We shall show this in the following. The idea is to show that

$$H_\mathrm{S} = \bigoplus_{w\in\mathbb{Z}_{1/2}} H_\mathrm{S}(w)$$

and $H_\mathrm{S}(w) \cong H_\mathrm{S}(-w)$ by symmetries hidden in H_S. This section is taken from [79]. Also refer to see [132] and [74, Sects. 3.8.4 and 3.8.9].

The polarization vectors $e^{\pm 1}$ are coherent polarization vectors in direction $n \in \mathrm{S}^2$ whenever there exists $z \in \mathbb{Z}$ such that for any $\phi \in [0, 2\pi)$ and any k with $k/|k| \neq n$,

$$\begin{pmatrix} e_\mu^{+1}(\mathcal{R}k) \\ e_\mu^{-1}(\mathcal{R}k) \end{pmatrix} = \begin{pmatrix} \cos(z\phi) & -\sin(z\phi) \\ \sin(z\phi) & \cos(z\phi) \end{pmatrix} \begin{pmatrix} (\mathcal{R}e^{+1}(k))_\mu \\ (\mathcal{R}e^{-1}(k))_\mu \end{pmatrix}, \quad \mu = 1, 2, 3,$$

where $\mathcal{R} = \mathcal{R}(n, \phi)$ is the matrix denoting the rotation around $n \in S^2$ with angle $\phi \in [0, 2\pi)$. Let $\mathfrak{J} : L^2(\mathbb{R}^3) \oplus L^2(\mathbb{R}^3) \to L^2(\mathbb{R}^3) \oplus L^2(\mathbb{R}^3)$ be given by

$$\mathfrak{J} = \mathrm{i}\begin{pmatrix} 0 & -\mathbb{1}_{L^2(\mathbb{R}^3)} \\ \mathbb{1}_{L^2(\mathbb{R}^3)} & 0 \end{pmatrix}.$$

We define

$$S_\mathrm{f} = \mathrm{d}\Gamma(z\mathfrak{J}).$$

Let $\ell_k = k \times (-\mathrm{i}\nabla_k)$ be the triplet of angular momentum. We define L_f by

$$L_{\mathrm{f}} = (L_{\mathrm{f},1}, L_{\mathrm{f},2}, L_{\mathrm{f},3}) = \mathrm{d}\Gamma(\ell_k). \tag{5.38}$$

S_{f} is called the helicity and L_{f} the angular momentum of the field. For $(n, z) \in \mathrm{S}^2 \times \mathbb{Z}$ define $J_{\mathrm{f}} = J_{\mathrm{f}}(n, z)$ by $J_{\mathrm{f}} = n \cdot L_{\mathrm{f}} + S_{\mathrm{f}}$ and $J_{\mathrm{p}} = J_{\mathrm{p}}(n)$ by $J_{\mathrm{p}} = n \cdot \ell_x - \frac{1}{2} n \cdot \sigma$. Write

$$\mathbb{J} = J_{\mathrm{p}} \otimes \mathbb{1} + \mathbb{1} \otimes J_{\mathrm{f}}.$$

Clearly, $\mathbb{J} = \mathbb{J}(n, z)$ is defined for each $(n, z) \in \mathrm{S}^2 \times \mathbb{Z}$.

Proposition 5.13 *If the polarization vectors are coherent in direction n, and $\hat{\varphi}$ and V are rotation-invariant, then $e^{\mathrm{i}\phi\mathbb{J}} H_{\mathrm{S}} e^{-\mathrm{i}\phi\mathbb{J}} = H_{\mathrm{S}}$ for every $\phi \in \mathbb{R}$.*

Proof Write $a^\sharp(\binom{f}{g})$ for $a^\sharp(f \oplus g)$. Notice that for a rotation-invariant f,

$$e^{\mathrm{i}\phi J_{\mathrm{f}}} a^* \left(f e^{-\mathrm{i}kx} \begin{pmatrix} e_\mu^+ \\ e_\mu^- \end{pmatrix} \right) e^{-\mathrm{i}\phi J_{\mathrm{f}}} = a^* \left(f e^{\mathrm{i}\phi(z\mathfrak{J} + n\ell_k)} e^{-\mathrm{i}kx} \begin{pmatrix} e_\mu^+ \\ e_\mu^- \end{pmatrix} \right).$$

Since the polarization vectors are coherent, we have

$$e^{\mathrm{i}\phi J_{\mathrm{f}}} a^* \left(f e^{-\mathrm{i}kx} \begin{pmatrix} e_\mu^+ \\ e_\mu^- \end{pmatrix} \right) e^{-\mathrm{i}\phi J_{\mathrm{f}}} = \sum_{\nu=1}^{3} \mathcal{R}_{\mu\nu} a^* \left(f e^{-\mathrm{i}k\mathcal{R}^{-1}x} \begin{pmatrix} e_\nu^+ \\ e_\nu^- \end{pmatrix} \right), \tag{5.39}$$

where $\mathcal{R} = \mathcal{R}(n, \phi) = (\mathcal{R}_{\mu\nu})_{1 \le \mu, \nu \le 3}$. By (5.39), we see that

$$(\mathrm{S}) \begin{cases} e^{\mathrm{i}\phi J_{\mathrm{f}}} H_{\mathrm{f}} e^{-\mathrm{i}\phi J_{\mathrm{f}}} = H_{\mathrm{f}}, \\ e^{\mathrm{i}\phi J_{\mathrm{f}}} A_\mu(x) e^{-\mathrm{i}\phi J_{\mathrm{f}}} = (\mathcal{R}A)_\mu(\mathcal{R}^{-1}x), \\ e^{\mathrm{i}\phi n\cdot\ell_x} x_\mu e^{-\mathrm{i}\phi n\cdot\ell_x} = (\mathcal{R}x)_\mu, \\ e^{\mathrm{i}\phi n\cdot\ell_x} (-\mathrm{i}\nabla_x)_\mu e^{-\mathrm{i}\phi n\cdot\ell_x} = (\mathcal{R}(-\mathrm{i}\nabla_x))_\mu, \\ e^{\mathrm{i}\phi n\cdot(1/2)\sigma} \sigma_\mu e^{-\mathrm{i}\phi n\cdot(1/2)\sigma} = (\mathcal{R}^{-1}\sigma)_\mu. \end{cases}$$

Then

$$e^{\mathrm{i}\phi\mathbb{J}} H_{\mathrm{S}} e^{-\mathrm{i}\phi\mathbb{J}} = \frac{1}{2}(\mathcal{R}\sigma \cdot (\mathcal{R}(-\mathrm{i}\nabla) - \mathcal{R}A(x)))^2 + H_{\mathrm{f}} + V(\mathcal{R}x) = H_{\mathrm{S}}.$$

QED

Denote the set of half integers by $\mathbb{Z}_{1/2} = \{w/2 | w \in \mathbb{Z}\}$. For each $(n, z) \in \mathrm{S}^2 \times \mathbb{Z}$, notice that $\sigma(n \cdot (\ell_x + \frac{1}{2}\sigma)) = \mathbb{Z}_{1/2}$, $\sigma(n \cdot L_{\mathrm{f}}) = \mathbb{Z}$ and $\sigma(S_{\mathrm{f}}) = \begin{cases} \mathbb{Z}, & z \neq 0, \\ 0, & z = 0. \end{cases}$ Thus for each $(n, z) \in \mathrm{S}^2 \times \mathbb{Z}$, it follows that

$$\sigma(\mathbb{J}) = \mathbb{Z}_{1/2}. \tag{5.40}$$

Theorem 5.14 *Suppose that the polarization vectors are coherent in direction n, and $\hat{\varphi}$ and V are rotation-invariant. Then $\mathcal{H}_S$ and H_S can be decomposed as*

$$\mathcal{H}_S = \bigoplus_{w\in\mathbb{Z}_{1/2}} \mathcal{H}_S(w), \quad H_S = \bigoplus_{w\in\mathbb{Z}_{1/2}} H_S(w). \tag{5.41}$$

Here $\mathcal{H}_S(w)$ is the subspace spanned by eigenvectors of $\mathbb{J}$ associated with eigenvalue $w \in \mathbb{Z}_{1/2}$ and $H_S(w) = H_S\lceil_{\mathcal{H}_S(w)}$.

Proof This follows from Proposition 5.13 and (5.40).

QED

The Pauli–Fierz Hamiltonians with different polarization vectors, however, are unitary equivalent. Denote the Pauli–Fierz Hamiltonian with polarization vectors $e^{\pm 1}$ by $H_S(e^{\pm 1})$. Let $\eta^{\pm 1}$ be arbitrary polarization vectors. Then

$$H_S(\eta^{\pm 1}) \cong \bigoplus_{w\in\mathbb{Z}_{1/2}} H_S(e^{\pm 1}, w).$$

By using symmetries of the Pauli–Fierz Hamiltonian with spin $1/2$ we can show the degeneracy of ground states of H_S. Assume that V is rotation-invariant and the polarization vectors $e^{\pm 1}$ are given by

$$e^{+1}(k) = \frac{(-k_2, k_1, 0)}{\sqrt{k_1^2 + k_2^2}}, \quad e^{-1}(k) = \hat{k} \times e^{+1}(k) = \frac{(-k_3 k_1, -k_2 k_3, k_1^2 + k_2^2)}{|k|\sqrt{k_1^2 + k_2^2}}. \tag{5.42}$$

These are coherent in direction $(0, 0, 1)$ and their helicity is zero. Let $\Lambda : \mathbb{R}^3 \to \mathbb{R}^3$ be the flip defined by $\Lambda \begin{pmatrix} k_1 \\ k_2 \\ k_3 \end{pmatrix} = \begin{pmatrix} k_1 \\ -k_2 \\ k_3 \end{pmatrix}$, and $\tilde{u}, u : L^2(\mathbb{R}^3) \oplus L^2(\mathbb{R}^3) \to L^2(\mathbb{R}^3) \oplus L^2(\mathbb{R}^3)$ by

$$\tilde{u} : \begin{pmatrix} f \\ g \end{pmatrix} \mapsto \begin{pmatrix} f \circ \Lambda \\ g \circ \Lambda \end{pmatrix}, \quad u : \begin{pmatrix} f \\ g \end{pmatrix} \mapsto \begin{pmatrix} -f \circ \Lambda \\ g \circ \Lambda \end{pmatrix}.$$

A computation gives

$$u^{\sharp -1} k_\mu u^\sharp = \begin{cases} k_\mu, & \mu = 1, 3, \\ -k_\mu, & \mu = 2, \end{cases} \quad u^{\sharp -1} \nabla_\mu u^\sharp = \begin{cases} \nabla_\mu, & \mu = 1, 3, \\ -\nabla_\mu, & \mu = 2, \end{cases}$$

where $u^\sharp = u$ or $\tilde{u}$. We have for rotation-invariant f and g,

$$u^{-1} \begin{pmatrix} e_\mu^{+1} f \\ e_\mu^{-1} g \end{pmatrix} = \begin{cases} \begin{pmatrix} e_\mu^{+1} f \\ e_\mu^{-1} g \end{pmatrix}, & \mu = 1, 3, \\ -\begin{pmatrix} e_\mu^{+1} f \\ e_\mu^{-1} g \end{pmatrix}, & \mu = 2. \end{cases}$$

Then the second quantization $\Gamma(u)$ of u induces the unitary operator on $\mathcal{F}$ and for rotation-invariant $\hat{\varphi}$ we obtain

$$\Gamma(u)^{-1}A_\mu(x)\Gamma(u) = \begin{cases} A_\mu(\Lambda x), & \mu = 1, 3, \\ -A_\mu(\Lambda x), & \mu = 2, \end{cases} \tag{5.43}$$

$$\Gamma(u)^{-1}H_{\mathrm{f}}\Gamma(u) = H_{\mathrm{f}}. \tag{5.44}$$

Next we consider the transformation on $\sigma = (\sigma_1, \sigma_2, \sigma_3)$ given by

$$\sigma_\mu \mapsto \sigma_2\sigma_\mu\sigma_2 = \begin{cases} -\sigma_\mu, & \mu = 1, 3, \\ \sigma_\mu, & \mu = 2. \end{cases} \tag{5.45}$$

Under the identification $L^2(\mathbb{R}^3) \oplus L^2(\mathbb{R}^3) \cong \mathbb{C}^2 \otimes L^2(\mathbb{R}^3)$, we define $\tau = \sigma_2 \otimes \tilde{u}$. This satisfies

$$\tau^{-1}(\sigma_\mu \otimes f)\tau = \begin{cases} -\sigma_\mu \otimes f \circ \Lambda^{-1}, & \mu = 1, 3, \\ \sigma_\mu \otimes f \circ \Lambda^{-1}, & \mu = 2, \end{cases} \tag{5.46}$$

$$\tau^{-1}(\sigma_\mu \otimes \nabla_\mu)\tau = -\sigma_\mu \otimes \nabla_\mu, \quad \mu = 1, 2, 3. \tag{5.47}$$

We finally define the unitary operator $\mathfrak{J} : \mathcal{H}_S \to \mathcal{H}_S$ by

$$\mathfrak{J} = \tau \otimes \Gamma(u).$$

Combining (5.43)–(5.47), for rotation-invariant $\hat{\varphi}$ and V

$$\begin{aligned} &\mathfrak{J}^{-1}\sigma_\mu(-\mathrm{i}\nabla_\mu - A_\mu)\mathfrak{J} = -\sigma_\mu(-\mathrm{i}\nabla_\mu - A_\mu), \\ &\mathfrak{J}^{-1}H_{\mathrm{f}}\mathfrak{J} = H_{\mathrm{f}}, \\ &\mathfrak{J}^{-1}V\mathfrak{J} = V \end{aligned}$$

are obtained. From these relations we can show the theorem below.

Theorem 5.15 *If $\hat{\varphi}$ and V are rotation-invariant, and the polarization vectors $e^{\pm 1}$ are given by* (5.42), *then $H_{\mathrm{S}}(z)$ and $H_{\mathrm{S}}(-z)$ are unitary equivalent.*

Proof Since $e^{\pm 1}$ is coherent in direction $(0, 0, 1)$ and its helicity is zero, $\mathbb{J}$ is of the form $\mathbb{J} = (\ell_{x,3} - \frac{1}{2}\sigma_3) \otimes \mathbb{1} + \mathbb{1} \otimes L_{\mathrm{f},3}$. It follows that $\mathfrak{J}^{-1}\mathbb{J}\mathfrak{J} = -\mathbb{J}$. This implies that $\mathfrak{J}$ maps $\mathcal{H}_{\mathrm{S}}(w)$ onto $\mathcal{H}_{\mathrm{S}}(-w)$. Furthermore, $\mathfrak{J}^{-1}H_{\mathrm{S}}\mathfrak{J} = H_{\mathrm{S}}$. Thus $\mathfrak{J}^{-1}H_{\mathrm{S}}(w)\mathfrak{J} = H_{\mathrm{S}}(-w)$ follows.

QED

An application of Theorem 5.15 is to estimate the multiplicity of eigenvalues of H_{S}.

Corollary 5.16 *Suppose that V and $\hat{\varphi}$ are rotation-invariant. Then the multiplicity of any eigenvalue is a non-negative even number. Moreover $\mathfrak{S}^{-1}e^{-tH_{\mathrm{S}}}\mathfrak{S}$ is not positivity improving.*

Proof We may suppose that the polarization vectors of H_S are given by (5.42). Thus $H_S = \bigoplus_{w \in \mathbb{Z}_{1/2}} H_S(w)$. Theorem 5.15 implies $H_S(w) \cong H_S(-w)$, and the multiplicity of any eigenvalue is even. The existence of the ground state is established in e.g., [8–13, 21, 45, 51, 97]. $\mathfrak{S}^{-1}e^{-tH_S}\mathfrak{S}$ can not be positivity improving, since the ground state is degenerate.

QED

6 Energy Comparison Inequalities

We show several inequalities derived from functional integral representations for the Pauli–Fierz type Hamiltonians. These inequalities are useful to study the properties of the Pauli–Fierz type Hamiltonians. See Table 7.

6.1 Pauli–Fierz Hamiltonian H and $H(p)$

An application of the functional integral representation of $(F, e^{-tH}G)$ is a diamagnetic inequality.

Corollary 6.1 *Under the conditions of Theorem* 3.3 *it follows that*

$$|(F, e^{-tH}G)| \le (|F|, e^{-t(H_p+H_f)}|G|). \tag{6.1}$$

Proof By the functional integral representation in Theorem 3.3 we have

$$|(F, e^{-tH}G)| \le \int_{\mathbb{R}^3} \mathbb{E}^x[e^{-\int_0^t V(B_s)ds}(\mathrm{J}_0|F(B_0)|, \mathrm{J}_t|G(B_t)|)]dx = (|F|, e^{-t(H_p+H_f)}|G|).$$

Here we used that $|\mathrm{J}_s F| \le \mathrm{J}_s|F|$, since J_s is positivity preserving.

QED

Table 7 Energy comparison inequality

	Energy comparison inequality
H	$E(0) \le E(e^2)$
$H(p)$	$E(e^2, 0) \le E(e^2, p)$
H_R	$E_R(0) \le E_R(e^2)$
$H_R(p)$	$E_R(e^2, 0) \le E_R(e^2, p)$
H_S	$\max_{(\alpha\beta\gamma)=(123),(231),(312)} E(0, \sqrt{\hat{B}_\alpha^2 + \hat{B}_\beta^2}, 0, \hat{B}_\gamma) \le E(e^2, \hat{B}_1, \hat{B}_2, \hat{B}_3)$
$H_S(p)$	$\max_{(\alpha\beta\gamma)=(123),(231),(312)} E(0, p, \sqrt{\hat{B}_\alpha^2 + \hat{B}_\beta^2}, 0, \hat{B}_\gamma) \le E(e^2, p, \hat{B}_1, \hat{B}_2, \hat{B}_3)$

Equation (6.1) is called the diamagnetic inequality. The diamagnetic inequality shows that coupling a particle to a quantized radiation field by minimal interaction increases the ground state energy of the non-interacting system. The exact statement is as follows. We introduce a coupling constant $e \in \mathbb{R}$ to the Pauli–Fierz Hamiltonian H as

$$H = \frac{1}{2}(-\mathrm{i}\nabla - e\hat{A})^2 + V + H_{\mathrm{f}}.$$

By a symmetry we can see that

$$\frac{1}{2}(-\mathrm{i}\nabla - e\hat{A})^2 + V + H_{\mathrm{f}} \cong \frac{1}{2}(-\mathrm{i}\nabla + e\hat{A})^2 + V + H_{\mathrm{f}}.$$

Then $\inf \sigma(H)$ depends on e^2. Let $E(e^2) = \inf \sigma(H)$.

Corollary 6.2 *We have* $E(0) \le E(e^2)$.

Proof Corollary 6.1 implies that

$$\sup_{F \neq 0} \frac{(F, e^{-tH} F)}{\|F\|^2} \le \sup_{F \neq 0} \frac{(|F|, e^{-t(H_{\mathrm{p}}+H_{\mathrm{f}})}|F|)}{\|F\|^2} \le \sup_{F \neq 0} \frac{(F, e^{-t(H_{\mathrm{p}}+H_{\mathrm{f}})} F)}{\|F\|^2}.$$

Hence $\|e^{-tH}\| \le \|e^{-t(H_{\mathrm{p}}+H_{\mathrm{f}})}\| = \|e^{-tH_{\mathrm{p}}}\|$. It implies that $\inf \sigma(H_{\mathrm{p}}) \le \inf \sigma(H)$. Thus the corollary follows.

QED

Another significant property of e^{-tH} is positivity improving. Unitary operator $e^{\mathrm{i}t(-\mathrm{i}\nabla_\mu)}$ is a shift operator on $L^2(\mathbb{R}^3)$, hence it is positivity preserving. While $e^{-t(-\Delta)}$ is positivity improving. I.e., $(f, e^{-t(-\Delta)}g) > 0$ for any non identically zero functions $f \ge 0$ and $g \ge 0$. By the integral kernel $e^{-\mathrm{i}\hat{A}(K_t)}$, in general $(F, e^{-tH}G) \in \mathbb{C}$ for $F \ge 0$ and $G \ge 0$. Let Ψ_{g} be the ground state of H. By Corollary 3.10 we see that $e^{-\mathrm{i}(\pi/2)N}\Psi_{\mathrm{g}}$ is strictly positive. Hence $(f \otimes \mathbb{1}, e^{-\mathrm{i}(\pi/2)N}\Psi_{\mathrm{g}}) \neq 0$ for any non identically zero $f \ge 0$. It is easy to see that $\lim_{t\to\infty}(f \otimes \mathbb{1}, e^{-t(H-E(e^2))} f \otimes \mathbb{1}) = |(f \otimes \mathbb{1}, \Psi_{\mathrm{g}})|^2 > 0$ and hence

$$\begin{aligned} &-\lim_{t\to\infty} \frac{1}{t} \log(f \otimes \mathbb{1}, e^{-tH} f \otimes \mathbb{1}) = -\lim_{t\to\infty} \frac{1}{t} \log \left\{ e^{-tE(e^2)} (f \otimes \mathbb{1}, e^{-t(H-E(e^2))} f \otimes \mathbb{1}) \right\} \\ &= E(e^2) - \lim_{t\to\infty} \frac{1}{t} \log(f \otimes \mathbb{1}, e^{-t(H-E(e^2))} f \otimes \mathbb{1}) = E(e^2). \end{aligned}$$

This allows to derive an expression of the ground state energy $E(e^2)$.

Corollary 6.3 *Suppose either Assumption 3.1 or Assumption 3.2. Then the function*

$$e^2 \mapsto E(e^2)$$

is monotonously increasing, continuous and concave.

Proof Notice that

$$(\mathbb{1}, e^{-\mathrm{i}e\hat{A}(K_t)}\mathbb{1}) = e^{-(e^2/2)\mathrm{q}(K_t,K_t)}.$$

For $f \geq 0$, we have

$$\begin{aligned} E(e^2) &= -\lim_{t\to\infty}\frac{1}{t}\log(f\otimes\mathbb{1}, e^{-tH}f\otimes\mathbb{1}) \\ &= -\lim_{t\to\infty}\frac{1}{t}\log\int_{\mathbb{R}^3}\mathbb{E}^x[f(B_0)f(B_t)e^{-\int_0^t V(B_s)\mathrm{d}s}e^{-(e^2/2)\mathrm{q}(K_t,K_t)}]\mathrm{d}x. \end{aligned}$$

As $e^{-(e^2/2)\mathrm{q}_\mathrm{E}(K_t,K_t)}$ is log-convex in e^2, $E(\cdot)$ is concave. Thus $E(e^2)$ is continuous on $(0,\infty)$. Since $E(e^2)$ is continuous at $e=0$, $E(e^2)$ can be expressed as

$$E(e^2) = E(0) + \int_0^{e^2}\rho(t)\mathrm{d}t$$

with a suitable non-negative function $\rho(t)$. This implies that $E(e^2)$ is monotonously increasing in e^2.

QED

Remark 6.1 The formula

$$-\lim_{t\to\infty}\frac{1}{t}\log(f\otimes\mathbb{1}, e^{-tH}f\otimes\mathbb{1}) = E(e^2)$$

holds for arbitrary strictly positive function f regardless of the existence of ground states. See [74, Lemma 1.56].

We derive a similar inequality for $e^{-tH(p)}$ to that for e^{-tH}.

Corollary 6.4 *It follows that*

$$|(\Psi, e^{-tH(p)}\Phi)| \leq (|\Psi|, e^{-t(\mathrm{P}_\mathrm{f}^2+H_\mathrm{f})}|\Phi|).$$

Proof By the functional integral representation of $(\Psi, e^{-tH(p)}\Phi)$ we have

$$|(\Psi, e^{-tH(p)}\Phi)| \leq \mathbb{E}[(\mathrm{J}_0|\Psi|, \mathrm{J}_t e^{-\mathrm{i}B_t\mathrm{P}_\mathrm{f}}|\Phi|] = (|\Psi|, e^{-t(\mathrm{P}_\mathrm{f}^2+H_\mathrm{f})}|\Phi|).$$

Here we used that $e^{-\mathrm{i}B_t\mathrm{P}_\mathrm{f}}$ is positivity preserving and hence $|e^{-\mathrm{i}B_t\mathrm{P}_\mathrm{f}}\Phi| \leq e^{-\mathrm{i}B_t\mathrm{P}_\mathrm{f}}|\Phi|$.

QED

From this diamagnetic inequality, we can only deduce the trivial energy comparison inequality $0 = \inf \sigma(\mathrm{P}_\mathrm{f}^2 + H_\mathrm{f}) \le \inf \sigma(H(p))$. However, combining the unitary transformation $e^{-\mathrm{i}(\pi/2)N}$ with the functional integral representation, we obtain an interesting result. Denote $E(e^2, p) = \inf \sigma(H(p))$.

Corollary 6.5 *It follows that*

(1) *Let* $p = 0$. *Then* $e^{\mathrm{i}(\pi/2)N} e^{-tH(0)} e^{-\mathrm{i}(\pi/2)N}$ *is positivity improving.*
(2) *The ground state of* $H(0)$ *is unique whenever it exists.*
(3) $E(e^2, 0) \le E(e^2, p)$ *holds.*
(4) *Map* $p \mapsto E(e^2, p)$ *is continuous and* $E(e^2, 0) = \inf \sigma(H_{V=0})$, *where* $H_{V=0}$ *is* H *with* $V = 0$.

Proof In the case of $p = 0$ we remark that $e^{\mathrm{i}pB_t} = 1$. Then we have

$$(\Psi, e^{\mathrm{i}(\pi/2)N} e^{-tH(0)} e^{-\mathrm{i}(\pi/2)N} \Phi) = \mathbb{E}^0[(\mathrm{J}_0 \Psi, e^{-\mathrm{i}\pi(K_t)} \mathrm{J}_t e^{-\mathrm{i}\mathrm{P}_\mathrm{f} B_t} \Phi)].$$

Since $\mathrm{J}_0^* e^{-\mathrm{i}\pi(K_t)} \mathrm{J}_t$ is positivity improving and $e^{-\mathrm{i}\mathrm{P}_\mathrm{f} B_t}$ is positivity preserving, (1) follows. (2) is implied by (1) and the Perron-Frobenius theorem [47]. We have

$$\begin{aligned} |(\Psi, e^{\mathrm{i}(\pi/2)N} e^{-tH(p)} e^{-\mathrm{i}(\pi/2)N} \Phi)| &\le \mathbb{E}^0[(\mathrm{J}_0 |\Psi|, e^{-\mathrm{i}\pi(K_t)} \mathrm{J}_t e^{-\mathrm{i}\mathrm{P}_\mathrm{f} B_t} |\Phi|)] \\ &= (|\Psi|, e^{\mathrm{i}(\pi/2)N} e^{-tH(0)} e^{-\mathrm{i}(\pi/2)N} |\Phi|). \end{aligned}$$

This yields (3). Finally we show (4). From $|(\Psi, (e^{-tH(p)} - e^{-tH(q)})\Phi)| \le c|p - q|$ the continuity follows. Let $E = \inf \sigma(H_{V=0})$ and we set $E(e^2, p) = E(p)$. We have

$$(\Psi, H\Psi) = \int_{\mathbb{R}^3} (\Psi(p), H(p)\Psi(p)) \mathrm{d}p \ge E(0) \|\Psi\|.$$

Then $E(0) \le E$ follows. Let $\Phi_\delta = \int_{\mathbb{R}^3}^{\oplus} \Phi(q) 1_{\{|p-q| \le \delta\}} \mathrm{d}q$, where we assume that $\Phi(q)$ satisfies

$$E(q) \le \frac{(\Phi(q), H(q)\Phi(q))}{\|\Phi(q)\|^2} < E, \quad |p - q| \le \delta.$$

We have

$$E\|\Phi_\delta\|^2 > \int_{|q-p| \le \delta} (\Phi(q), H(q)\Phi(q)) \mathrm{d}q = (\Phi_\delta, H\Phi_\delta).$$

This contradicts that E is the bottom of the spectrum of $H_{V=0}$. Hence $E(0) \le E \le E(p)$ for all $p \in \mathbb{R}^3$. From the continuity of $E(p)$, it follows that $E(0) = E$.

QED

6.2 Relativistic Pauli–Fierz Hamiltonian H_{R} and $H_{\mathrm{R}}(p)$

For H_{R} by using the functional integral representation we can obtain similar results to those of H. Let $E_{\mathrm{R}}(e^2) = \inf \sigma(H_{\mathrm{R}})$ and $E_{\mathrm{R}}(e^2, p) = \inf \sigma(H_{\mathrm{R}}(p))$.

Corollary 6.6 *Suppose Assumption 4.1.*
(1) It follows that

$$|(F, e^{-tH_{\mathrm{R}}}G)| \le \left(|F|, e^{-t((-\Delta+m^2)^{1/2}-m+V+H_{\mathrm{f}})}|G|\right).$$

(2) The function $e^2 \mapsto E_{\mathrm{R}}(e^2)$ is monotonously increasing, continuous and concave.
(3) It follows that

$$|(\Psi, e^{-tH_{\mathrm{R}}(p)}\Phi)| \le (|\Psi|, e^{-t(\sqrt{\mathrm{P}_{\mathrm{f}}^2+m^2}-m+H_{\mathrm{f}})}|\Phi|).$$

(4) Statements (1)–(4) below are satisfied.

(1) *Let $p = 0$. Then $e^{\mathrm{i}(\pi/2)N}e^{-tH_{\mathrm{R}}(0)}e^{-\mathrm{i}(\pi/2)N}$ is positivity improving.*
(2) *The ground state of $H_{\mathrm{R}}(0)$ is unique whenever it exists.*
(3) *Energy comparison inequality $E_{\mathrm{R}}(0) \le E_{\mathrm{R}}(p)$ holds.*
(4) *Map $p \mapsto E_{\mathrm{R}}(p)$ is continuous and $E_{\mathrm{R}}(0) = \inf \sigma(H_{\mathrm{R}\,V=0})$, where $H_{\mathrm{R}\,V=0}$ is H_{R} with $V = 0$.*

6.3 Pauli–Fierz Hamiltonian with Spin 1/2 $H_{\mathbb{Z}_2}$ and $H_{\mathbb{Z}_2}(p)$

Let us consider $H_{\mathbb{Z}_2}$. By using the functional integral representation we can estimate the ground state energy of $H_{\mathbb{Z}_2}$. Write

$$E(e^2, \hat{B}_1, \hat{B}_2, \hat{B}_3) = \inf \sigma(H_{\mathbb{Z}_2}).$$

For the spinless Pauli–Fierz Hamiltonian H we have $\inf \sigma(H) = E(e^2, 0, 0, 0)$ and the diamagnetic inequality $E(0, 0, 0, 0) \le E(e^2, 0, 0, 0)$ was already seen. We extend this inequality to $H_{\mathbb{Z}_2}$. Define

$$(H_{\mathbb{Z}_2}(0)F)(\theta) = (H_{\mathrm{p}} + H_{\mathrm{f}} + \hat{H}_{\mathrm{d}}(\theta))F(\theta) - |\hat{H}_{\mathrm{od}}(-\theta)|F(-\theta),$$

where

$$|\hat{H}_{\mathrm{od}}(-\theta)| = \frac{1}{2}\sqrt{\hat{B}_1^2(x) + \hat{B}_2^2(x)}$$

is independent of $\theta \in \mathbb{Z}_2$. $H_{\mathbb{Z}_2}(0)$ corresponds to

$$H_{\rm p}+H_{\rm f}-\frac{1}{2}\begin{pmatrix}\hat{B}_3 & \sqrt{\hat{B}_1^2+\hat{B}_2^2}\\ \sqrt{\hat{B}_1^2+\hat{B}_2^2} & -\hat{B}_3\end{pmatrix}$$

acting in $L^2(\mathbb{R}^3;\mathbb{C}^2)\otimes L^2(\mathrm{Q})$. The functional integral representation of $e^{-tH_{\mathbb{Z}_2}(0)}$ is given by

$$(F,e^{-tH_{\mathbb{Z}_2}(0)}G)=\lim_{\varepsilon\downarrow 0}e^t\sum_{\alpha\in\mathbb{Z}_2}\int_{\mathbb{R}^3}\mathbb{E}^{x,\alpha}[e^{-\int_0^t V(B_s)\mathrm{d}s}(\mathrm{J}_0F(X_0),e^{X_t^\perp(\varepsilon)}\mathrm{J}_tG(X_t))]\mathrm{d}x,$$

where

$$X_t^\perp(\varepsilon)=-\int_0^t\mathbb{H}_{\rm d}(B_s,\theta_{N_s},s)\mathrm{d}s+\int_0^{t+}\log(\Psi_\varepsilon(|\mathbb{H}_{\rm od}(B_s,s)|))\mathrm{d}N_s. \tag{6.2}$$

Corollary 6.7 *It follows that*

$$|(F,e^{-tH_{\mathbb{Z}_2}}G)|\le(|F|,e^{-tH_{\mathbb{Z}_2}(0)}|G|) \tag{6.3}$$

and

$$\max_{(\alpha\beta\gamma)=(123),(231),(312)}E(0,\sqrt{\hat{B}_\alpha^2+\hat{B}_\beta^2},0,\hat{B}_\gamma)\le E(e^2,\hat{B}_1,\hat{B}_2,\hat{B}_3). \tag{6.4}$$

Proof Since $|\Psi_\varepsilon(\mathbb{H}_{\rm od})|\le\Psi_\varepsilon(|\mathbb{H}_{\rm od}|)$, $|e^{Z_t(\varepsilon)}|\le e^{X_t^\perp(\varepsilon)}$ and $|\mathrm{J}_tG|\le \mathrm{J}_t|G|$, by the functional integral representation of $e^{-tH_{\mathbb{Z}_2}}$ we have (6.3). From this,

$$E(0,\sqrt{\hat{B}_1^2+\hat{B}_2^2},0,\hat{B}_3)\le E(e^2,\hat{B}_1,\hat{B}_2,\hat{B}_3)$$

is obtained. We will show that

$$E(e^2,\hat{B}_1,\hat{B}_2,\hat{B}_3)=E(e^2,\hat{B}_2,\hat{B}_3,\hat{B}_1)=E(e^2,\hat{B}_3,\hat{B}_1,\hat{B}_2) \tag{6.5}$$

by SU(2)-symmetry. Let $\mathcal{R}\in O(3)$ be such that $\mathcal{R}\begin{pmatrix}x_1\\x_2\\x_3\end{pmatrix}=\begin{pmatrix}x_2\\x_3\\x_1\end{pmatrix}$. Then there exists $(n,\phi)\in S^2\times\mathbb{R}$ such that $\mathcal{R}=\mathcal{R}(n,\phi)$. Here $\mathcal{R}(n,\phi)$ describes the rotation around $n\in\mathbb{R}^3$ by angle $\phi\in[0,2\pi)$. Hence we see that

$$e^{\mathrm{i}\phi n\cdot(1/2)\sigma}\sigma_\mu e^{-\mathrm{i}\phi n\cdot(1/2)\sigma}=(\mathcal{R}^{-1}\sigma)_\mu.$$

Now we write $H_{\mathbb{Z}_2}$ by $H_{\mathbb{Z}_2}(e,\hat{B}_1,\hat{B}_2,\hat{B}_3)$. Thus we obtain that

$$e^{\mathrm{i}\phi n\cdot(1/2)\sigma}H_{\mathbb{Z}_2}(e,\hat{B}_1,\hat{B}_2,\hat{B}_3)e^{-\mathrm{i}\phi n\cdot(1/2)\sigma}=H_{\mathbb{Z}_2}(e,\hat{B}_2,\hat{B}_3,\hat{B}_1)$$

which implies the first equality in (6.5). The second one is proven in the same way.

QED

Write

$$E(p, \hat{A}, \hat{B}_1, \hat{B}_2, \hat{B}_3) = \inf \sigma(H_{\mathbb{Z}_2}(p)),$$

and define $H_{\mathbb{Z}_2}(p, 0)$ by

$$(H_{\mathbb{Z}_2}(p, 0)\Psi)(\theta) = \left(\frac{1}{2}(p - \mathrm{P_f})^2 + \hat{H}_{\mathrm{d}}(0, \theta)\right)\Psi(\theta) - |\hat{H}_{\mathrm{od}}(0, -\theta)|\Psi(-\theta),$$

where $|\hat{H}_{\mathrm{od}}(0, -\theta)| = \frac{1}{2}\sqrt{\hat{B}_1(0)^2 + \hat{B}_2(0)^2}$. This corresponds to

$$\frac{1}{2}(p - \mathrm{P_f})^2 + H_{\mathrm{f}} - \frac{1}{2}\begin{pmatrix} \hat{B}_3(0) & \sqrt{\hat{B}_1(0)^2 + \hat{B}_2(0)^2} \\ \sqrt{\hat{B}_1(0)^2 + \hat{B}_2(0)^2} & -\hat{B}_3(0) \end{pmatrix}$$

in $\mathbb{C}^2 \otimes L^2(\mathrm{Q})$. Note that $|\hat{H}_{\mathrm{od}}(0, -\theta)|$ is independent of θ.

Corollary 6.8 *It follows that*

$$|(\Phi, e^{-tH_{\mathbb{Z}_2}(p)}\Psi)| \le (|\Phi|, e^{-tH_{\mathbb{Z}_2}(p,0)}|\Psi|) \tag{6.6}$$

and

$$\max_{(\alpha\beta\gamma)=(123),(231),(312)} E(0, 0, \sqrt{\hat{B}_\alpha^2 + \hat{B}_\beta^2}, 0, \hat{B}_\gamma) \le E(p, \hat{A}, \hat{B}_1, \hat{B}_2, \hat{B}_3). \tag{6.7}$$

Proof We have

$$|(\Phi, e^{-tH_{\mathrm{S}}(p)}\Psi)| \le e^t \lim_{\varepsilon\downarrow 0} \sum_{\theta\in\mathbb{Z}_2} \mathbb{E}^{0,\alpha}[(\mathrm{J}_0|\Phi(\theta_\alpha)|, e^{X_t^\perp(\varepsilon)}\mathrm{J}_t e^{-\mathrm{i}\mathrm{P_f}B_t}|\Phi(\theta_{N_t})|)] = \mathrm{RHS}\ (6.6),$$

where $X_t^\perp(\varepsilon)$ is given by (6.2). Equation (6.7) is immediate from (6.6) and a similar argument to (6.4).

QED

7 Concluding Remarks

(1) The positivity improvingness of the semigroups generated by translation invariant Hamiltonians is introduced in this paper only for the case $p = 0$. We can also show the positivity improvingness for $p \neq 0$ by functional integral representation. We refer to see [70]. This is also prove in [110] in the different manners.

(2) To find an invariant domain by using functional integral representations can be extended to more general cases. We refer to see [69].
(3) The Nelson model without cutoff function can be defined through a renormalization. The Nelson model with a cutoff function is defined by

$$H = H_{\mathrm{p}} \otimes \mathbb{1} + \mathbb{1} \otimes H_{\mathrm{f}} + \phi(\tilde{\varphi}(\cdot - x)),$$

where $\phi(f)$ for $f \in L^2(\mathbb{R}^3)$ denotes a scalar field satisfying $\mathbb{E}[e^{z\phi(f)}] = e^{z^2\|f\|^2/2}$ for $z \in \mathbb{C}$, $\mathbb{E}[\phi(f)\phi(g)] = \frac{1}{2}(f, g)$ and $\mathbb{E}[\phi(f)] = 0$. The renormalization was succeeded in by Edward Nelson [111, 112], and gave a proof in terms of a path measure by [49, 107]. The spectral properties of the renormalized Nelson Hamiltonian H_∞ is studied in [60, 75, 76]. In particular in [75] a Gibbs measure [14, 17] associated with the ground state Ψ_{g} of H_∞ is constructed, and $(\Psi_{\mathrm{g}}, \mathscr{O}\Psi_{\mathrm{g}})$ for $\mathscr{O} = e^{\beta N}, e^{\phi(f)^2}$ are expressed in terms of the Gibbs measure.

For the Gibbs measure associated with the ground state of the Pauli–Fierz Hamiltonian is constructed in [66], but there is no useful applications.

Acknowledgements FH is financially supported by JSPS KAKENHI 20K20886 and JSPS KAKENHI 20H01808.

Appendix A Abstract Boson Fock Space

A.1 Annihilation Operators and Creation Operators

In this Appendix we introduce general tools concerning with abstract boson Fock spaces and second quantization. Instead of physical space-time dimension we work in general d dimension.

Let $\mathfrak{H}$ be a separable Hilbert space over $\mathbb{C}$. Consider the operation $\otimes_{\mathrm{s}}^n$ of n-fold symmetric tensor product defined through the symmetrization operator

$$S_n(f_1 \otimes \cdots \otimes f_n) = \frac{1}{n!}\sum_{\pi \in \mathfrak{S}_n} f_{\pi(1)} \otimes \cdots \otimes f_{\pi(n)}, \quad n \geq 1,$$

where $f_1, \ldots, f_n \in \mathfrak{H}$ and $\mathfrak{S}_n$ denotes the permutation group of order n. Define

$$\mathcal{F}^{(n)} = \begin{cases} S_n(\otimes^n \mathfrak{H}), & n \geq 1, \\ \mathbb{C}, & n = 0. \end{cases}$$

The space

$$\mathcal{F} = \mathcal{F}(\mathfrak{H}) = \left\{ \bigoplus_{n=0}^{\infty} \mathcal{F}^{(n)} \ni \bigoplus_{n=0}^{\infty} \Psi^{(n)} \,\middle|\, \|\Psi\|_{\mathcal{F}}^2 = \sum_{n=0}^{\infty} \|\Psi^{(n)}\|_{\mathcal{F}^{(n)}}^2 < \infty \right\}$$

is called the boson Fock space over $\mathfrak{H}$. The boson Fock space $\mathcal{F}$ can be identified with the space of ℓ_2-sequences $(\Psi^{(n)})_{n\geq 0}$ such that $\Psi^{(n)} \in \mathcal{F}^{(n)}$ and $\sum_{n=0}^{\infty} \|\Psi^{(n)}\|^2_{\mathcal{F}^{(n)}} < \infty$. $\mathcal{F}$ is a Hilbert space endowed with the scalar product

$$(\Psi, \Phi)_{\mathcal{F}} = \sum_{n=0}^{\infty} (\Psi^{(n)}, \Phi^{(n)})_{\mathcal{F}^{(n)}}.$$

The vector $\Omega = (1, 0, 0, \dots)$ is called Fock vacuum. The subspace $\mathcal{F}^{(n)}$ can be interpreted as consisting of the states of the quantum field having exactly n boson particles, while the permutation symmetry corresponds to the fact that the particles are indistinguishable.

There are two fundamental operators, the creation operator denoted by $a^*(f)$, $f \in \mathfrak{H}$, and the annihilation operator by $a(f)$ defined by

$$(a^*(f)\Psi)^{(n)} = \begin{cases} \sqrt{n} S_n(f \otimes \Psi^{(n-1)}), & n \geq 1, \\ 0, & n = 0 \end{cases}$$

with domain

$$D(a^*(f)) = \left\{ (\Psi^{(n)})_{n\geq 0} \in \mathcal{F} \,\middle|\, \sum_{n=1}^{\infty} n \|S_n(f \otimes \Psi^{(n-1)})\|^2_{\mathcal{F}^{(n)}} < \infty \right\},$$

and

$$a(f) = (a^*(\bar{f}))^*.$$

Note that $a^\sharp(f)$ is linear in f by the definitions.

As the terminology suggests, the action of $a^*(f)$ increases the number of bosons by one, while $a(f)$ decreases it by one. The relation

$$(\Phi, a(f)\Psi) = (a^*(\bar{f})\Phi, \Psi)$$

holds. Furthermore, since both operators are closable by the dense definition of their adjoints, we use and denote their closed extensions by the same symbols. Let $D \subset \mathfrak{H}$ be a dense subset. It is known that

$$\mathrm{LH}\{a^*(f_1) \cdots a^*(f_n)\Omega, \Omega | \, f_j \in D, \, j = 1, \dots, n, n \geq 1\}$$

is dense in $\mathcal{F}(\mathfrak{H})$. The space

$$\mathcal{F}_{\text{fin}} = \{(\Psi^{(n)})_{n\geq 0} \in \mathcal{F} | \, \Psi^{(m)} = 0 \text{ for all } m \geq M \text{ with some } M\}$$

is called the finite particle subspace. The field operators $a(f)$ and $a^*(f)$ leave $\mathcal{F}_{\text{fin}}$ invariant and satisfy the canonical commutation relations

$$[a(f), a^*(g)] = (\bar{f}, g)\mathbb{1}, \quad [a(f), a(g)] = 0 = [a^*(f), a^*(g)]$$

on $\mathcal{F}_{\mathrm{fin}}$. We introduce several technical estimates. Let

$$\Phi(f) = \frac{1}{\sqrt{2}}(a^*(f) + a(\bar{f})).$$

Then $\Phi(f)$ is essentially self-adjoint on $\mathcal{F}_{\mathrm{fin}}$. The conjugate momentum operator is also defined by

$$\Pi(f) = \frac{\mathrm{i}}{\sqrt{2}}(a^*(f) - a(\bar{f})).$$

They satisfy that

$$[\Phi(f), \Pi(g)] = i\Re(f, g), \quad [\Pi(f), \Pi(g)] = i\Im(f, g), \quad [\Phi(f), \Phi(g)] = i\Im(f, g).$$

Products $\prod_{j=1}^m \Phi(f_j)$ can be represented by the sum of Wick ordered operators. We see this in the lemma below. We simply write

$$c^m_{j_1,\dots,j_k} = {:}\Phi(f_1)\cdots\widehat{\Phi(f_{j_1})}\cdots\cdots\widehat{\Phi(f_{j_k})}\cdots\Phi(f_m){:},$$

where $\widehat{\Phi(f)}$ describes neglecting $\Phi(f)$. We also set

$$c^m = {:}\Phi(f_1)\cdots\Phi(f_m){:}.$$

Lemma A.1 *Let $f_j \in \mathfrak{H}$, $j = 1, \dots, n$. Then $\prod_{j=1}^m \Phi(f_j)$ can be represented in terms of Wick ordered operators as follows.*
$(m = 2n)$

$$\prod_{j=1}^{2n} \Phi(f_j) = c^{2n} + \sum_{k=1}^{n} \frac{1}{2^k} \sum_{j_1,\dots,j_{2k}}^{\mathrm{pair}} \left(\prod_{i=1}^{k} (f_{j_{2i-1}}, f_{j_{2i}})\right) c^{2n}_{j_1,\dots,j_{2k}}, \tag{A.1}$$

where $\sum_{j_1,\dots,j_{2k}}^{\mathrm{pair}}$ describes the summation over all k-pairs chosen from $\{1, \dots, 2n\}$.
$(m = 2n + 1)$

$$\prod_{j=1}^{2n+1} \Phi(f_j) = c^{2n+1} + \sum_{k=1}^{n} \frac{1}{2^k} \sum_{j_1,\dots,j_{2k}}^{\mathrm{pair}} \left(\prod_{i=1}^{k} (f_{j_{2i-1}}, f_{j_{2i}})\right) c^{2n+1}_{j_1,\dots,j_{2k}}, \tag{A.2}$$

where $\sum_{j_1,\dots,j_{2k}}^{\mathrm{pair}}$ also describes the summation over all k-pairs chosen from $\{1, \dots, 2n + 1\}$.

Proof From the definition of Wick products the lemma directly follows.

QED

Lemma A.2 *Let $f \in \mathfrak{H}$. Then*

$$\Phi(f)^n = n! \sum_{k=0}^{[n/2]} \sum_{l+m+2k=n} \frac{(\frac{\|f\|^2}{4})^k}{k!} \frac{(\frac{1}{\sqrt{2}}a^*(f))^l}{l!} \frac{(\frac{1}{\sqrt{2}}a(\bar{f}))^m}{m!}. \tag{A.3}$$

Here $[m]$ is the integer part of m, i.e., $[2n/2] = n$ and $[(2n+1)/2] = n$.

Proof We see that by Lemma A.1

$$\Phi(f)^n = \sum_{k=0}^{[n/2]} \frac{n!}{k!(n-2k)!} \left(\frac{\|f\|^2}{4}\right)^k :\Phi(f)^{n-2k}:,$$

$$:\Phi(f)^m: = 2^{-m/2} \sum_{k=0}^{m} \binom{m}{k} a^*(f)^k a(\bar{f})^{m-k}.$$

Together with them (A.3) follows. Then the lemma is proven.

QED

A.2 Second Quantization

Given a bounded operator T on $\mathfrak{H}$, the second quantization of T is the operator $\Gamma(T)$ on $\mathcal{F}$ defined by

$$\Gamma(T) = \bigoplus_{n=0}^{\infty} (\otimes^n T).$$

Here it is understood that $\otimes^0 T = \mathbb{1}$. In most cases $\Gamma(T)$ is an unbounded operator (Fig. 14).

However, for a contraction operator T, the second quantization $\Gamma(T)$ is also a contraction on $\mathcal{F}$, or equivalently, Γ is a functor

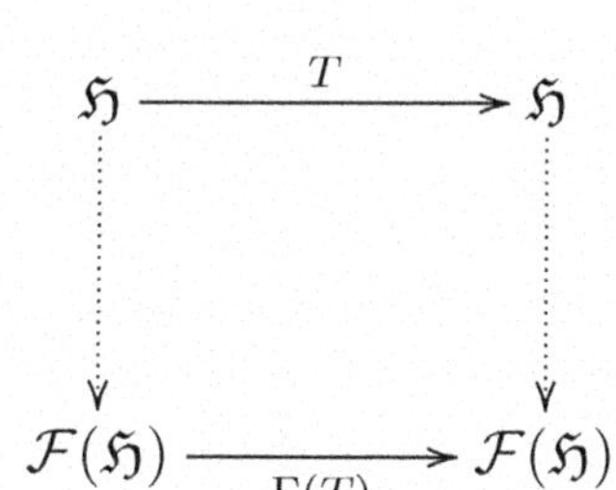

Fig. 14 Functor Γ

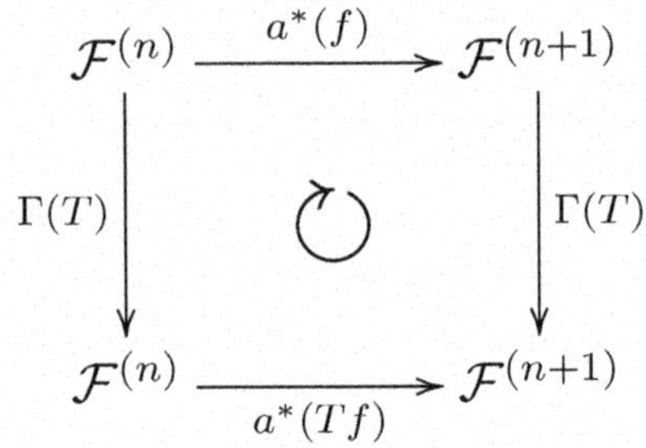

Fig. 15 $a^*(f)$ and $\Gamma(T)$

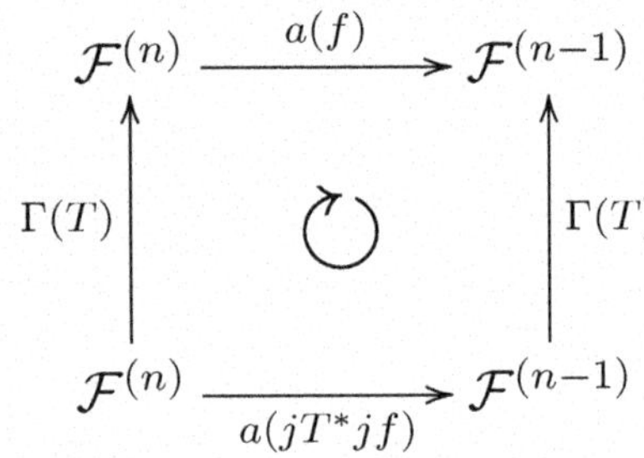

Fig. 16 $a(f)$ and $\Gamma(T)$

$$\Gamma : \mathscr{C}(\mathfrak{H} \to \mathfrak{H}) \to \mathscr{C}(\mathcal{F} \to \mathcal{F}),$$

of the set $\mathscr{C}(X \to Y)$ of contraction operators from X to Y. The functor Γ has the semigroup property, while $\mathscr{C}(\mathfrak{H} \to \mathfrak{H})$ is a $*$-algebra with respect to operator multiplication and conjugation $*$. The map Γ pulls this structure over to $\mathcal{F}$ so that

$$\Gamma(S)\Gamma(T) = \Gamma(ST), \quad \Gamma(S)^* = \Gamma(S^*), \quad \Gamma(\mathbb{1}) = \mathbb{1}, \tag{A.4}$$

for $S, T \in \mathscr{C}(\mathfrak{H} \to \mathfrak{H})$ (Figs. 15 and 16).

We can see relationship between $a^\sharp(f)$ and $\Gamma(T)$. Intertwining properties

$$\Gamma(T)a^*(f) = a^*(Tf)\Gamma(T), \tag{A.5}$$
$$a(f)\Gamma(T) = \Gamma(T)a(jT^*jf) \tag{A.6}$$

can be checked directly and from this we can derive commutation relations

$$[\Gamma(T), a^*(f)] = a^*((T - \mathbb{1})f)\Gamma(T), \tag{A.7}$$
$$[\Gamma(T), a(f)] = \Gamma(T)a(j(\mathbb{1} - T^*)jf), \tag{A.8}$$

where $jf = \bar{f}$ denotes the complex conjugate of f. Suppose in particular that T satisfies that $T^*T = \mathbb{1}$ and $TT^* = E$. Then (Fig. 17) (A.5) and (A.6) yield that

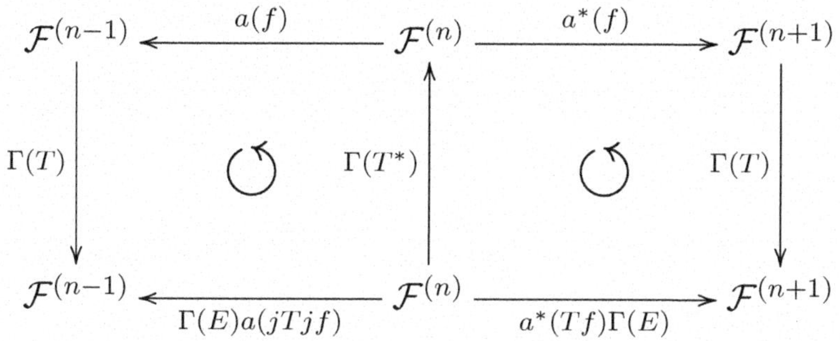

Fig. 17 Creation and annihilation operators, and second quantization

$$\Gamma(T)a^*(f)\Gamma(T^*) = a^*(Tf)\Gamma(E), \tag{A.9}$$

$$\Gamma(T)a(f)\Gamma(T^*) = \Gamma(E)a(jTjf). \tag{A.10}$$

A.3 Differential Second Quantization

For a self-adjoint operator h on $\mathfrak{H}$ the structure relations (A.4) imply in particular that $\{\Gamma(e^{ith}) : t \in \mathbb{R}\}$ is a strongly continuous one-parameter unitary group on $\mathcal{F}$. Then by the Stone's theorem there exists a unique self-adjoint operator $\mathrm{d}\Gamma(h)$ on $\mathcal{F}$ such that

$$\Gamma(e^{ith}) = e^{it\mathrm{d}\Gamma(h)}, \quad t \in \mathbb{R}.$$

The operator $\mathrm{d}\Gamma(h)$ is called the differential second quantizationof h or simply second quantization of h. Thus we have

$$e^{it\mathrm{d}\Gamma(h)}a^*(f)e^{-it\mathrm{d}\Gamma(h)} = a^*(e^{ith}f), \tag{A.11}$$

$$e^{it\mathrm{d}\Gamma(h)}a(f)e^{-it\mathrm{d}\Gamma(h)} = a(je^{ith}jf). \tag{A.12}$$

Since $\mathrm{d}\Gamma(h) = -\mathrm{i}\frac{d}{dt}\Gamma(e^{ith})\lceil_{t=0}$ on some domain, we have

$$\mathrm{d}\Gamma(h) = 0 \oplus \left[\bigoplus_{n=1}^{\infty}\left(\sum_{j=1}^{n}\underbrace{\mathbb{1}\otimes\cdots\otimes\overset{j}{h}\otimes\cdots\otimes\mathbb{1}}_{n}\right)\right], \tag{A.13}$$

where j on top of h indicates its position in the product. Thus the action of $\mathrm{d}\Gamma(h)$ is given by

$$\mathrm{d}\Gamma(h)\Omega = 0,$$

$$\mathrm{d}\Gamma(h)a^*(f_1)\cdots a^*(f_n)\Omega = \sum_{j=1}^{n} a^*(f_1)\cdots a^*(hf_j)\cdots a^*(f_n)\Omega.$$

Then it holds that

$$[\mathrm{d}\Gamma(h), a^*(f)] = a^*(hf), \quad [\mathrm{d}\Gamma(h), a(f)] = -a(hf).$$

It can be also seen by (A.13) that

$$\sigma(\mathrm{d}\Gamma(h)) = \overline{\{\lambda_1 + \cdots + \lambda_n | \lambda_j \in \sigma(h), j = 1, \ldots, n, n \geq 1\} \cup \{0\}},$$
$$\sigma_{\mathrm{p}}(\mathrm{d}\Gamma(h)) = \{\lambda_1 + \cdots + \lambda_n | \lambda_j \in \sigma_{\mathrm{p}}(h), j = 1, \ldots, n, n \geq 1\} \cup \{0\}.$$

If $0 \notin \sigma_{\mathrm{p}}(h)$ and $h \geq 0$, the multiplicity of 0 in $\sigma_{\mathrm{p}}(\mathrm{d}\Gamma(h))$ is one. A crucial operator in quantum field theory is the number operator defined by the second quantization of the identity operator on $\mathfrak{H}$:

$$N = \mathrm{d}\Gamma(\mathbb{1}).$$

Let h be a positive self-adjoint operator and $f \in D(h)$. The following bound is a fundamental. Let $f \in D(h^{-1/2})$ and $\Psi \in D(\mathrm{d}\Gamma(h)^{1/2})$. Then $\Psi \in D(a^\sharp(f))$ and

$$\|a(f)\Psi\| \leq \|h^{-1/2}f\|\|\mathrm{d}\Gamma(h)^{1/2}\Psi\|, \tag{A.14}$$
$$\|a^*(f)\Psi\| \leq \|h^{-1/2}f\|\|\mathrm{d}\Gamma(h)^{1/2}\Psi\| + \|f\|\|\Psi\|. \tag{A.15}$$

In particular, $D(\mathrm{d}\Gamma(h)^{1/2}) \subset D(a^\sharp(f))$, whenever $f \in D(h^{-1/2})$.

Appendix B The Case of $\mathcal{H} = L^2(\mathbb{R}^3)$

This section is taken from [74, Sect. 1.2].

B.1 Useful Bounds

We set $\mathcal{H} = L^2(\mathbb{R}^3)$ and $H_{\mathrm{f}} = \mathrm{d}\Gamma(\omega)$. Then $\mathcal{F}^{(n)} \ni \Phi$ is a function $\Phi(k_1 \ldots, k_n)$ which is symmetric with respect to any permutation of $k_1, \ldots, k_n$. Thus the annihilation operator and the creation operator are defined by

$$(a(f)\Psi)^{(n)}(k_1,\ldots,k_n)=\sqrt{n+1}\int_{\mathbb{R}^3} f(k)\Phi^{(n+1)}(k,k_1\ldots,k_n)\mathrm{d}k,$$

$$(a^*(f)\Psi)^{(n)}(k_1,\ldots,k_n)=\frac{1}{\sqrt{n}}\sum_{j=1}^n f(k_j)\Phi^{(n-1)}(k_1\ldots,\hat{k}_j\ldots,k_n).$$

Furthermore

$$H_{\mathrm{f}}\Psi^{(n)}(k_1,\ldots,k_n)=\left(\sum_{j=1}^n \omega(k_j)\right)\Psi^{(n)}(k_1,\ldots,k_n).$$

We show some useful inequalities on $a^\sharp$ and H_{f}.

Lemma B.3 *Let $\psi\in D(H_{\mathrm{f}}^{m/2})$. Suppose that $\|f_j/\sqrt{\omega}\|<\infty$ for $j=1,\ldots,m$. Then $\psi\in D(\prod_{j=1}^n a(f_j))$ and*

$$\left\|\prod_{j=1}^m a(f_j)\psi\right\|\leq\prod_{j=1}^m\|f_j/\sqrt{\omega}\|\|H_{\mathrm{f}}^{m/2}\psi\|.$$

Proof Let $n>m$. Since $\psi^{(n)}$ is symmetric, by the definition of H_{f} we have

$$(\psi^{(n)},(H_{\mathrm{f}}^m\psi)^{(n)})=n(n-1)\cdots(n-m+1)\int_{\mathbb{R}^{dn}}|\psi^{(n)}(k_1,\ldots,k_n)|^2\prod_{j=1}^n\omega(k_j)\mathrm{d}k_1\cdots\mathrm{d}k_n.$$

While by the definition of $a(f)$ we can see that

$$\begin{aligned}&\left\|(\prod_{j=1}^m a(f_j)\psi)^{(n-m)}\right\|^2\\ &\leq n(n-1)\cdots(n-m+1)\prod_{j=1}^m\|f_j/\sqrt{\omega}\|\int_{\mathbb{R}^{dn}}|\psi^{(n)}(k_1,\ldots,k_n)|^2\prod_{j=1}^m\omega(k_j)\mathrm{d}k_1\cdots\mathrm{d}k_n.\end{aligned}$$

Thus we have

$$\left\|(\prod_{j=1}^m a(f_j)\psi)^{(n-m)}\right\|^2\leq\prod_{j=1}^m\|f_j/\sqrt{\omega}\|(\psi^{(n)},(H_{\mathrm{f}}^m\psi)^{(n)})$$

for $m\leq n$. For $m>n$, the above inequality also holds true since the left hand side is zero. Summing over n on both sides, we conclude the desired results.

QED

Let us consider to evaluate the product of creation operators $\|\prod_{j=1}^m a^*(f_j)\Phi\|$. Since we know a bound of the product of annihilation operators $\|\prod_{j=1}^m a(f_j)\Phi\|$, we can also evaluate $\| \prod_{j=1}^m a^*(f_j)\Phi\|$, but which is rather technically complicated than that of $\| \prod_{j=1}^m a(f_j)\Phi\|$. To see this we use the fact

$$\left\| \prod_{j=1}^m a^*(f_j)\Phi \right\|^2 = \left(\Phi, \prod_{j=1}^m a(\bar{f}_j) \prod_{j=1}^m a^*(f_j)\Phi \right)$$

and compute the commutation relation $\left[\prod_{j=1}^m a(\bar{f}_j), \prod_{j=1}^m a^*(f_j)\right]$. We then conclude that

$$\left\| \prod_{j=1}^m a^*(f_j)\Phi \right\|^2 = \left\| \prod_{j=1}^m a(\bar{f}_j)\Phi \right\|^2 + \left(\Phi, \left[\prod_{j=1}^m a(\bar{f}_j), \prod_{j=1}^m a^*(f_j) \right] \Phi \right).$$

Evaluating the second term of the right-hand side above, we can have a bound of $\left\| \prod_{j=1}^m a^*(f_j)\Phi \right\|$.

Lemma B.4 *Let $f_i, g_j \in D(1/\sqrt{\omega})$ for $i, j = 1, \ldots, n$ and $\Phi \in D(H_{\mathrm{f}}^{n/2})$. Then*

$$\left| \left(\prod_{j=1}^n a^*(g_j)\Phi, \prod_{j=1}^n a^*(f_j)\Phi \right) \right| \leq n!2^n \left(\prod_{l=1}^n \|f_l\|_\omega \|g_l\|_\omega \right) \sum_{m=0}^n \frac{1}{m!} \|H_{\mathrm{f}}^{m/2}\Phi\|^2,$$

where $\|f\|_\omega = \|f\| + \|f/\sqrt{\omega}\|$.

By Lemma B.4 we have useful bounds for products of annihilation operators and creation operators. We summarize them as follows. Suppose that $f_j \in D(1/\sqrt{\omega})$ for $j = 1, \ldots, n$. Then for $\Phi \in D(H_{\mathrm{f}}^{n/2})$, we have

$$\left\| \prod_{j=1}^n a(f_j)\Phi \right\| \leq \left(\prod_{j=1}^n \|f_j/\sqrt{\omega}\| \right) \|H_{\mathrm{f}}^{n/2}\Phi\|, \tag{B.1}$$

$$\left\| \prod_{j=1}^n a^*(f_j)\Phi \right\| \leq \sqrt{n!}2^{n/2} \left(\prod_{l=1}^n \|f_l\|_\omega \right) \left(\sum_{m=0}^n \frac{1}{m!} \|H_{\mathrm{f}}^{m/2}\Phi\|^2 \right)^{1/2}. \tag{B.2}$$

B.2 Exponent of Annihilation Operators and Creation Operators

Although exponential operator $e^{a^*(f)}$ is unbounded, it can be seen in the proposition below that $e^{a^*(f)}e^{-\frac{t}{2}H_{\mathrm{f}}}$ is bounded for any $t > 0$.

Proposition B.5 *Let $t > 0$. (1) Let $f \in D(1/\sqrt{\omega})$. Then both $\overline{e^{a^*(f)}e^{-\frac{t}{2}H_f}}$ and $\overline{e^{-\frac{t}{2}H_f}e^{a(f)}}$ are bounded. (2) Let $f \in L^2(\mathbb{R}^3)$. Then both $e^{a^*(f)}e^{-\frac{t}{2}N}$ and $\overline{e^{-\frac{t}{2}N}e^{a(f)}}$ are bounded.*

Proof Suppose that $t < 1$. For any $s < t$ we have for $\Phi \in \mathcal{F}$,

$$\left\| \sum_{n=0}^{m} \frac{1}{n!} a^*(f)^n e^{-\frac{t}{2}H_f}\Phi \right\| \leq \sum_{n=0}^{m} \frac{1}{\sqrt{n!}} 2^{n/2} s^{-n/2} \|f\|_{\omega}^{n} \left(\sum_{k=0}^{n} \frac{1}{k!} \|(sH_f)^{k/2} e^{-\frac{t}{2}H_f}\Phi\|^2 \right)^{1/2}.$$

We can see that sequence $\{\sum_{n=0}^{m} \frac{1}{n!} a^*(f)^n e^{-\frac{t}{2}H_f}\Phi\}_{m=0}^{\infty}$ is a Cauchy sequence in $\mathcal{F}$. Hence $e^{-\frac{t}{2}H_f}\Phi \in D(e^{a^*(f)})$ and as $m \to \infty$ on both sides above and we have

$$\|e^{a^*(f)}e^{-\frac{t}{2}H_f}\Phi\| \leq A(f,s)\|e^{-\frac{1}{2}(t-s)H_f}\Phi\|,$$

where

$$A(f,s) = \sum_{n=0}^{\infty} \frac{1}{\sqrt{n!}} 2^{n/2} s^{-n/2} \|f\|_{\omega}^{n}. \tag{B.3}$$

Choosing s such that $s < t$, we can see that $\|e^{-\frac{1}{2}(t-s)H_f}\Phi\| \leq \|\Phi\|$ and $e^{a^*(f)}e^{-\frac{t}{2}H_f}$ for $t < 1$ is bounded. Suppose $1 \leq t$. Choosing $s = 1$ in the above discussion, we have

$$\|e^{a^*(f)}e^{-\frac{t}{2}H_f}\Phi\| \leq A(f,1)\|e^{-\frac{1}{2}(t-1)H_f}\Phi\| \leq A(f,1)\|\Phi\|.$$

Thus $e^{a^*(f)}e^{-\frac{t}{2}H_f}$ for $t \geq 1$ is bounded. Finally since $\left(e^{-\frac{t}{2}H_f}e^{a(f)}\right)^* \supset e^{a^*(\bar{f})}e^{-\frac{t}{2}H_f}$, the second statement follows. Then (1) follows. (2) is similarly proven by replacing ω with the identity $\mathbb{1}$.

QED

B.3 Technical Estimates I

Let $X = e^{a^*(g)}e^{a(\bar{g})}$ for $g \in L^2(\mathbb{R}^3)$. We already see that $e^{a^*(g)}e^{-\frac{t}{2}H_f}e^{a(\bar{g})}$ is bounded for any $t > 0$. We extend this to more general cases. By (A.3) we know the explicit form of Wick ordering of $\Phi(f)^n$, and see that

$$X\Phi(f)^n\Psi = n! \sum_{k=0}^{[n/2]} \sum_{l+m+2k=n} \frac{(\frac{\|f\|^2}{4})^k}{k!} \frac{(\frac{1}{\sqrt{2}}a^*(f)+y)^l}{l!} X \frac{(\frac{1}{\sqrt{2}}a(\bar{f}))^m}{m!}\Psi. \tag{B.4}$$

Here $y = (g, f)/\sqrt{2}$. Operator $X\Phi(f)^n$ is not bounded, but we are interested in estimating operator $Z(t, n)$ which is defined by inserting e^{-tH_f} into $X\Phi(f)^n$ as

$$Z(t,n) = n! \sum_{k=0}^{[n/2]} \sum_{l+m+2k=n} \frac{(\frac{\|f\|^2}{4})^k}{k!} \frac{(\frac{1}{\sqrt{2}} a^*(f) + y)^l}{l!} e^{a^*(\bar{g})} e^{-tH_f} e^{a(\bar{g})} \frac{(\frac{1}{\sqrt{2}} a(\bar{f}))^m}{m!}. \tag{B.5}$$

Lemma B.6 *Let $f \in D(1/\sqrt{\omega})$. Then $Z(t, n)$ is bounded for $t > 0$, and the bound of $\|Z(t, n)\Psi\|$ is given as follows.*
$(t \geq 1)$

$$\|Z(t,n)\Psi\| \leq z_t(n) \left\| e^{-\frac{1}{2}(t-1)H_f} \right\|^2 \|\Psi\|,$$

where

$$z_t(n) = n! \sum_{k=0}^{[n/2]} \sum_{l+m+2k=n} \frac{\left(\frac{\|f\|^2}{4}\right)^k (\|f\|_\omega + |y|)^l \|f\|_\omega^m}{k!l!m!} \xi(l,m)$$

with $y = (g, f)/\sqrt{2}$ and

$$\xi(l,m) = \sum_{j,j'=0}^{\infty} \frac{(\sqrt{2}\|g\|_\omega)^{j+j''} \sqrt{(j+l)!}\sqrt{(j'+m)!}}{j!j'!}.$$

$(t < 1)$ For any $0 < s < t$,

$$\|Z(t,n)\Psi\| \leq z_t(s,n) \left\| e^{-\frac{1}{2}(t-s)H_f} \right\|^2 \|\Psi\|,$$

where

$$z_t(s,n) = n! \sum_{k=0}^{[n/2]} \sum_{l+m+2k=n} \frac{\left(\frac{\|f\|^2}{4}\right)^k (\|f\|_\omega + |y|)^l \|f\|_\omega^m}{k!l!m!} \xi_s(l,m)$$

and

$$\xi_s(l,m) = \sum_{j,j'=0}^{\infty} \frac{(\sqrt{2}\|g\|_\omega)^{j+j''} \sqrt{(j+l)!}\sqrt{(j'+m)!}}{s^{(j+l)/2} s^{(j'+m)/2} j!j'!}.$$

Proof In this proof we set $s = 1$ for $t \geq 1$ and $0 < s < t$ for $t < 1$. Since

$$\left\| \sum_{n=0}^{N} \frac{a^*(f)^n}{n!} \Phi \right\| \leq \sum_{n=0}^{N} \frac{(\sqrt{2}\|f\|_\omega)^n s^{-n/2}}{\sqrt{n!}} \left\| e^{\frac{1}{2}sH_f} \Phi \right\|,$$

and

$$\|(a^*(f)+y)^m a^*(g)^n\Psi\| \le \sum_{k=0}^{m}\binom{m}{k}|y|^k\|a^*(f)^{m-k}a^*(g)^n\Psi\|$$
$$\le s^{-(n+m)/2}(\sqrt{2}\|f\|+|y|)^m(\sqrt{2}\|g\|_\omega)^n\sqrt{(n+m)!}\|e^{\frac{1}{2}sH_{\rm f}}\Psi\|,$$

we have for any non-negative integers $m \ge 0$

$$\left\|(a^*(g)+y)^m e^{a^*(f)}\Phi\right\| \le \sum_{n=0}^{\infty}\frac{(\sqrt{2}\|f\|_\omega+|y|)^m(\sqrt{2}\|g\|_\omega)^n\sqrt{(n+m)!}}{s^{(n+m)/2}n!}\left\|e^{\frac{1}{2}sH_{\rm f}}\Phi\right\|.$$

We have

$$\|Z(t,n)\Psi\| \le \sum_{k=0}^{[n/2]}\sum_{l+m+2k=n}\frac{n!\left(\frac{\|f\|^2}{4}\right)^k}{k!l!m!}\left\|\left(\frac{a^*(f)}{\sqrt{2}}+y\right)^l e^{a^*(g)}e^{-\frac{t}{2}H_{\rm f}}\right\|\left\|e^{-\frac{t}{2}H_{\rm f}}e^{a(\bar g)}\left(\frac{a(\bar f)}{\sqrt{2}}\right)^m\right\|\|\Psi\|. \tag{B.6}$$

Inserting inequalities:

$$\left\|\left(\frac{a^*(f)}{\sqrt{2}}+|y|\right)^l e^{a^*(g)}e^{-\frac{t}{2}H_{\rm f}}\right\| \le \sum_{j=0}^{\infty}\frac{(\sqrt{2}\|g\|_\omega)^j(\|f\|_\omega+|y|)^l\sqrt{(j+l)!}}{s^{(j+l)/2}j!}A,$$
$$\left\|e^{-\frac{t}{2}H_{\rm f}}e^{a(\bar g)}\left(\frac{a(\bar f)}{\sqrt{2}}\right)^m\right\| \le \sum_{j=0}^{\infty}\frac{(\sqrt{2}\|g\|_\omega)^j\|f\|_\omega^m\sqrt{(j+m)!}}{s^{(j+m)/2}j!}A$$

with $A = \left\|e^{-\frac{1}{2}(t-s)H_{\rm f}}\right\|$ into (B.6), we prove the lemma.

QED

Next we estimate operator $Z(t)$ defined by

$$Z(t) = \sum_{n=0}^{\infty}\frac{t^n}{n!}Z(t,n). \tag{B.7}$$

Lemma B.7 *Let $f, g \in D(1/\sqrt{\omega})$. Then $Z(t)$ for $t > 0$ is bounded with*

$$\|Z(t)\| \le \frac{e^{\frac{\|f\|^2}{4}}\left\|e^{-\frac{t-s}{2}H_{\rm f}}\right\|^2 C(h)^2}{\left(1-\frac{h}{\sqrt{s}}(\sqrt{2}\|g\|_\omega+\|f\|_\omega+y)\right)\left(1-\frac{h}{\sqrt{s}}(\sqrt{2}\|g\|_\omega+\|f\|_\omega)\right)}. \tag{B.8}$$

Here $s = 1$ for $t \ge 1$ and $0 < s < t$ for $t < 1$, and $C(h)$ is a constant depending on $h > 0$ such that $1 - \frac{h}{\sqrt{s}}(\sqrt{2}\|g\|_\omega + \|f\|_\omega + y) > 0$ for $y = (g, f)/\sqrt{2}$.

Proof In this proof we set $s = 1$ for $t \geq 1$ and $0 < s < t$ for $t < 1$. By Lemma B.6 and the fact that $Z(t) = \sum_{n=0}^{\infty} \frac{t^n}{n!} Z(t,n)$ we have

$$\|Z(t)\Psi\| \leq z_t \left\| e^{-\frac{1}{2}(t-1)H_f} \right\|^2 \|\Psi\|,$$

where

$$z_t = \sum_{n=0}^{\infty} \sum_{l+m+2k=n} \frac{\left(\frac{\|tf\|^2}{4}\right)^k (\|tf\|_\omega + t|y|)^l \|tf\|_\omega^m}{k!l!m!} \xi_s(l,m).$$

Then we replace tf and $t|y|$ by f and $|y|$, respectively in what follows. We have

$$z_t = e^{\|f\|^2/4} \xi_s(y) \xi_s(0), \tag{B.9}$$

where

$$\xi_s(y) = \sum_{j=0}^{\infty} \sum_{l=0}^{\infty} \frac{a^j b^l \sqrt{(j+l)!}}{l!j!}. \tag{B.10}$$

with $a = \sqrt{2}\|g\|_\omega / \sqrt{s}$ and $b = (\|f\|_\omega + y)/\sqrt{s}$. Let $0 < h$ be such that $h < 1/(a+b)$. Then

$$\xi_s(y) = \sum_{j,l=0}^{\infty} \frac{a^j b^l h^{j+l} (j+l)!}{l!j!} \frac{(1/h)^j (1/h)^l}{\sqrt{(j+l)!}} \leq \frac{C}{1 - h(a+b)} < \infty.$$

Here $C = C(h) = \sup_{(j,l) \in \mathbb{N} \times \mathbb{N}} \frac{(1/h)^j (1/h)^l}{\sqrt{(j+l)!}}$. It can be also shown that $\xi(0)$ is finite. Hence we have (B.8). Then z_t is finite, and the lemma is proven.

QED

B.4 Technical Estimates II

Operator $Z(t,n)$ is defined through commutation relations: $[X, \Phi(f)^n]$, where $X = e^{a^*(g)} e^{a(\bar{g})}$. We also need to see operator $W(t,n)$ defined through commutation relations:

$$\left[X, \prod_{j=1}^{n} \Phi(f_j) \right]. \tag{B.11}$$

To compute the commutation relation (B.11) explicitly we apply Wick's theorem to product $\prod_{j=1}^n \Phi(f_j)$:

$$\prod_{j=1}^{n} \Phi(f_j) = \sum_{k=0}^{[n/2]} \frac{1}{2^k} \sum_{j_1,\dots,j_{2k}}^{\text{pair}} \left(\prod_{i=1}^{k} (f_{j_{2i-1}}, f_{j_{2i}}) \right) c^n_{j_1,\dots,j_{2k}}.$$

Since

$$:\prod_{j=1}^{n} \Phi(f_j): = 2^{-n/2} \sum_{k=0}^{m} \sum_{\{j_1,\dots,j_k\}\subset\{1,\dots,n\}} \prod_{j\in\{j_1,\dots,j_k\}} a^*(f_j) \prod_{i\in\{j_1,\dots,j_k\}^c} a(\bar{f}_i),$$

we can compute $c^n_{j_1,\dots,j_{2k}}$ as

$$c^n_{j_1,\dots,j_{2k}} = 2^{-(n-2k)/2} \sum_{l=0}^{n-2k} \sum_{\{i_1,\dots,i_l\}\subset\{j_1,\dots,j_{2k}\}^c} \prod_{j\in\{i_1,\dots,i_l\}} a^*(f_j) \prod_{i\in\{i_1,\dots,i_l\}^c\cap\{j_1,\dots,j_{2k}\}^c} a(\bar{f}_i).$$

Hence

$$X\prod_{j=1}^{n} \Phi(f_j) = \sum_{k=0}^{[n/2]} \frac{1}{2^k} \sum_{j_1,\dots,j_{2k}}^{\text{pair}} \left(\prod_{i=1}^{k} (f_{j_{2i-1}}, f_{j_{2i}}) \right) \sum_{l=0}^{n-2k} \sum_{\{i_1,\dots,i_l\}\subset\{j_1,\dots,j_{2k}\}^c} AXB$$

where

$$A = \prod_{j\in\{i_1,\dots,i_l\}} \left(\frac{a^*(f_j)}{\sqrt{2}} + y_j \right), \quad B = \prod_{i\in\{i_1,\dots,i_l\}^c\cap\{j_1,\dots,j_{2k}\}^c} \frac{a(\bar{f}_i)}{\sqrt{2}} \tag{B.12}$$

with $y_j = (g, f_j)/\sqrt{2}$. Since $X\prod_{j=1}^m \Phi(f_j)$ is not bounded, we then define $W(t,n)$ by inserting $e^{-tH_{\rm f}}$ into $X\prod_{j=1}^m \Phi(f_j)$ as

$$W(t,n) = \sum_{k=0}^{[n/2]} \frac{1}{2^k} \sum_{j_1,\dots,j_{2k}}^{\text{pair}} \sum_{l=0}^{n-2k} \sum_{\{i_1,\dots,i_l\}\subset\{j_1,\dots,j_{2k}\}^c} \left(\prod_{i=1}^{k} (f_{j_{2i-1}}, f_{j_{2i}}) \right) Ae^{a^*(g)}e^{-tH_{\rm f}}e^{a(\bar{g})}B. \tag{B.13}$$

Furthermore we set

$$W(t) = \sum_{n=0}^{\infty} \frac{t^n}{n!} W(t,n). \tag{B.14}$$

$W(t,n)$ and $W(t)$ are extensions of $Z(t,n)$ and $Z(t)$ discussed in (B.5) and (B.7), respectively.

Formula proved in Theorem B.8 below is used to estimate integral kernels of semigroup e^{-tL} in Sect. 5.4.3. The integral kernel of e^{-tL} is

$$\mathrm{J}_0^* e^Y \mathrm{J}_t = \mathrm{J}_0^* e^{\frac{B^\dagger}{\sqrt{2}}+\frac{B}{\sqrt{2}}} \prod_{j=1}^N \left(\left| \frac{C_j^\dagger}{\sqrt{2}} + \frac{C_j}{\sqrt{2}} \right| + \varepsilon\psi_\varepsilon \left(\left| \frac{C_j^\dagger}{\sqrt{2}} + \frac{C_j}{\sqrt{2}} \right| \right) \right) \mathrm{J}_t. \tag{B.15}$$

See (5.22). (B.15) is of the form of $W(t)$ given by (B.14). Theorem B.8 tells that the integral kernel of e^{-tL} is bounded.

Theorem B.8 *Let $f_j \in D(1/\sqrt{\omega})$, $j \in \mathbb{N}$. Suppose that there exists $\alpha > 0$ such that $\|f_j\|_\omega \le \alpha$ for any j. Then $W(t, n)$ and $W(t)$ are bounded for $t > 0$ with*

$$\|W(t)\| \le \frac{e^{\frac{\alpha^2}{4}} \left\| e^{-\frac{t-s}{2}H_{\mathrm{f}}} \right\|^2 C(h)^2}{\left(1 - \frac{h}{\sqrt{s}}(\sqrt{2}\|g\|_\omega + \alpha + y)\right)\left(1 - \frac{h}{\sqrt{s}}(\sqrt{2}\|g\|_\omega + \alpha)\right)}.$$

Here $s = 1$ for $t \ge 1$ and $0 < s < t$ for $t < 1$, $y = \alpha\|g\|/\sqrt{2}$, and $C(h)$ is a constant depending on $h > 0$ such that $1 - \frac{h}{\sqrt{s}}(\sqrt{2}\|g\|_\omega + \alpha + y) > 0$.

Proof In this proof we set $s = 1$ for $t \ge 1$ and $0 < s < t$ for $t < 1$. Let A and B be (B.12). We can estimate $\left\| A e^{a^*(g)} e^{-tH_{\mathrm{f}}} e^{a(\bar{g})} B \right\|$ as

$$\left\| A e^{a^*(g)} e^{-tH_{\mathrm{f}}} e^{a(\bar{g})} B \right\| \le \left\| A e^{a^*(g)} e^{-\frac{t}{2}H_{\mathrm{f}}} \right\| \left\| e^{-\frac{t}{2}H_{\mathrm{f}}} e^{a(\bar{g})} B \right\|,$$

with

$$\left\| A e^{a^*(g)} e^{-\frac{t}{2}H_{\mathrm{f}}} \Phi \right\| \le \sum_{j=0}^\infty \frac{(\alpha+y)^l (\sqrt{2}\|g\|_\omega)^j \sqrt{(l+j)!}}{s^{(j+l)/2} j!} \left\| e^{-\frac{1}{2}(t-s)H_{\mathrm{f}}} \right\|,$$

$$\left\| e^{-\frac{t}{2}H_{\mathrm{f}}} e^{a(\bar{g})} B \right\| \le \sum_{j=0}^\infty \frac{\alpha^{n-2k-l} (\sqrt{2}\|g\|_\omega)^j \sqrt{(n-2k-l+j)!}}{s^{(j+n-2k-l)/2} j!} \left\| e^{-\frac{1}{2}(t-s)H_{\mathrm{f}}} \right\|.$$

Let $y = \alpha\|g\|/\sqrt{2}$. Then $|y_j| \le y$. In a similar manner to the proof of Lemma B.6 we have bounds below:
$(t \ge 1)$

$$\|W(t, n)\Psi\| \le w_t(n) \left\| e^{-\frac{1}{2}(t-1)H_{\mathrm{f}}} \right\|^2 \|\Psi\|,$$

where

$$w_t(n) = n! \sum_{k=0}^{[n/2]} \sum_{l+m+2k=n} \frac{\left(\frac{\alpha^2}{4}\right)^k (\alpha+y)^l \alpha^m}{k! l! m!} \xi(l, m).$$

($t<1$) For any $0<s<t$,

$$\|W(t,n)\Psi\| \le w_t(s,n)\left\|e^{-\frac{1}{2}(t-s)H_{\mathrm{f}}}\right\|^2 \|\Psi\|,$$

where

$$w_t(s,n) = n!\sum_{k=0}^{[n/2]}\sum_{l+m+2k=n}\frac{\left(\frac{\alpha^2}{4}\right)^k(\alpha+y)^l\alpha^m}{k!l!m!}\xi_s(l,m).$$

Here $\xi(l,m)$ and $\xi_s(l,m)$ defined in Lemma B.6. Hence $W(t,n)$ is bounded. We shall show the boundedness of $W(t)$. The proof is also similar to that of Lemma B.7. We have

$$\|W(t)\Psi\| \le w_t\left\|e^{-\frac{1}{2}(t-1)H_{\mathrm{f}}}\right\|^2 \|\Psi\|,$$

where

$$w_t = \sum_{n=0}^{\infty}\sum_{l+m+2k=n}\frac{\left(\frac{t^2\alpha^2}{4}\right)^k(t\alpha+ty)^l(t\alpha)^m}{k!l!m!}\xi_s(l,m).$$

Then we replace $t\alpha$ and ty by α and y, respectively in what follows. We have

$$w_t = \sum_{n=0}^{\infty}\sum_{l+m+2k=n}\frac{\left(\frac{\alpha^2}{4}\right)^k}{k!}\frac{(\alpha+y)^l}{l!}\frac{\alpha^m}{m!}\xi_s(l,m) = e^{\alpha^2/4}\xi_s(y)\xi_s(0),$$

where

$$\xi_s(y) = \sum_{j=0}^{\infty}\sum_{l=0}^{\infty}\frac{(\sqrt{2}\|g\|_\omega)^j(\alpha+y)^l\sqrt{(j+l)!}}{s^{(j+l)/2}l!j!}.$$

The right-hand side above is finite which is shown in Lemma B.7. Then the proof is complete.

QED

Appendix C Proofs of (1) and (2) in the Proof of Theorem 5.8

Lemma C.9 *Statement (1) of the proof of Theorem 5.8 is true.*

Proof For each $(x,\alpha,w,m)\in\mathbb{R}^3\times\mathbb{Z}_2\times\mathscr{X}\times\mathscr{X}_\mu$, it can be directly seen that $F,G\in D(e^{Z_t(\varepsilon)}\mathrm{J}_0)\cap D(e^{Z_t^n(\varepsilon)}\mathrm{J}_0)$ for any $n\in\mathbb{N}$. We see the inequality: $\|Y_t(2)\Phi\|\le\gamma\sqrt{m+1}\|\Phi\|$ for $\Phi\in L^2_m(\mathrm{Q})$, where $\gamma=\sqrt{2}t\||k|\hat{\varphi}/\sqrt{\omega}\|$. Since $L^2_{\mathrm{fin}}(\mathrm{Q})$ is the set of analytic vectors for $e^{Y_t^n(2)}$, for $F\in L^2_{\mathrm{fin}}(\mathrm{Q})$ we have $e^{Y_t^n(2)}F=\sum_{j=0}^{\infty}\frac{Y_t^n(2)^j}{j!}F$. We assume that $F \in L^2_{m'}(\mathrm{Q})$ and $G \in L^2_m(\mathrm{Q})$, hence $NF = m'F$ and $NG = mG$ for the number operator N. Then

$$\|e^{Y_t^n(2)} F\| \le \sum_{j=0}^{\infty} \frac{\sqrt{(m+j-1)\cdots(m+1)m}}{j!} \gamma^j \|F\|.$$

On the other hand by the definition of $Y_t^n(3,\varepsilon)$ we have

$$e^{Y_t^n(3,\varepsilon)} = \prod_{i=1}^n \exp\left(\int_{\frac{(i-1)t}{n}}^{\frac{it}{n}+} \log\left(-\Psi_\varepsilon(\mathbb{H}_{\mathrm{od}}(B_s, -\theta_{N_{s-}}, \frac{(i-1)t}{n}\right)\mathrm{d}N_s\right). \tag{C.1}$$

For every $m \in \mathscr{X}_\mu$ there exists $N = N(m) \in \mathbb{N}$ such that map $t \mapsto N_t(m)$ is not continuous at points $s_1 = s_1(m), \ldots, s_N = s_N(m)$. For sufficiently large n the number of discontinuous points in $(\frac{(i-1)t}{n}, \frac{it}{n}]$ is at most one. Then by taking n large enough and putting $(n(s_i), n(s_i) + t/n]$ for the interval containing s_i, we get

$$e^{Y_t^n(3,\varepsilon)} = \prod_{i=1}^N (-\phi_i - \varepsilon\psi_\varepsilon(\phi_i)),$$

where $\phi_i = \mathbb{H}_{\mathrm{od}}(B_{s_i}, -\theta_{N_{s_i-}}, n(s_i))$. Fix $m \in \mathscr{X}_\mu$ above. We have

$$\prod_{i=1}^N (-\phi_i - \varepsilon\psi_\varepsilon(\phi_i)) = (-1)^N \sum_{p=0}^N \varepsilon^{N-p} \sum_{\{j_1,\ldots,j_p\}\subset[N]} \prod_{i\notin\{j_1,\ldots,j_p\}} \psi_i \cdot \prod_{i=1}^p \phi_{j_i},$$

where $\psi_i = \psi_\varepsilon(\phi_{j_i})$ and $[N] = \{1, \ldots, N\}$, and

$$\begin{aligned}
&(F, \mathrm{J}_0^* e^{Y_t^n(1)} e^{Y_t^n(2)} e^{Y_t^n(3,\varepsilon)} \mathrm{J}_t G) \\
&= \sum_{p=0}^N \varepsilon^{N-p} \sum_{\{j_1,\ldots,j_p\}\subset[N]} \left(e^{-Y_t^n(1)} \mathrm{J}_0 F, e^{Y_t^n(2)} \prod_{i\notin\{j_1,\ldots,j_p\}} \psi_i \cdot \prod_{i=1}^p \phi_{j_i} \mathrm{J}_t G\right).
\end{aligned}$$

Notice again that $\|\phi_{j_i}\Phi\| \le 2\gamma\sqrt{m+1}\|\Phi\|$ for $\Phi \in L^2(\mathrm{Q})$ and $i = 1, \ldots, p$. Since $\psi_\varepsilon \le 1$, each terms on the right-hand side above can be estimated as

$$\begin{aligned}
&\left|\left(e^{-Y_t^n(1)} \mathrm{J}_0 F, e^{Y_t^n(2)} \prod_{i\notin\{j_1,\ldots,j_p\}} \psi_i \cdot \prod_{i=1}^p \phi_{j_i} \mathrm{J}_t G\right)\right| \\
&\le \sum_{j=0}^{\infty} \varepsilon^{N-p} \frac{\sqrt{(j+p+m-1)\cdots(m+1)m}}{j!} (2\gamma)^{j+p-m+1} \|F\|\|G\|.
\end{aligned}$$

Thus

$$
\begin{aligned}
&|(F, \mathrm{J}_0^* e^{Y_t^n(1)} e^{Y_t^n(2)} e^{Y_t^n(3,\varepsilon)} \mathrm{J}_t G)| \\
&\leq \sum_{p=0}^{N} \frac{\varepsilon^{N-p} N!}{p!(N-p)!} \sum_{j=0}^{\infty} \frac{\sqrt{(j+p+m-1)\cdots(m+1)m}}{j!} (2\gamma)^{j+p-m+1} \|F\|\|G\| \\
&\leq A_t(N)\|F\|\|G\|,
\end{aligned}
$$

where

$$
A_t(N) = \sum_{j=0}^{\infty} (1+\varepsilon)^N \frac{\sqrt{(j+N+m-1)\cdots(m+1)m}}{j!} (2\gamma)^{j+N-m+1}.
$$

The number N depends on $m \in \mathscr{X}_\mu$, then $A_t(N(m))$ turns to be a random process on $(\mathscr{X}_\mu, \mathcal{B}_\mu, \mu)$. We set $C_t(m) = A_t(N(m))$. For each $(x, \alpha, w, m) \in \mathbb{R}^3 \times \mathbb{Z}_2 \times \mathscr{X} \times \mathscr{X}_\mu$, $|(F, \mathrm{J}_0^* e^{Z_t^n(\varepsilon)} \mathrm{J}_t G)|$ is finite with the bound $|(F, \mathrm{J}_0^* e^{Z_t^n(\varepsilon)} \mathrm{J}_t G)| \leq C_t(m)$ and

$$
\begin{aligned}
&\mathbb{E}_\mu[C_t^2] = \sum_{N=0}^{\infty} \frac{t^N A_t(N)^2}{N!} e^{-t} \\
&\leq e^{-t} \sum_{N=0}^{\infty} \left(\sum_{j=0}^{\infty} (1+\varepsilon)^N (2\gamma)^{j+N-m+1} \frac{\sqrt{(j+N+m-1)\cdots(m+1)m}}{N!j!} \right)^2 < \infty.
\end{aligned}
$$

Thus (1) follows.

QED

Lemma C.10 *Statement (2) of the proof of Theorem 5.8 is true.*

Proof We show the convergence of $\mathrm{J}_0^* e^{Z_t^n(\varepsilon)} \mathrm{J}_t$ as $n \to \infty$. Let $F \in L^2_{\mathrm{fin}}(\mathrm{Q})$. We show the convergences of $e^{Y_t^n(1)}$, $e^{Y_t^n(2)}$ and $e^{Y_t^n(3,\varepsilon)}$, separately. Thus

$$
\|e^{-Y_t(1)} \mathrm{J}_0 F - e^{-Y_t^n(1)} \mathrm{J}_0 F\| \leq \left\| \bigoplus_{\mu=1}^{3} \sum_{j=1}^{n} \int_{t_{j-1}}^{t_j} (\mathrm{j}_{t_{j-1}} - \mathrm{j}_s) \tilde{\varphi}(\cdot - B_s) \mathrm{d}B_s^\mu \right\| \|(N+\mathbb{1})^{1/2} F\|.
$$

Then it follows that

$$
\begin{aligned}
&\mathbb{E}_{\mathcal{W}}^x[\|e^{-Y_t(1)} \mathrm{J}_0 F - e^{-Y_t^n(1)} \mathrm{J}_0 F\|^2] \leq 3 \sum_{j=1}^{n} \int_{t_{j-1}}^{t_j} \|(\mathrm{j}_{t_{j-1}} - \mathrm{j}_s) \tilde{\varphi}(\cdot - B_s)\|^2 \mathrm{d}s \|(N+\mathbb{1})^{1/2} F\|^2 \\
&= 3 \sum_{j=1}^{n} \int_{t_{j-1}}^{t_j} \left(\frac{\hat{\varphi}}{\sqrt{\omega}}, (1 - e^{-|s-t_{j-1}|\omega}) \frac{\hat{\varphi}}{\sqrt{\omega}} \right) \mathrm{d}s \|(N+\mathbb{1})^{1/2} F\|^2 \\
&\leq 3 \frac{t}{n} \|\sqrt{\omega} \hat{\varphi}\|^2 \|(N+\mathbb{1})^{1/2} F\|^2.
\end{aligned}
$$

Hence we have

$$\lim_{n\to\infty} \mathbb{E}^x_{\mathscr{W}}[\|e^{-Y_t(1)}\mathrm{J}_0 F - e^{-Y_t^n(1)}\mathrm{J}_0 F\|] = 0. \tag{C.2}$$

On the other hand in the Fock representation $Y_t(2)$ and $Y_t^n(2)$ are represented as

$$Y_t(2) = -\frac{1}{\sqrt{2}}\sum_{j=\pm 1}\left\{a_{\mathrm{E}}^*\left(\int_0^t b_j^-(s)\mathrm{d}s, j\right) - a_{\mathrm{E}}\left(\int_0^t b_j^+(s)\mathrm{d}s, j\right)\right\},$$

$$Y_t^n(2) = -\frac{1}{\sqrt{2}}\sum_{j=\pm 1}\sum_{i=1}^n\left\{a_{\mathrm{E}}^*\left(\int_{t_{i-1}}^{t_i} b_j^-(t_{i-1})\mathrm{d}s, j\right) - a_{\mathrm{E}}\left(\int_{t_{i-1}}^{t_i} b_j^+(t_{i-1})\mathrm{d}s, j\right)\right\}.$$

Here $b_j^{\pm}(s) = S_j(\theta_{N_s})e^{\pm \mathrm{i}kB_s}\hat{\mathrm{j}}_s\frac{\hat{\varphi}}{\sqrt{\omega}}$ and $b_j^{\pm}(t_{i-1}) = S_j(\theta_{N_s})e^{\pm \mathrm{i}kB_s}\hat{\mathrm{j}}_{t_{i-1}}\frac{\hat{\varphi}}{\sqrt{\omega}}$. Then the distance between test functions of $Y_t(2)$ and $Y_t^n(2)$ can be estimated as

$$\left\|\int_0^t b_j^-(s)\mathrm{d}s - \sum_{i=1}^n\int_{t_{i-1}}^{t_i} b_j^-(t_{i-1})\mathrm{d}s\right\|^2 \le t\sum_{i=1}^n\int_{t_{i-1}}^{t_i}\left\|S_j(\theta_{N_s})e^{-\mathrm{i}kB_s}\left(\hat{\mathrm{j}}_s\frac{\hat{\varphi}}{\sqrt{\omega}} - \hat{\mathrm{j}}_{t_{i-1}}\frac{\hat{\varphi}}{\sqrt{\omega}}\right)\right\|^2 \mathrm{d}s.$$

Since $\left\|S_j(\theta_{N_s})e^{-\mathrm{i}kB_s}(\hat{\mathrm{j}}_s\frac{\hat{\varphi}}{\sqrt{\omega}} - \hat{\mathrm{j}}_{t_{i-1}}\frac{\hat{\varphi}}{\sqrt{\omega}})\right\|^2 \le 2|s - t_{i-1}|\||k|\hat{\varphi}\|^2$, we see that

$$\left\|\int_0^t b_j^-(s)\mathrm{d}s - \sum_{i=1}^n\int_{t_{i-1}}^{t_i} b_j^-(t_{i-1})\mathrm{d}s\right\|^2 \le \frac{2}{n}t^2\||k|\hat{\varphi}\|^2.$$

It is concluded that

$$\lim_{n\to\infty}\left\|\int_0^t b_j^-(s)\mathrm{d}s - \sum_{i=1}^n\int_{t_{i-1}}^{t_i} b_j^-(t_{i-1})\mathrm{d}s\right\| = 0. \tag{C.3}$$

Furthermore by (C.3) it can be straightforwardly seen that

$$e^{Y_t^n(2)}F \to e^{Y_t(2)}F \tag{C.4}$$

by the expansion $e^{Y_t^n(2)}F = \sum_{j=0}^\infty \frac{Y_t^n(2)^j}{j!}F$.

In the same argument as in the proof of Lemma C.9 for every $m \in \mathscr{X}_\mu$ there exists $N = N(m) \in \mathbb{N}$ such that map $t \mapsto N_t(m)$ is not continuous at points $s_1 = s_1(m), \ldots, s_N = s_N(m)$. We suppose that s_i is in $(n(s_i), n(s_i) + t/n]$ for sufficiently large n. We get

$$e^{Y_t^n(3,\varepsilon)} = \prod_{i=1}^N(-\phi_i - \varepsilon\psi_\varepsilon(\phi_i)),$$

where $\phi_i = \mathbb{H}_{\mathrm{od}}(B_{s_i}, -\theta_{N_{s_i-}}, n(s_i))$. In Fock representation we can also have

$$\mathbb{H}_{\rm od}(B_{s_i}, -\theta_{N_{s_i-}}, s_i) \cong \frac{1}{\sqrt{2}} \sum_{j=\pm 1} \left\{ a_{\rm E}^* \left(c_j^-(s_i), j \right) - a_{\rm E} \left(c_j^+(s_i), j \right) \right\},$$

$$\mathbb{H}_{\rm od}(B_{s_i}, -\theta_{N_{s_i-}}, n(s_i)) \cong \frac{1}{\sqrt{2}} \sum_{j=\pm 1} \left\{ a_{\rm E}^* \left(c_j^-(n(s_i)), j \right) - a_{\rm E} \left(c_j^+(n(s_i)), j \right) \right\},$$

where $c_j^{\pm}(t) = T_j(-\theta_{N_{s_i-}})(\hat{\jmath}_t \frac{\hat{\varphi}}{\sqrt{\omega}})e^{\pm ikB_{s_i}}$. We then estimate the distance between their test functions as

$$\left\| c_j^-(n(s_i)) - c_j^-(s_i) \right\|^2 \leq 2 \left(\frac{|k|\hat{\varphi}}{\sqrt{\omega}}, (1 - e^{-|n(s_i)-s_i|\omega}) \frac{|k|\hat{\varphi}}{\sqrt{\omega}} \right).$$

Clearly, $n(s_i) \to s_i$ as $n \to \infty$. Then

$$\lim_{n\to\infty} \mathbb{H}_{\rm od}(B_{s_i}, -\theta_{N_{s_i-}}, n(s_i)) = \mathbb{H}_{\rm od}(B_{s_i}, -\theta_{N_{s_i-}}, s_i),$$
$$\lim_{n\to\infty} \psi_\varepsilon(\mathbb{H}_{\rm od}(B_{s_i}, -\theta_{N_{s_i-}}, n(s_i))) = \psi_\varepsilon(\mathbb{H}_{\rm od}(B_{s_i}, -\theta_{N_{s_i-}}, s_i))$$

on $\mathcal{K}_\infty$ and we have

$$\lim_{n\to\infty} e^{Y_t^n(3,\varepsilon)} = \prod_{i=1}^{N} (-\Psi_\varepsilon(\mathbb{H}_{\rm od}(B_{s_i}, -\theta_{N_{s_i-}}, s_i))). \tag{C.5}$$

Then by (C.2), (C.4) and (C.5) we can see that for each $(x, \alpha, m) \in \mathbb{R}^3 \times \mathbb{Z}_2 \times \mathscr{X}_\mu$,

$$\mathbb{E}_{\mathcal{W}}^x \left[\left(F(q_0), \mathrm{J}_0^* e^{Z_t^n(\varepsilon)} \mathrm{J}_t G(q_t) \right) \right] \to \mathbb{E}_{\mathcal{W}}^x \left[\left(F(q_0), \mathrm{J}_0^* e^{Z_t(\varepsilon)} \mathrm{J}_t G(q_t) \right) \right]$$

as $n \to \infty$ and $|\mathbb{E}_{\mathcal{W}}^x[(F(q_0), \mathrm{J}_0^* e^{Z_t^n(\varepsilon)} \mathrm{J}_t G(q_t))]| \leq C_t(m) \mathbb{E}_W^x[\|F(q_0)\| \|G(q_t)\|]$ and the dominated function $C_t(m)\|F(q_0)\|\|G(q_t)\|$ is integrable. The Lebesgue dominated convergence theorem yields that

$$\int_{\mathbb{R}^3} \sum_{\alpha\in\mathbb{Z}_2} \mathbb{E}^{x,\alpha} \left[e^{-\int_0^t V(B_r) dr} \left(F(q_0), \mathrm{J}_0^* e^{Z_t^n(\varepsilon)} \mathrm{J}_t G(q_t) \right) \right] dx$$
$$\to \int_{\mathbb{R}^3} \sum_{\alpha\in\mathbb{Z}_2} \mathbb{E}^{x,\alpha} \left[e^{-\int_0^t V(B_r) dr} \left(F(q_0), \mathrm{J}_0^* e^{Z_t^n(\varepsilon)} \mathrm{J}_t G(q_t) \right) \right] dx$$

for $F, G \in \mathcal{K}_\infty$ as $n \to \infty$. Hence the proof of (2) is complete.

QED

References

1. M. Aizenman, B. Simon, Brownian motion and Harnack's inequality for Schrödinger operators. Commun. Pure Appl. Math. **35**, 209–271 (1982)
2. S. Albeverio, R. Høegh-Krohn. *Mathematical Theory of Feynman Path Integrals* (Springer, 1976)
3. Z. Ammari, M. Falconi, F. Hiroshima, Towards a derivation of classical electrodynamics of charges and fields from qed (2022). arXiv:2202.05015
4. D.B. Applebaum. *Lévy Processes and Stochastic Calculus* (Cambridge University Press, 2009)
5. A. Arai, Rigorous theory of spectra and radiation for a model in quantum electrodynamics. J. Math. Phys. **24**, 1896–1910 (1983)
6. A. Arai, Noninvertible bogoliubov transformations and instability of embedded eigenvalues. J. Math. Phys. **32**, 1838–1846 (1991)
7. A. Arai, *Analysis of Fock Spaces and Mathematical Theory of Quantum Fields* (World Scientific, 2018)
8. V. Bach, J. Fröhlich, A. Pizzo, Infrared-finite algorithms in QED II. The expansion of the groundstate of an atom interacting with the quantized radiation field. Adv. Math. **220**, 1023–1074 (2006)
9. V. Bach, J. Fröhlich, A. Pizzo, Infrared-finite algorithms in QED: the ground state of an atom interacting with the quantized radiation field. Commun. Math. Phys. **264**, 145–165 (2006)
10. V. Bach, J. Fröhlich, I.M. Sigal, Mathematical theory of nonrelativistic matter and radiation. Lett. Math. Phys. **34**, 183–201 (1995)
11. V. Bach, J. Fröhlich, I.M. Sigal, Quantum electrodynamics of confined nonrelativistic particles. Adv. Math. **137**, 299–395 (1998)
12. V. Bach, J. Fröhlich, I.M. Sigal, Renormalization group analysis of spectral problems in quantum field theory. Adv. Math. **137**, 205–298 (1998)
13. V. Bach, J. Fröhlich, I.M. Sigal, Spectral analysis for systems of atoms and molecules coupled to the quantized radiation field. Commun. Math. Phys. **207**, 249–290 (1999)
14. V. Betz, Existence of Gibbs measures relative to Brownian motion. Markov Proc. Rel. Fields **9**, 85–102 (2003)
15. V. Betz, F. Hiroshima, Gibbs measures with double stochastic integrals on path space. Inf. Dim. Anal. Relat. Top. **12**, 135–152 (2009)
16. V. Betz, F. Hiroshima, J. Lörinczi, R.A. Minlos, H. Spohn, Ground state properties of the Nelson Hamiltonian - a Gibbs measure-based approach. Rev. Math. Phys. **14**, 173–198 (2002)
17. V. Betz, H. Spohn, A central limit theorem for Gibbs measures relative to Brownian motion. Probab. Theory Rel. Fields **131**, 459–478 (2005)
18. R.H. Cameron, A family of integrals serving to connect the Wiener and Feynman integrals. J. Math. Phys. **39**, 126–141 (1960)
19. R. Carmona, Pointwise bounds for Schrödinger eigenstates. Commun. Math. Phys. **62**, 97–106 (1978)
20. R. Carmona, W.C. Masters, B. Simon, Relativistic Schrödinger operators: asymptotic behavior of the eigenfunctions. J. Funct. Anal. **91**, 117–142 (1990)
21. I. Catto, P. Exner, C. Hainzl, Enhanced binding revisited for a spinless particle in nonrelativistic QED. J. Math. Phys. **45**, 4174–4185 (2004)
22. P. Chernoff, Note on product formulas for operator semigroups. J. Funct. Anal. **2**, 238–242 (1968)
23. H.L. Cycon, R.G. Froese, W. Kirsch, B. Simon, *Schrödinger Operators. With Application to Quantum Mechanics and Global Geometry* (Springer, 1987)
24. I. Daubechies, One electron molecules with relativistic kinetic energy: properties of the discrete spectrum. Commun. Math. Phys. **94**, 523–535 (1984)
25. E.B. Davies. *Heat Kernels and Spectral Theory* (Cambridge University Press, 1989)
26. E.B. Davies, Intrinsic ultracontractivity and the Dirichlet Laplacian. J. Funct. Anal. **100**, 163–180 (1991)

27. E.B. Davies, B. Simon, Ultracontractivity and the heat kernel for Schrödinger operators and Dirichlet Laplacians. J. Funct. Anal. **59**, 335–395 (1984)
28. G.F. De Angelis, Stochastic mechanics of relativistic spinless particle. J. Math. Phys. **31**, 1408–1412 (1990)
29. G.F. De Angelis, G. Jona-Lasinio, M. Serva, Stochastic mechanics of a Dirac particle in two spacetime dimensions. J. Phys. A **19**, 865–871 (1986)
30. G.F. De Angelis, G. Jona-Lasinio, V. Sidoravicius, Berezin integrals and Poisson processes. J. Phys. A **31**, 289–308 (1998)
31. G.F. De Angelis, G. Jona-Lasinio, M. Sirugue, Probabilistic solution of Pauli type equations. J. Phys. A **16**, 2433–2444 (1983)
32. G.F. De Angelis, A. Rinaldi, M. Serva, Imaginary-time path integral for a relativistic spin-(1/2) particle in a magnetic field. Europhys. Lett. **14**, 95–100 (1991)
33. H. Ezawa, J. Klauder, L.A. Shepp, On the divergence of certain integrals of the wiener progress. Ann. Inst. Fourier (Grenoble) **24**, 189–193 (1974)
34. H. Ezawa, J. Klauder, L.A. Shepp, Vestigial effects of singular potentials in diffusion theory and quantum mechanics. J. Math. Phys. **16**, 783–799 (1975)
35. M. Falconi, Self-adjointness criterion for operators in Fock spaces. Math. Phys. Anal. Geom. **18**, 18 (2015)
36. W.G. Faris, Invariant cones and uniqueness of the ground state for Fermion systems. J. Math. Phys. **13**, 1285–1290 (1972)
37. W.G. Faris, B. Simon, Degenerate and nondegenerate ground states for Schrödinger operators. Duke Math. J. **42**, 559–567 (1975)
38. R.P. Feynman, Space-time approach to non-relativistic quantum mechanics. Rev. Mod. Phys. **20**, 367–387 (1948)
39. R.P. Feynman, A.R. Hibbs, *Quantum Mechanics and Path Integrals* (McGraw-Hill, 1965)
40. J. Fröhlich, Y.M. Park, Correlation inequalities and thermodynamic limit for classical and quantum continuous systems II. Bose-Einstein and Fermi-Dirac statistics. J. Stat. Phys. **23**, 701–753 (1980)
41. D. Fujiwara, A construction of the fundamental solution for the Schrödinger equation. J. Anal. Math. **35**, 41–96 (1979)
42. B. Gaveau, J. Vauthier, Intégrales oscillantes stochastiques: l'équation de Pauli. J. Funct. Anal. **44**, 388–400 (1981)
43. J. Glimm, A. Jaffe, The $\lambda(\phi^4)_2$ quantum field theory without cutoffs: II. The field operators and the approximate vacuum. Ann. Math. **91**, 362–401 (1970)
44. J. Glimm, A. Jaffe, *Quantum Physics: A Functional Integral Point of View* (Springer, 1981)
45. M. Griesemer, E. Lieb, M. Loss, Ground states in non-relativistic quantum electrodynamics. Invent. Math. **145**, 557–595 (2001)
46. C. Grosche, F. Steiner, *Handbook of Feynman Path Integrals* (Springer, 1998)
47. L. Gross, A noncommutative extension of the Perron-Frobenius theorem. Bull. Amer. Math. Soc. **77**, 343–347 (1971)
48. L. Gross, The relativistic polaron without cutoffs. Commun. Math. Phys. **31**, 25–73 (1973)
49. M. Gubinelli, F. Hiroshima, J. Lörinczi, Ultraviolet renormalization of the Nelson Hamiltonian through functional integration. J. Funct. Anal. **267**, 3125–3153 (2014)
50. Z. Haba, *Feynman Integral and Random Dynamics in Quantum Physics: A Probabilistic Approach to Quantum Dynamics* (Kluwer Academic, 1999)
51. C. Hainzl, R. Seiringer, Mass renormalization and energy level shift in non-relativistic QED. Adv. Theor. Math. Phys. **6**, 847–881 (2002)
52. D. Hasler, I. Herbst, On the self-adjointness and domain of Pauli-Fierz type Hamiltonians. Rev. Math. Phys. **20**, 787–800 (2008)
53. I.W. Herbst, Spectral theory of the operator $(p^2 + m^2)^{1/2} - Ze^2/r$. Commun. Math. Phys. **53**, 285–294 (1977)
54. T. Hidaka, F. Hiroshima, Pauli-Fierz model with Kato class potential and exponential decay. Rev. Math. Phys. **22**, 1181–1208 (2010)

55. T. Hidaka, F. Hiroshima, Self-adjointness of semi-relativistic Pauli-Fierz models. Rev. Math. Phys. **27**, 1550015, 18 p. (2015)
56. T. Hidaka, F. Hiroshima, Spectrum of semi-relativistic Pauli-Fierz Hamiltonian I. J. Math. Anal. Appl. **437**, 330–349 (2016)
57. T. Hidaka, F. Hiroshima, I. Sasaki, Spectrum of semi-relativistic Pauli-Fierz Hamiltonian II. J. Spectral Theory **11**, 1779–1830 (2021)
58. M. Hirokawa, F. Hiroshima, Absence of energy level crossing for the ground state energy of the Rabi model. Commun. Stoch. Anal. **8**, 551–560 (2014)
59. M. Hirokawa, F. Hiroshima, J. Lőrinczi, Spin-boson model through a Poisson driven stochastic process. Math. Zeitschrift **277**, 1165–1198 (2014)
60. M. Hirokawa, F. Hiroshima, H. Spohn, Ground state for point particles interacting through a massless scalar bose field. Adv. Math. **191**, 339–392 (2005)
61. F. Hiroshima, Functional integral representation of a model in quantum electrodynamics. Rev. Math. Phys. **9**, 489–530 (1997)
62. F. Hiroshima, Essential self-adjointness of translation-invariant quantum field models for arbitrary coupling constants. Commun. Math. Phys. **211**, 585–613 (2000)
63. F. Hiroshima, Ground states of a model in nonrelativistic quantum electrodynamics II. J. Math. Phys. **41**, 661–674 (2000)
64. F. Hiroshima, Self-adjointness of the Pauli-Fierz Hamiltonian for arbitrary values of coupling constants. Ann. Henri Poincaré **3**, 171–201 (2002)
65. F. Hiroshima, Fiber Hamiltonians in the non-relativistic quantum electrodynamics. J. Funct. Anal. **252**, 314–355 (2007)
66. F. Hiroshima, Functional integral approach to semi-relativistic Pauli-Fierz models. Adv. Math. **259**, 784–840 (2014)
67. F. Hiroshima, *Ground States in Quantum Field Models* (Springer, 2019)
68. F. Hiroshima, Non-relativistic limit of the semi-relativistic Pauli-Fierz model. RIMS Kokyuroku **2187**, 1–10 (2021)
69. F. Hiroshima, Invariant domains by function a integrations. In preparation (2022)
70. B. Hinrichs, F. Hiroshima, Positivity of semigroup generated by translation invariant Nelson Hamiltonian. In preparation (2024)
71. F. Hiroshima, T. Ichinose, J. Lörinczi, Path integral representation for Schrödinger operators with Bernstein functions of the Laplacian. Rev. Math. Phys. **24**, 1250013, 40 p. (2012)
72. F. Hiroshima, T. Ichinose, J. Lörinczi, Probabilistic representation and fall-off of bound states of relativistic Schrödinger operators with spin 1/2. Publ. RIMS Kyoto **49**, 189–214 (2013)
73. F. Hiroshima, J. Lörinczi, Functional integral representations of the Pauli-Fierz model with spin 1/2. J. Funct. Anal. **254**, 2127–2185 (2008)
74. F. Hiroshima, J. Lőrinczi, Feynman-Kac Type Theorems and Its Applications, vol. 2, 2nd edn. (De Gruyter, 2020)
75. F. Hiroshima, O. Matte, Ground states and their associated Gibbs measures in the renormalized nelson model. Rev. Math. Phys. **33**, 2250002 (84 p.) (2021)
76. F. Hiroshima, O. Matte, Point-wise spatial decay of eigenvectors in the Nelson model. In preparation (2024)
77. F. Hiroshima, I. Sasaki, H. Spohn, A. Suzuki, Enhanced Biding in Quantum Field Theory. IMI Lecture Note 38 (Kyushu University, 2012)
78. F. Hiroshima, H. Spohn, Enhanced binding through coupling to a quantum field. Ann. Henri Poincaré **2**, 1159–1187 (2001)
79. F. Hiroshima, H. Spohn, Ground state degeneracy of the Pauli-Fierz model with spin. Adv. Theor. Math. Phys. **5**, 1091–1104 (2001)
80. K. Itô, Stochastic integral. Proc. Imp. Acad. Tokyo **20**, 519–524 (1944)
81. G.W. Johnson, M.L. Lapidus, *The Feynman Integral and Feynman's Operational Calculus* (Oxford University Press, 2000)
82. M. Kac, On distributions of certain Wiener functionals. Trans. Amer. Math. Soc. **65**, 1–13 (1949)

83. M. Kac, On some connections between probability theory and differential and integral equations, in *Proceedings of Second Berkeley Symposium* (1951), pp. 189–215
84. M. Kac, Wiener and integration in function spaces. Bull. Amer. Math. Soc. **72**, 52–68 (1966)
85. T. Kato, Fundamental properties of Hamiltonian operators of Schrödinger type. Trans. Amer. Math. Soc. **70**, 195–211 (1951)
86. T. Kato, *Perturbation Theory for Linear Operators* (Springer, 1966)
87. T. Kato, On the Trotter-Lie product formula. Proc. Japan Acad. **50**, 694–698 (1974)
88. T. Kato, Trotter's product formula for an arbitrary pair of self-adjoint contraction semigroups. Adv. Math. Suppl. Stud. **3**, 185–195 (1978)
89. T. Kato, K. Masuda, Trotter's product formula for nonlinear semigroups generated by the subdifferentiables of convex functionals. J. Math. Soc. Japan **30**, 169–178 (1978)
90. M. Könenberg, O. Matte, Ground states of semi-relativistic Pauli-Fierz and no-pair Hamiltonians in QED at critical Coulomb coupling. J. Operator Theory **70**, 211–237 (2013)
91. M. Könenberg, O. Matte, On enhanced binding and related effects in the non- and semi-relativistic Pauli-Fierz models. Commun. Math. Phys. **323**, 635–661 (2013)
92. M. Könenberg, O. Matte, E. Stockmeyer, Existence of ground states of hydrogen-like atoms in relativistic QED I: the semi-relativistic Pauli-Fierz operator. Rev. Math. Phys. **23**, 635–661 (2011)
93. M. Könenberg, O. Matte, E. Stockmeyer, Existence of ground states of hydrogen-like atoms in relativistic quantum electrodynamics. II. The no-pair operator. J. Math. Phys. **52**, 123501 (34 p.) (2011)
94. E. Lieb, Bounds on the eigenvalues of the Laplacian and Schrödinger operator. Bull. Amer. Math. Soc. **82**, 751–753 (1976)
95. E. Lieb, The number of bound states of one-body Schrödinger operators and the Weyl problem. Proc. Amer. Math. Soc. Symp. Pure Math. **36**, 241–252 (1980)
96. E. Lieb, M. Loss, *Analysis* (Academic, 2001)
97. E. Lieb, M. Loss, Existence of atoms and molecules in non-relativistic quantum electrodynamics. Adv. Theor. Math. Phys. **7**, 667–710 (2003)
98. E. Lieb, W.E. Thirring, Bound for the kinetic energy of fermions which prove the stability of matters. Phys. Rev. Lett. **35**, 687–689 (1975)
99. E. Lieb, W.E. Thirring. *Inequalities for the Moments of the Eigenvalues of the Schrödinger Hamiltonian and Their Relation to Sobolev Inequalities* (Princeton University Press, 1976), pp. 269–303
100. J. Lőrinczi, F. Hiroshima, V. Betz, *Feynman-Kac Type Theorems and Its Applications*, vol. 1, 2nd edn. (De Gruyter, 2020)
101. M. Loss, T. Miyao, H. Spohn, Lowest energy states in nonrelativistic QED: atoms and ions in motion. J. Funct. Anal. **243**, 353–393 (2007)
102. M. Loss, T. Miyao, H. Spohn, Kramers degeneracy theorem in nonrelativistic QED. Lett. Math. Phys. **89**, 21–31 (2009)
103. S. Machihara, K. Nakanishi, T. Ozawa, Nonrelativistic limit in the energy space for nonlinear klein-gordon equations. Math. Ann. **322**, 603–621 (2002)
104. S. Machihara, K. Nakanishi, T. Ozawa, Small global solutions and the nonrelativistic limit for the nonlinear dirac equation. Rev. Mat. Iberoamericana **19**, 179–194 (2003)
105. O. Matte, Continuity properties of the semi-group and its integral kernel in non-relativistic QED. Rev. Math. Phys. **28**, 1650011 (90 p.) (2016)
106. O. Matte, Pauli-Fierz type operators with singular electromagnetic potentials on general domains. Math. Phys. Anal. Geom. **20**, 41 (2017)
107. O. Matte, J. Møller, Feynman-Kac formulas for the ultra-violet renormalized Nelson model. Astérisque **404**, vi+110 (2018)
108. O. Matte, E. Stockmeyer, Exponential localization of hydrogen-like atoms in relativistic quantum electrodynamics. Commun. Math. Phys. **295**, 551–583 (2010)
109. T. Miyao, Nondegeneracy of ground states in nonrelativistic quantum field theory. J. Oper. Theory **64**, 207–241 (2010)

110. T. Miyao, On the semigroup generated by the renormalized nelson hamiltonian. J. Funct. Anal. **276**, 1948–1977 (2019)
111. E. Nelson, Feynman integrals and the Schrödinger equation. J. Math. Phys. **5**, 332–343 (1964)
112. E. Nelson, Interaction of nonrelativistic particles with a quantized scalar field. J. Math. Phys. **5**, 1990–1997 (1964)
113. E. Nelson, Schrödinger particles interacting with a quantized scalar field, in *Proceedings of a Conference on Analysis in Function Space*, ed. by W.T. Martin, I. Segal (MIT Press, 1964), p. 87
114. E. Nelson, Construction of quantum fields from Markoff fields. J. Funct. Anal. **12**, 97–112 (1973)
115. E. Nelson, Probability theory and Euclidean quantum field theory. *Lecture Note in Physics*, vol. 25, pp. 94–124 (1973)
116. E. Nelson, The free Markoff field. J. Funct. Anal. **12**, 211–227 (1973)
117. W. Pauli, M. Fierz, Zur Theorie der Emission langwelliger Lichtquanten. Nuovo Cimento **15**, 167–188 (1938)
118. M. Reed, B. Simon, *Methods of Modern Mathematical Physics IV* (Academic, 1978)
119. S.N.M. Rujusenaars, On Bogoliubov transformtions for systems of relativistic charged particles. J. Math. Phys. **18**, 517–526 (1977)
120. S.N.M. Rujusenaars, On Bogoliubo transformtions. II. The general case. Ann. Phys. **116**, 105–134 (1978)
121. K.I. Sato, *Lévy Processes and Infinite Divisibility* (Cambridge University Press, 1999)
122. R. Schilling, R. Song, Z. Vondracek, *Bernstein Functions. Theory and Applications* (Walter de Gruyter, 2010)
123. I. Segal, Tensor algebras over Hilbert spaces, I. Trans. Amer. Math. Soc. **81**, 106–134 (1956)
124. I. Segal, Tensor algebras over Hilbert spaces, II. Ann. Math. **63**, 160–175 (1956)
125. B. Simon, Quadratic forms and Klauder's phenomenon: a remark on very singular perturbations. J. Funct. Anal. **14**, 295–298 (1973)
126. B. Simon, *The $P(\phi)_2$ Euclidean (Quantum) Field Theory* (Princeton University Press, 1974)
127. B. Simon, A canonical decomposition for quadratic forms with applications to monotone convergence theorems. J. Funct. Anal. **9**, 377–385 (1978)
128. B. Simon. *Functional Integrations and Quantum Physics* (Academic, 1979) (2nd edn, 2005)
129. B. Simon, Schrödinger semigroups. Bull. Amer. Math. Soc. **7**, 447–526 (1982)
130. A.D. Sloan, A nonperturbative approach to nondegeneracy of ground states in quantum field theory: polaron models. J. Funct. Anal. **16**, 161–191 (1974)
131. H. Spohn, Effective mass of the polaron: a functional integral approach. Ann. Phys. **175**, 278–318 (1987)
132. H. Spohn, *Dynamics of Charged Particles and their Radiation Field* (Cambridge University Press, 2004)
133. B. Thaller, *The Dirac Equation* (Springer, 1992)
134. H.F. Trotter, On the product of semi-groups of operators. Proc. Amer. Math. Soc. **10**, 545–551 (1959)

The manufacturer's authorised representative in the EU is Springer Nature Customer Service Centre GmbH, Europaplatz 3, 69115 Heidelberg, Germany. If you have any concerns regarding our products, please contact ProductSafety@springernature.com

Printed and bound by CPI Group (UK) Ltd, Croydon, CR0 4YY

30/09/2026

02225298-0015